Experientia Supplementum

Volume 108

More information about this series at http://www.springer.com/series/4822

Mario D. Cordero • Elísabet Alcocer-Gómez
Editors

Inflammasomes: Clinical and Therapeutic Implications

Springer

Editors
Mario D. Cordero
Department of Physiology
Institute of Nutrition and Food Technology
"José Mataix", Biomedical Research
Center (CIBM), University of Granada
Armilla, Spain

Elísabet Alcocer-Gómez
Departamento de Psicología Experimental
Facultad de Psicología
Universidad de Sevilla
Seville, Spain

ISSN 1664-431X ISSN 2504-3692 (electronic)
Experientia Supplementum
ISBN 978-3-030-07758-7 ISBN 978-3-319-89390-7 (eBook)
https://doi.org/10.1007/978-3-319-89390-7

This Springer imprint is published by the registered company Springer Nature Switzerland AG
The registered company address is: Gewerbestrasse 11, 6330 Cham, Switzerland

Acknowledgments

The editors want to thank all the authors for the tremendous effort and dedication to the development of this book as well as the high level of chapters.

Contents

Chapter 1
Inflammasomes in Clinical Practice: A Brief Introduction

Elísabet Alcocer-Gómez, Beatriz Castejón-Vega, Macarena López-Sánchez, and Mario D. Cordero

Contents

Abstract Inflammasomes are multiprotein complexes formed and activated after exposure to pathogenic microbes and host danger signals that control the maturation and production of IL-1β and IL-18. Their implication in different diseases such as cardiovascular, neurodegenerative, psychiatric, and metabolic diseases opens a door to developing new therapeutic perspectives. However, the rapid increase in the

E. Alcocer-Gómez (✉)
Departamento de Psicología Experimental, Facultad de Psicología, Universidad de Sevilla, Sevilla, Spain

B. Castejón-Vega
Departamento de Psicología Experimental, Facultad de Psicología, University of Sevilla, Sevilla, Spain

Research Laboratory, Oral Medicine Department, University of Sevilla, Sevilla, Spain

M. López-Sánchez
Research Laboratory, Oral Medicine Department, University of Sevilla, Sevilla, Spain

Centro Andaluz de Biología del Desarrollo (CABD), Universidad Pablo de Olavide-CSIC-Junta de Andalucía, Sevilla, Spain

M. D. Cordero (✉)
Centro Andaluz de Biología del Desarrollo (CABD), Universidad Pablo de Olavide-CSIC-Junta de Andalucía, Sevilla, Spain

Department of Physiology, Institute of Nutrition and Food Technology "José Mataix", Biomedical Research Center (CIBM), University of Granada, Armilla, Spain
e-mail: mdcormor@ugr.es

M. D. Cordero, E. Alcocer-Gómez (eds.), *Inflammasomes: Clinical and Therapeutic Implications*, Experientia Supplementum 108,
https://doi.org/10.1007/978-3-319-89390-7_1

knowledge about inflammasomes is associated with their involvement in clinical practice. Two topics open the way to future lines of research: a clinical trial with the new specific inhibitors and the development of diagnostic tools.

Keywords Inflammasomes · NLRP3 · Clinical medicine · Diagnosis · Pharmacological treatment

1.1 Introduction

The aim of medicine and clinical research is to understand the biological processes of health and the development of tools for the treatment and diagnosis of diseases. Thus, we need a balance between basic and clinical research in a translational form. Different examples of this are the development of drugs and diagnostic methods for the identification of molecular pathways and targets involved in the pathophysiology of the diseases.

For decades, inflammation has been the focus of researchers' attention because of its direct and indirect participation in the pathophysiology of many diseases. From an immunity viewpoint, after an infection, microorganisms are initially sensed by pattern-recognition receptors (PRRs) of the innate immune system. These PRRs are expressed in various immune cells such as macrophages, dendritic cells, epithelial cells, neutrophils, and adaptive immune cells. The Toll-like receptors (TLRs), a particular type of PRRs, are expressed on the cell membrane surface and can be activated by different external signals with a pathogen profile known as pathogen-associated molecular patterns (PAMPs). The PRRs can also be triggered during sterile inflammatory diseases, in the absence of microbes, by different damaging profiles, suggesting the crucial role of danger signals that are host-derived, and that are known as danger-associated molecular patterns (DAMPs). PAMPs and DAMPs can also trespass on the plasma membrane and trigger intracellular innate immune receptors directly, for example, via recognition of DNA or RNA in the cytoplasm. In the loss of homeostasis conditions, some cytosolic innate immune sensors can also be indirectly activated. Among these sensors are included NOD-like receptors or nucleotide-binding oligomerization domain-like receptors (NLRs) that possess a pyrin domain (PYD) or caspase activation and recruitment domain (CARD) in their N-terminal regions, which have partial homology in man and mouse (Fig. 1.1a) and which have specific ligands (Fig. 1.1b). Of these, the NLRP3 is the best described, which are composed of a carboxy-terminal LRR domain, a central NACHT and NAD domain (NBD), and an amino-terminal PYD (Fig. 1.2a) (Lamkanfi and Dixit 2014), with a similar structure between humans and the mouse, both the most studied species with regard to inflammasomes (Fig. 1.2b). After an assembly stimulus, NLR is associated with the adapter molecule apoptosis-associated speck-like protein that contains a CARD domain (ASD) in the N-terminal regions, and the effector protein caspase-1 (Fig. 1.3a), and forms the named inflammasome complex (Lamkanfi and Dixit 2014). These inflammasome complexes regulate the activation of caspase-1, which in turn regulates the cleavage of cytokines interleukin-1beta (IL-1β) and interleukin-18 (IL-18). The NLRP3

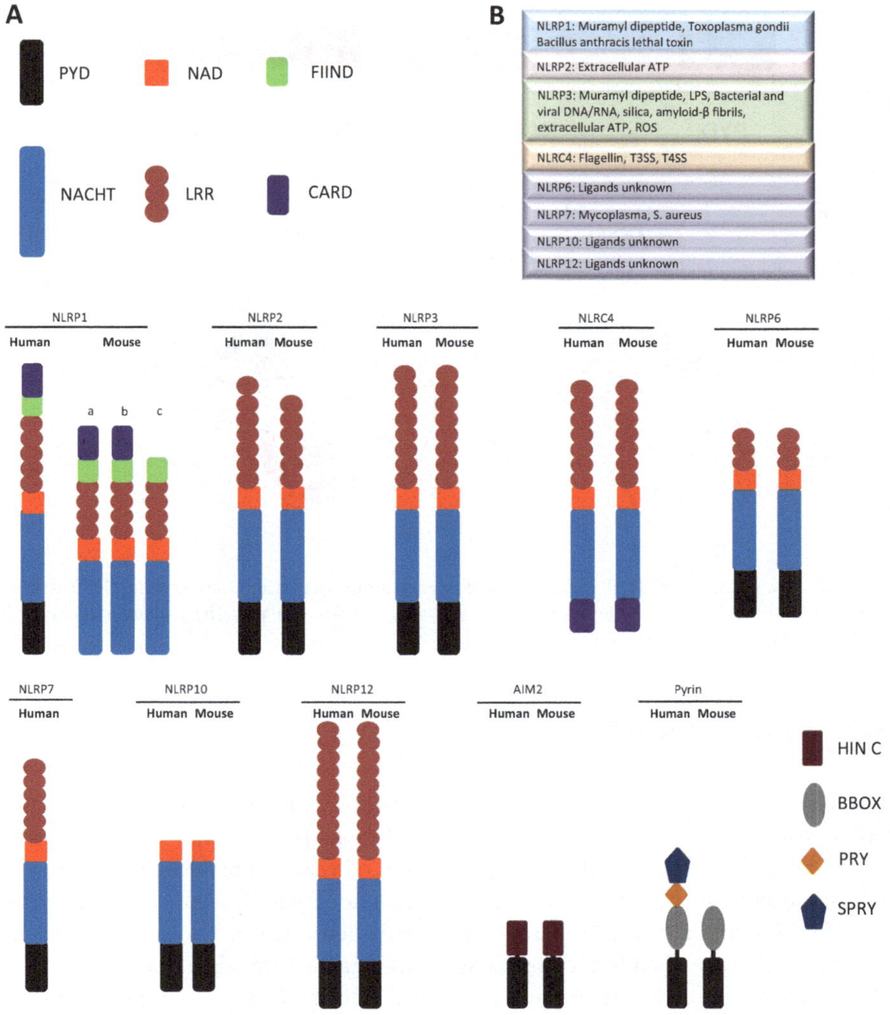

Fig. 1.1 Overview of the NLRP family and domain organization of inflammasomes. The principal identified components of the NOD-like receptor (NLR) family and ligands. (**a**) They all contain a nucleotide-binding domain (NBD) with a NACHT and NAD, carboxy-terminal leucine-rich repeat (LRR) and may contain either a pyrin domain (PYD) or a caspase activation domain and recruitment domain (CARD) or both. (**b**) Several of the described ligands of the inflammasome complexes NLRP1, NLRP2, NLRP3, NLRC4, NLRP6, NLRP7, NLRP10, and NLRP12

inflammasome is the most frequently studied because is activated in response to a large array of endogenous danger signals, such as extracellular adenosine triphosphate (ATP) (Sadatomi et al. 2017), uric acid crystals (MSU) (Braga et al. 2017), amyloid-β fibrils (Heneka et al. 2013), cholesterol crystals (Duewell et al. 2010), and reactive oxygen species (ROS) (Abderrazak et al. 2015) (Fig. 1.3b).

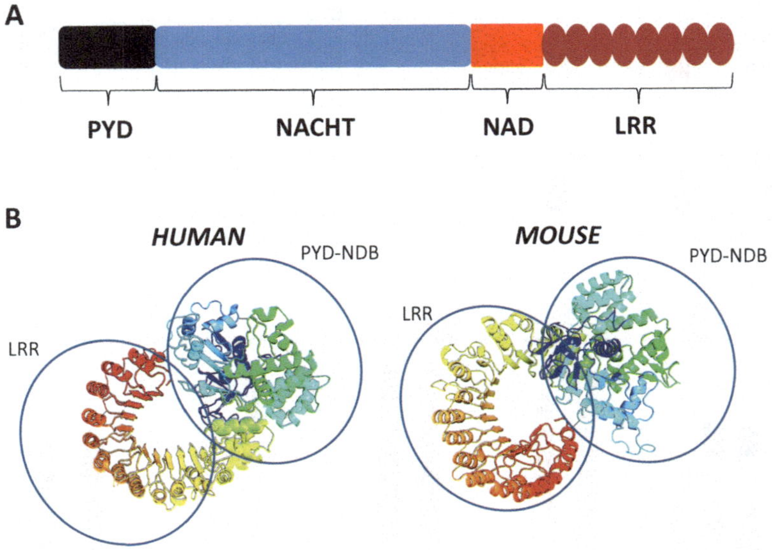

Fig. 1.2 Overview of NLRP3 inflammasome organization. (**a**) NLRP3 contains a NBD with a NACHT and NAD, carboxy-terminal LRR, and may contain a PYD. (**b**) Crystal structure of NLRP3 in humans and the mouse

1.2 Clinical Aspect

From a clinical perspective, many researchers have focused their studies on the inflammasomes because of the variety of molecular patterns they can recognize, which have been linked to the pathophysiology of various prevalent disorders and where the inflammasome is shown as a therapeutic target. Spontaneous inflammatory diseases with abnormal inflammasome activation, such as cryopyrin-associated periodic syndrome (CAPS) are caused by NLRP3 gain-of-function mutations, which do not require an activating stimulus (Yang and Chiang 2015). Numerous mutations with gain-of-function have been described in the different subunits of the inflammasomes, which trigger auto-inflammatory syndromes in the absence of cognate ligands. Extensive evidence has indicated that NLRP3 inflammasome is activated in several psychiatric and neurodegenerative disorders such depression (Alcocer-Gómez et al. 2013), Alzheimer's disease (Masters and Selkoe 2012), Parkinson's disease (Zhou et al. 2016), amyotrophic lateral sclerosis (ALS), Huntington's disease, and multiple sclerosis (MS) (Heneka et al. 2014; Glass et al. 2010) and prion diseases (Hafner-Bratkovic et al. 2012) among others. The NLRP3 inflammasome is also up-regulated after myocardial infarction (Ajdukovic 2015) and recently, NLRP3 and inflammatory cytokines have also been proposed as new biomarkers of cardiovascular risk (Bullón et al. 2017). The NLRP3 inflammasome has also been shown to play a role in metabolic diseases such as obesity, diabetes, gout, and various age-associated diseases (Cordero et al. 2018), and to further the implication in different diseases, the NLRP3 inflammasome has been proposed as a

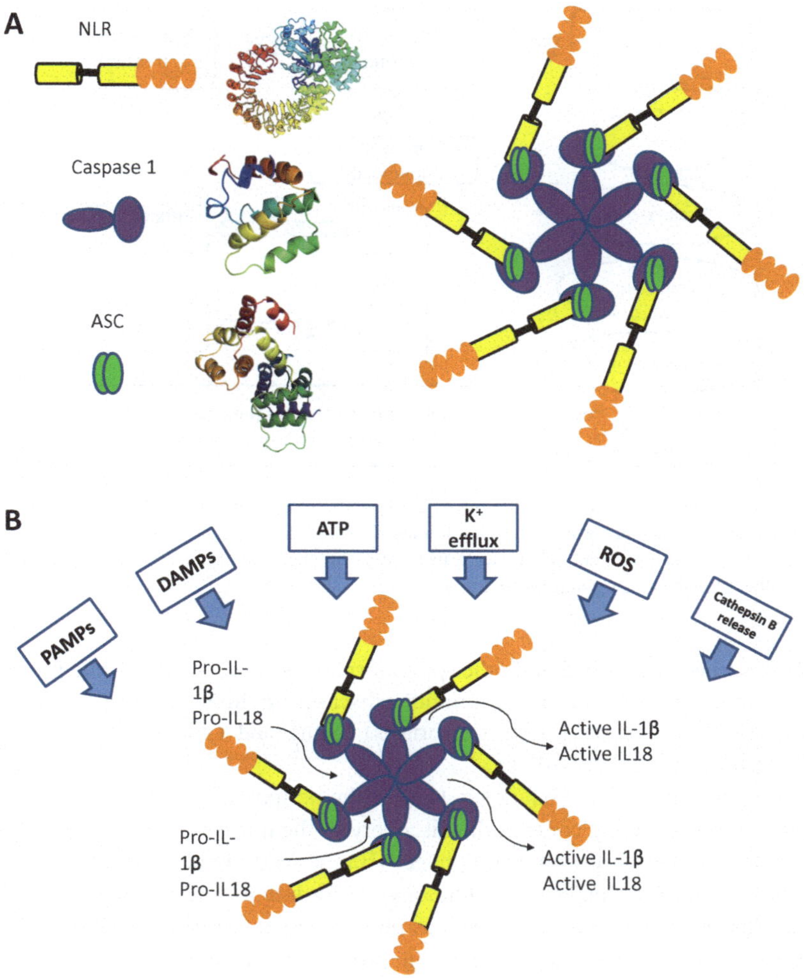

Fig. 1.3 Overview of the NLRP3 inflammasome complex and activation. (**a**) The NLRP3 inflammasome complex is composed of an NLR, the effector protein caspase-1, and the adapter molecule apoptosis-associated speck-like protein that contains a CARD (ASC). (**b**) PAMPs and DAMPs are thought to activate the NLRP3 inflammasome by reducing intracellular K+ concentrations, cytosolic release of lysosomal cathepsins such as cathepsin B, or by inducing mitochondrial damage, which induces the production of ROS

common nexus among several diseases, such as cardiovascular conditions and major depression (Alcocer-Gómez and Cordero 2017).

1.2.1 Diagnostic Tools

The implication of the different inflammasomes in so many diseases is a clear sign of its importance in medical practice. However, we need to find clear objectives for its utility in the development of clinical tools. Because many mutations in different

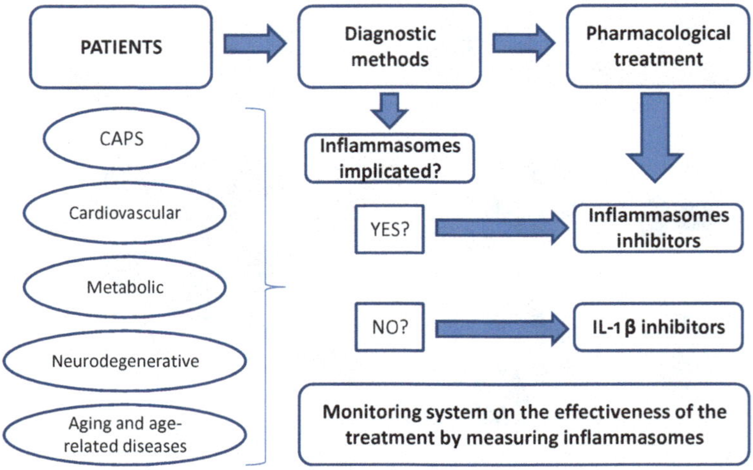

Fig. 1.4 Overview of the utility of the inflammasomes in clinical practice. Clinical methodology for evaluating the implication of the inflammasomes in the pathophysiology of patients and selecting the correct treatment. Then, we will need to monitor the effectiveness of the inhibitors of the inflammasomes using diagnostic tools

NLRs have been described and other polymorphisms associated with the predisposition of other diseases such as autoimmune diseases, multiple sclerosis, rheumatoid arthritis or skin diseases (vitiligo, psoriasis) (Yang and Chiang 2015), we have evidence for genetic diagnostics.

A great advance of the diagnostic tools is the peripheral determination of metabolic biomarkers. In this context, the discovery of the release of oligomeric NLRP3 inflammasome particles from macrophages (Baroja-Mazo et al. 2014), opens a new path in the design of diagnostic technology. According to this, different technological approaches have been made, such as bioluminescent resonance energy transfer (BRET), a methodology for monitoring protein interactions, to detect NLRP3 oligomerization in living cells before and during NLRP3 inflammasome activation (Martín-Sánchez et al. 2016). Recently, an optimized whole blood assay to determine the expression of different inflammasomes such as NLRP3, NLRC4, and AIM2 has been shown, in which a reduced sample was used, 140 µl of blood per well (Grinstein et al. 2017). All these methods provide the first step toward to future diagnostic methods and contribute to the knowledge of the efficacy of pharmacological treatments by determination in real time of inflammasome reduction (Fig. 1.4). This is very important, for example, in treatment-resistant depression.

1.2.2 Pharmacological Treatments

The other area in which inflammasomes are important in clinical practice is pharmacology. In the last few years, several specific inhibitors of NLRP3 have been studied and the efficacy has been demonstrated in different experimental models of diseases. Recently, a study identified the small molecule MCC950, which prevents the formation of the NLRP3 inflammasome complex, and in parallel, the ketone metabolite β-hydroxybutyrate as a specific inhibitor of the NLRP3 inflammasome was also investigated (Coll et al. 2015; Youm et al. 2015). Since then, new compounds have been explored; however, the specific inhibitors have only been studied in experimental models. Thus, in the near future, new trials in human diseases with inflammasome inhibitors will show an important clinical field.

1.3 Conclusions

The rapid increase in knowledge about the biology of inflammasomes and their implications in the pathophysiology of human diseases offers several utilities for clinical practice, which will lead to a new future in medicine. Next, there will be a specific inhibitor for treatment in humans and we will have to make a special effort to develop diagnostic mechanisms.

References

Abderrazak A, Syrovets T, Couchie D, El Hadri K, Friguet B, Simmet T, Rouis M (2015) NLRP3 inflammasome: from a danger signal sensor to a regulatory node of oxidative stress and inflammatory diseases. Redox Biol 4:296–307

Ajdukovic J (2015) The role of NLRP3 inflammasome in cardiovascular diseases. J Clin Exp Cardiol 6:399

Alcocer-Gómez E, Cordero MD (2017) NLRP3 inflammasome: common nexus between depression and cardiovascular diseases. Nat Rev Cardiol 14:124

Alcocer-Gómez E, de Miguel M, Casas-Barquero N, Núñez-Vasco J, Sánchez-Alcazar JA, Fernández-Rodríguez A, Cordero MD (2013) NLRP3 inflammasome is activated in mononuclear blood cells from patients with major depressive disorder. Brain Behav Immun 36:111–117

Baroja-Mazo A, Martín-Sánchez F, Gomez AI, Martínez CM, Amores-Iniesta J, Compan V, Barberà-Cremades M, Yagüe J, Ruiz-Ortiz E, Antón J, Buján S, Couillin I, Brough D, Arostegui JI, Pelegrín P (2014) The NLRP3 inflammasome is released as a particulate danger signal that amplifies the inflammatory response. Nat Immunol 15:738–748

Braga TT, Forni MF, Correa-Costa M, Ramos RN, Barbuto JA, Branco P, Castoldi A, Hiyane MI, Davanso MR, Latz E, Franklin BS, Kowaltowski AJ, Camara NO (2017) Soluble uric acid activates the NLRP3 inflammasome. Sci Rep 7:39884

Bullón P, Cano-García FJ, Alcocer-Gómez E, Varela-López A, Roman-Malo L, Ruiz-Salmerón RJ, Quiles JL, Navarro-Pando JM, Battino M, Ruiz-Cabello J, Jiménez-Borreguero LJ, Cordero MD (2017) Could NLRP3-inflammasome be a cardiovascular risk biomarkers in acute myocardial infarction patients? Antioxid Redox Signal 27:269–275

Coll RC, Robertson AA, Chae JJ, Higgins SC, Muñoz-Planillo R, Inserra MC, Vetter I, Dungan LS, Monks BG, Stutz A, Croker DE, Butler MS, Haneklaus M, Sutton CE, Núñez G, Latz E, Kastner DL, Mills KH, Masters SL, Schroder K, Cooper MA, O'Neill LA (2015) A small-molecule inhibitor of the NLRP3 inflammasome for the treatment of inflammatory diseases. Nat Med 21:248–255

Cordero MD, Williams MR, Ryffel B (2018) AMP-activated protein kinase regulation of the NLRP3 inflammasome during aging. Trends Endocrinol Metab 29:8–17

Duewell P, Kono H, Rayner KJ, Sirois CM, Vladimer G, Bauernfeind FG, Abela GS, Franchi L, Nuñez G, Schnurr M, Espevik T, Lien E, Fitzgerald KA, Rock KL, Moore KJ, Wright SD, Hornung V, Latz E (2010) NLRP3 inflammasomes are required for atherogenesis and activated by cholesterol crystals. Nature 464:1357–1361

Glass CK, Saijo K, Winner B, Marchetto MC, Gage FH (2010) Mechanisms underlying inflammation in neurodegeneration. Cell 140:918–934

Grinstein L, Endter K, Hedrich CM, Reinke S, Luksch H, Schulze F, Robertson AAB, Cooper MA, Rösen-Wolff A, Winkler S (2017) An optimized whole blood assay measuring expression and activity of NLRP3, NLRC4 and AIM2 inflammasomes. Clin Immunol. pii: S1521-6616(17) 30179-1

Hafner-Bratkovic I, Bencina M, Fitzgerald KA, Golenbock D, Jerala R (2012) NLRP3 inflammasome activation in macrophage cell lines by prion protein fibrils as the source of IL-1beta and neuronal toxicity. Cell Mol Life Sci 69:4215–4228

Heneka MT, Kummer MP, Stutz A, Delekate A, Schwartz S, Vieira-Saecker A, Griep A, Axt D, Remus A, Tzeng TC, Gelpi E, Halle A, Korte M, Latz E, Golenbock DT (2013) NLRP3 is activated in Alzheimer's disease and contributes to pathology in APP/PS1 mice. Nature 493:674–678

Heneka MT, Kummer MP, Latz E (2014) Innate immune activation in neurodegenerative disease. Nat Rev Immunol 14:463–477

Lamkanfi M, Dixit VM (2014) Mechanisms and functions of inflammasomes. Cell 157:1013–1022

Martín-Sánchez F, Compan V, Pelegrín P (2016) Measuring NLR oligomerization III: detection of NLRP3 complex by bioluminescence resonance energy transfer. Methods Mol Biol 1417:159–168

Masters CL, Selkoe DJ (2012) Biochemistry of amyloid β-protein and amyloid deposits in Alzheimer disease. Cold Spring Harb Perspect Med 2:a006262

Sadatomi D, Nakashioya K, Mamiya S, Honda S, Kameyama Y, Yamamura Y, Tanimura S, Takeda K (2017) Mitochondrial function is required for extracellular ATP-induced NLRP3 inflammasome activation. J Biochem 161:503–512

Yang CA, Chiang BL (2015) Inflammasomes and human autoimmunity: a comprehensive review. J Autoimmun 61:1–8

Youm YH, Nguyen KY, Grant RW, Goldberg EL, Bodogai M, Kim D, D'Agostino D, Planavsky N, Lupfer C, Kanneganti TD, Kang S, Horvath TL, Fahmy TM, Crawford PA, Biragyn A, Alnemri E, Dixit VD (2015) The ketone metabolite β-hydroxybutyrate blocks NLRP3 inflammasome-mediated inflammatory disease. Nat Med 21:263–269

Zhou Y, Lu M, Du RH, Qiao C, Jiang CY, Zhang KZ, Ding JH, Hu G (2016) MicroRNA-7 targets Nod-like receptor protein 3 inflammasome to modulate neuroinflammation in the pathogenesis of Parkinson's disease. Mol Neurodegener 11:28

Chapter 2
The Inflammasomes in Cardiovascular Disease

Gerardus P. J. van Hout and Lena Bosch

Contents

G. P. J. van Hout (✉)
Department of Cardiology, Utrecht University Medical Center, Utrecht, The Netherlands
e-mail: G.P.J.vanHout-2@umcutrecht.nl

L. Bosch
Department of Experimental Cardiology, Utrecht University Medical Center, Utrecht, The Netherlands
e-mail: L.bosch@umcutrecht.nl

M. D. Cordero, E. Alcocer-Gómez (eds.), *Inflammasomes: Clinical and Therapeutic Implications*, Experientia Supplementum 108,
https://doi.org/10.1007/978-3-319-89390-7_2

Abstract Cardiovascular disease (CVD) is the number one cause of death worldwide. The pathogenesis of various disease entities that comprise the area of CVD is complex and multifactorial. Inflammation serves a central role in these complex aetiologies. The inflammasomes are intracellular protein complexes activated by danger-associated molecular patterns (DAMPs) present in CVD such as atherosclerosis and myocardial infarction (MI). After a two-step process of priming and activation, inflammasomes are responsible for the formation of pro-inflammatory cytokines interleukin-1β and interleukin-18, inducing a signal transduction cascade resulting in a strong immune response that culminates in disease progression. In the past few years, increased interest has been raised regarding the inflammasomes in CVD. Inflammasome activation is thought to be involved in the pathogenesis of various disease entities such as atherosclerosis, MI and heart failure (HF). Interference with inflammasome-mediated signalling could reduce inflammation and attenuate the severity of disease. In this chapter we provide an overview of the current literature available on the role of inflammasome inhibition as a therapeutic intervention and the possible clinical implications for CVD.

Keywords Cardiovascular disease · Inflammasome · Atherosclerosis · Heart failure · Myocardial infarction · Inflammation

2.1 Cardiovascular Disease and the Inflammasome

Cardiovascular disease (CVD) comprises all disease entities of the heart and blood vessels. Together they are the primary cause of death worldwide, supporting the intensive investigation of the mechanisms that play a central role in CVD pathogenesis.[1] Identification of these mechanisms will enable the development of novel therapies that can hamper disease progression and decrease the burden on society.

The most abundant disease of the cardiovascular system is the formation of lipid-rich plaques in the arterial vessel wall, named atherosclerosis. Atherosclerosis can occur anywhere in the human body, yet it often develops at certain locations, such as the carotid and coronary arteries. It becomes clinically manifest when a stable plaque is significant enough to decrease blood flow or when a vulnerable plaque ruptures, thereby inducing thrombus formation leading to vessel occlusion. This results in ischemia of the tissue downstream of the occluded blood vessel. In the heart this leads to myocardial infarction (MI) and possibly heart failure (HF). Occlusion of vessels in the brain will lead to stroke. Apart from these organs, peripheral artery disease can result in ischemic damage to other parts of the body.

Inflammasomes are pattern recognition receptors (PRRs) that are formed in response to a multitude of stimuli. Inflammasome-based signalling seems to play a crucial role during the development of atherosclerosis and acute infarction of the heart. The most frequently studied inflammasome in CVD is the NLRP3

[1]http://www.who.int/mediacentre/factsheets/fs317/en

inflammasome, composed of NLRP3 (nod-like receptor protein 3) an adaptor protein ASC (Apoptosis-associated speck-like protein containing a CARD), and the protease caspase-1. The activation of this multimeric complex initiates downstream responses including the maturation of the pro-inflammatory cytokines interleukin-1β (IL-1β) and interleukin-18 (IL-18) and inflammation-related cell death named pyroptosis (Tschopp et al. 2003; van Hout et al. 2016). Apart from atherosclerosis and acute MI, the role of the inflammasome in other inflammation-driven cardiovascular disease entities has been established. With regard to the heart, the most import diseases are affecting the heart muscle (cardiomyopathies) or involve inflammation of the myocardium (myocarditis) or pericardium (pericarditis). Concerning the blood vessels, the most important pathologies are dilation of the aorta (aortic aneurysm) and inflammation of the vessel wall (vasculitis).

In the current chapter, we will elaborate on the role of the inflammasome in CVD, especially focussing on atherosclerosis (leading to coronary artery disease), and its major consequence (MI). Additionally we will discuss the role of the inflammasome in HF and the less prevalent myocarditis and pericarditis, finishing with the evidence of inflammasome signalling in non-atherosclerotic vascular disease.

2.2 The Inflammasome in Atherosclerosis

2.2.1 Inflammation and Atherosclerosis

Atherosclerosis is the main cause of ischemic heart disease and stroke. It is characterized by the gradual development of lipid-rich plaques in the vessel wall. Lipoproteins, such as low-density lipoproteins (LDL), passively diffuse through the endothelial layer into the intima of the vessel wall (Lusis 2000). In the vessel wall, a complex set of biochemical reactions results in the oxidation of LDL. This so-called oxidized LDL, or oxLDL, serves as a pro-inflammatory mediator. OxLDL stimulates endothelial cells to produce pro-inflammatory molecules and increases the expression of adhesion factors on their cell surface (such as intercellular adhesion molecule (ICAM) and vascular cell adhesion molecule 1 (VCAM-1)) to attract monocytes and lymphocytes to the vessel wall. Adhered monocytes transmigrate into the intima, differentiate into macrophages and phagocytose oxLDL, thereby forming foam cells. In contrast to LDL, high-density lipoprotein (HDL) is protective against atherosclerosis by removing cholesterol and inhibiting lipoprotein oxidation. The continuous accumulation of oxLDL particles and the phagocytosis of these particles by macrophages through scavenger receptors like CD36 lead to the formation of an early atherosclerotic plaque or fatty streak. From this stage onwards, the plaque progresses, as more immune cells infiltrate, smooth muscle cells (SMC) start to proliferate and calcium depositions take place (Lusis 2000).

Inflammation is the driving force behind the progression from a simple fatty streak to a complex, instable atherosclerotic plaque. The accumulated macrophages and lymphocytes produce inflammatory cytokines, such as TNFα, IL-1β, IL-6 and

IFNγ, inducing a positive feedback loop by attracting more circulating cells to the newly formed plaque. SMCs migrate from the media of the vessel wall to the surface of the plaque and start to form a fibrous cap. SMCs secrete fibrous extracellular matrix (ECM) further enhancing plaque growth, which eventually leads to expansion of the plaque into the vessel lumen (Libby and Hansson 2015). When the size of the plaque increases, blood vessels start to infiltrate the plaque to enable central oxygenation. However, this neovascularization leads to frequent intraplaque haemorrhage, thereby damaging and destabilizing the plaque. Moreover, since adequate blood supply fails, central necrosis of the plaque occurs. The local production of cytokines results in thinning of the fibrous cap, further destabilizing the plaque, making it prone to rupture. When the plaque ruptures, acute lumen occlusion due to thrombus formation occurs, leading to clinical complications such as MI. The pathogenesis of plaque formation is not only supported by experimental studies but also by observational data in which human vulnerable plaques generally have an increased number of inflammatory cells, a large necrotic core and a thin fibrous cap. Furthermore, plaque instability is associated with more intraplaque haemorrhage and more calcifications (Lusis 2000; Janoudi et al. 2016).

As outlined above, the key processes that drive the formation of atherosclerosis are inflammation on the one hand and lipid metabolism on the other (Gistera and Hansson 2017). Recently, many studies have shown that the inflammasome may serve as a central signalling structure in these processes and may be the link between cholesterol metabolism and immune activation (Janoudi et al. 2016). The expression, activation and role of the inflammasome in the pathogenesis of atherosclerosis are described below. In addition, inhibition of the inflammasome in preclinical and clinical studies is discussed.

2.2.2 Inflammasome Expression in Atherosclerosis

Many studies have shown that NLRP3 inflammasome activity is increased in atherosclerosis and that the level of activation correlates with the severity of disease. The inflammasome components (NLRP3, ASC, caspase-1) and its downstream effector molecules (IL-1β and IL-18) are present in human atherosclerotic lesions, with increased expression levels compared to healthy arteries (Paramel Varghese et al. 2016; Mallat et al. 2001). Vulnerable plaques (characterized by erosions, bleeding or ulcers) also show increased expression of the inflammasome (on histology and mRNA level) compared to morphologically stable plaques (Shi et al. 2015). The same is seen in symptomatic plaques, expressing higher IL-18 mRNA levels compared to asymptomatic plaques (Mallat et al. 2001). The inflammasome is localized in different areas of the human atherosclerotic plaque. IL-1β, IL-18 and the IL-18 receptors are mainly localized in macrophage-rich regions (Folco et al. 2014), and the IL-18 receptor is also expressed in endothelial cells (Mallat et al. 2001). In circulating human peripheral blood mononuclear cells (PBMCs), NLRP3 protein levels correlate with the severity of coronary artery disease (on coronary angiography assessed by SYNTAX scores) and with clinical risk scores (GRACE score) (Afrasyab et al. 2016). Apart from local expression in the vessel wall and in circulating mononuclear cells, the levels of NLRP3, IL-1β and IL-18 (mRNA) in subcutaneous adipose tissue also correlate with and are independent predictors of the severity of coronary atherosclerosis (GEMINI and SYNTAX scores) (Bando et al. 2015).

2.2.3 *Inflammasome Activators in Atherosclerosis*

As mentioned, the inflammasome is considered an important link between lipid metabolism and inflammation (Janoudi et al. 2016). In the setting of atherosclerosis, multiple players in lipid metabolism are able to induce inflammasome activation. Cholesterol crystals, oxLDL and oxHDL can all activate the inflammasome and induce the secretion of IL-1β and IL-18 in human monocytes and macrophages in vitro (L'Homme et al. 2013; Thacker et al. 2016; Bleda et al. 2016; Xiao et al. 2013). OxLDL is recognized by CD36 receptors on recruited monocytes, leading through lysosomal pathways to NLRP3 inflammasome activation (Chen et al. 2014). In LPS-primed human monocytes, saturated fatty acids can also induce the release of IL-1β, whereas unsaturated fatty acids cannot (L'Homme et al. 2013). In contrast, HDL is able to suppress inflammasome activity in response to cholesterol crystals in human monocyte-derived macrophages (Thacker et al. 2016). Like monocytes, endothelial cells in vitro also show NLRP1 inflammasome activation after stimulation with plasma from patients with high triglyceride and cholesterol levels (Bleda et al. 2016).

Apart from lipid metabolism, other mechanisms are thought to be involved in inflammasome activation in atherosclerosis. Hypoxia, generally present in atherosclerotic plaques, is able to increase NLRP3 expression in human macrophages and limit degradation of pro-IL-1β, thereby prolonging its half-life (Folco et al. 2014). Atheroprone oscillatory shear blood flow is also able to induce NLRP3 inflammasome activation in endothelial cells. This is thought to happen via sterol regulatory element-binding protein 2 (SREBP2) (Xiao et al. 2013; Chen et al. 2014). The process of autophagy (controlled intracellular degradation of cell content) is often dysfunctional in atherosclerotic plaques. Mice lacking ATG5 (a protein important for autophagy) in macrophages showed decreased autophagy, resulting in an increased inflammasome activation and plaque size. Caspase-1 inhibition in these autophagy-deficient ATG5−/− macrophages reduced the IL-1β response. These results indicate that just like lipid products, hypoxia and oscillatory flow, dysfunctional autophagy can lead to inflammasome activation in atherosclerotic plaques (Razani et al. 2012).

2.2.4 The Inflammasome in the Pathogenesis of Atherosclerosis

Apart from associative evidence in human atherosclerotic plaques, most research on the role of the inflammasome in atherosclerosis is performed in knock-out mouse models to establish a causative role for the inflammasome. Widely used mouse models include LDLr−/− and ApoE−/− mice. Both mice develop atherosclerosis in a matter of weeks on a high-fat diet. The advantage of the ApoE−/− model is that complex vascular lesions also develop on a normal diet, but a high-fat diet results in more rapid lesion development with more foam cells present in the plaque. A downside of the ApoE−/− mice model is that ApoE has pleiotropic effects apart from plasma lipid levels. For instance, ApoE is described to have a function in macrophages and adrenal cells. The advantage of the LDLr−/− mice is that the LDL receptor does not have multiple functions as described for ApoE. However, on a normal diet, limited lesion development occurs in the LDLr−/− mice (Getz and Reardon 2015).

In LDLr−/− mice on a high-cholesterol diet, haematopoietic deletion of NLRP3, ASC or IL-1a/IL-β resulted in markedly decreased atherosclerosis and a reduction of inflammasome-dependent IL-18 levels (Duewell et al. 2010). Haematopoietic deletion of caspase-1/11 in LDLr−/− mice also resulted in a strong reduction in atherosclerotic plaque size with a reduced necrotic core (Hendrikx et al. 2015). Earlier studies in ApoE−/− mice indicated that IL-1 plays a role in fatty streak formation (Elhage et al. 1998; Kirii et al. 2003). In addition, the role of IL-18 was established in atherosclerosis development in ApoE−/− mice. ApoE−/− IL-18−/− double knock-out mice showed reduced IFN-γ responses and increased α-smooth muscle actin$^+$ (αSMA) SMCs, indicating a more stable plaque phenotype. Surprisingly, the serum cholesterol and triglyceride levels were higher in the IL-18-deficient mice (Elhage et al. 2003; Whitman et al. 2002). ApoE−/− caspase-1−/− mice or ApoE−/− mice in which the NLRP3 gene was silenced also showed smaller atherosclerotic plaque areas compared to ApoE−/− alone, with less pro-inflammatory cytokine production (such as IL-1β) and reduced macrophage numbers in the plaque. Silencing of NLRP3 increased SMCs and collagen, leading to a more stabilized plaque phenotype (Zheng et al. 2014). In contrast, Usui et al. showed that in ApoE−/− caspase-1−/− mice, the amount of vascular SMCs in the plaques was reduced (Usui et al. 2012). Surprisingly, a study by Menu et al. was unable to show a difference in plaque progression, stability or infiltration of macrophages in ApoE−/− NLRP3−/− and ApoE−/− ASC−/− and ApoE−/− caspase-1−/− double-deficient mice compared to ApoE−/− (Menu et al. 2011). The reason for this discrepancy is unclear; the only difference with the study by Usui et al. (2012) is the amount of cholesterol (0.15% or 1.25%) the food pellets contained. Mice deficient in the $P2X_7$ receptor (ApoE−/− $P2X_7r$−/− mice), a receptor involved in activation of the NLRP3 inflammasome, showed less aortic atherosclerosis compared to ApoE−/− mice (Peng et al. 2015; Hansson and Klareskog 2011; Stachon et al. 2017). Apart from LDLr−/− and ApoE−/− models of atherosclerosis, the role of the inflammasome was also studied in a model of vascular injury and neointima formation. This study showed that bone marrow-derived ASC is critical for neointima formation after vascular injury (Yajima et al. 2008).

These in vivo mice data support a role for the inflammasome and its downstream molecules IL-1β and IL-18 in the development of atherosclerosis. Different cell types, such as macrophages, SMCs and endothelial cells are proposed to be involved in this proatherogenic role of the inflammasome. Macrophages show reduced migratory capacity and increased susceptibility to lipid deposition after NLRP3 inflammasome activation in vitro; this can facilitate retention in the arterial wall and foam cell formation (Li et al. 2014). In vitro stimulation of SMCs with IL-1β induces VCAM-1 and monocyte chemoattractant protein-1 (MCP-1) expression and can in this way facilitate the recruitment of inflammatory cells to the atherosclerotic lesions (Wang et al. 1995). Inflammasome activation in SMCs can also lead to calcifications, a process involved in plaque progression (Wen et al. 2013). As mentioned before, endothelial cells can activate the inflammasome upon atheroprone oscillatory shear flow. This activation of the innate immune response can result in endothelial dysfunction, an important first step of atherogenesis (Xiao et al. 2013). Figure 2.1 summarizes the role of the inflammasome in atherosclerosis development.

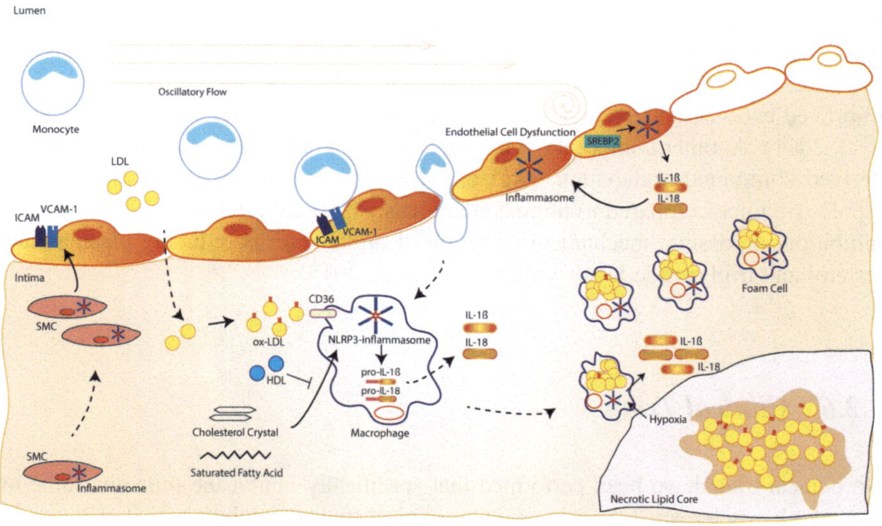

Fig. 2.1 Inflammasome activation in the pathogenesis of atherosclerotic plaque development. Low-density lipoprotein (LDL) migrates through the endothelial layer into the intima of the vessel wall, where it is oxidized, forming oxLDL. OxLDL and oscillatory blood flow activate the NLRP3 inflammasome in endothelial cells (the latter through sterol regulatory element-binding protein (SREBP2)), stimulating the expression of the adhesion molecules ICAM and VCAM-1. This facilitates monocyte adherence and migration to the intima of the vessel. OxLDL, cholesterol crystals and saturated fatty acids again induce NLRP3 inflammasome activation in macrophages present in the vessel wall. This results in the formation of foam cells and the production of IL-1β and IL-18. This results in local inflammation, inducing SMC migration and the formation of a fibrous cap. Activation of the inflammasome in SMCs, macrophages and endothelial cells initiates a vicious circle of endothelial dysfunction, monocyte recruitment and more foam cell formation, leading to the formation of a hypoxic, necrotic lipid core, intraplaque haemorrhage and eventual thinning of the fibrous cap, making the plaque prone to rupture

2.2.5 *Inflammasome Inhibition in Atherosclerosis*

Since genetic mouse models suggest a role for the inflammasome in the development of atherosclerosis, various inflammasome inhibitors for the prevention and treatment of atherosclerosis have been proposed. Injection of an anti-IL-1β monoclonal antibody in ApoE−/− mice inhibited the formation of atherosclerotic lesions and is associated with lower plasma non-HDL/HDL cholesterol ratios (Bhaskar et al. 2011). The most specific NLRP3 inflammasome inhibitor described today, MCC950 (Coll et al. 2015), has been tested in an ageing mouse model of atherosclerosis. Mice with TET2−/− bone marrow (causing somatic mutations in haematopoietic cells representing ageing) on an LDLr−/− background showed increased levels of inflammasome activity compared to LDLr−/− on a high-fat diet alone. The NLRP3 inflammasome inhibitor MCC950 significantly reduced atherosclerotic plaque size by 50% in the TET2−/− LDLr−/− mice compared to saline. In the LDLr−/− mice alone, a nonstatistically significant 20% reduction of plaque size by MCC950 was witnessed (Fuster et al. 2017). DPP-4 inhibitors, widely used in the treatment of patients with type-2 diabetes, can suppress NLRP3 activation and IL-1β release in a human monocyte cell line. LDLr−/− mice treated for 12 weeks with a DPP-4 inhibitor showed markedly decreased aortic plaque size with a reduced plaque macrophage content. Furthermore, DPP-4 inhibition in atherosclerosis led to reduced proliferation of vascular smooth muscle cells, inflammatory reaction, improved endothelial function and reduced thrombogenesis (Dai et al. 2014; Shah et al. 2011). Another compound that influences inflammasome activity is the plant-derived compound arglabation. Arglabation is able to attenuate atherosclerosis in ApoE−/− mice compared to untreated animals. Apart from NLRP3 inflammasome inhibition, a possible mechanism of action of arglabation is reducing plasma cholesterol and triglyceride levels (Abderrazak et al. 2015).

2.2.6 *Clinical Trials*

No clinical trials have been performed that specifically inhibit the inflammasome in human atherosclerosis. However, a major clinical trial on inhibition of IL-1β, one of the downstream effector molecules of the inflammasome, has recently produced very interesting results. The CANTOS (The Canakinumab Antiinflammatory Thrombosis Outcome Study) trial showed that the IL-1β antibody canakinumab in patients with stable coronary artery disease led to a significant lower rate of recurrent cardiovascular events than placebo. This double-blinded randomized trial enrolled 10,061 patients with coronary artery disease. Patients with elevated high-sensitive C-reactive protein (CRP) (>2 mg/L), despite contemporary secondary prevention, were included. CANTOS proves the hypothesis that after sufficient lipid lowering, there remains a 'residual inflammatory risk' and shows for the first time that targeting inflammatory processes in patients with cardiovascular disease significantly improves outcome

(Ridker et al. 2017). Blocking IL-1β is not the same as inflammasome inhibition. Direct inflammasome inhibition can theoretically prevent a broader range of potentially pathologic processes such as IL-18 signalling and pyropotosis. Moreover, IL-1beta signalling is not completely abolished by inflammasome inhibition, thereby presumably reducing severe infections, which were major adverse events in CANTOS.

Colchicine was also tested for its effects on recurrence rates in MI patients. Colchicine is a drug long known for its effectiveness in gout. Apart from other mechanisms, colchicine has shown to exert its effects through upstream inhibition of the NLRP3 inflammasome, thereby reducing the secretion of IL-1β and IL-18 (Martinon et al. 2006). In this trial, low-dose colchicine (LoDoCo) was tested in patients with stable coronary artery disease. The LoDoCo trial showed a strong effect of colchicine on the primary composite endpoint of acute coronary syndrome, out-of-hospital cardiac arrest or ischemic stroke (Nidorf et al. 2013). These promising effects of colchicine will be validated in a large, multicentre, double-blind LoDoCo2 trial which is currently being conducted.

2.2.7 Conclusion

Numerous animal and human studies show associative evidence for the inflammasome and the formation and severity of atherosclerosis. Mechanistic experiments have identified, e.g. oxidized lipid products and cholesterol crystals as the inducers of inflammasome activation in macrophages, leading to foam cell formation and IL-1β and IL-18 production. This results in a pro-inflammatory milieu that induces a vicious circle of endothelial dysfunction, recruitment of more inflammatory cells and additional foam cell formation.

Mouse models of atherosclerosis have established a key role for inflammasome activation in plaque development. These findings have led to clinical testing with the anti-IL-1β antibody canakinumab in patients with stable coronary artery disease in the CANTOS trial. This large phase III clinical trial showed for the first time that targeting inflammatory processes in patients with atherosclerosis significantly improves outcome and sets the stage for future trials specifically targeting the inflammasome in the setting of cardiovascular disease.

2.3 The Inflammasome in Myocardial Infarction and Heart Failure

2.3.1 From Myocardial Infarction to Heart Failure

Myocardial infarction (MI) is a major consequence of progressive atherosclerosis. MI occurs when an instable atherosclerotic plaque ruptures. This enables clot formation, leading to sudden coronary artery occlusion, thereby hampering nutrients and oxygen delivery to the myocardium. In turn, this results in irreversible ischemic cell death and life-threatening deterioration of cardiac function. The primary treatment for MI should be immediate to confine as much damage as possible. Therapies have evolved rapidly in the past few decades from conservative approaches to minimally invasive low-risk percutaneous coronary interventions.

The ultimate treatment goal is to salvage viable myocardium from ischemic damage by re-establishing coronary perfusion (reperfusion) as soon as possible, thereby limiting infarct size and preserving cardiac function. It has been unequivocally proven that this rapid reperfusion is very beneficial to patients and reduces mortality (Steg et al. 2012). Paradoxically, a large body of evidence suggest that reperfusion itself can also damage viable myocardium (Vander Heide and Steenbergen 2013). Among other mechanisms, the damage induced by ischemia and reperfusion (ischemia-reperfusion injury or IRI) in the heart is due to activation of an exaggerated inflammatory reaction.

This acute phase of reperfusion injury (minutes to hours) is followed by a more chronic phase (days to weeks), in which the heart adapts to the loss of contractile function. A collagen-based scar is formed, and alteration of regional contractility, pressure and volume leads to geometric adaptations of the left ventricle. This process is termed adverse cardiac remodelling and is a major risk factor for the development of HF.

The syndrome of HF results from various structural and functional impairments of the cardiac muscle. Not only MI but also different aetiologies like hypertension, infections and genetic causes can lead to persistent cardiac damage and subsequent HF. The disease remains a major cause of morbidity and mortality, and despite improvements in therapy, overall prognosis continues to be poor (mortality rates approaching 50% in 5 years) (Braunwald 2015).

2.3.2 Inflammation as a Key Process

The post-MI inflammatory response is to some extent essential since the irreversibly damaged myocardium should be replaced by a strong collagen-based scar to prevent cardiac muscle rupture. However, inflammation after MI also results in additional cardiomyocyte death and further deterioration of cardiac function in the acute (minutes to hours), subacute (days to weeks) and chronic (weeks to months or years) phase after MI (van der Laan et al. 2012a, b; Maekawa et al. 2002; Takahashi et al. 2008).

Following cardiac ischemia, intracellular molecules such as ATP, mitochondrial DNA and high levels of potassium are released from the damaged cardiomyocytes. Reperfusion amplifies this effect, and as a result, these molecules are swiftly released into the systemic circulation. These so-called danger molecules or danger-associated molecular patterns (DAMPs) can be recognized by certain PRRs. These PRRs then induce a pro-inflammatory state by activation of intranuclear transcription factors leading to activation of cytokines.

The activation of PRRs results in the induction of a pro-inflammatory state and the subsequent influx of circulating leucocytes. These circulating cells, of which neutrophils and monocytes are the first responders, cause injury to endothelial cells due to the excretion of reactive oxygen species (ROS), cytokines and proteases (Vinten-Johansen 2004). This will lead to the expression of cell adhesion molecules by endothelial cells, enabling the transmigration of circulating leucocytes, into the damaged myocardium. Here these inflammatory cells generate more ROS and proteases resulting into tissue breakdown and cellular clearance, thereby directly contributing to myocardial IRI (Arslan et al. 2011). Additionally, neutrophils produce matrix metalloproteinases (MMPs) that degrade the intermyocyte collagen struts thereby destabilizing the ventricular wall. This leads to infarct expansion, a process that is known to occur within hours from initial myocyte injury (Fig. 2.2) (Sutton and Sharpe 2000).

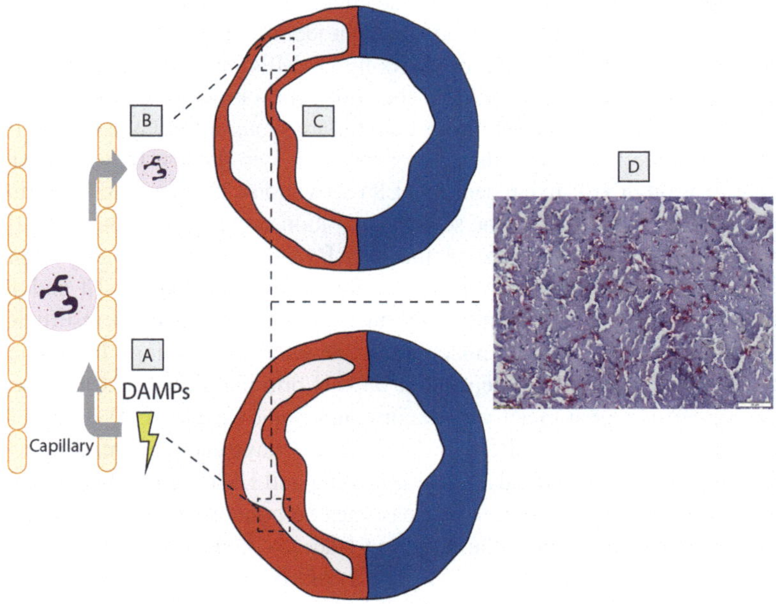

Fig. 2.2 Schematic overview of the occurrence of infarct extension after MI. Due to ischaemia-reperfusion injury, the infarct becomes larger according to a wave-front principle, with early subendocardial involvement and progression towards a transmural infarction. Due to ischaemia-reperfusion, DAMPs are released from the damaged myocardium and enter the systemic circulation (**a**). Here they activate circulating leucocytes, which subsequently transmigrate out of the circulation into the damaged tissue (**b**). Once resided, these inflammatory cells cause damage, thereby inducing infarct enlargement (**c**). (**d**) Histological sample of infarct tissue containing neutrophils in porcine myocardium 72 h after ischaemia-reperfusion injury (van Hout et al. 2016)

Depending on the size and location of the infarct, the loss of contractile myocardium leads to an alteration of ventricular pressure and increased wall stress. The increased wall stress again results in DAMP release and chronic low-grade inflammation. This gradually induces cardiomyocyte slippage in the left ventricle. Together with infarct expansion, this leads to further left ventricular wall thinning and progressive dilatation within the first days to weeks after MI. This process, referred to as adverse cardiac remodelling, initiates systemic neurohormonal adaptations and eventually culminates in HF (Sutton and Sharpe 2000; Heusch et al. 2014).

As in cardiac IRI and adverse cardiac remodelling, different chronic HF animal models indicate a role for the innate immune system in the pathophysiology of HF. In this phase of cardiac disease, DAMPs also play a central role and modulate interstitial cardiac fibrosis, cardiomyocyte apoptosis and hypertrophy. Low-grade chronic inflammation is present in HF, and pro-inflammatory cytokines (such as TNFα, IL-6, IL-1β, CRP) are increased in patients, and their levels relate to HF severity and prognosis (Hofmann and Frantz 2013; Butts et al. 2015).

The exaggerated inflammatory response after MI thus enhances acute IRI and is associated with an increased risk of adverse remodelling, HF and a worse prognosis. As outlined above, the activation of PRRs by DAMPs plays a central role in both these processes. Among these PRRs, the inflammasomes, of which the NLRP3 inflammasome has been studied in most detail, are thought to play a crucial role in MI. Inflammasome-based signalling involves a two-step process by which it is first primed (e.g. through Toll-like receptor (TLR) activation). This leads to the formation of inactive NLRP3 and IL-1β. The second activation step results in the formation of the inflammasome by adherence of the NLRP3 protein to ASC and caspase-1. Cleavage of caspase-1 then results in the formation of active IL-1β and IL-18. Importantly, triggering of the inflammasome alone in the heart is insufficient to induce cardiac dysfunction in mice in the absence of priming. Inflammasome formation in the heart is thus dependent on this priming signal and a subsequent separate triggering signal to activate the inflammasome (Toldo et al. 2015).

The evidence that the (NLRP3) inflammasome plays such a central role in the inflammatory reaction in MI, adverse remodelling and HF is generally derived from the observation that (1) different inflammasome components are upregulated, (2) the absence or inhibition of the inflammasome leads to damage reduction or (3) the effector molecules (IL-1β and IL-18) play a role in the pathogenesis of MI and HF. These three observations will be discussed here in more detail.

2.3.3 Inflammasome Upregulation in Myocardial Infarction and Heart Failure

The first direct evidence implicating the NLRP3 inflammasome as a key component in post-MI inflammation was provided by Kawaguchi et al. (2011). The investigators observed an upregulation of the inflammasome component ASC in human cardiac tissue after MI. This was confirmed in a murine model of IRI. Importantly, both ASC−/− and caspase-1−/− mice subjected to 30 min of IR of the left coronary artery had smaller infarcts as a percentage of the area at risk at 48 h of reperfusion compared to wild-type mice. These hearts also showed less neutrophil and macrophage infiltration. Chimer experiments pointed to an upregulation of the inflammasome in both cardiac resident cells (fibroblast) and circulating cells, equally contributing to myocardial IRI.

In the same year, Mezzaroma et al. also provided strong evidence on the role of the NLRP3 inflammasome in MI (Mezzaroma et al. 2011). This study revealed increased inflammasome activation in isolated mouse cardiomyocytes when exposed to either ischemic conditions or a combination of the TLR primer LPS and ATP, a $P2X_7$ receptor activator. The investigators also observed a significantly increased expression of ASC in mice subjected to permanent coronary artery ligation. ASC activation was upregulated in cardiomyocytes, as well as cardiac fibroblast and infiltrated leucocytes, both after 3 and 7 days following MI.

It was revealed that, although cardiomyocytes do not secrete IL-1β or IL-18, the NLRP3 inflammasome is most certainly activated in this cell type. Instead of cytokine secretion, in cardiomyocytes, the activation of the NLRP3 inflammasome directly leads to pyroptosis (Mezzaroma et al. 2011). These experiments showed that the key role of the inflammasome in MI is not only due to an indirect effect, which is the secretion of the pro-inflammatory cytokines IL-1β or IL-18, but also directly on the viable myocardium by inducing cell death (Fig. 2.3).

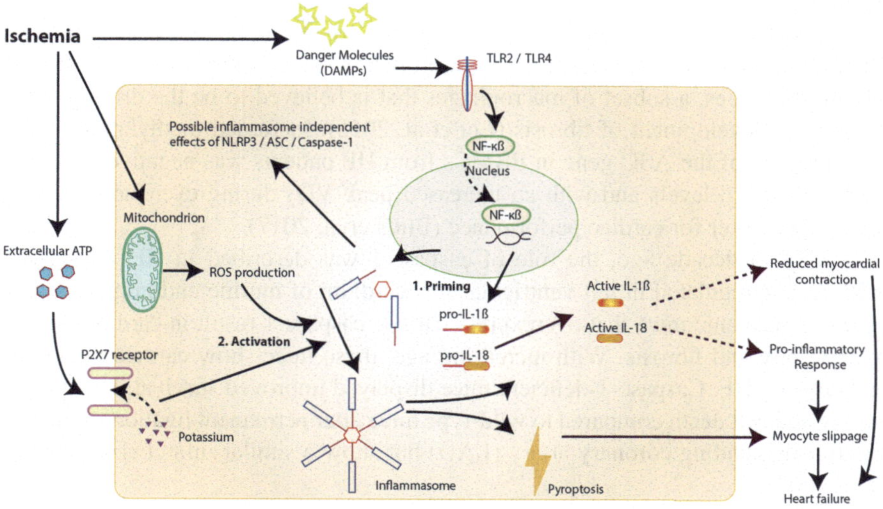

Fig. 2.3 Simplified schematic overview of inflammasome activation after myocardial infarction. Ischaemia results in the release of DAMPs that activate PRRs (e.g. Toll-like receptors (TLRs)) and induce nuclear migration of NF-κβ, resulting in priming of the inflammasome and production of pro-IL-1β. The inflammasome is then activated, for example, by activation of the $P2X_7$ receptor or reactive oxygen species (ROS) production after mitochondrial damage. This leads to the release of active forms of IL-1β and IL-18 and pyroptosis. Figure is adapted from (van Hout et al. 2016)

In addition to these data, clinical evidence indicates that certain polymorphisms in the NLRP3 gene protect against the development of MI, especially in women (Varghese et al. 2013). Another murine study showed that NLRP3−/− hearts were not susceptible to ischemic preconditioning, while wild-type and interestingly also ASC−/− hearts did show an infarct size reduction tested ex vivo in murine hearts (Zuurbier et al. 2012), indicating that the NLRP3 protein may be essential in cardioprotection. Moreover, when subjected to ex vivo global IR, NLRP3−/− hearts also show a preservation of cardiac function compared to ASC−/− hearts subjected to the same conditions (Sandanger et al. 2013). Interestingly, these data are contradictory with studies that showed that a lack of ASC was protective in mice subjected to MI.

Importantly, also negative results on the role of the NLRP3 inflammasome have been reported. In a closed-chest murine model of IR, NLRP3−/− mice did not show a reduction in infarct size compared to wild-type mice, possibly indicating that the NLRP3 inflammasome fulfils a role in the subacute and not acute time frame after MI (Jong et al. 2014). In extension to these findings, one study also reported larger infarcts as percentage of the area at risk in NLRP3−/− mice subjected to MI compared to wild-type or ASC−/− mice after both 3 and 24 h (Sandanger et al. 2016). These observations suggest a complex role for the inflammasome and its different components (NLRP3, ASC, caspase-1). It has been postulated that especially NLRP3 may have inflammasome-dependent and inflammasome-independent effects (Mezzaroma et al. 2014). These results also imply that the role of the inflammasome in MI is greatly dependent on the animal model and experimental conditions.

The NLRP3 inflammasome also plays a role in adverse remodelling and HF. The NLRP3 inflammasome enhanced fibrosis through increased expression in infiltrated M1 macrophages, a subset of macrophages that is believed to be the driving force behind the development of fibrosis (Liu et al. 2015). Increased methylation of the intron region of the ASC gene in PBMCs from HF patients was negatively associated with IL-1β levels and with an increased peak VO_2 during exercise testing, a surrogate marker for cardiac performance (Butts et al. 2017).

Already a decade ago, the role of caspase-1 was described in HF. Caspase-1 mRNA is upregulated in left ventricular myocardium of murine and human failing hearts. Transgenic mice that overexpress cardiac caspase-1 result in cardiomyocyte hypertrophy and fibrosis. With increasing age, these mice show cardiac dilatation and develop HF. Caspase-1-deficient mice displayed improved survival, less hypertrophy and cell death compared to wild-type mice after permanent ligation of the left anterior descending coronary artery (LAD) but show a similar infarct size (Merkle et al. 2007).

2.3.4 Inhibition of the Inflammasome

Apart from observational data, mechanistic experiments and studies with knock-out mice, important evidence for the role of the NLRP3 inflammasome in MI, also come from the pharmacological inhibition of inflammasome formation in preclinical MI models. Mice subjected to permanent coronary artery ligation showed a marked infarct size reduction when pretreated with silencing RNA for either NLRP3 or the $P2X_7$ receptor that, after activation by ATP, opens a cation channel allowing for potassium efflux that can lead to NLRP3 inflammasome activation in MI (Mezzaroma et al. 2011). Similar results were obtained when a pharmacological inhibitor of $P2X_7$ was administered.

A newly developed small-molecule inhibitor named 16,673-34-0 has also shown to decrease infarct size in a pretreatment mouse model of permanent coronary artery ligation (Marchetti et al. 2014). These findings were later confirmed in both a permanent ligation and IR model of MI (Marchetti et al. 2015). Importantly a recent study revealed that when administering this compound up to 1 h after reperfusion, it could still effectively decrease the infarct size in a murine model of 30-min transient coronary artery occlusion. This effect could be detected no sooner than 24 h, suggesting that NLRP3 inflammasome inhibition in the first hours is not sufficient to render a clinically significant effect (Toldo et al. 2016).

Recent evidence has also revealed that reperfusion therapy with recombinant human relaxin-2 (serelaxin) reduces infarct size in a murine model of IRI through inhibition of the NLRP3 inflammasome (Valle Raleigh et al. 2017). Interestingly, also L5-LDL, an electronegative LDL particle that is increased in patients suffering from MI, is able to activate the NLRP3 inflammasome and could therefore be a clinically relevant DAMP of the inflammasome in acute MI (Yang et al. 2017). Moreover colchicine administered to mice undergoing permanent coronary artery ligation leads to a reduction in inflammasome expression as well as infarct size compared to these parameters in control animals (Fujisue et al. 2017). In addition, in patients with an acute coronary syndrome, colchicine is able to inhibit the production of IL-1β and IL-18 (Martinez et al. 2015).

Calcineurin-transgenic (CNTg) mice develop progressive cardiac dysfunction. NLRP3−/− CNTg double-deficient mice show improved cardiac function assessed by fractional shortening compared to CNTg mice, indicating a role for the inflammasome in this HF model. IL-1 receptor antagonism for 2 weeks in CNTg mice resulted in significantly reduced left ventricular dilatation and an improved FS compared to saline. Mononuclear cell infiltrate was reduced in the treated mice, but no changes were observed at the level of hypertrophy (Bracey et al. 2013).

To translate these findings to a clinical application, a study with a highly translational pig model of MI with a clinically feasible treatment protocol has recently been performed. In this study pigs were subjected to a 75-min transient coronary artery occlusion. The selective NLRP3 inflammasome inhibitor MCC950 showed a dose-dependent effect on both infarct size and cardiac function. Moreover, this resulted in decreased inflammasome expression and a reduction of cardiac inflammation (van Hout et al. 2017).

2.3.5 *Targeting the Downstream Cytokines IL-1β and IL-18*

Activation of the NLRP3 inflammasome leads to the formation and subsequent release of active IL-1β and IL-18. The role of het NLRP3 inflammasome in cardiac IRI and adverse remodelling could therefore not only be determined by direct inhibition of the inflammasome but also through interference with signalling of both these cytokines.

2.3.5.1 Interleukin-1β

IL-1β is thought to play a central role in post-MI inflammation (Frangogiannis 2015). Interference with IL-1β signalling by genetic deletion of the IL-1 receptor (IL-1R1) protected against both cardiac IR and permanent ligation in mice (Bujak et al. 2008; Abbate et al. 2011). Overexpression of the naturally occurring IL-1R antagonist also resulted in a preservation of cardiac function, and deletion of this antagonist culminates in deterioration of cardiac function in mice (Abbate et al. 2011; Suzuki et al. 2001). Moreover, administration of the IL-1 receptor blocker anakinra resulted in enhanced cardiac performance and reduced cardiomyocyte apoptosis, presumably independent of infarct size (Abbate et al. 2008; Salloum et al. 2009). Additionally, pretreatment with anakinra also led to infarct size reduction in a mouse model of MI (Feng et al. 2010).

After the development of anakinra, another IL-1 inhibitor was developed, consisting of the IL-1 receptor, the IL-1 receptor-associated protein and the Fc fragment of an immunoglobulin (Van Tassell et al. 2010). This recombinant protein, named the 'IL-1 trap' (rilonacept), showed to have beneficial effects on remodelling in a mouse model of MI. These data suggest that the role of IL-1β is pivotal, since it not only deteriorates cardiac function through directly decreasing cardiac contractility but also through the enhancement of myocardial infarct size.

Importantly, by blocking the IL-receptor, IL-1α signalling is also hampered, so part of the effect that was seen in these studies could be due to interference with signalling of this IL-1 isoform (Van Tassell et al. 2013a, 2015). To further investigate this, several IL-1β antibodies have been developed, enabling identification of the specific role of IL-1β in cardiac remodelling. The first study on an antibody directed at IL-1β reported impaired healing of the heart and favoured cardiac rupture in a permanent MI model (Hwang et al. 2001). Since studies that investigated inference with combined IL-1β and IL-1α signalling showed opposite results, it was suggested that selective IL-1β blockade would induce adverse effects. Recently, more thoroughly characterized antibodies, specially developed for in vivo use, were tested. In both of these studies, beneficial effects were seen in mice subjected to MI and treated with these compounds (Toldo et al. 2013; Abbate et al. 2010a). The adverse effects seen in the first study were therefore believed to be caused by pleiotropic effects of the antibody.

Since anakinra has been registered for clinical usage in rheumatoid arthritis, off-target testing of these compounds was feasible, and clinical evidence on the role of IL-1β in the healing process of MI is also available. Two pilot studies (VCU-ART and VCU-ART2) have been performed (Abbate et al. 2010b, 2013). Both of these studies included ST-segment elevation MI patients that were clinically stable and had undergone percutaneous coronary intervention with successful reperfusion. Patients were treated with anakinra and were followed up for 3 months. Although no significant results were seen regarding major adverse cardiac events, the anakinra-treated patients did show lower levels of CRP at 72 h and were less likely to develop new-onset HF. Although not significant when corrected for baseline differences, a trend towards improved left ventricular geometry was also seen in these patients. Future phase III trials should further investigate if anakinra is effective in MI patients.

In HF patients, IL-1β also seems to play an important role. In patients with idiopathic dilated cardiomyopathy, plasma Il-1β levels correlate with left ventricular mass and severity of mitral valve regurgitation. In these patients, IL-1β is also a predictor of outcome (death or cardiac transplantation) (Aleksova et al. 2017). From mice models we know that IL-1β negatively influences myocardial contractility. Injection of IL-1β (3 μg/kg) in healthy mice reduces cardiac function measured by left ventricular fractional shortening (LVFS) already at 4 h after injection. Stressing these mice with β-receptor stimulation (using isoproterenol), an impaired contractile reserve with a right shift of the dose-response curve was revealed. After stopping IL-1β injections, LVFS returned to baseline levels. This indicates that IL-1β is able to induce a reversible contractile dysfunction (Van Tassell et al. 2013b). Two pilot studies investigated the effect of anakinra on cardiopulmonary exercise testing performance in patients with HF with a reduced ejection fraction (HFrEF) (Van Tassell et al. 2012) and in patients with HF with a preserved ejection fraction (HFpEF) (Van Tassell et al. 2014). In both HF groups, anakinra led to an improvement of peak oxygen consumption (VO_2) and resulted in a reduction of CRP. The improved aerobic exercise capacity in these patients by anakinra could be predicted by baseline exercise capacity and not by baseline CRP or BNP (Canada et al. 2014).

2.3.5.2 Interleukin-18

Similar to IL-1β, IL-18 is also secreted after activation of the (NLRP3) inflammasome as a result of caspase-1 cleavage. Unlike IL-1β, however, the inactive precursor of IL-18 is not formed through cellular priming (e.g. by TLR activation) but is abundantly present in inactivated cells of almost every cell type. Also similar to IL-1β, the activity of IL-18 is balanced by its counterpart, the IL-18-binding protein (Dinarello et al. 2013). Inflammation caused by the downstream effects of both IL-1β and IL-18 therefore not only depends on inflammasome activation but also on the balance between these cytokines and their naturally occurring antagonists (Dinarello and van der Meer 2013).

Several experimental studies have been performed, showing that blocking IL-18 signalling is protective in cardiac injury. Administration of IL-18 in mice results in left ventricular hypertrophy and increased collagen formation, both predictors of long-term cardiac failure (Platis et al. 2008; Woldbaek et al. 2005). In another study, mice infused with the well-characterized danger molecule lipopolysaccharide showed a depressed cardiac function. When IL-18 signalling was neutralized in this study, animals showed a preserved cardiac function, presumably through decreased release and expression of TNFα and adhesion molecules (Raeburn et al. 2002).

Interestingly, IL-18 and IL-1β not only enhance myocardial damage and suppress cardiac function separately but also work in a synergistic way (Toldo et al. 2014a). In these experiments, mice lacking IL-18 did not show decreased cardiac contractility when treated with recombinant IL-1β, whereas the control group did. Importantly, downstream IL-6-mediated signalling was not affected in this study. This suggests that IL-18 is essential for IL-1β-mediated reduced contractility, but not for IL-1-β-mediated inflammation, the two processes by which these cytokines directly and indirectly induce cardiac dysfunction. Inflammasome inhibition could therefore be more effective than blocking either one of these cytokines by both directly preserving cardiac contractility as well as attenuation of the inflammatory response.

Human studies also show evidence for an important role of IL-18 in relationship to cardiac damage and contractility. HF patients have elevated levels of IL-18, and a correlation between these levels and mortality exists (Mallat et al. 2004). Experiments with ex vivo human atrial muscle strips revealed increased contractility and increased intracellular tissue creatine kinase when IL-18-binding protein was added to the perfusate after inducing cardiac ischaemia (Pomerantz et al. 2001). In a large cohort of 1229 patients with a median follow-up of almost 4 years, levels of plasma IL-18 correlated with future cardiac events and mortality (Blankenberg et al. 2002).

2.4 Inflammasome in Myocarditis

Myocarditis is characterized by an acute or chronic inflammatory response of the heart to environmental (such as viruses) or endogenous triggers (such as autoimmune myocarditis). The pathogenesis of myocarditis varies per trigger. In virus-mediated myocarditis, within hours after viral entry in the cardiomyocyte, type 1 interferon is produced leading to myocyte cell death. The second phase evolves after hours to days, involving activation of innate immune responses, including the inflammasome. The inflammatory response in myocarditis can rapidly escalate into an auto-inflammatory cycle, leading to chronic autoantigen-driven inflammation. This can progress in dilated cardiomyopathy and HF (Heymans et al. 2016).

2.4.1 Inflammasome Activation in Myocarditis

Endomyocardial biopsies from patients with acute lymphocytic myocarditis or myocarditis diagnosed in post-mortem samples showed signs of inflammasome activation by ASC aggregation in leucocytes, cardiomyocytes, fibroblasts and endothelial cells, whereas in control samples, these aggregates were absent. The number of inflammasome-activated cells was higher in patients presenting with severe HF (NYHA III–IV compared to I–II) and in patients with no recovery of LVEF after 6 months (Toldo et al. 2014b). In vitro experiments with cardiomyocytes exposed to Coxsackie B (CVB3) virus (a well-known trigger for myocarditis) reveal an upregulation of inflammasome activity. In a mouse model of CVB3-induced viral myocarditis, levels of ASC, caspase-1 and IL-1β were upregulated in cardiac tissue. Importantly, IL-1β production correlated positively with myocarditis severity (Wang et al. 2014).

2.4.2 *Inflammasome Inhibition in Myocarditis*

CVB3-induced viral myocarditis mice treated with a caspase-1 inhibitor (Ac-YVAD-CHO) or an IL-1β blocking antibody showed less severe myocarditis expressed by creatine kinase levels and increased cardiac LVEF compared to placebo (Wang et al. 2014). In humans only case reports of inflammasome pathway inhibition by IL-1β blockade using anakinra have been described in fulminant, viral myocarditis. Standard clinical management for these patients includes mechanical support, but no specific treatments are available. A patient with fulminant myocarditis that developed severe biventricular dysfunction with systemic inflammation leading to cardiogenic shock received anakinra. Already 24 h after initiation of anakinra (100 mg/day), clinical improvement was witnessed with fever reduction, lowering of infection parameters and improvement of LVEF (Cavalli et al. 2017; Noji 2016). Another report describes a similar case with fulminant myocarditis treated with anakinra; within 4 days of treatment, clinical improvement and weaning from mechanical support were achieved (Cavalli et al. 2016). These case reports might indicate that IL-1 blockade is effective for the treatment of fulminant myocarditis, although further confirmation in the setting of clinical trials is needed. No reports on the role of the inflammasome in autoimmune myocarditis are available.

2.5 Inflammasome in Pericarditis

Pericarditis is inflammation of the pericardium. Most cases (80–90%) are thought to be idiopathic, although unidentified viral infection may to some extent be responsible. Among severe complications is recurrent pericarditis. Recurrent pericarditis affects up to 30% of patients after a first episode of acute pericarditis and is a difficult clinical problem. The cause of recurrent pericarditis is unknown but appears to be autoimmune mediated. The primary treatment for pericarditis is colchicine as an adjunctive therapy to NSAIDs. One of the working mechanisms of colchicine is through upstream inhibition of the NLRP3 inflammasome (Stack et al. 2015).

Colchicine effectively reduces recurrence rates in patients with recurrent pericarditis or acute pericarditis (Alabed et al. 2014). However, there are patients with recurrent pericarditis with colchicine resistance and corticosteroid dependence. In these patients, the effect of anakinra was studied in two small clinical trials. These preliminary reports appear promising. However, further larger randomized controlled trials are required (Baskar et al. 2016; Brucato et al. 2016).

2.6 The Inflammasome in Abdominal Aortic Aneurysms

Abdominal aortic aneurysms (AAA) are permanent and localized aortic dilations that mainly develop below the renal arteries. Often the aneurysms remain asymptomatic and undiagnosed, but with increasing size, the risk of rupture dramatically increases. Histological features include chronic medial and adventitial inflammatory cell infiltration (neutrophils, T- and B-cells, macrophages, mast cells, NK cells) and elastin degeneration. Chronic inflammation is a driving force in the pathogenesis of AAA, resulting in progressive remodelling and deterioration of the aortic wall (Libby and Hansson 2015; Shimizu et al. 2006). The inflammasome is thought to play a role in these inflammatory pathways. This is suggested by increased plasma IL-1β levels in AAA patients compared to controls (Wu et al. 2016). Immunohistochemistry also revealed higher expression of NLRP3, ASC, caspase-1 and caspase-5 and AIM2 in AAA compared to control aortas (Dihlmann et al. 2014; Wu et al. 2017). In PBMCs isolated from AAA patients, caspase-1 and IL-1β mRNA levels were also increased compared to controls. These differences were especially pronounced in males with AAA and not present in females. In contrast, PBMC AIM2, NLRP3 and ASC mRNA levels did not differ between AAA patients and controls (Wu et al. 2016).

The structural integrity of the aortic wall depends on vascular smooth muscle cells (SMCs) and the extracellular matrix. SMC contractile dysfunction is thought to play a role in aortic aneurysm and dissection development. Stressing SMCs leads to tropomyosin and myosin heavy chain degradation. Caspase-1 is able to cleave these contractile proteins. Reduction of inflammasome activity (by siRNA or pharmacological inhibition) prevents the degradation of tropomyosin and myosin heavy chain in SMCs in vitro. These findings indicate that the inflammasome might be involved in SMC dysfunction.

In vivo mice models are available to study the role of the inflammasome in AAA. Mice on a high-fat diet receiving angiotensin II (AngII) infusion develop aortic aneurysms and dissections. In these aortic lesions, inflammasome activity is increased compared to healthy arteries. AngII infusion in NLRP3−/− and caspase-1−/− mice resulted in a preserved aortic structure and reduced aortic enlargement and dissection development compared to the WT phenotype (Wu et al. 2017). The same reduction was seen in AngII-infused double knock-out Apoe−/−NLRP3−/−, Apoe−/− ASC−/− and Apoe−/−casp-1−/− mice, compared to Apoe−/− alone. One of the possible mechanisms may be inhibition of mitochondria-derived ROS that is stimulated by AngII infusion in the inflammasome-deficient animals (Usui et al. 2015). In aortic rings from these aneurysm-developing mice, the contractile response to phenylephrine is reduced. NLRP3−/− and caspase-1−/− aortic rings showed preserved contractile response to phenylephrine (Wu et al. 2017). In another mice model of AAA development (by perivascular calcium phosphate treatment), lentiviral silencing of NLRP3 resulted in a smaller diameter of the aorta compared to empty virus (Sun et al. 2015).

2.7 The Inflammasome in Vasculitis

Vasculitides are heterogeneous clinical entities all characterized by inflammation of the vessel wall. In most cases, the cause of vasculitis is unknown, but often autoimmune processes are thought to play a role. Therapy depends on the specific type of vasculitis but most of the time includes immune suppression. The different types of vasculitis are grouped by size of the affected blood vessels: large, medium or small. The role of the inflammasome is described in a subset of vasculitides and will be summarized below (Ramirez et al. 2014).

2.7.1 Giant-Cell Arteritis

Giant-cell arteritis (GCA) is a chronic systemic vasculitis affecting large- and medium-sized arteries. It is the most predominant vasculitis in Western countries mainly affecting females. Pro-inflammatory cytokines play a major role in the pathogenesis of GCA, and the inflammasome may be involved in this complex polygenic disease. IL-1β and IL-18 are expressed in temporal arteries from patients with GCA (Hernandez-Rodriguez et al. 2004; Blain et al. 2002), but IL-18 expression does not correlate with clinical manifestations or haematological parameters (Shahriar Nabili et al. 2008). IL-18 gene polymorphisms (rs1946518) have been described to be associated with GCA susceptibility, but again not with clinical manifestations (Palomino-Morales et al. 2010). Also NLRP1 gene polymorphisms (rs8182352) are associated with biopsy-proven GCA (Serrano et al. 2013).

2.7.2 Kawasaki Disease

Kawasaki disease (KD), a medium-sized vessel vasculitis, is the most common cause of acute vasculitis in children. It is the main cause of acquired heart disease affecting mainly children below the age of 5. The disease characterizes itself by coronary arteritis with inflammatory cell infiltration and extracellular matrix distraction. Consequences of KD are coronary artery aneurysms, MI and sudden cardiac death. Standard therapy includes intravenous immunoglobulins, but up to a third of patients fail to respond to therapy. These children are at increased risk for development of coronary abnormalities. The cause of KD is unknown, but abnormal immune responses to infectious agents could be involved in the pathophysiology. Many cytokines and chemokines are elevated during the acute phase of the disease, including IL-1β (Takahashi et al. 2014). Peripheral blood mononuclear cells (PBMCs) from KD patients spontaneously release IL-1β in greater levels compared to healthy controls (Suzuki et al. 1996). Higher IL-1β release is seen in patients that failed to respond to standard therapy with intravenous immunoglobulins and in patients that develop coronary artery abnormalities (Leung et al. 1989).

Genetic data show that polymorphisms in the IL-1β gene that are related to increased IL-1β production are associated with intravenous immunoglobulin resistance (Weng et al. 2010). In patients unresponsive to immunoglobulins, transcript abundance for IL-1 pathway genes is found to be higher compared to responsive KD patients (Fury et al. 2010). To study the causative role of the inflammasome in KD, mice models are used. A frequently used model is the lactobacillus casei cell wall extract (LCWE)-induced model of coronary arteritis. A single injection of LCWE reproducibly induces proximal coronary arteritis with histopathologic characteristics very similar to the coronary arteritis observed in human KD (Lehman et al. 1985). In this model, IL-1β increased in an inflammasome-dependent manner. NLRP3−/−, caspase-1−/−, IL-1β−/− and IL-1R−/− mice were all protected from LCWE-induced vasculitis and coronary arthritis and showed less vascular inflammation (Lee et al. 2012, 2015). LCWE injection induced NLRP3 activity in endothelial cells and in $CD11^{+}$ macrophages in the vascular lesions, resulting in increased caspase-1 activity and IL-1β production (Chen et al. 2015). Experiments with chimeric mice showed that stromal IL-1β signalling is required for LCWE-induced vasculitis and coronary arteritis and that IL-1 signalling is not required in haematopoietic cells (Lee et al. 2015). In LCWE-induced vasculitis, anakinra (IL-1R antagonist) was able to block development of coronary lesions and myocarditis (Lee et al. 2012). Quercetin, an antioxidant (found in fruits, vegetables and nuts), was able to inhibit both the NLRP3 and AIM2 inflammasome by preventing ASC oligomerization. Intraperitoneal injection with quercetin for 7 days following LCWE injection reduced coronary arteritis and aneurysm formation. Quercetin inhibited local caspase-1 activity in vascular lesions resulting in less intimal and myofibroblast proliferation (Domiciano et al. 2017).

2.7.3 *Inflammasome in Behçet*

Behçet's syndrome is a systemic inflammatory disorder with multiple disease manifestations including vasculitis, affecting both small and large vessels. In Behçet patients with vascular involvement, IL-1β production after whole blood LPS stimulation was significantly increased compared to healthy controls and to patients without vascular involvement (Yuksel et al. 2014). For the treatment of Behçet's, anakinra and canakinumab have proven to be safe and efficacious in refractory Behçet's disease, strengthening the hypothesis that Behçet's may be considered an IL-1-mediated disease (Emmi et al. 2016).

References

Abbate A, Salloum FN, Vecile E, Das A, Hoke NN, Straino S, Biondi-Zoccai GG, Houser JE, Qureshi IZ, Ownby ED, Gustini E, Biasucci LM, Severino A, Capogrossi MC, Vetrovec GW, Crea F, Baldi A, Kukreja RC, Dobrina A (2008) Anakinra, a recombinant human interleukin-1 receptor antagonist, inhibits apoptosis in experimental acute myocardial infarction. Circulation 117(20):2670–2683. https://doi.org/10.1161/circulationaha.107.740233

Abbate A, Van Tassell BW, Seropian IM, Toldo S, Robati R, Varma A, Salloum FN, Smithson L, Dinarello CA (2010a) Interleukin-1beta modulation using a genetically engineered antibody prevents adverse cardiac remodelling following acute myocardial infarction in the mouse. Eur J Heart Fail 12(4):319–322. https://doi.org/10.1093/eurjhf/hfq017

Abbate A, Kontos MC, Grizzard JD, Biondi-Zoccai GG, Van Tassell BW, Robati R, Roach LM, Arena RA, Roberts CS, Varma A, Gelwix CC, Salloum FN, Hastillo A, Dinarello CA, Vetrovec GW (2010b) Interleukin-1 blockade with anakinra to prevent adverse cardiac remodeling after acute myocardial infarction (Virginia Commonwealth University Anakinra Remodeling Trial [VCU-ART] Pilot study). Am J Cardiol 105(10):1371–1377.e1371. https://doi.org/10.1016/j.amjcard.2009.12.059

Abbate A, Salloum FN, Van Tassell BW, Vecile E, Toldo S, Seropian I, Mezzaroma E, Dobrina A (2011) Alterations in the interleukin-1/interleukin-1 receptor antagonist balance modulate cardiac remodeling following myocardial infarction in the mouse. PLoS One 6(11):e27923. https://doi.org/10.1371/journal.pone.0027923

Abbate A, Van Tassell BW, Biondi-Zoccai G, Kontos MC, Grizzard JD, Spillman DW, Oddi C, Roberts CS, Melchior RD, Mueller GH, Abouzaki NA, Rengel LR, Varma A, Gambill ML, Falcao RA, Voelkel NF, Dinarello CA, Vetrovec GW (2013) Effects of interleukin-1 blockade with anakinra on adverse cardiac remodeling and heart failure after acute myocardial infarction [from the Virginia Commonwealth University-Anakinra Remodeling Trial (2) (VCU-ART2) pilot study]. Am J Cardiol 111(10):1394–1400. https://doi.org/10.1016/j.amjcard.2013.01.287

Abderrazak A, Couchie D, Mahmood DF, Elhage R, Vindis C, Laffargue M, Mateo V, Buchele B, Ayala MR, El Gaafary M, Syrovets T, Slimane MN, Friguet B, Fulop T, Simmet T, El Hadri K, Rouis M (2015) Anti-inflammatory and antiatherogenic effects of the NLRP3 inflammasome inhibitor arglabin in ApoE2.Ki mice fed a high-fat diet. Circulation 131(12):1061–1070. https://doi.org/10.1161/circulationaha.114.013730

Afrasyab A, Qu P, Zhao Y, Peng K, Wang H, Lou D, Niu N, Yuan D (2016) Correlation of NLRP3 with severity and prognosis of coronary atherosclerosis in acute coronary syndrome patients. Heart Vessel 31(8):1218–1229. https://doi.org/10.1007/s00380-015-0723-8

Alabed S, Cabello JB, Irving GJ, Qintar M, Burls A (2014) Colchicine for pericarditis. Cochrane Database Syst Rev (8):Cd010652. https://doi.org/10.1002/14651858.CD010652.pub2

Aleksova A, Beltrami AP, Carriere C, Barbati G, Lesizza P, Perrieri-Montanino M, Isola M, Gentile P, Salvioni E, Not T, Agostoni P, Sinagra G (2017) Interleukin-1beta levels predict long-term mortality and need for heart transplantation in ambulatory patients affected by idiopathic dilated cardiomyopathy. Oncotarget 8(15):25131–25140. https://doi.org/10.18632/oncotarget.15349

Arslan F, de Kleijn DP, Pasterkamp G (2011) Innate immune signaling in cardiac ischemia. Nat Rev Cardiol 8(5):292–300. https://doi.org/10.1038/nrcardio.2011.38

Bando S, Fukuda D, Soeki T, Nishimoto S, Uematsu E, Matsuura T, Ise T, Tobiume T, Yamaguchi K, Yagi S, Iwase T, Yamada H, Wakatsuki T, Shimabukuro M, Sata M (2015)

Expression of NLRP3 in subcutaneous adipose tissue is associated with coronary atherosclerosis. Atherosclerosis 242(2):407–414. https://doi.org/10.1016/j.atherosclerosis.2015.07.043
Baskar S, Klein AL, Zeft A (2016) The use of IL-1 receptor antagonist (anakinra) in idiopathic recurrent pericarditis: a narrative review. Cardiol Res Pract 2016:7840724. https://doi.org/10.1155/2016/7840724
Bhaskar V, Yin J, Mirza AM, Phan D, Vanegas S, Issafras H, Michelson K, Hunter JJ, Kantak SS (2011) Monoclonal antibodies targeting IL-1 beta reduce biomarkers of atherosclerosis in vitro and inhibit atherosclerotic plaque formation in apolipoprotein E-deficient mice. Atherosclerosis 216(2):313–320. https://doi.org/10.1016/j.atherosclerosis.2011.02.026
Blain H, Abdelmouttaleb I, Belmin J, Blain A, Floquet J, Gueant JL, Jeandel C (2002) Arterial wall production of cytokines in giant cell arteritis: results of a pilot study using human temporal artery cultures. J Gerontol A Biol Sci Med Sci 57(4):M241–M245
Blankenberg S, Tiret L, Bickel C, Peetz D, Cambien F, Meyer J, Rupprecht HJ (2002) Interleukin-18 is a strong predictor of cardiovascular death in stable and unstable angina. Circulation 106 (1):24–30
Bleda S, de Haro J, Varela C, Ferruelo A, Acin F (2016) Elevated levels of triglycerides and vldl-cholesterol provoke activation of nlrp1 inflammasome in endothelial cells. Int J Cardiol 220:52–55. https://doi.org/10.1016/j.ijcard.2016.06.193
Bracey NA, Beck PL, Muruve DA, Hirota SA, Guo J, Jabagi H, Wright JR Jr, Macdonald JA, Lees-Miller JP, Roach D, Semeniuk LM, Duff HJ (2013) The Nlrp3 inflammasome promotes myocardial dysfunction in structural cardiomyopathy through interleukin-1beta. Exp Physiol 98(2):462–472. https://doi.org/10.1113/expphysiol.2012.068338
Braunwald E (2015) The war against heart failure: the Lancet lecture. Lancet (London, England) 385(9970):812–824. https://doi.org/10.1016/s0140-6736(14)61889-4
Brucato A, Imazio M, Gattorno M, Lazaros G, Maestroni S, Carraro M, Finetti M, Cumetti D, Carobbio A, Ruperto N, Marcolongo R, Lorini M, Rimini A, Valenti A, Erre GL, Sormani MP, Belli R, Gaita F, Martini A (2016) Effect of anakinra on recurrent pericarditis among patients with colchicine resistance and corticosteroid dependence: the AIRTRIP randomized clinical trial. JAMA 316(18):1906–1912. https://doi.org/10.1001/jama.2016.15826
Bujak M, Dobaczewski M, Chatila K, Mendoza LH, Li N, Reddy A, Frangogiannis NG (2008) Interleukin-1 receptor type I signaling critically regulates infarct healing and cardiac remodeling. Am J Pathol 173(1):57–67. https://doi.org/10.2353/ajpath.2008.070974
Butts B, Gary RA, Dunbar SB, Butler J (2015) The importance of NLRP3 inflammasome in heart failure. J Card Fail 21(7):586–593. https://doi.org/10.1016/j.cardfail.2015.04.014
Butts B, Butler J, Dunbar SB, Corwin E, Gary RA (2017) ASC methylation and interleukin-1beta are associated with aerobic capacity in heart failure. Med Sci Sports Exerc 49(6):1072–1078. https://doi.org/10.1249/mss.0000000000001200
Canada JM, Van Tassell BW, Christopher S, Oddi C, Abouzaki NA, Gambill ML, Mueller G, Melchior R, Shah KB, Dinarello CA, Abbate A, Arena R (2014) Clinical predictors of response to anakinra in patients with heart failure. Int J Cardiol 173(3):537–539. https://doi.org/10.1016/j.ijcard.2014.03.023
Cavalli G, Pappalardo F, Mangieri A, Dinarello CA, Dagna L, Tresoldi M (2016) Treating life-threatening myocarditis by blocking interleukin-1. Crit Care Med 44(8):e751–e754. https://doi.org/10.1097/ccm.0000000000001654
Cavalli G, Foppoli M, Cabrini L, Dinarello CA, Tresoldi M, Dagna L (2017) Interleukin-1 receptor blockade rescues myocarditis-associated end-stage heart failure. Front Immunol 8:131. https://doi.org/10.3389/fimmu.2017.00131
Chen Z, Martin M, Li Z, Shyy JY (2014) Endothelial dysfunction: the role of sterol regulatory element-binding protein-induced NOD-like receptor family pyrin domain-containing protein 3 inflammasome in atherosclerosis. Curr Opin Lipidol 25(5):339–349. https://doi.org/10.1097/mol.0000000000000107

Chen Y, Li X, Boini KM, Pitzer AL, Gulbins E, Zhang Y, Li PL (2015) Endothelial Nlrp3 inflammasome activation associated with lysosomal destabilization during coronary arteritis. Biochim Biophys Acta 1853(2):396–408. https://doi.org/10.1016/j.bbamcr.2014.11.012

Coll RC, Robertson AA, Chae JJ, Higgins SC, Munoz-Planillo R, Inserra MC, Vetter I, Dungan LS, Monks BG, Stutz A, Croker DE, Butler MS, Haneklaus M, Sutton CE, Nunez G, Latz E, Kastner DL, Mills KH, Masters SL, Schroder K, Cooper MA, O'Neill LA (2015) A small-molecule inhibitor of the NLRP3 inflammasome for the treatment of inflammatory diseases. Nat Med 21(3):248–255. https://doi.org/10.1038/nm.3806

Dai Y, Dai D, Wang X, Ding Z, Mehta JL (2014) DPP-4 inhibitors repress NLRP3 inflammasome and interleukin-1beta via GLP-1 receptor in macrophages through protein kinase C pathway. Cardiovasc Drugs Ther 28(5):425–432. https://doi.org/10.1007/s10557-014-6539-4

Dihlmann S, Erhart P, Mehrabi A, Nickkholgh A, Lasitschka F, Bockler D, Hakimi M (2014) Increased expression and activation of absent in melanoma 2 inflammasome components in lymphocytic infiltrates of abdominal aortic aneurysms. Mol Med (Cambridge, MA) 20:230–237. https://doi.org/10.2119/molmed.2013.00162

Dinarello CA, van der Meer JW (2013) Treating inflammation by blocking interleukin-1 in humans. Semin Immunol 25(6):469–484. https://doi.org/10.1016/j.smim.2013.10.008

Dinarello CA, Novick D, Kim S, Kaplanski G (2013) Interleukin-18 and IL-18 binding protein. Front Immunol 4:289. https://doi.org/10.3389/fimmu.2013.00289

Domiciano TP, Wakita D, Jones HD, Crother TR, Verri WA Jr, Arditi M, Shimada K (2017) Quercetin inhibits inflammasome activation by interfering with ASC oligomerization and prevents interleukin-1 mediated mouse vasculitis. Sci Rep 7:41539. https://doi.org/10.1038/srep41539

Duewell P, Kono H, Rayner KJ, Sirois CM, Vladimer G, Bauernfeind FG, Abela GS, Franchi L, Nunez G, Schnurr M, Espevik T, Lien E, Fitzgerald KA, Rock KL, Moore KJ, Wright SD, Hornung V, Latz E (2010) NLRP3 inflammasomes are required for atherogenesis and activated by cholesterol crystals. Nature 464(7293):1357–1361. https://doi.org/10.1038/nature08938

Elhage R, Maret A, Pieraggi MT, Thiers JC, Arnal JF, Bayard F (1998) Differential effects of interleukin-1 receptor antagonist and tumor necrosis factor binding protein on fatty-streak formation in apolipoprotein E-deficient mice. Circulation 97(3):242–244

Elhage R, Jawien J, Rudling M, Ljunggren HG, Takeda K, Akira S, Bayard F, Hansson GK (2003) Reduced atherosclerosis in interleukin-18 deficient apolipoprotein E-knockout mice. Cardiovasc Res 59(1):234–240

Emmi G, Talarico R, Lopalco G, Cimaz R, Cantini F, Viapiana O, Olivieri I, Goldoni M, Vitale A, Silvestri E, Prisco D, Lapadula G, Galeazzi M, Iannone F, Cantarini L (2016) Efficacy and safety profile of anti-interleukin-1 treatment in Behcet's disease: a multicenter retrospective study. Clin Rheumatol 35(5):1281–1286. https://doi.org/10.1007/s10067-015-3004-0

Feng Y, Zou L, Si R, Nagasaka Y, Chao W (2010) Bone marrow MyD88 signaling modulates neutrophil function and ischemic myocardial injury. Am J Physiol Cell Physiol 299(4):C760–C769. https://doi.org/10.1152/ajpcell.00155.2010

Folco EJ, Sukhova GK, Quillard T, Libby P (2014) Moderate hypoxia potentiates interleukin-1beta production in activated human macrophages. Circ Res 115(10):875–883. https://doi.org/10.1161/circresaha.115.304437

Frangogiannis NG (2015) Interleukin-1 in cardiac injury, repair, and remodeling: pathophysiologic and translational concepts. Discoveries (Craiova, Romania) 3(1):e41. https://doi.org/10.15190/d.2015.33

Fujisue K, Sugamura K, Kurokawa H, Matsubara J, Ishii M, Izumiya Y, Kaikita K, Sugiyama S (2017) Colchicine improves survival, left ventricular remodeling, and chronic cardiac function after acute myocardial infarction. Circ J 81(8):1174–1182. https://doi.org/10.1253/circj.CJ-16-0949

Fury W, Tremoulet AH, Watson VE, Best BM, Shimizu C, Hamilton J, Kanegaye JT, Wei Y, Kao C, Mellis S, Lin C, Burns JC (2010) Transcript abundance patterns in Kawasaki disease

patients with intravenous immunoglobulin resistance. Hum Immunol 71(9):865–873. https://doi.org/10.1016/j.humimm.2010.06.008
Fuster JJ, MacLauchlan S, Zuriaga MA, Polackal MN, Ostriker AC, Chakraborty R, Wu CL, Sano S, Muralidharan S, Rius C, Vuong J, Jacob S, Muralidhar V, Robertson AA, Cooper MA, Andres V, Hirschi KK, Martin KA, Walsh K (2017) Clonal hematopoiesis associated with Tet2 deficiency accelerates atherosclerosis development in mice. Science (New York, NY) 355 (6327):842–847. https://doi.org/10.1126/science.aag1381
Getz GS, Reardon CA (2015) Use of mouse models in atherosclerosis research. Methods Mol Biol (Clifton, NJ) 1339:1–16. https://doi.org/10.1007/978-1-4939-2929-0_1
Gistera A, Hansson GK (2017) The immunology of atherosclerosis. Nat Rev Nephrol 13 (6):368–380. https://doi.org/10.1038/nrneph.2017.51
Hansson GK, Klareskog L (2011) Pulling down the plug on atherosclerosis: cooling down the inflammasome. Nat Med 17(7):790–791. https://doi.org/10.1038/nm0711-790
Hendrikx T, Jeurissen ML, van Gorp PJ, Gijbels MJ, Walenbergh SM, Houben T, van Gorp R, Pottgens CC, Stienstra R, Netea MG, Hofker MH, Donners MM, Shiri-Sverdlov R (2015) Bone marrow-specific caspase-1/11 deficiency inhibits atherosclerosis development in Ldlr(−/−) mice. FEBS J 282(12):2327–2338. https://doi.org/10.1111/febs.13279
Hernandez-Rodriguez J, Segarra M, Vilardell C, Sanchez M, Garcia-Martinez A, Esteban MJ, Queralt C, Grau JM, Urbano-Marquez A, Palacin A, Colomer D, Cid MC (2004) Tissue production of pro-inflammatory cytokines (IL-1beta, TNFalpha and IL-6) correlates with the intensity of the systemic inflammatory response and with corticosteroid requirements in giant-cell arteritis. Rheumatology (Oxford) 43(3):294–301. https://doi.org/10.1093/rheumatology/keh058
Heusch G, Libby P, Gersh B, Yellon D, Bohm M, Lopaschuk G, Opie L (2014) Cardiovascular remodelling in coronary artery disease and heart failure. Lancet (London, England) 383 (9932):1933–1943. https://doi.org/10.1016/s0140-6736(14)60107-0
Heymans S, Eriksson U, Lehtonen J, Cooper LT Jr (2016) The quest for new approaches in myocarditis and inflammatory cardiomyopathy. J Am Coll Cardiol 68(21):2348–2364. https://doi.org/10.1016/j.jacc.2016.09.937
Hofmann U, Frantz S (2013) How can we cure a heart "in flame"? A translational view on inflammation in heart failure. Basic Res Cardiol 108(4):356. https://doi.org/10.1007/s00395-013-0356-y
Hwang MW, Matsumori A, Furukawa Y, Ono K, Okada M, Iwasaki A, Hara M, Miyamoto T, Touma M, Sasayama S (2001) Neutralization of interleukin-1beta in the acute phase of myocardial infarction promotes the progression of left ventricular remodeling. J Am Coll Cardiol 38(5):1546–1553
Janoudi A, Shamoun FE, Kalavakunta JK, Abela GS (2016) Cholesterol crystal induced arterial inflammation and destabilization of atherosclerotic plaque. Eur Heart J 37(25):1959–1967. https://doi.org/10.1093/eurheartj/ehv653
Jong WM, Leemans JC, Weber NC, Juffermans NP, Schultz MJ, Hollmann MW, Zuurbier CJ (2014) Nlrp3 plays no role in acute cardiac infarction due to low cardiac expression. Int J Cardiol 177(1):41–43. https://doi.org/10.1016/j.ijcard.2014.09.148
Kawaguchi M, Takahashi M, Hata T, Kashima Y, Usui F, Morimoto H, Izawa A, Takahashi Y, Masumoto J, Koyama J, Hongo M, Noda T, Nakayama J, Sagara J, Taniguchi S, Ikeda U (2011) Inflammasome activation of cardiac fibroblasts is essential for myocardial ischemia/reperfusion injury. Circulation 123(6):594–604. https://doi.org/10.1161/circulationaha.110.982777
Kirii H, Niwa T, Yamada Y, Wada H, Saito K, Iwakura Y, Asano M, Moriwaki H, Seishima M (2003) Lack of interleukin-1beta decreases the severity of atherosclerosis in ApoE-deficient mice. Arterioscler Thromb Vasc Biol 23(4):656–660. https://doi.org/10.1161/01.atv.0000064374.15232.c3
L'Homme L, Esser N, Riva L, Scheen A, Paquot N, Piette J, Legrand-Poels S (2013) Unsaturated fatty acids prevent activation of NLRP3 inflammasome in human monocytes/macrophages. J Lipid Res 54(11):2998–3008. https://doi.org/10.1194/jlr.M037861

Lee Y, Schulte DJ, Shimada K, Chen S, Crother TR, Chiba N, Fishbein MC, Lehman TJ, Arditi M (2012) Interleukin-1beta is crucial for the induction of coronary artery inflammation in a mouse model of Kawasaki disease. Circulation 125(12):1542–1550. https://doi.org/10.1161/circulationaha.111.072769

Lee Y, Wakita D, Dagvadorj J, Shimada K, Chen S, Huang G, Lehman TJ, Fishbein MC, Hoffman HM, Crother TR, Arditi M (2015) IL-1 signaling is critically required in stromal cells in kawasaki disease vasculitis mouse model: role of both IL-1alpha and IL-1beta. Arterioscler Thromb Vasc Biol 35(12):2605–2616. https://doi.org/10.1161/atvbaha.115.306475

Lehman TJ, Walker SM, Mahnovski V, McCurdy D (1985) Coronary arteritis in mice following the systemic injection of group B Lactobacillus casei cell walls in aqueous suspension. Arthritis Rheum 28(6):652–659

Leung DY, Cotran RS, Kurt-Jones E, Burns JC, Newburger JW, Pober JS (1989) Endothelial cell activation and high interleukin-1 secretion in the pathogenesis of acute Kawasaki disease. Lancet (London, England) 2(8675):1298–1302

Li X, Zhang Y, Xia M, Gulbins E, Boini KM, Li PL (2014) Activation of Nlrp3 inflammasomes enhances macrophage lipid-deposition and migration: implication of a novel role of inflammasome in atherogenesis. PLoS One 9(1):e87552. https://doi.org/10.1371/journal.pone.0087552

Libby P, Hansson GK (2015) Inflammation and immunity in diseases of the arterial tree: players and layers. Circ Res 116(2):307–311. https://doi.org/10.1161/circresaha.116.301313

Liu W, Zhang X, Zhao M, Zhang X, Chi J, Liu Y, Lin F, Fu Y, Ma D, Yin X (2015) Activation in M1 but not M2 macrophages contributes to cardiac remodeling after myocardial infarction in rats: a critical role of the calcium sensing receptor/NRLP3 inflammasome. Cell Physiol Biochem 35(6):2483–2500. https://doi.org/10.1159/000374048

Lusis AJ (2000) Atherosclerosis. Nature 407(6801):233–241. https://doi.org/10.1038/35025203

Maekawa Y, Anzai T, Yoshikawa T, Asakura Y, Takahashi T, Ishikawa S, Mitamura H, Ogawa S (2002) Prognostic significance of peripheral monocytosis after reperfused acute myocardial infarction: a possible role for left ventricular remodeling. J Am Coll Cardiol 39(2):241–246

Mallat Z, Corbaz A, Scoazec A, Besnard S, Leseche G, Chvatchko Y, Tedgui A (2001) Expression of interleukin-18 in human atherosclerotic plaques and relation to plaque instability. Circulation 104(14):1598–1603

Mallat Z, Heymes C, Corbaz A, Logeart D, Alouani S, Cohen-Solal A, Seidler T, Hasenfuss G, Chvatchko Y, Shah AM, Tedgui A (2004) Evidence for altered interleukin 18 (IL)-18 pathway in human heart failure. FASEB J 18(14):1752–1754. https://doi.org/10.1096/fj.04-2426fje

Marchetti C, Chojnacki J, Toldo S, Mezzaroma E, Tranchida N, Rose SW, Federici M, Van Tassell BW, Zhang S, Abbate A (2014) A novel pharmacologic inhibitor of the NLRP3 inflammasome limits myocardial injury after ischemia-reperfusion in the mouse. J Cardiovasc Pharmacol 63(4):316–322. https://doi.org/10.1097/fjc.0000000000000053

Marchetti C, Toldo S, Chojnacki J, Mezzaroma E, Liu K, Salloum FN, Nordio A, Carbone S, Mauro AG, Das A, Zalavadia AA, Halquist MS, Federici M, Van Tassell BW, Zhang S, Abbate A (2015) Pharmacologic inhibition of the NLRP3 inflammasome preserves cardiac function after ischemic and nonischemic injury in the mouse. J Cardiovasc Pharmacol 66(1):1–8. https://doi.org/10.1097/fjc.0000000000000247

Martinez GJ, Robertson S, Barraclough J, Xia Q, Mallat Z, Bursill C, Celermajer DS, Patel S (2015) Colchicine acutely suppresses local cardiac production of inflammatory cytokines in patients with an acute coronary syndrome. J Am Heart Assoc 4(8):e002128. https://doi.org/10.1161/jaha.115.002128

Martinon F, Petrilli V, Mayor A, Tardivel A, Tschopp J (2006) Gout-associated uric acid crystals activate the NALP3 inflammasome. Nature 440(7081):237–241. https://doi.org/10.1038/nature04516

Menu P, Pellegrin M, Aubert JF, Bouzourene K, Tardivel A, Mazzolai L, Tschopp J (2011) Atherosclerosis in ApoE-deficient mice progresses independently of the NLRP3 inflammasome. Cell Death Dis 2:e137. https://doi.org/10.1038/cddis.2011.18

Merkle S, Frantz S, Schon MP, Bauersachs J, Buitrago M, Frost RJ, Schmitteckert EM, Lohse MJ, Engelhardt S (2007) A role for caspase-1 in heart failure. Circ Res 100(5):645–653. https://doi.org/10.1161/01.res.0000260203.55077.61

Mezzaroma E, Toldo S, Farkas D, Seropian IM, Van Tassell BW, Salloum FN, Kannan HR, Menna AC, Voelkel NF, Abbate A (2011) The inflammasome promotes adverse cardiac remodeling following acute myocardial infarction in the mouse. Proc Natl Acad Sci USA 108 (49):19725–19730. https://doi.org/10.1073/pnas.1108586108

Mezzaroma E, Marchetti C, Toldo S (2014) Letter by Mezzaroma, et al regarding article, "NLRP3 inflammasome as a therapeutic target in myocardial infarction". Int Heart J 55(4):379

Nidorf SM, Eikelboom JW, Budgeon CA, Thompson PL (2013) Low-dose colchicine for secondary prevention of cardiovascular disease. J Am Coll Cardiol 61(4):404–410. https://doi.org/10.1016/j.jacc.2012.10.027

Noji Y (2016) Anakinra in fulminant myocarditis: targeting interleukin-1 and the inflammasome formation. Crit Care Med 44(8):1630–1631. https://doi.org/10.1097/ccm.0000000000001769

Palomino-Morales RJ, Vazquez-Rodriguez TR, Torres O, Morado IC, Castaneda S, Miranda-Filloy JA, Callejas-Rubio JL, Fernandez-Gutierrez B, Gonzalez-Gay MA, Martin J (2010) Association between IL-18 gene polymorphisms and biopsy-proven giant cell arteritis. Arthritis Res Ther 12 (2):R51. https://doi.org/10.1186/ar2962

Paramel Varghese G, Folkersen L, Strawbridge RJ, Halvorsen B, Yndestad A, Ranheim T, Krohg-Sorensen K, Skjelland M, Espevik T, Aukrust P, Lengquist M, Hedin U, Jansson JH, Fransen K, Hansson GK, Eriksson P, Sirsjo A (2016) NLRP3 inflammasome expression and activation in human atherosclerosis. J Am Heart Assoc 5(5):e003031. https://doi.org/10.1161/jaha.115.003031

Peng K, Liu L, Wei D, Lv Y, Wang G, Xiong W, Wang X, Altaf A, Wang L, He D, Wang H, Qu P (2015) P2X7R is involved in the progression of atherosclerosis by promoting NLRP3 inflammasome activation. Int J Mol Med 35(5):1179–1188. https://doi.org/10.3892/ijmm.2015.2129

Platis A, Yu Q, Moore D, Khojeini E, Tsau P, Larson D (2008) The effect of daily administration of IL-18 on cardiac structure and function. Perfusion 23(4):237–242. https://doi.org/10.1177/0267659108101511

Pomerantz BJ, Reznikov LL, Harken AH, Dinarello CA (2001) Inhibition of caspase 1 reduces human myocardial ischemic dysfunction via inhibition of IL-18 and IL-1beta. Proc Natl Acad Sci USA 98(5):2871–2876. https://doi.org/10.1073/pnas.041611398

Raeburn CD, Dinarello CA, Zimmerman MA, Calkins CM, Pomerantz BJ, McIntyre RC Jr, Harken AH, Meng X (2002) Neutralization of IL-18 attenuates lipopolysaccharide-induced myocardial dysfunction. Am J Phys Heart Circ Phys 283(2):H650–H657. https://doi.org/10.1152/ajpheart.00043.2002

Ramirez GA, Maugeri N, Sabbadini MG, Rovere-Querini P, Manfredi AA (2014) Intravascular immunity as a key to systemic vasculitis: a work in progress, gaining momentum. Clin Exp Immunol 175(2):150–166. https://doi.org/10.1111/cei.12223

Razani B, Feng C, Coleman T, Emanuel R, Wen H, Hwang S, Ting JP, Virgin HW, Kastan MB, Semenkovich CF (2012) Autophagy links inflammasomes to atherosclerotic progression. Cell Metab 15(4):534–544. https://doi.org/10.1016/j.cmet.2012.02.011

Ridker PM, Everett BM, Thuren T, MacFadyen JG, Chang WH, Ballantyne C, Fonseca F, Nicolau J, Koenig W, Anker SD, Kastelein JJP, Cornel JH, Pais P, Pella D, Genest J, Cifkova R, Lorenzatti A, Forster T, Kobalava Z, Vida-Simiti L, Flather M, Shimokawa H, Ogawa H, Dellborg M, Rossi PRF, Troquay RPT, Libby P, Glynn RJ (2017) Antiinflammatory therapy with canakinumab for atherosclerotic disease. N Engl J Med 377(12):1119–1131. https://doi.org/10.1056/NEJMoa1707914

Salloum FN, Chau V, Varma A, Hoke NN, Toldo S, Biondi-Zoccai GG, Crea F, Vetrovec GW, Abbate A (2009) Anakinra in experimental acute myocardial infarction – does dosage or duration of treatment matter? Cardiovasc Drugs Ther 23(2):129–135. https://doi.org/10.1007/s10557-008-6154-3

Sandanger O, Ranheim T, Vinge LE, Bliksoen M, Alfsnes K, Finsen AV, Dahl CP, Askevold ET, Florholmen G, Christensen G, Fitzgerald KA, Lien E, Valen G, Espevik T, Aukrust P, Yndestad A (2013) The NLRP3 inflammasome is up-regulated in cardiac fibroblasts and mediates myocardial ischaemia-reperfusion injury. Cardiovasc Res 99(1):164–174. https://doi.org/10.1093/cvr/cvt091

Sandanger O, Gao E, Ranheim T, Bliksoen M, Kaasboll OJ, Alfsnes K, Nymo SH, Rashidi A, Ohm IK, Attramadal H, Aukrust P, Vinge LE, Yndestad A (2016) NLRP3 inflammasome activation during myocardial ischemia reperfusion is cardioprotective. Biochem Biophys Res Commun 469(4):1012–1020. https://doi.org/10.1016/j.bbrc.2015.12.051

Serrano A, Carmona FD, Castaneda S, Solans R, Hernandez-Rodriguez J, Cid MC, Prieto-Gonzalez S, Miranda-Filloy JA, Rodriguez-Rodriguez L, Morado IC, Gomez-Vaquero C, Blanco R, Sopena B, Ortego-Centeno N, Unzurrunzaga A, Mari-Alfonso B, Sanchez-Martin J, Garcia-Villanueva MJ, Hidalgo-Conde A, Pazzola G, Boiardi L, Salvarani C, Gonzalez-Gay MA, Martin J (2013) Evidence of association of the NLRP1 gene with giant cell arteritis. Ann Rheum Dis 72(4):628–630. https://doi.org/10.1136/annrheumdis-2012-202609

Shah Z, Kampfrath T, Deiuliis JA, Zhong J, Pineda C, Ying Z, Xu X, Lu B, Moffatt-Bruce S, Durairaj R, Sun Q, Mihai G, Maiseyeu A, Rajagopalan S (2011) Long-term dipeptidyl-peptidase 4 inhibition reduces atherosclerosis and inflammation via effects on monocyte recruitment and chemotaxis. Circulation 124(21):2338–2349. https://doi.org/10.1161/circulationaha.111.041418

Shahriar Nabili PB, Roberts F, Gracie A, McFadzean R (2008) Local expression of IL-18 in the temporal artery does not correlate with clinical manifestations of giant cell arteritis. Neuro-Ophthalmology 32(1):3–6

Shi X, Xie WL, Kong WW, Chen D, Qu P (2015) Expression of the NLRP3 inflammasome in carotid atherosclerosis. J Stroke Cerebrovasc Dis 24(11):2455–2466. https://doi.org/10.1016/j.jstrokecerebrovasdis.2015.03.024

Shimizu K, Mitchell RN, Libby P (2006) Inflammation and cellular immune responses in abdominal aortic aneurysms. Arterioscler Thromb Vasc Biol 26(5):987–994. https://doi.org/10.1161/01.ATV.0000214999.12921.4f

Stachon P, Heidenreich A, Merz J, Hilgendorf I, Wolf D, Willecke F, von Garlen S, Albrecht P, Hardtner C, Ehrat N, Hoppe N, Reinohl J, von Zur Muhlen C, Bode C, Idzko M, Zirlik A (2017) P2X7 deficiency blocks lesional inflammasome activity and ameliorates atherosclerosis in mice. Circulation 135(25):2524–2533. https://doi.org/10.1161/circulationaha.117.027400

Stack J, Ryan J, McCarthy G (2015) Colchicine: new insights to an old drug. Am J Ther 22(5):e151–e157. https://doi.org/10.1097/01.mjt.0000433937.07244.e1

Steg PG, James SK, Atar D, Badano LP, Blomstrom-Lundqvist C, Borger MA, Di Mario C, Dickstein K, Ducrocq G, Fernandez-Aviles F, Gershlick AH, Giannuzzi P, Halvorsen S, Huber K, Juni P, Kastrati A, Knuuti J, Lenzen MJ, Mahaffey KW, Valgimigli M, van 't Hof A, Widimsky P, Zahger D (2012) ESC guidelines for the management of acute myocardial infarction in patients presenting with ST-segment elevation. Eur Heart J 33(20):2569–2619. https://doi.org/10.1093/eurheartj/ehs215

Sun W, Pang Y, Liu Z, Sun L, Liu B, Xu M, Dong Y, Feng J, Jiang C, Kong W, Wang X (2015) Macrophage inflammasome mediates hyperhomocysteinemia-aggravated abdominal aortic aneurysm. J Mol Cell Cardiol 81:96–106. https://doi.org/10.1016/j.yjmcc.2015.02.005

Sutton MG, Sharpe N (2000) Left ventricular remodeling after myocardial infarction: pathophysiology and therapy. Circulation 101(25):2981–2988

Suzuki H, Uemura S, Tone S, Iizuka T, Koike M, Hirayama K, Maeda J (1996) Effects of immunoglobulin and gamma-interferon on the production of tumour necrosis factor-alpha and interleukin-1 beta by peripheral blood monocytes in the acute phase of Kawasaki disease. Eur J Pediatr 155(4):291–296

Suzuki K, Murtuza B, Smolenski RT, Sammut IA, Suzuki N, Kaneda Y, Yacoub MH (2001) Overexpression of interleukin-1 receptor antagonist provides cardioprotection against ischemia-reperfusion injury associated with reduction in apoptosis. Circulation 104(12 Suppl 1):I308–I303

Takahashi T, Hiasa Y, Ohara Y, Miyazaki S, Ogura R, Suzuki N, Hosokawa S, Kishi K, Ohtani R (2008) Relationship of admission neutrophil count to microvascular injury, left ventricular dilation, and long-term outcome in patients treated with primary angioplasty for acute myocardial infarction. Circ J 72(6):867–872

Takahashi K, Oharaseki T, Yokouchi Y (2014) Update on etio and immunopathogenesis of Kawasaki disease. Curr Opin Rheumatol 26(1):31–36. https://doi.org/10.1097/bor.0000000000000010

Thacker SG, Zarzour A, Chen Y, Alcicek MS, Freeman LA, Sviridov DO, Demosky SJ Jr, Remaley AT (2016) High-density lipoprotein reduces inflammation from cholesterol crystals by inhibiting inflammasome activation. Immunology 149(3):306–319. https://doi.org/10.1111/imm.12638

Toldo S, Mezzaroma E, Van Tassell BW, Farkas D, Marchetti C, Voelkel NF, Abbate A (2013) Interleukin-1beta blockade improves cardiac remodelling after myocardial infarction without interrupting the inflammasome in the mouse. Exp Physiol 98(3):734–745. https://doi.org/10.1113/expphysiol.2012.069831

Toldo S, Mezzaroma E, O'Brien L, Marchetti C, Seropian IM, Voelkel NF, Van Tassell BW, Dinarello CA, Abbate A (2014a) Interleukin-18 mediates interleukin-1-induced cardiac dysfunction. Am J Phys Heart Circ Phys 306(7):H1025–H1031. https://doi.org/10.1152/ajpheart.00795.2013

Toldo S, Kannan H, Bussani R, Anzini M, Sonnino C, Sinagra G, Merlo M, Mezzaroma E, De-Giorgio F, Silvestri F, Van Tassell BW, Baldi A, Abbate A (2014b) Formation of the inflammasome in acute myocarditis. Int J Cardiol 171(3):e119–e121. https://doi.org/10.1016/j.ijcard.2013.12.137

Toldo S, Mezzaroma E, McGeough MD, Pena CA, Marchetti C, Sonnino C, Van Tassell BW, Salloum FN, Voelkel NF, Hoffman HM, Abbate A (2015) Independent roles of the priming and the triggering of the NLRP3 inflammasome in the heart. Cardiovasc Res 105(2):203–212. https://doi.org/10.1093/cvr/cvu259

Toldo S, Marchetti C, Mauro AG, Chojnacki J, Mezzaroma E, Carbone S, Zhang S, Van Tassell B, Salloum FN, Abbate A (2016) Inhibition of the NLRP3 inflammasome limits the inflammatory injury following myocardial ischemia-reperfusion in the mouse. Int J Cardiol 209:215–220. https://doi.org/10.1016/j.ijcard.2016.02.043

Tschopp J, Martinon F, Burns K (2003) NALPs: a novel protein family involved in inflammation. Nat Rev Mol Cell Biol 4(2):95–104. https://doi.org/10.1038/nrm1019

Usui F, Shirasuna K, Kimura H, Tatsumi K, Kawashima A, Karasawa T, Hida S, Sagara J, Taniguchi S, Takahashi M (2012) Critical role of caspase-1 in vascular inflammation and development of atherosclerosis in Western diet-fed apolipoprotein E-deficient mice. Biochem Biophys Res Commun 425(2):162–168. https://doi.org/10.1016/j.bbrc.2012.07.058

Usui F, Shirasuna K, Kimura H, Tatsumi K, Kawashima A, Karasawa T, Yoshimura K, Aoki H, Tsutsui H, Noda T, Sagara J, Taniguchi S, Takahashi M (2015) Inflammasome activation by mitochondrial oxidative stress in macrophages leads to the development of angiotensin II-induced aortic aneurysm. Arterioscler Thromb Vasc Biol 35(1):127–136. https://doi.org/10.1161/atvbaha.114.303763

Valle Raleigh J, Mauro AG, Devarakonda T, Marchetti C, He J, Kim E, Filippone S, Das A, Toldo S, Abbate A, Salloum FN (2017) Reperfusion therapy with recombinant human relaxin-2 (Serelaxin) attenuates myocardial infarct size and NLRP3 inflammasome following ischemia/reperfusion injury via eNOS-dependent mechanism. Cardiovasc Res 113(6):609–619. https://doi.org/10.1093/cvr/cvw246

van der Laan AM, Nahrendorf M, Piek JJ (2012a) Healing and adverse remodelling after acute myocardial infarction: role of the cellular immune response. Heart 98(18):1384–1390. https://doi.org/10.1136/heartjnl-2012-301623

van der Laan AM, Hirsch A, Robbers LF, Nijveldt R, Lommerse I, Delewi R, van der Vleuten PA, Biemond BJ, Zwaginga JJ, van der Giessen WJ, Zijlstra F, van Rossum AC, Voermans C, van der Schoot CE, Piek JJ (2012b) A proinflammatory monocyte response is associated with myocardial injury and impaired functional outcome in patients with ST-segment elevation myocardial infarction: monocytes and myocardial infarction. Am Heart J 163(1):57–65.e52. https://doi.org/10.1016/j.ahj.2011.09.002

van Hout GP, Arslan F, Pasterkamp G, Hoefer IE (2016) Targeting danger-associated molecular patterns after myocardial infarction. Expert Opin Ther Targets 20(2):223–239. https://doi.org/10.1517/14728222.2016.1088005

van Hout GP, Bosch L, Ellenbroek GH, de Haan JJ, van Solinge WW, Cooper MA, Arslan F, de Jager SC, Robertson AA, Pasterkamp G, Hoefer IE (2017) The selective NLRP3-inflammasome inhibitor MCC950 reduces infarct size and preserves cardiac function in a pig model of myocardial infarction. Eur Heart J 38(11):828–836. https://doi.org/10.1093/eurheartj/ehw247

Van Tassell BW, Varma A, Salloum FN, Das A, Seropian IM, Toldo S, Smithson L, Hoke NN, Chau VQ, Robati R, Abbate A (2010) Interleukin-1 trap attenuates cardiac remodeling after experimental acute myocardial infarction in mice. J Cardiovasc Pharmacol 55(2):117–122. https://doi.org/10.1097/FJC.0b013e3181c87e53

Van Tassell BW, Arena RA, Toldo S, Mezzaroma E, Azam T, Seropian IM, Shah K, Canada J, Voelkel NF, Dinarello CA, Abbate A (2012) Enhanced interleukin-1 activity contributes to exercise intolerance in patients with systolic heart failure. PLoS One 7(3):e33438. https://doi.org/10.1371/journal.pone.0033438

Van Tassell BW, Toldo S, Mezzaroma E, Abbate A (2013a) Targeting interleukin-1 in heart disease. Circulation 128(17):1910–1923. https://doi.org/10.1161/circulationaha.113.003199

Van Tassell BW, Seropian IM, Toldo S, Mezzaroma E, Abbate A (2013b) Interleukin-1beta induces a reversible cardiomyopathy in the mouse. Inflamm Res 62(7):637–640. https://doi.org/10.1007/s00011-013-0625-0

Van Tassell BW, Arena R, Biondi-Zoccai G, McNair Canada J, Oddi C, Abouzaki NA, Jahangiri A, Falcao RA, Kontos MC, Shah KB, Voelkel NF, Dinarello CA, Abbate A (2014) Effects of interleukin-1 blockade with anakinra on aerobic exercise capacity in patients with heart failure and preserved ejection fraction (from the D-HART pilot study). Am J Cardiol 113(2):321–327. https://doi.org/10.1016/j.amjcard.2013.08.047

Van Tassell BW, Raleigh JM, Abbate A (2015) Targeting interleukin-1 in heart failure and inflammatory heart disease. Curr Heart Fail Rep 12(1):33–41. https://doi.org/10.1007/s11897-014-0231-7

Vander Heide RS, Steenbergen C (2013) Cardioprotection and myocardial reperfusion: pitfalls to clinical application. Circ Res 113(4):464–477. https://doi.org/10.1161/circresaha.113.300765

Varghese GP, Fransen K, Hurtig-Wennlof A, Bengtsson T, Jansson JH, Sirsjo A (2013) Q705K variant in NLRP3 gene confers protection against myocardial infarction in female individuals. Biomed Rep 1(6):879–882. https://doi.org/10.3892/br.2013.155

Vinten-Johansen J (2004) Involvement of neutrophils in the pathogenesis of lethal myocardial reperfusion injury. Cardiovasc Res 61(3):481–497. https://doi.org/10.1016/j.cardiores.2003.10.011

Wang X, Feuerstein GZ, Gu JL, Lysko PG, Yue TL (1995) Interleukin-1 beta induces expression of adhesion molecules in human vascular smooth muscle cells and enhances adhesion of leukocytes to smooth muscle cells. Atherosclerosis 115(1):89–98

Wang Y, Gao B, Xiong S (2014) Involvement of NLRP3 inflammasome in CVB3-induced viral myocarditis. Am J Phys Heart Circ Phys 307(10):H1438–H1447. https://doi.org/10.1152/ajpheart.00441.2014

Wen C, Yang X, Yan Z, Zhao M, Yue X, Cheng X, Zheng Z, Guan K, Dou J, Xu T, Zhang Y, Song T, Wei C, Zhong H (2013) Nalp3 inflammasome is activated and required for vascular smooth muscle cell calcification. Int J Cardiol 168(3):2242–2247. https://doi.org/10.1016/j.ijcard.2013.01.211

Weng KP, Hsieh KS, Ho TY, Huang SH, Lai CR, Chiu YT, Huang SC, Lin CC, Hwang YT, Ger LP (2010) IL-1B polymorphism in association with initial intravenous immunoglobulin treatment failure in Taiwanese children with Kawasaki disease. Circ J 74(3):544–551

Whitman SC, Ravisankar P, Daugherty A (2002) Interleukin-18 enhances atherosclerosis in apolipoprotein E(−/−) mice through release of interferon-gamma. Circ Res 90(2):E34–E38

Woldbaek PR, Sande JB, Stromme TA, Lunde PK, Djurovic S, Lyberg T, Christensen G, Tonnessen T (2005) Daily administration of interleukin-18 causes myocardial dysfunction in healthy mice. Am J Phys Heart Circ Phys 289(2):H708–H714. https://doi.org/10.1152/ajpheart.01179.2004

Wu X, Cakmak S, Wortmann M, Hakimi M, Zhang J, Bockler D, Dihlmann S (2016) Sex- and disease-specific inflammasome signatures in circulating blood leukocytes of patients with abdominal aortic aneurysm. Mol Med (Cambridge, MA) 22:508–518. https://doi.org/10.2119/molmed.2016.00035

Wu D, Ren P, Zheng Y, Zhang L, Xu G, Xie W, Lloyd EE, Zhang S, Zhang Q, Curci JA, Coselli JS, Milewicz DM, Shen YH, LeMaire SA (2017) NLRP3 (nucleotide oligomerization domain-like receptor family, pyrin domain containing 3)-caspase-1 inflammasome degrades contractile proteins: implications for aortic biomechanical dysfunction and aneurysm and dissection formation. Arterioscler Thromb Vasc Biol 37(4):694–706. https://doi.org/10.1161/atvbaha.116.307648

Xiao H, Lu M, Lin TY, Chen Z, Chen G, Wang WC, Marin T, Shentu TP, Wen L, Gongol B, Sun W, Liang X, Chen J, Huang HD, Pedra JH, Johnson DA, Shyy JY (2013) Sterol regulatory element binding protein 2 activation of NLRP3 inflammasome in endothelium mediates hemodynamic-induced atherosclerosis susceptibility. Circulation 128(6):632–642. https://doi.org/10.1161/circulationaha.113.002714

Yajima N, Takahashi M, Morimoto H, Shiba Y, Takahashi Y, Masumoto J, Ise H, Sagara J, Nakayama J, Taniguchi S, Ikeda U (2008) Critical role of bone marrow apoptosis-associated speck-like protein, an inflammasome adaptor molecule, in neointimal formation after vascular injury in mice. Circulation 117(24):3079–3087. https://doi.org/10.1161/circulationaha.107.746453

Yang TC, Chang PY, Lu SC (2017) L5-LDL from ST-elevation myocardial infarction patients induces IL-1beta production via LOX-1 and NLRP3 inflammasome activation in macrophages. Am J Phys Heart Circ Phys 312(2):H265–h274. https://doi.org/10.1152/ajpheart.00509.2016

Yuksel S, Eren E, Hatemi G, Sahillioglu AC, Gultekin Y, Demiroz D, Akdis C, Fresko I, Ozoren N (2014) Novel NLRP3/cryopyrin mutations and pro-inflammatory cytokine profiles in Behcet's syndrome patients. Int Immunol 26(2):71–81. https://doi.org/10.1093/intimm/dxt046

Zheng F, Xing S, Gong Z, Mu W, Xing Q (2014) Silence of NLRP3 suppresses atherosclerosis and stabilizes plaques in apolipoprotein E-deficient mice. Mediat Inflamm 2014:507208. https://doi.org/10.1155/2014/507208

Zuurbier CJ, Jong WM, Eerbeek O, Koeman A, Pulskens WP, Butter LM, Leemans JC, Hollmann MW (2012) Deletion of the innate immune NLRP3 receptor abolishes cardiac ischemic preconditioning and is associated with decreased Il-6/STAT3 signaling. PLoS One 7(7):e40643. https://doi.org/10.1371/journal.pone.0040643

Chapter 3
Inflammasomes in CNS Diseases

Eduardo A. Albornoz, Trent M. Woodruff, and Richard Gordon

Contents

Abstract Neuroinflammation is a common pathological feature in almost all neurological diseases and is a response triggered as a consequence of the chronic activation of the innate immune response in the CNS against a variety of stimuli, including infection, traumatic brain injury, toxic metabolites, aggregated proteins, or autoimmunity. Crucial mediators of this neurinflammatory process are the intracellular protein complexes known as inflammasomes which can be triggered by pathogens as well as pathogen-associated molecular patterns (PAMPs) and damage-

E. A. Albornoz · T. M. Woodruff, PhD
School of Biomedical Sciences, Faculty of Medicine, University of Queensland, Brisbane, QLD, Australia

R. Gordon, PhD, DABT (✉)
School of Biomedical Sciences, Faculty of Medicine, University of Queensland, Brisbane, QLD, Australia

UQ Centre for Clinical Research, Faculty of Medicine, University of Queensland, Brisbane, QLD, Australia
e-mail: r.gordon1@uq.edu.au

M. D. Cordero, E. Alcocer-Gómez (eds.), *Inflammasomes: Clinical and Therapeutic Implications*, Experientia Supplementum 108,
https://doi.org/10.1007/978-3-319-89390-7_3

associated molecular patterns (DAMPs). However, chronic inflammasome activation can eventually result in cellular death and tissue damage, leading to the release of DAMPs that can reactivate the inflammasome, thereby propagating a vicious cycle of inflammation. The primary cells involved in CNS inflammasome activation are the immunocompetent microglia and the infiltrating macrophages into the CNS. However, astrocytes and neurons also express inflammasomes, and the understanding of how they are engaged in the pathogenesis of a variety of neurological diseases is crucial to develop effective therapeutic approaches for CNS pathologies that are propagated by chronic inflammasome activation. This chapter covers the activation mechanisms of relevant inflammasomes in the brain and summarizes their roles in the pathogenesis and progression of different neurological conditions.

Keywords Neuroinflammation · Neurodegeneration · NLRP3 · Inflammasome · Microglia · IL-1β · ASC

3.1 Introduction

Inflammation of the central nervous system (CNS) or neuroinflammation is a common underlying pathological feature of most neurological disorders. Chronic neuroinflammation is evident in progressive neurodegenerative diseases like Alzheimer's and Parkinson's disease (Heneka et al. 2014) and in autoimmune disorders such as multiple sclerosis (Barclay and Shinohara 2017). Neuroinflammation is also present in psychiatric illnesses such as depression (Alcocer-Gomez et al. 2014) and a consequence of direct damage to the CNS in the form of brain injuries such as stroke (Barrington et al. 2017) or traumatic brain injury (Wallisch et al. 2017). Innate immune activation is also evident as an acute response against viral or bacterial infections in the CNS (Klein et al. 2017). The neuroinflammatory process is a response to a variety of pathogenic signals, and its primary role is to maintain homeostasis in the brain conferring neuroprotection and promoting remyelination and even axonal regeneration (Wee Yong 2010). Nevertheless, the delicate balance between the benefit and harm that is conferred by activation of the innate immune system is directly related to length, reactivation, and spread of the inflammatory response. This mechanism of defense in the innate immune response occurs through pattern recognition receptors (PRRs) that can recognize infections through pathogen-specific proteins (PAMPs) and internal cellular disturbances through damage-associated proteins (DAMPs) (Kigerl et al. 2014). Primarily expressed in the CNS by glial cells (microglia and astrocytes) and also in neurons, these receptors can be located on extracellular membranes or within the cytosol such as the NOD-like receptors (NLRs). NLR activation leads to the oligomerization of cytosolic protein complex known as inflammasomes that regulate the activation of caspase-1 (Martinon et al. 2002). This active caspase cleaves the precursor forms of the pro-inflammatory cytokines of the IL-1 family, IL-1β, IL-18, and IL-33, into their active forms that exert their biological effects. The expression and release of the IL-1 family cytokines in the normal brain is normally at low levels and is dramatically upregulated in response to local or systemic disease and injury (Rothwell and Luheshi 2000). In the CNS, multiple cell types express receptors for IL-1β and IL-18 making them particularly sensitive to these cytokines. (Allan et al. 2005; Alboni et al. 2010). The primary role of the inflammasome activation in the healthy brain is to limit and clear any pathogenic or metaboloic damage, but when this activation becomes persistent, it entails other associated process that can propagate and spread the inflammatory response. These include pyroptotic cell death and the release of ASC specks, NLRP3, and DAMPs that can propagate the inflammatory response leading to tissue damage (Franklin et al. 2014; Baroja-Mazo et al. 2014).

Recent advances in our understanding of inflammasome triggers and activation mechanisms in the CNS, as well as its regulation by endogenous inhibitors such as dopamine and the ketone body β-hydroxybutyrate (Youm et al. 2015) could have therapeutic applications for CNS diseases. Indeed, the development of highly specific NLRP3 inhibitors such as MCC950 (Coll et al. 2015) and caspase-1/ICE inhibitors have opened up an exciting new area of translational research with the potential for disease modification and therapeutic targeting of progressive neurodegenerative diseases (Lipinska et al. 2014; Boxer et al. 2010; Ross et al. 2007).

3.2 Inflammasomes in the Healthy Brain

Until recently, the CNS was considered an immune-privileged site, protected by the blood-brain barrier (BBB), creating an isolated environment between the CNS and the peripheral immune system (Carson et al. 2006). However, the recent paradigm shifts in our understanding has revealed the far-reaching scope and intricate communication between the innate immune response in the CNS and the peripheral immune system in maintaining brain homeostasis (Louveau et al. 2015a, b). The immunocompetent microglia are primarily responsible for sensing a broad range of exogenous pathogenic stimuli, such as fungal, bacterial, and viral components, or endogenous such as aggregated and misfolded proteins, extracellular ATP, and reactive oxygen species (ROS) among others (Walsh et al. 2014). The microglia express and can activate NLRC4 and NLRP3 inflammasomes, the best characterized of which is NLRP3, considered in the recent years as a key pathway in the development of neuroinflammation and neurodegeneration (Song et al. 2017). NLRP3 activation requires two signals: the first is a priming step (such as TLR4 activation) which drives the transcription and translation of inflammasome components including IL-1β, caspase-1, NLRP3, and ASC. A secondary signal or activation step (i.e., pathogen infiltration or aggregated proteins) is then required to trigger the formation of the inflammasome complex (NRLP3-ASC-caspase-1) leading to cluster-dependent caspase-1 activation and the cleavage and release of IL1-β (Guo et al. 2015). There is believed to be an association of this pro-inflammatory phenotype or M1-like microglia with the expression of IL-1β, IL-6, IL-18, tumor necrosis factor-α (TNFα), the production of superoxide, reactive oxygen species (ROS), and

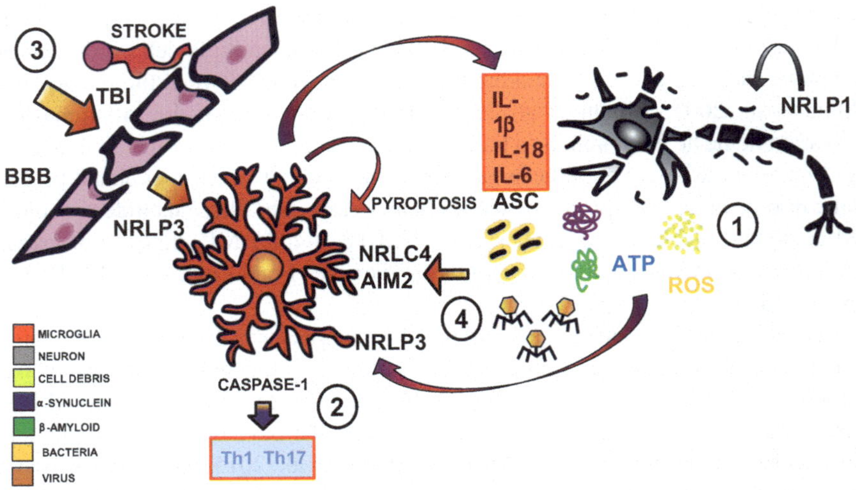

Fig. 3.1 Overview of inflammasome activation in neurological diseases. Microglial inflammasomes can be chronically activated in progressive neurodegenerative diseases such as Alzheimer and Parkinson's disease where they drive neuropathology (**1**). Inflammasome activation has also been documented in autoimmune diseases such as Multiple Sclerosis where it contributes to T cell dysregulation and nerve damage (**2**). Emerging evidence also indicates a pathogenic role for inflammasome activation in Traumatic Brain Injury (TBI) and stroke (**3**), as well as CNS infection by bacteria and viruses (**4**). The pathological contributions of systemic versus CNS inflammasome activation remains to be determined in these diseases and is an active area of research

nitric oxide (NO), and an impaired phagocytic capacity (Tang and Le 2016). The microglia also expresses NLRC4 a key sensor for bacterial infection, sensing bacterial flagellin related with the cause of meningitis in the CNS (Wu et al. 2010). Unlike NLRP3, NLRC4 can be activated without the ASC adapter protein (Latz et al. 2013), and bacterial activation can lead to a dual recruitment of NLRC4 and NLRP3 to the same macromolecular complex (Man et al. 2014). In contrast, in the healthy brain, the microglia interact with the neighboring neurons, remodeling synapses and secreting soluble neurotrophic factors, such as brain-derived neurotrophic factor (BDNF) and transforming growing factor-β (TGF-β). Microglial cells in the healthy brain are involved in regulation of synaptic strength, neuronal pruning, protease secretion to maintain extracellular matrices, and building and maintaining proper neuronal network functions and phagocytic activity to remove accumulating cell debris and misfolded proteins to maintain brain homeostasis (Fig. 3.1) (Waisman et al. 2015). This neuroprotective phenotype or M2-like is associated with the production of anti-inflammatory cytokines such as IL-4, IL-13, and IL-10 (Kabba et al. 2018). The balance between microglial phenotypes (M1/M2) depends on the disease stage and severity, and the type of response may determine whether microglial activation can be beneficial or leads to chronic neuroinflammation (Fig. 3.1) (Tang and Le 2016). The astrocytes also play a significant role in the brain homeostasis, including regulating neurotransmitter and growth factor release, forming the BBB, and regulating the immune response (Dong and Benveniste 2001). The role of astrocytes in innate immunity is crucial; these cells express an array of receptors including Toll-like receptors (TLR), double-stranded RNA-dependent protein kinase, scavenger receptors, components of the complement system, and NOD-like receptors like NLRP3 (Farina et al. 2007; Johann et al. 2015). Recently, it was shown that human astrocytes also express NLRP2 inflammasome (Minkiewicz et al. 2013). ATP can activate the inflammasome forming a multiprotein complex with ASC and caspase-1 and also interacts with the P2X7 receptor and pannexin 1 channel, leading to the release of IL-1β (Minkiewicz et al. 2013). Furthermore, increasing evidence has shown the expression of inflammasome-forming NLRs in non-myeloid cells such as neurons; the most characterized and implicated in some pathologies is NRLP1 (Kaushal et al. 2015). Human NLRP1 is unique compared to the rest of the inflammasomes due to the complexity of its domain structure, possessing two interaction domains, an N-terminal PYD and a C-terminal CARD, and interestingly the adaptor protein ASC is not necessary for activating the pathway, but it might enhance NLRP1 activation (Faustin et al. 2007). Furthermore, cerebral pericytes express inflammasomes, and Toll-like receptors, which are involved in controlling key neurovascular functions and BBB permeability and peripheral leukocyte trafficking (Nyul-Toth et al. 2017). Emerging evidence suggests that inflammasomes could have previously unknown roles in normal brain function beyond neuronal protection and maintaining homeostasis. It can therefore be expected that new roles for the various inflammasome components will emerge as these pathways are studied in more detail in the context of the central nervous system.

3.3 Inflammasomes in Neurodegenerative Diseases

There is extensive accumulating evidence for the pathogenic role of chronic inflammasome activation in propagating neuroinflammation, a key underlying feature of neurodegenerative disorders such as Alzheimer's and Parkinson's disease (Heneka 2017; Mao et al. 2017). Despite the different factors and mediators that can trigger inflammasome activation in this sterile environment, aging is a common factor in most progressive neurodegenerative diseases (Wyss-Coray 2016). Recent studies correlate microglia and enhanced sensitivity to inflammatory stimuli with aging, demonstrating in senescence-accelerated mice a "primed microglial" phenotype, characterized by increased production of pro-inflammatory cytokines and ROS, leading to elevated basal inflammasome activation that can influence neurodegeneration in more elderly populations (Luo et al. 2010; Spittau 2017). Furthermore, it has recently been demonstrated that the expression of specific inflammasome gene modules can accurately classify older populations in two groups, one with constitutive expression of IL-1β and the other without this expression, showing a correlation in the first group with an increased mortality and related inflammatory disease state, a common hallmark of neurodegenerative diseases

(Furman et al. 2017). We describe below, the role of the inflammasome and their therapeutic approaches in these two most common neurodegenerative diseases.

3.3.1 Alzheimer's Disease

Alzheimer's disease (AD) is the most prevalent chronic and progressive neurodegenerative disease. The first symptoms are short-term memory loss and worsen over time leading to progressive cognitive impairment and dementia in 70% of the cases (Tarawneh and Holtzman 2012). Currently, there are no treatments to stop the progression of the disease, only to ameliorate some symptoms (Yiannopoulou and Papageorgiou 2013). The primary hallmark of the development of AD is the deposition of amyloid β (Aβ) aggregates in the hippocampus (Friedrich et al. 2010), correlated with a chronic activation of the innate immune system and microglial activation (Doens and Fernandez 2014). The microglia can bind soluble Aβ oligomers and fibrils via cell-surface receptors including CD14, CD36, CD47, and TLRs among others (Bamberger et al. 2003). In AD, NLRP3 inflammasome activation proceeds via phagosomal disruption or cell surface K^+ channels (Salminen et al. 2009). Increasing evidence from AD mouse models such as APP/PS1 transgenic model and clinical studies have shown the relationship between innate immune mechanisms and neurodegenerative process involving the persistent activation of NLRP3 in microglia and peripheral macrophages and NLRP1 inflammasome in neurons (Saresella et al. 2016). NLRP3 activation has been shown to enhance AD pathology and may be involved in synaptic dysfunction, cognitive impairments, and the restriction of microglial clearance functions (Heneka et al. 2013). Aβ fibrils can activate NLRP3 in mouse microglia, and higher levels of active caspase-1 and IL-1β in the brains of patients with AD and APP/PS1 mice have been detected compared to healthy patients and wild-type (WT) control mice. Furthermore, there is a protective phenotype in caspase-1 or NLRP3 knockout (KO) in terms of spatial memory impairments and loss of hippocampal neurons, associated with behavioral disturbances present in AD, showing a neuroprotective M2-like phenotype (Heneka et al. 2013; Heneka 2017). In neurons NLRP1 levels are upregulated in APP/PS1 mice, and in vitro results have shown that silencing NLRP1 reduces Aβ-induced neuronal pyroptotic cell death, positioning NLRP1 activation in neurons as a new factor relevant to neurodegeneration in AD (Tan et al. 2014; Kaushal et al. 2015). Recently the NLRC4 inflammasome was shown to be upregulated in AD brains; it has been shown that NLRC4 can be activated in response to fatty acid palmitate in astrocytes (Kaushal et al. 2015); this is correlated with a higher fatty acid content in AD brains compared with healthy brains (Cutler et al. 2004). Based on these results, introducing pharmacologic treatments targeting the NLRP3 inflammasome at different levels in the pathway may have beneficial effects in patients with AD (White et al. 2017). The first successful study targeting NLRP3 pharmacologically in a transgenic mouse model of AD was recently published (Daniels et al. 2016). This study has shown that several clinically approved and widely used nonsteroidal anti-inflammatory drugs (NSAIDs) of the fenamate class are efficient and selective inhibitors of the NLRP3 inflammasome via reversible blockade of volume-regulated anion channel (VRAC) in the plasma membrane of macrophages. Inhibiting cognitive impairments in 3 × TgAD transgenic mice (Daniels et al. 2016), this treatment completely abated AD-related neuroinflammation with reduced levels of microglial activation and IL-1β expression compared to WT mice, opening an exciting new translational field to repurposition these drugs as Alzheimer's disease therapeutics (Venegas et al. 2017). A potentially key paradigm shift in our understanding of inflammasome pathology in driving AD is that microglial ASC specks can cross-seed amyloid-β during the course of the disease and thereby contribute to the spread of amyloid-β pathology.

3.3.2 *Parkinson's Disease*

Parkinson's disease (PD) is the most prevalent synucleinopathy and the second most common neurodegenerative disorder worldwide after AD (Mhyre et al. 2012). PD is a chronic neurodegenerative disease of the CNS, and its pathological hallmark is a profound loss of nigrostriatal dopaminergic neurons that is preceded by the accumulation and spread of characteristic Lewy bodies, consisting primarily of misfolded fibrillar α-synuclein (Syn) (Obeso et al. 2010). There is a correlation between lack of dopamine in the CNS with debilitating motor symptoms including tremor, rigidity, and slowness of the movements and in advanced-stage dementia (Lotharius and Brundin 2002). The current treatments target the symptoms but similar to AD do not stop the progression of the disease due to the lack of treatments to prevent chronic neuroinflammation that leads to neurodegeneration (Schapira et al. 2006). In recent years, accumulating evidence suggests that the innate immune system specifically NLRP3 is involved in the prominent neuroinflammatory response observed in PD (Codolo et al. 2013; Walsh et al. 2014; Guo et al. 2015; Mao et al. 2017). It has been shown that Syn fibrils can activate NLRP3 inflammasome in macrophages and microglia, acting as an endogenous trigger of the inflammasome in PD (Codolo et al. 2013; Guo et al. 2015; Gustot et al. 2015). The mechanism of Syn activation of NLRP3 is not fully elucidated, but recent evidence suggests that it could be through microglial endocytosis and subsequent lysosomal cathepsin B release (Zhou et al. 2016) and deficiency of caspase-1, significantly inhibited Syn-induced microglia activation, and IL-1β production in vitro (Zhou et al. 2016). Clinical studies have found higher levels of IL-1β and an upregulation of NLRP3 in PD patients (Zhou et al. 2016; Zhang et al. 2016). Additionally, it has been demonstrated that dopamine can inhibit systemic NLRP3 activation through dopamine D1 receptor (DDR1) via a second messenger cyclic adenosine monophosphate (cAMP), which binds to NLRP3 and promote its ubiquitination and degradation via the E3 ubiquitin ligase MARCH7 (Yan et al. 2015), suggesting an important novel endogenous regulatory role for dopamine that correlates with an increased NLRP3 activation in PD patients who inevitably have reduced levels of dopamine over the course of the disease.

Although the etiology of PD is unknown, pesticide exposure is well recognized as an environmental risk factor to acquire this disease (Hancock et al. 2008). In fact chronic infusion of the pesticide, rotenone, has been used as a rodent model of PD, since this model reproduces many relevant features of the disease (Panov et al. 2005). In this model it has also been demonstrated that microglial NLRP3 inflammasome can be activated by rotenone exposure (Liang et al. 2015). Microglial NLRP3 can be activated with rotenone via the ROS/c-abl/NLRP3 signaling pathway, affecting the auto-lysosomal system. In this study it was demonstrated that targeting oxidative stress-induced c-Abl activation in the microglia can diminish microglial activation associated with neurodegeneration in PD (Lawana et al. 2017). The effect of rotenone also was observed in neurons in an animal model showing that pathological pesticide exposure could activate neuronal inflammasome in the substantia nigra and promote the expression of NLRP3, ASC, and caspase-1 and the secretion of IL-1β and IL-18 in a dose- and time-dependent manner (Zhang et al. 2016). Moreover, inflammasome activation components were detected in cerebrospinal fluid of PD patients, showing that cyclin-dependent kinase 5 (Cdk5) is necessary for NLRP3 activation in neurons and its pharmacological inhibition with roscovitine a Cdk5 inhibitor or Cdk5-targeted deletion could efficiently block neuronal inflammasome activation in 1,2,3,6-methyl-phenyl-tetrahydropyridine (MPTP) and Syn transgenic mouse PD models (Zhang et al. 2016). Besides, a central component of the inflammasome, caspase-1, causes truncation and aggregation of Syn, and this was recently demonstrated in a neuronal cell model of PD, dmeonstrating another potentially pathogenic role of the NLRP3 inflammasome in PD etiology (Wang et al. 2016). NLRP3 upregulation also was observed in a rat model of PD, the 6-hydroxydopamine (6-OHDA) model, showing high levels of mRNA and protein expression of NLRP3 components in the 6-OHDA injected side. In this study, microinjections of different doses of caspase-1 inhibitor (Ac-YVAD-CMK) in the striatum were performed. This treatment showed an inhibition of the mRNA and protein expression of NLRP3 components and an improvement in the rotational behavior and the number of dopamine neurons in the substantia nigra, indicating that NLRP3 inflammasome participates in the pathogenesis of PD and downstream inhibition of the inflammasome can alleviate the occurrence of PD symptoms (Mao et al. 2017). Astrocytes also play an important role in neurodegeneration, regulating ROS production through uncoupling protein 2 (UCP2), maintaining the proper levels of oxidative stress in the brain (Lu et al. 2014). It has been shown that UCP2 knockout mice exhibit exacerbated dopaminergic neuron loss in MPTP mouse model, with NLRP3 inflammasome activation in astrocytes (Lu et al. 2014). Moreover, targeting NLRP3 to inhibit the inflammatory process also has been approached with gene therapy. In a recent study, it was shown that NLRP3 is a target gene of microRNA-7 (miR-7) and microglial transfection with miR-7 produces an inhibition of microglial NLRP3 inflammasome activation, whereas anti-miR-7 aggravated inflammasome activation in vitro (Zhou et al. 2016). Also, this was demonstrated in vivo with stereotaxic injections of miR-7 into mouse striatum, decreasing dopaminergic neuron loss accompanied by the amelioration of microglial activation in MPTP mouse model (Zhou et al. 2016). Collectively, these exciting new findings place NLRP3 inflammasome as a central pathway in the progression of the neuroinflammatory process in PD and an attractive disease-modifying therapeutic target.

3.4 Inflammasome in CNS Autoimmune Disease: Multiple Sclerosis

Multiple sclerosis (MS) is an autoimmune inflammatory demyelinating disease of the CNS that is thought to be mediated by myelin-specific autoreactive T cells, characterized by an increased microglial activation associated with extensive and chronic neurodegeneration (Compston and Coles 2008). This neuroinflammatory damage disrupts the proper transmission of the nerve impulse, resulting in a range of debilitating signs and symptoms including fatigue, ataxia, cognitive impairment, and depression among others (Bruck 2005). Although the etiology of the disease remains unknown, several clinical studies have reported the association of the elevated expression of caspase-1, IL-1β, and IL-18 with the susceptibility, progression, and severity of MS patients (Losy and Niezgoda 2001; Ming et al. 2002; Huang et al. 2004; Heidary et al. 2014). This finding is linked with an increasing number of reports that strongly suggest the involvement of NLRP3 inflammasome in the pathophysiology of MS (Gris et al. 2010; Inoue and Shinohara 2013; Barclay and Shinohara 2017; Guo et al. 2017). Recent studies have shown that NLRP3 plays a critical role in the induction and progression of experimental autoimmune encephalomyelitis (EAE), an animal model of MS. This is through direct effects on caspase-1-dependent cytokines which influence Th1 and Th17 responses (Gris et al. 2010), showing a protective phenotype in NLRP3 KO mice with EAE, due to a reduction in the severity of the disease with a significant decrease of the inflammatory infiltrates including macrophages, dendritic cells, CD4, and $CD8^+$ T cells in the spinal cord and with a reduction in the destruction of myelin and astrogliosis (Gris et al. 2010). Additionally, high levels of cytoplasmic caspase-1 in resident oligodendrocytes of MS lesions have been reported. In this study oligodendrocytes were exposed to a cytokine challenge, observing a blockage in cell death induction by the caspase-1inhibitor Z-YVAD-FMK, suggesting that caspase-1 may play a key role in the inflammatory and pyroptotic processes associated with MS pathogenesis (Ming et al. 2002). Furthermore, evidence in patients with MS suggests that inflammasome activation occurs during MS progression finding caspase-1 and IL-1β in MS plaques and an increase of IL-1β and IL-18 in peripheral blood mononuclear cells (PBMCs) of MS patients (Huang et al. 2004; Inoue and Shinohara 2013). Moreover, pharmacological studies with the specific NLRP3 inhibitor, the small molecule MCC950, were performed investigating the possibility that MCC950 may suppress the T-cell response that mediates autoimmune diseases (Coll et al. 2015). Pretreatment of EAE mice with MCC950 delayed the onset and reduced the severity of EAE, with a reduction in serum concentration of IL-1β and IL-6. Furthermore, analysis of brain mononuclear cells from mice culled on day 22 after treatment showed modestly reduced frequencies of IL-17- and IFN-γ-producing $CD3^+$ T cells in MCC950-treated mice in comparison with vehicle-treated mice, showing an attenuation in the severity and progression of the disease, positioning this drug as a potential therapeutic for NLRP3-associated syndromes including auto-inflammatory and autoimmune diseases (Coll et al. 2015). Similar results were observed with a novel small molecule, a hydroxyl sulfonamide analog JC-171, delaying the progression and severity of EAE in prophylactic and therapeutic settings and blocking IL-1β production and Th17 response (Guo et al. 2017). Collectively, all these novel therapeutic approaches suggest that sustained NLRP3 inflammasome activation is an important mechanism in MS pathophysiology and a potential therapeutic target for the treatment.

3.5 Inflammasomes in Brain Injury

The leading causes of brain injury related to mortality and morbidity are acute ischemic stroke and traumatic brain injury (Feigin et al. 2014; Levin and Diaz-Arrastia 2015). The first is due to an insufficient supply of blood to regions of the brain producing damage and death of tissue (Donnan et al. 2008). The second, TBI, is a consequence of a mechanical trauma to the CNS such as head or spinal cord injury (Ghajar 2000). Both forms of injury result in an acute necrotic and apoptotic loss of neuronal and some glial populations driven by inflammatory cascades leading to the overactivation of innate immune responses. Below we describe the role of the inflammasome activation in stroke and TBI.

3.5.1 Stroke

Stroke is one of the most frequent causes of death and disability worldwide (Donnan et al. 2008; Kuklina et al. 2012). The most common form is ischemic stroke and is caused when a blood clot slows or interrupts the normal blood flow to a region of the brain leading to inflammation and tissue damage (Feigin et al. 2003). Increasing evidence indicates that NLRP3 inflammasome plays a significant role in the pathogenesis and progression of stroke (Barrington et al. 2017; Ye et al. 2017). Increased levels of NLRP3 protein has been found after experimental ischemic stroke, accompanied by elevated levels of IL-1β and IL-18 and extensive neuronal and glial cell death (Lammerding et al. 2016). Preclinical studies in animal models have also found an increase in NLRP3 and NLRP1 expression and activation in primary cortical neurons and cerebral tissue under in vitro and in vivo ischemic conditions. This activation is through NF-κB and MAPK signaling pathways, showing high levels of IL-1β, IL-18, and caspase-3 triggering neuronal apoptosis (Fann et al. 2018). Furthermore, in the same study, treatment with intravenous immunoglobulin

(IVIg) was reported to reduce the activation of the NF-κB and MAPK pathways resulting in decreased expression and activation of NLRP1 and NLRP3 in primary cortical neurons under ischemic conditions, suggesting that therapeutic interventions targeting inflammasome activation in neurons may provide new treatments for ischemic stroke. Moreover, NLRP3 deficiency ameliorated cerebral injury in mice after ischemic stroke by reducing infarcts and BBB damage through NOX2-mediated oxidative stress (Yang et al. 2014). Also, the specific selective inhibition of NLRP3 with MCC950 or with P2X7R antagonist BBG post stroke has shown a reduction in neuronal apoptosis, infarction volume, and neurological impairment demonstrating that P2X7R/NLRP3 pathway plays a critical role in caspase-3-dependent neuronal apoptosis after ischemic stroke (Ye et al. 2017). The increase of caspase-1 expression has also been described in neurons and astrocytes after thromboembolic stroke and observed 24 h later in microglia (Abulafia et al. 2009). Therapeutic caspase-1 inhibition has been evaluated pharmacologically and with transgenic mice, showing a reduction of brain damage in knockout mice compared with WT controls after experimental stroke (Friedlander et al. 1997), as well as a protective effect with the intracerebroventricular administration of the caspase-1 inhibitors Ac-YVAD-cmk or VRT-018858 in experimental stroke models (Rabuffetti et al. 2000; Ross et al. 2007). Collectively, these findings suggest that the NLRP3 inflammasome could be a potential novel therapeutic target for stroke.

3.5.2 Traumatic Brain Injury

Traumatic brain injury (TBI), or mechanical trauma to the CNS, results in the disruption of the cellular microenvironment leading to massive necrotic and apoptotic loss of neuronal and glial populations. This loss is accompanied by an acute production of ROS and activation of the innate immune system triggering inflammation with the release of pro-inflammatory cytokines leading to neuron damage and death (Werner and Engelhard 2007). Growing evidence indicates that TBI could activate the inflammasome, specifically NLRP3 with increased levels of ASC, activation of caspase-1, and release of IL-1β and IL-18 in humans and murine models of TBI (Adamczak et al. 2012; Liu et al. 2013). The first study in human patients has shown increased levels of NLRP1 and NLRP3 in cerebrospinal fluid (CFS) after severe TBI in children correlating NLRP3 as a marker of TBI severity (Adamczak et al. 2012). Also, neuronal NLRP1 constitutes an important component of the innate immune response after TBI, demonstrated in a rat model neutralizing ASC with anti-ASC antibodies, showing a reduced caspase-1 activation and processing of IL-1β resulting in a significant decrease in contusion volume after injury (de Rivero Vaccari et al. 2009). To study if NLRP3 is involved in the outcome of TBI, NLRP3 activation was targeted pharmacologically in a rat model of blast-induced TBI (bTBI) with the administration of propofol. Propofol is a lipid-soluble intravenous anesthetic, which has been shown to possess therapeutic benefit during neuroinflammation on various brain injury models, indicating an inhibition of the inflammatory response and a reduction in brain injury by inhibiting ROS. This inhibition leads to a decrease in NLRP3 activation and pro-inflammatory cytokines release in the cerebral cortex of bTBI rats with an amelioration of cerebral cortex damage (Ma et al. 2016). Moreover, similar results were obtained with the Chinese medicine mangiferin, showing a neuroprotective effect in rats treated with mangiferin in a bTBI model, suppressing the activation and expression of NLRP3 through the inhibition of oxidative stress and pro-inflammatory cytokines production in the cerebral cortex, alleviating brain damage, and positioning this approach as a potential therapeutic strategy for bTBI (Fan et al. 2017). Another therapeutic approach has been achieved, targeting NLRP3 in TBI with the use of omega-3 fatty acids ($\omega-3$ FAs). This treatment prevents NLRP3 mitochondrial localization and caspase-1 cleavage through G protein-coupled receptor 40 (GPR40) reducing the release of IL-1β ameliorating neuronal death and behavioral deficits after TBI in rats (Lin et al. 2017), positioning the innate immune response, specifically NLRP3, as a major contributor to the neuroinflammatory process that leads to the severity of TBI and as an important novel therapeutic target for the treatment of this debilitating neurological disease.

3.6 Inflammasome Activation in CNS Infections

Infections of the CNS can be caused primarily by viruses, bacteria, and fungi as well as pathogenic prions (Koyuncu et al. 2013; Coureuil et al. 2017; Shi and Mody 2016). Some acute infections can lead to meningitis or encephalitis and have the potential to contribute to massive acute inflammasome activation. This response against a particular pathogen can be distinct in the CNS compared to systemic responses (Walsh et al. 2014). Therefore it is important to understand how the inflammasome can be distinctly activated in the CNS to develop novel and effective therapeutic strategies against CNS infections; below we outline the role of the inflammasomes in bacterial and viral infections in the CNS based on recent evidence.

3.6.1 Bacterial Infection

Streptococcus pneumoniae is a common cause of bacterial meningitis which occurs when these bacteria invade the CSF leading to an inflammatory process and brain damage (Geldhoff et al. 2013). This process is orchestrated by a variety of innate immune responses, including through acute NLRP3 inflammasome activation. As recently demonstrated in murine models and human patients, NLRP3 could play an important role in the pathologic progression of pneumococcal meningitis infection (Hoegen et al. 2011; Geldhoff et al. 2013). In a murine model of pneumococcal meningitis, ASC and NLRP3 were shown to be involved in modulating the severity of the disease through the release of IL-1β and IL-18. In this study using differentiated human macrophages, THP-1 cells were utilized to confirm that the pneumococcal pore-forming toxin pneumolysin is an essential inducer of IL-1β expression and inflammasome activation upon pneumococcal challenge, through the release of ATP, lysosomal destabilization, and cathepsin B activation (Hoegen et al. 2011). In murine microglia pneumolysin can induce caspase-1-dependent pyroptotic cell death with NLRP3 being critical for caspase-1 activation during the process, and the induction of autophagy could transiently protect microglia from pyroptosis boosting the infective mechanism (Kim et al. 2015). In patients with bacterial meningitis, CSF levels of IL-1β and IL-18 were correlated with severity of the disease. Inversely in ASC and NLRP3 KO mice, a decreased systemic inflammatory response and bacterial outgrowth is observed in blood and brain homogenates compared to WT mice. Moreover, the NLRP3 deficiency was associated with an increase in cerebral neutrophil infiltration and cerebral hemorrhages in comparison to WT controls (Geldhoff et al. 2013). Another common etiological agent of brain abscesses is *Staphylococcus aureus* (*S. aureus*) which is characterized by widespread inflammation and necrosis. *S. aureus* also induces NLRP3 inflammasome activation in microglia in an ATP and cathepsin B-dependent manner, with significantly reduced IL-1β production in NLRP3 and ASC KO microglia following exposure to *S. aureus* (Hanamsagar et al. 2011). *Listeria monocytogenes* (LM) infection, or listeriosis, is a common foodborne disease that can lead to severe and potentially fatal cases of bacteremia and meningitis (Thonnings et al. 2016). Infection with LM can activate caspase-1 and the processing of IL-1β and IL-18 and pyroptosis through the activation of multiple inflammasomes such as AIM2, NLRC4, and NLRP3 and collectively orchestrate a robust pro-inflammatory response (Wu et al. 2010).

3.6.2 Viral Infections

Viral infections typically begin in the peripheral tissues and rarely invade the CNS; however, some viruses can infect the CNS triggering the innate immune response (Koyuncu et al. 2013). Japanese encephalitis virus (JEV) represents a common cause of acute viral encephalitis; JEV can invade the CNS and consequently induce acute neuroinflammation, which is characterized by neurodegeneration, astrogliosis. JEV infection has recently been linked with microglial NLRP3 activation resulting in the production of IL-1β and IL-18 (Kaushik et al. 2012). JEV activates NLRP3 through K^+ efflux and ROS production in mouse microglia, and depletion of NLRP3 results in the reduction of caspase-1 activity and cytokine release (Kaushik et al. 2012). Another related flavivirus that causes viral encephalitis; West Nile virus (WNV) has been shown to interact with the innate immune response. In contrast with JEV, NLRP3 inflammasome activation has shown to be a protective response during encephalitis caused with WNV. IL-1β production is a key host restriction factor involved in WNV control, as shown in animals lacking the IL-1 receptor or components involved in inflammasome pathway a higher susceptibility to WNV pathogenesis. In this study it was demonstrated that IL-1β production is essential for the development of an effective host immunity against WNV, revealing a novel role for IL-1β in antiviral action that restricts virus replication in neurons (Ramos et al. 2012). Furthermore, ASC-deficient mice exhibited increased susceptibility to WNV infection, associated with a reduced survival with enhanced virus replication in the peripheral tissues and CNS. However, brains from ASC KO mice displayed unrestrained inflammation, including elevated levels of pro-inflammatory cytokines and chemokines, correlated with astrogliosis and enhanced infiltration of peripheral immune cells in the CNS (Kumar et al. 2013). Recently neurological complications related to Zika virus (ZIKV) infection have emerged as a significant threat to public health worldwide (Russo et al. 2017). Several studies have identified microglial nodules, gliosis, neuronal and glial cell degeneration, and necrosis in the brain of ZIKV-infected infants. This suggests that ZIKV could play a role in these neurological disorders through neuroinflammation and microglial dysfunction (Tricarico et al. 2017). ZIKV has been shown to replicate in infective cells causing increased ROS leading to NLRP3 activation and IL-1β release in the human primary glioblastoma cell line U87-MG process that culminates in cell death, implicating inflammasome as a novel and relevant pathogenic factor in ZIKV infection and associated neuropathology (Tricarico et al. 2017).

3.7 Concluding Remarks

In the past decade, crucial advances have been made in our understanding of functional roles of inflammasomes in the healthy CNS and more significantly, in understading the pathological consequences of chronic inflammasome activation during neurological disease. Indeed, sustained innate immune activation more broadly has emerged as a key pathological mechanism during progressive neurodegeneration. While insufficient activation causes the host to become vulnerable to PAMPs and DAMPs, on the contrary chronic and sustained inflammasome activation can drive unfavorable outcomes in almost all progressive neurological diseases that have been studied (Singhal et al. 2014; Song et al. 2017). The current understanding of the specific pathological mechanisms driving inflammasome activation remains insufficient to develop effective therapeutic strategies for complex and challening CNS diseases with multifactorial etioligies. Specifically, the mechanisms by which inflammasome activation is terminated and CNS homeostasis is restored remain to be defined. Likewise the contribution of systemic inflammasome activation to CNS pathology has not been elucidated. Therefore, a more in-depth understanding of the regulation of this key pathway in different neurological contexts could lead to novel and more effective disease-modifying therapeutic strategies that could slow or halt disease progression and potentially restore homeostasis in the CNS.

References

Abulafia DP, de Rivero Vaccari JP, Lozano JD, Lotocki G, Keane RW, Dietrich WD (2009) Inhibition of the inflammasome complex reduces the inflammatory response after thromboembolic stroke in mice. J Cereb Blood Flow Metab 29:534–544

Adamczak S, Dale G, de Rivero Vaccari JP, Bullock MR, Dietrich WD, Keane RW (2012) Inflammasome proteins in cerebrospinal fluid of brain-injured patients as biomarkers of functional outcome: clinical article. J Neurosurg 117:1119–1125

Alboni S, Cervia D, Sugama S, Conti B (2010) Interleukin 18 in the CNS. J Neuroinflammation 7:9

Alcocer-Gomez E, de Miguel M, Casas-Barquero N, Nunez-Vasco J, Sanchez-Alcazar JA, Fernandez-Rodriguez A, Cordero MD (2014) NLRP3 inflammasome is activated in mononuclear blood cells from patients with major depressive disorder. Brain Behav Immun 36:111–117

Allan SM, Tyrrell PJ, Rothwell NJ (2005) Interleukin-1 and neuronal injury. Nat Rev Immunol 5:629–640

Bamberger ME, Harris ME, Mcdonald DR, Husemann J, Landreth GE (2003) A cell surface receptor complex for fibrillar beta-amyloid mediates microglial activation. J Neurosci 23:2665–2674

Barclay W, Shinohara ML (2017) Inflammasome activation in multiple sclerosis and experimental autoimmune encephalomyelitis (EAE). Brain Pathol 27:213–219

Baroja-Mazo A, Martin-Sanchez F, Gomez AI, Martinez CM, Amores-Iniesta J, Compan V, Barbera-Cremades M, Yague J, Ruiz-Ortiz E, Anton J, Bujan S, Couillin I, Brough D, Arostegui JI, Pelegrin P (2014) The NLRP3 inflammasome is released as a particulate danger signal that amplifies the inflammatory response. Nat Immunol 15:738–748

Barrington J, Lemarchand E, Allan SM (2017) A brain in flame; do inflammasomes and pyroptosis influence stroke pathology? Brain Pathol 27:205–212

Boxer MB, Quinn AM, Shen M, Jadhav A, Leister W, Simeonov A, Auld DS, Thomas CJ (2010) A highly potent and selective caspase 1 inhibitor that utilizes a key 3-cyanopropanoic acid moiety. ChemMedChem 5:730–738

Bruck W (2005) Clinical implications of neuropathological findings in multiple sclerosis. J Neurol 252(Suppl 3):iii10–iii14

Carson MJ, Doose JM, Melchior B, Schmid CD, Ploix CC (2006) CNS immune privilege: hiding in plain sight. Immunol Rev 213:48–65

Codolo G, Plotegher N, Pozzobon T, Brucale M, Tessari I, Bubacco L, de Bernard M (2013) Triggering of inflammasome by aggregated alpha-synuclein, an inflammatory response in synucleinopathies. PLoS One 8:e55375

Coll RC, Robertson AA, Chae JJ, Higgins SC, Munoz-Planillo R, Inserra MC, Vetter I, Dungan LS, Monks BG, Stutz A, Croker DE, Butler MS, Haneklaus M, Sutton CE, Nunez G, Latz E, Kastner DL, Mills KH, Masters SL, Schroder K, Cooper MA, O'Neill LA (2015) A small-molecule inhibitor of the NLRP3 inflammasome for the treatment of inflammatory diseases. Nat Med 21:248–255

Compston A, Coles A (2008) Multiple sclerosis. Lancet 372:1502–1517

Coureuil M, Lecuyer H, Bourdoulous S, Nassif X (2017) A journey into the brain: insight into how bacterial pathogens cross blood-brain barriers. Nat Rev Microbiol 15:149–159

Cutler RG, Kelly J, Storie K, Pedersen WA, Tammara A, Hatanpaa K, Troncoso JC, Mattson MP (2004) Involvement of oxidative stress-induced abnormalities in ceramide and cholesterol metabolism in brain aging and Alzheimer's disease. Proc Natl Acad Sci U S A 101:2070–2075

Daniels MJ, Rivers-Auty J, Schilling T, Spencer NG, Watremez W, Fasolino V, Booth SJ, White CS, Baldwin AG, Freeman S, Wong R, Latta C, Yu S, Jackson J, Fischer N, Koziel V, Pillot T, Bagnall J, Allan SM, Paszek P, Galea J, Harte MK, Eder C, Lawrence CB, Brough D (2016) Fenamate NSAIDs inhibit the NLRP3 inflammasome and protect against Alzheimer's disease in rodent models. Nat Commun 7:12504

de Rivero Vaccari JP, Lotocki G, Alonso OF, Bramlett HM, Dietrich WD, Keane RW (2009) Therapeutic neutralization of the NLRP1 inflammasome reduces the innate immune response

and improves histopathology after traumatic brain injury. J Cereb Blood Flow Metab 29:1251–1261
Doens D, Fernandez PL (2014) Microglia receptors and their implications in the response to amyloid beta for Alzheimer's disease pathogenesis. J Neuroinflammation 11:48
Dong Y, Benveniste EN (2001) Immune function of astrocytes. Glia 36:180–190
Donnan GA, Fisher M, Macleod M, Davis SM (2008) Stroke. Lancet 371:1612–1623
Fan K, Ma J, Xiao W, Chen J, Wu J, Ren J, Hou J, Hu Y, Gu J, Yu B (2017) Mangiferin attenuates blast-induced traumatic brain injury via inhibiting NLRP3 inflammasome. Chem Biol Interact 271:15–23
Fann DY, Lim YA, Cheng YL, Lok KZ, Chunduri P, Baik SH, Drummond GR, Dheen ST, Sobey CG, Jo DG, Chen CL, Arumugam TV (2018) Evidence that NF-kappaB and MAPK signaling promotes NLRP inflammasome activation in neurons following ischemic stroke. Mol Neurobiol 55(2):1082–1096
Farina C, Aloisi F, Meinl E (2007) Astrocytes are active players in cerebral innate immunity. Trends Immunol 28:138–145
Faustin B, Lartigue L, Bruey JM, Luciano F, Sergienko E, Bailly-Maitre B, Volkmann N, Hanein D, Rouiller I, Reed JC (2007) Reconstituted NALP1 inflammasome reveals two-step mechanism of caspase-1 activation. Mol Cell 25:713–724
Feigin VL, Lawes CM, Bennett DA, Anderson CS (2003) Stroke epidemiology: a review of population-based studies of incidence, prevalence, and case-fatality in the late 20th century. Lancet Neurol 2:43–53
Feigin VL, Forouzanfar MH, Krishnamurthi R, Mensah GA, Connor M, Bennett DA, Moran AE, Sacco RL, Anderson L, Truelsen T, O'Donnell M, Venketasubramanian N, Barker-Collo S, Lawes CM, Wang W, Shinohara Y, Witt E, Ezzati M, Naghavi M, Murray C, Global Burden of Diseases I, Risk Factors S, The, G. B. D. S. E. G (2014) Global and regional burden of stroke during 1990–2010: findings from the Global Burden of Disease Study 2010. Lancet 383:245–254
Franklin BS, Bossaller L, De Nardo D, Ratter JM, Stutz A, Engels G, Brenker C, Nordhoff M, Mirandola SR, Al-Amoudi A, Mangan MS, Zimmer S, Monks BG, Fricke M, Schmidt RE, Espevik T, Jones B, Jarnicki AG, Hansbro PM, Busto P, Marshak-Rothstein A, Hornemann S, Aguzzi A, Kastenmuller W, Latz E (2014) The adaptor ASC has extracellular and 'prionoid' activities that propagate inflammation. Nat Immunol 15:727–737
Friedlander RM, Gagliardini V, Hara H, Fink KB, Li W, Macdonald G, Fishman MC, Greenberg AH, Moskowitz MA, Yuan J (1997) Expression of a dominant negative mutant of interleukin-1 beta converting enzyme in transgenic mice prevents neuronal cell death induced by trophic factor withdrawal and ischemic brain injury. J Exp Med 185:933–940
Friedrich RP, Tepper K, Ronicke R, Soom M, Westermann M, Reymann K, Kaether C, Fandrich M (2010) Mechanism of amyloid plaque formation suggests an intracellular basis of Abeta pathogenicity. Proc Natl Acad Sci U S A 107:1942–1947
Furman D, Chang J, Lartigue L, Bolen CR, Haddad F, Gaudilliere B, Ganio EA, Fragiadakis GK, Spitzer MH, Douchet I, Daburon S, Moreau JF, Nolan GP, Blanco P, Dechanet-Merville J, Dekker CL, Jojic V, Kuo CJ, Davis MM, Faustin B (2017) Expression of specific inflammasome gene modules stratifies older individuals into two extreme clinical and immunological states. Nat Med 23:174–184
Geldhoff M, Mook-Kanamori BB, Brouwer MC, Troost D, Leemans JC, Flavell RA, van der Ende A, van der Poll T, van de Beek D (2013) Inflammasome activation mediates inflammation and outcome in humans and mice with pneumococcal meningitis. BMC Infect Dis 13:358
Ghajar J (2000) Traumatic brain injury. Lancet 356:923–929
Gris D, Ye Z, Iocca HA, Wen H, Craven RR, Gris P, Huang M, Schneider M, Miller SD, Ting JP (2010) NLRP3 plays a critical role in the development of experimental autoimmune encephalomyelitis by mediating Th1 and Th17 responses. J Immunol 185:974–981
Guo H, Callaway JB, Ting JP (2015) Inflammasomes: mechanism of action, role in disease, and therapeutics. Nat Med 21:677–687

Guo C, Fulp JW, Jiang Y, Li X, Chojnacki JE, Wu J, Wang XY, Zhang S (2017) Development and characterization of a hydroxyl-sulfonamide analogue, 5-chloro-N-[2-(4-hydroxysulfamoyl-phenyl)-ethyl]-2-methoxy-benzamide, as a novel NLRP3 inflammasome inhibitor for potential treatment of multiple sclerosis. ACS Chem Neurosci 8(10):2194–2201

Gustot A, Gallea JI, Sarroukh R, Celej MS, Ruysschaert JM, Raussens V (2015) Amyloid fibrils are the molecular trigger of inflammation in Parkinson's disease. Biochem J 471:323–333

Hanamsagar R, Torres V, Kielian T (2011) Inflammasome activation and IL-1beta/IL-18 processing are influenced by distinct pathways in microglia. J Neurochem 119:736–748

Hancock DB, Martin ER, Mayhew GM, Stajich JM, Jewett R, Stacy MA, Scott BL, Vance JM, Scott WK (2008) Pesticide exposure and risk of Parkinson's disease: a family-based case-control study. BMC Neurol 8:6

Heidary M, Rakhshi N, Pahlevan Kakhki M, Behmanesh M, Sanati MH, Sanadgol N, Kamaladini H, Nikravesh A (2014) The analysis of correlation between IL-1B gene expression and genotyping in multiple sclerosis patients. J Neurol Sci 343:41–45

Heneka MT (2017) Inflammasome activation and innate immunity in Alzheimer's disease. Brain Pathol 27:220–222

Heneka MT, Kummer MP, Stutz A, Delekate A, Schwartz S, Vieira-Saecker A, Griep A, Axt D, Remus A, Tzeng TC, Gelpi E, Halle A, Korte M, Latz E, Golenbock DT (2013) NLRP3 is activated in Alzheimer's disease and contributes to pathology in APP/PS1 mice. Nature 493:674–678

Heneka MT, Kummer MP, Latz E (2014) Innate immune activation in neurodegenerative disease. Nat Rev Immunol 14:463–477

Hoegen T, Tremel N, Klein M, Angele B, Wagner H, Kirschning C, Pfister HW, Fontana A, Hammerschmidt S, Koedel U (2011) The NLRP3 inflammasome contributes to brain injury in pneumococcal meningitis and is activated through ATP-dependent lysosomal cathepsin B release. J Immunol 187:5440–5451

Huang WX, Huang P, Hillert J (2004) Increased expression of caspase-1 and interleukin-18 in peripheral blood mononuclear cells in patients with multiple sclerosis. Mult Scler 10:482–487

Inoue M, Shinohara ML (2013) NLRP3 inflammasome and MS/EAE. Autoimmune Dis 2013:859145

Johann S, Heitzer M, Kanagaratnam M, Goswami A, Rizo T, Weis J, Troost D, Beyer C (2015) NLRP3 inflammasome is expressed by astrocytes in the SOD1 mouse model of ALS and in human sporadic ALS patients. Glia 63:2260–2273

Kabba JA, Xu Y, Christian H, Ruan W, Chenai K, Xiang Y, Zhang L, Saavedra JM, Pang T (2018) Microglia: housekeeper of the central nervous system. Cell Mol Neurobiol 38(1):53–71

Kaushal V, Dye R, Pakavathkumar P, Foveau B, Flores J, Hyman B, Ghetti B, Koller BH, Leblanc AC (2015) Neuronal NLRP1 inflammasome activation of Caspase-1 coordinately regulates inflammatory interleukin-1-beta production and axonal degeneration-associated Caspase-6 activation. Cell Death Differ 22:1676–1686

Kaushik DK, Gupta M, Kumawat KL, Basu A (2012) NLRP3 inflammasome: key mediator of neuroinflammation in murine Japanese encephalitis. PLoS One 7:e32270

Kigerl KA, de Rivero Vaccari JP, Dietrich WD, Popovich PG, Keane RW (2014) Pattern recognition receptors and central nervous system repair. Exp Neurol 258:5–16

Kim JY, Paton JC, Briles DE, Rhee DK, Pyo S (2015) Streptococcus pneumoniae induces pyroptosis through the regulation of autophagy in murine microglia. Oncotarget 6:44161–44178

Klein RS, Garber C, Howard N (2017) Infectious immunity in the central nervous system and brain function. Nat Immunol 18:132–141

Koyuncu OO, Hogue IB, Enquist LW (2013) Virus infections in the nervous system. Cell Host Microbe 13:379–393

Kuklina EV, Tong X, George MG, Bansil P (2012) Epidemiology and prevention of stroke: a worldwide perspective. Expert Rev Neurother 12:199–208

Kumar M, Roe K, Orillo B, Muruve DA, Nerurkar VR, Gale M Jr, Verma S (2013) Inflammasome adaptor protein apoptosis-associated speck-like protein containing CARD (ASC) is critical for the immune response and survival in west Nile virus encephalitis. J Virol 87:3655–3667

Lammerding L, Slowik A, Johann S, Beyer C, Zendedel A (2016) Poststroke inflammasome expression and regulation in the peri-infarct area by gonadal steroids after transient focal ischemia in the rat brain. Neuroendocrinology 103:460–475

Latz E, Xiao TS, Stutz A (2013) Activation and regulation of the inflammasomes. Nat Rev Immunol 13:397–411

Lawana V, Singh N, Sarkar S, Charli A, Jin H, Anantharam V, Kanthasamy AG, Kanthasamy A (2017) Involvement of c-Abl kinase in microglial activation of NLRP3 inflammasome and impairment in autolysosomal system. J Neuroimmune Pharmacol 12(4):624–660

Levin HS, Diaz-Arrastia RR (2015) Diagnosis, prognosis, and clinical management of mild traumatic brain injury. Lancet Neurol 14:506–517

Liang Y, Jing X, Zeng Z, Bi W, Chen Y, Wu X, Yang L, Liu J, Xiao S, Liu S, Lin D, Tao E (2015) Rifampicin attenuates rotenone-induced inflammation via suppressing NLRP3 inflammasome activation in microglia. Brain Res 1622:43–50

Lin C, Chao H, Li Z, Xu X, Liu Y, Bao Z, Hou L, Liu Y, Wang X, You Y, Liu N, Ji J (2017) Omega-3 fatty acids regulate NLRP3 inflammasome activation and prevent behavior deficits after traumatic brain injury. Exp Neurol 290:115–122

Lipinska K, Malone KE, Moerland M, Kluft C (2014) Applying caspase-1 inhibitors for inflammasome assays in human whole blood. J Immunol Methods 411:66–69

Liu HD, Li W, Chen ZR, Hu YC, Zhang DD, Shen W, Zhou ML, Zhu L, Hang CH (2013) Expression of the NLRP3 inflammasome in cerebral cortex after traumatic brain injury in a rat model. Neurochem Res 38:2072–2083

Losy J, Niezgoda A (2001) IL-18 in patients with multiple sclerosis. Acta Neurol Scand 104:171–173

Lotharius J, Brundin P (2002) Pathogenesis of Parkinson's disease: dopamine, vesicles and alpha-synuclein. Nat Rev Neurosci 3:932–942

Louveau A, Harris TH, Kipnis J (2015a) Revisiting the mechanisms of CNS immune privilege. Trends Immunol 36:569–577

Louveau A, Smirnov I, Keyes TJ, Eccles JD, Rouhani SJ, Peske JD, Derecki NC, Castle D, Mandell JW, Lee KS, Harris TH, Kipnis J (2015b) Structural and functional features of central nervous system lymphatic vessels. Nature 523:337–341

Lu M, Sun XL, Qiao C, Liu Y, Ding JH, Hu G (2014) Uncoupling protein 2 deficiency aggravates astrocytic endoplasmic reticulum stress and nod-like receptor protein 3 inflammasome activation. Neurobiol Aging 35:421–430

Luo XG, Ding JQ, Chen SD (2010) Microglia in the aging brain: relevance to neurodegeneration. Mol Neurodegener 5:12

Ma J, Xiao W, Wang J, Wu J, Ren J, Hou J, Gu J, Fan K, Yu B (2016) Propofol inhibits NLRP3 inflammasome and attenuates blast-induced traumatic brain injury in rats. Inflammation 39:2094–2103

Man SM, Hopkins LJ, Nugent E, Cox S, Gluck IM, Tourlomousis P, Wright JA, Cicuta P, Monie TP, Bryant CE (2014) Inflammasome activation causes dual recruitment of NLRC4 and NLRP3 to the same macromolecular complex. Proc Natl Acad Sci U S A 111:7403–7408

Mao Z, Liu C, Ji S, Yang Q, Ye H, Han H, Xue Z (2017) The NLRP3 inflammasome is involved in the pathogenesis of Parkinson's disease in rats. Neurochem Res 42:1104–1115

Martinon F, Burns K, Tschopp J (2002) The inflammasome: a molecular platform triggering activation of inflammatory caspases and processing of proIL-beta. Mol Cell 10:417–426

Mhyre TR, Boyd JT, Hamill RW, Maguire-Zeiss KA (2012) Parkinson's disease. Subcell Biochem 65:389–455

Ming X, Li W, Maeda Y, Blumberg B, Raval S, Cook SD, Dowling PC (2002) Caspase-1 expression in multiple sclerosis plaques and cultured glial cells. J Neurol Sci 197:9–18

Minkiewicz J, De Rivero Vaccari JP, Keane RW (2013) Human astrocytes express a novel NLRP2 inflammasome. Glia 61:1113–1121

Nyul-Toth A, Kozma M, Nagyoszi P, Nagy K, Fazakas C, Hasko J, Molnar K, Farkas AE, Vegh AG, Varo G, Galajda P, Wilhelm I, Krizbai IA (2017) Expression of pattern recognition receptors and activation of the non-canonical inflammasome pathway in brain pericytes. Brain Behav Immun 64:220–231

Obeso JA, Rodriguez-Oroz MC, Goetz CG, Marin C, Kordower JH, Rodriguez M, Hirsch EC, Farrer M, Schapira AH, Halliday G (2010) Missing pieces in the Parkinson's disease puzzle. Nat Med 16:653–661

Panov A, Dikalov S, Shalbuyeva N, Taylor G, Sherer T, Greenamyre JT (2005) Rotenone model of Parkinson disease: multiple brain mitochondria dysfunctions after short term systemic rotenone intoxication. J Biol Chem 280:42026–42035

Rabuffetti M, Sciorati C, Tarozzo G, Clementi E, Manfredi AA, Beltramo M (2000) Inhibition of caspase-1-like activity by Ac-Tyr-Val-Ala-Asp-chloromethyl ketone induces long-lasting neuroprotection in cerebral ischemia through apoptosis reduction and decrease of proinflammatory cytokines. J Neurosci 20:4398–4404

Ramos HJ, Lanteri MC, Blahnik G, Negash A, Suthar MS, Brassil MM, Sodhi K, Treuting PM, Busch MP, Norris PJ, Gale M Jr (2012) IL-1beta signaling promotes CNS-intrinsic immune control of West Nile virus infection. PLoS Pathog 8:e1003039

Ross J, Brough D, Gibson RM, Loddick SA, Rothwell NJ (2007) A selective, non-peptide caspase-1 inhibitor, VRT-018858, markedly reduces brain damage induced by transient ischemia in the rat. Neuropharmacology 53:638–642

Rothwell NJ, Luheshi GN (2000) Interleukin 1 in the brain: biology, pathology and therapeutic target. Trends Neurosci 23:618–625

Russo FB, Jungmann P, Beltrao-Braga PCB (2017) Zika infection and the development of neurological defects. Cell Microbiol 19:e12744

Salminen A, Ojala J, Kauppinen A, Kaarniranta K, Suuronen T (2009) Inflammation in Alzheimer's disease: amyloid-beta oligomers trigger innate immunity defence via pattern recognition receptors. Prog Neurobiol 87:181–194

Saresella M, La Rosa F, Piancone F, Zoppis M, Marventano I, Calabrese E, Rainone V, Nemni R, Mancuso R, Clerici M (2016) The NLRP3 and NLRP1 inflammasomes are activated in Alzheimer's disease. Mol Neurodegener 11:23

Schapira AH, Bezard E, Brotchie J, Calon F, Collingridge GL, Ferger B, Hengerer B, Hirsch E, Jenner P, Le Novere N, Obeso JA, Schwarzschild MA, Spampinato U, Davidai G (2006) Novel pharmacological targets for the treatment of Parkinson's disease. Nat Rev Drug Discov 5:845–854

Shi M, Mody CH (2016) Fungal infection in the brain: what we learned from intravital imaging. Front Immunol 7:292

Singhal G, Jaehne EJ, Corrigan F, Toben C, Baune BT (2014) Inflammasomes in neuroinflammation and changes in brain function: a focused review. Front Neurosci 8:315

Song L, Pei L, Yao S, Wu Y, Shang Y (2017) NLRP3 inflammasome in neurological diseases, from functions to therapies. Front Cell Neurosci 11:63

Spittau B (2017) Aging microglia-phenotypes, functions and implications for age-related neurodegenerative diseases. Front Aging Neurosci 9:194

Tan MS, Tan L, Jiang T, Zhu XC, Wang HF, Jia CD, Yu JT (2014) Amyloid-beta induces NLRP1-dependent neuronal pyroptosis in models of Alzheimer's disease. Cell Death Dis e1382:5

Tang Y, Le W (2016) Differential roles of M1 and M2 microglia in neurodegenerative diseases. Mol Neurobiol 53:1181–1194

Tarawneh R, Holtzman DM (2012) The clinical problem of symptomatic Alzheimer disease and mild cognitive impairment. Cold Spring Harb Perspect Med 2:a006148

Thonnings S, Knudsen JD, Schonheyder HC, Sogaard M, Arpi M, Gradel KO, Ostergaard C, Danish Collaborative Bacteraemia, N (2016) Antibiotic treatment and mortality in patients with Listeria monocytogenes meningitis or bacteraemia. Clin Microbiol Infect 22:725–730

Tricarico PM, Caracciolo I, Crovella S, D'Agaro P (2017) Zika virus induces inflammasome activation in the glial cell line U87-MG. Biochem Biophys Res Commun 492(4):597–602

Venegas C, Kumar S, Franklin BS, Dierkes T, Brinkschulte R, Tejera D, Vieira-Saecker A, Schwartz S, Santarelli F, Kummer MP, Griep A, Gelpi E, Beilharz M, Riedel D, Golenbock DT, Geyer M, Walter J, Latz E, Heneka MT (2017) Microglia-derived ASC specks cross-seed amyloid-β in Alzheimer's disease. Nature 552(7685):355–361. https://doi.org/10.1038/nature25158

Waisman A, Ginhoux F, Greter M, Bruttger J (2015) Homeostasis of microglia in the adult brain: review of novel microglia depletion systems. Trends Immunol 36:625–636

Wallisch JS, Simon DW, Bayir H, Bell MJ, Kochanek PM, Clark RS (2017) Cerebrospinal fluid NLRP3 is increased after severe traumatic brain injury in infants and children. Neurocrit Care 27 (1):44–50

Walsh JG, Muruve DA, Power C (2014) Inflammasomes in the CNS. Nat Rev Neurosci 15:84–97

Wang W, Nguyen LT, Burlak C, Chegini F, Guo F, Chataway T, Ju S, Fisher OS, Miller DW, Datta D, Wu F, Wu CX, Landeru A, Wells JA, Cookson MR, Boxer MB, Thomas CJ, Gai WP, Ringe D, Petsko GA, Hoang QQ (2016) Caspase-1 causes truncation and aggregation of the Parkinson's disease-associated protein alpha-synuclein. Proc Natl Acad Sci U S A 113:9587–9592

Wee Yong V (2010) Inflammation in neurological disorders: a help or a hindrance? Neuroscientist 16:408–420

Werner C, Engelhard K (2007) Pathophysiology of traumatic brain injury. Br J Anaesth 99:4–9

White CS, Lawrence CB, Brough D, Rivers-Auty J (2017) Inflammasomes as therapeutic targets for Alzheimer's disease. Brain Pathol 27:223–234

Wu J, Fernandes-Alnemri T, Alnemri ES (2010) Involvement of the AIM2, NLRC4, and NLRP3 inflammasomes in caspase-1 activation by *Listeria monocytogenes*. J Clin Immunol 30:693–702

Wyss-Coray T (2016) Ageing, neurodegeneration and brain rejuvenation. Nature 539:180–186

Yan Y, Jiang W, Liu L, Wang X, Ding C, Tian Z, Zhou R (2015) Dopamine controls systemic inflammation through inhibition of NLRP3 inflammasome. Cell 160:62–73

Yang F, Wang Z, Wei X, Han H, Meng X, Zhang Y, Shi W, Li F, Xin T, Pang Q, Yi F (2014) NLRP3 deficiency ameliorates neurovascular damage in experimental ischemic stroke. J Cereb Blood Flow Metab 34:660–667

Ye X, Shen T, Hu J, Zhang L, Zhang Y, Bao L, Cui C, Jin G, Zan K, Zhang Z, Yang X, Shi H, Zu J, Yu M, Song C, Wang Y, Qi S, Cui G (2017) Purinergic 2X7 receptor/NLRP3 pathway triggers neuronal apoptosis after ischemic stroke in the mouse. Exp Neurol 292:46–55

Yiannopoulou KG, Papageorgiou SG (2013) Current and future treatments for Alzheimer's disease. Ther Adv Neurol Disord 6:19–33

Youm YH, Nguyen KY, Grant RW, Goldberg EL, Bodogai M, Kim D, D'Agostino D, Planavsky N, Lupfer C, Kanneganti TD, Kang S, Horvath TL, Fahmy TM, Crawford PA, Biragyn A, Alnemri E, Dixit VD (2015) The ketone metabolite beta-hydroxybutyrate blocks NLRP3 inflammasome-mediated inflammatory disease. Nat Med 21:263–269

Zhang P, Shao XY, Qi GJ, Chen Q, Bu LL, Chen LJ, Shi J, Ming J, Tian B (2016) Cdk5-dependent activation of neuronal inflammasomes in Parkinson's disease. Mov Disord 31:366–376

Zhou Y, Lu M, Du RH, Qiao C, Jiang CY, Zhang KZ, Ding JH, Hu G (2016) MicroRNA-7 targets Nod-like receptor protein 3 inflammasome to modulate neuroinflammation in the pathogenesis of Parkinson's disease. Mol Neurodegener 11:28

Chapter 4
Lung Diseases

Saleela M. Ruwanpura, Sarah Rosli, and Michelle D. Tate

Contents

Abstract Inflammasomes are large innate cytoplasmic complexes that play a major role in promoting inflammation in the lung in response to a range of environmental and infectious stimuli. Inflammasomes are critical for driving acute innate immune responses that resolve infection and maintain tissue homeostasis. However, dysregulated or excessive inflammasome activation can be detrimental. Here, we discuss the plethora of recent data from clinical studies and small animal disease models that implicate excessive inflammasome responses in the pathogenesis of a

S. M. Ruwanpura · S. Rosli · M. D. Tate (✉)
Centre for Innate Immunity and Infectious Diseases, Hudson Institute of Medical Research, Clayton, VIC, Australia

Department of Molecular Translational Science, Faculty of Medicine, Nursing and Health Sciences, Monash University, Clayton, VIC, Australia
e-mail: saleela.ruwanpura@hudson.org.au; sarah.rosli@hudson.org.au; michelle.tate@hudson.org.au

M. D. Cordero, E. Alcocer-Gómez (eds.), *Inflammasomes: Clinical and Therapeutic Implications*, Experientia Supplementum 108,
https://doi.org/10.1007/978-3-319-89390-7_4

number of acute and chronic respiratory inflammatory diseases. Understanding of the role of inflammasomes in lung disease is of great therapeutic interest.

Keywords Inflammasomes · Lung disease · Inflammation · Disease pathogenesis

4.1 Introduction

The innate immune system plays a pivotal role in restoring homeostasis in the lung following an insult such as infection, cellular stress or injury. However, excessive or chronic activation of the immune system can contribute to the development of a number of inflammatory diseases such as acute respiratory distress syndrome (ARDS), asthma, cystic fibrosis (CF) and chronic obstructive pulmonary disease (COPD). Current treatments for such diseases are limited and ineffective, and new treatments are required to reduce morbidity and mortality. It is of great therapeutic interest that the mechanisms involved in the progression and persistence of immunopathology in the lung be delineated in greater detail.

Inflammasomes are large innate cytoplasmic complexes that play a major role in promoting inflammation in the lung, by enzymatically maturating the inactive pro-inflammatory cytokine precursors, pro-IL-1β and pro-IL-18 into bioactive IL-1β and IL-18, respectively. Inflammasomes are critical for driving acute innate immune responses that resolve infection and maintain tissue homeostasis. However, as discussed in this chapter, there is increasing evidence that excessive inflammasome activation can lead to lung disease.

4.1.1 Inflammasome Activation in the Lung

The lung is continuously exposed to potentially noxious stimuli, which include exogenous signals such as microbial (bacteria, viruses) and environmental antigens (smoke, silica, asbestos, allergens), as well as a plethora of host-derived endogenous danger signals. Innate immune responses produced within the host recognise these noxious stimuli through the tightly coordinated activation of a series of extracellular and cytosolic receptors called pattern recognition receptors (PRRs), which are widely expressed in both immune (e.g. alveolar macrophage, neutrophils) and non-immune (e.g. epithelial) cells in the lung (Bals and Hiemstra 2004). PRRs are classified into several families such as Toll-like receptors (TLRs), absent in melanoma 2 (AIM2)-like receptors (ALRs) and nucleotide-binding oligomerisation domain-containing (NOD)-like receptors (NLRs) (Kawai and Akira 2010; Kersse 2011; Ratsimandresy et al. 2013) (Fig. 4.1). Collectively, PRRs trigger inflammatory responses following recognition of a diverse range of ligands comprising microbial motifs called pathogen-associated molecular patterns (PAMPs), or danger-associated molecular patterns (DAMPs), which can involve endogenous host-derived signals or exogenous stimulants, such as smoke or silica, as above.

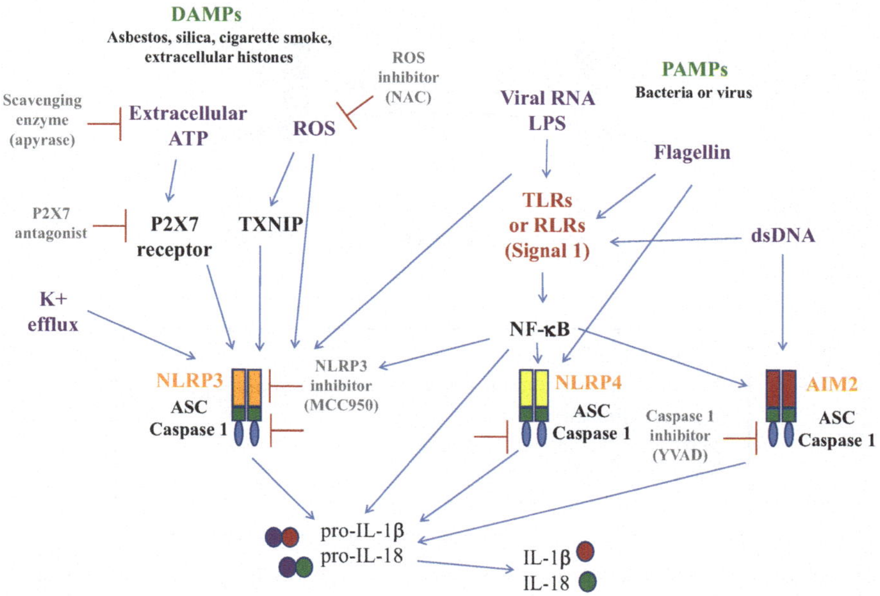

Fig. 4.1 Schematic of inflammasome responses in the lung. The lung is continuously exposed to exogenous signals such as microbial (bacteria, viruses) and environmental (smoke, silica, asbestos, allergens), as well as a plethora of host-derived danger signals. Activation of inflammasomes NLRP3, NLRC4 and AIM2 requires two signals. Signal 1 involves recognition of PAMPs (e.g. viral RNA or bacterial LPS) by PRRs such as TLRs and RLRs, inducing the expression of inflammasome components and pro-IL-1β/18. The second signal activates the inflammasome complexes NLRP3, NLRC4 or AIM2 in response to DAMPS (e.g. extracellular adenosine triphosphate (ATP) and reactive oxygen species (ROS)) or specific PAMPs (e.g. NLRP3, viral RNA/proteins; NLRC4, bacterial flagellin; AIM2, dsDNA). Inflammasome activation initiates the processing of pro-IL-1β and pro-IL-18 into their bioactive forms IL-1β and IL-18, by caspase-1

The most well-characterised PRR families are TLRs, which are transmembrane proteins associated with host cell surfaces and endosomes, and the cytosolic NLRs and ALRs (Fritz et al. 2006; Zuo et al. 2015). Signalling by TLRs, with the exception of TLR3, is dependent on the adaptor MyD88 and downstream activation of nuclear factor kappa-light-chain-enhancer of activated B cells (NF-κB) (Schnare et al. 2000) (Fig. 4.1). Members of the NLR family, such as NLRP3, NLRC4 (NLR CARD domain containing) (De Nardo et al. 2014) and AIM2 (a cytosolic DNA sensor in the ALR family) (Hornung et al. 2009), form the core of distinct inflammasomes, which are multiprotein complexes regulating the release of bioactive pro-inflammatory cytokines IL-1β and IL-18, in a two-step process. The first "priming" step involves the induced expression of biologically inactive pro-IL-1β and pro-IL-18 precursors, as well as inflammasome components via a PRR-mediated signal (e.g. lipopolysaccharide (LPS)-induced activation of TLR4/NF-κB). The second step involves the sensing of a specific DAMP or PAMP by each NLR or AIM2, leading to the recruitment and oligomerisation of the key adaptor protein, apoptosis-related speck-like protein

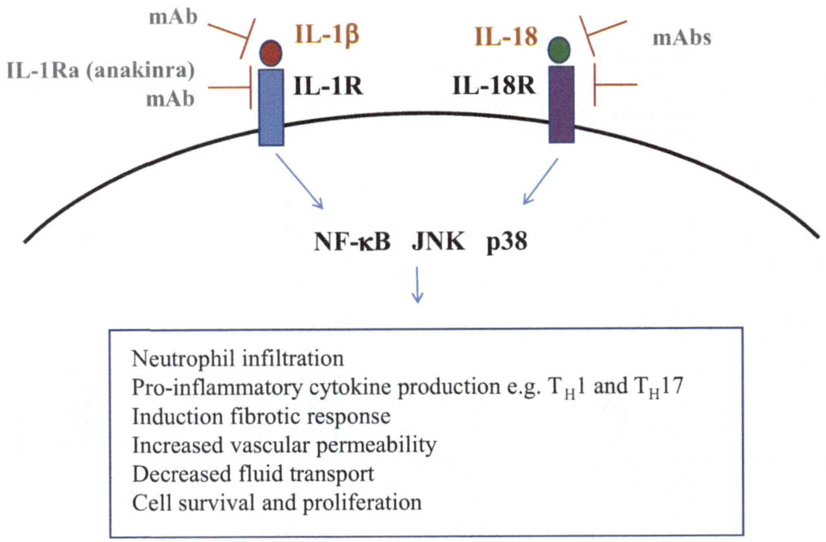

Fig. 4.2 Biological effects of inflammasome-dependent cytokines IL-1β and IL-18. Following inflammasome activation, mature IL-1β and IL-18 are secreted and can then bind their cell surface receptors IL-1R and IL-18R, respectively, which are expressed on a range of cell types. This results in a signalling cascade involving NF-κB, p38 and JNK, leading to a range of biological outcomes, such as pro-inflammatory cytokine secretion, neutrophil infiltration and increased vascular permeability, which have been implicated in a range of respiratory diseases

containing a CARD (ASC, also known as PYCARD), into large filamentous scaffolds called "specks" (De Nardo et al. 2014; Franklin et al. 2014). These ASC speck structures then facilitate the subsequent recruitment and activation of caspase-1, which in turn catalyses the maturation of pro-IL-1β or pro-IL-18 proteins into secreted bioactive cytokines (Vanaja et al. 2015; Franklin et al. 2014), which potently promote inflammatory host responses such as neutrophil infiltration and cytokine production (Latz et al. 2013) (Figs. 4.1 and 4.2). IL-1β and IL-18 mediate their biological effects following binding to cell surface receptors IL-1R and IL-18R, respectively, activating a signalling cascade involving NF-κB, p38 and Jun N-terminal kinase (JNK) (Fig. 4.2).

As described above, inflammasome activation depends on the recognition of PAMPs and DAMPs by PRRs. All cells expressing PRRs immediately identify PAMP-expressing microbes and act as the front line of host defence against infection in the lung. The membrane-bound TLRs scan the extracellular milieu and the endosomal compartment for PAMPs. Bacterial LPS, endotoxins found on the cell membrane of Gram-negative bacteria, and viral RNA are considered to be major PAMPs. LPS is specifically recognised by TLR4. Models of LPS-induced inflammation are widely employed to investigate both host responses in the lung and specific diseases such as acute lung injury (ALI) (Andonegui et al. 2003; Grailer et al. 2014; Jiang et al. 2016), asthma (Kim et al. 2014; Tran et al. 2012) and idiopathic pulmonary fibrosis (IPF) (Lasithiotaki et al. 2016). Mouse models in the above disease settings have shown evidence of neutrophil infiltration, production of IL-1β and IL-18 and most importantly in NLRP3 inflammasome activation in the lung following LPS stimulation (Andonegui et al. 2003; Grailer et al. 2014; Jiang et al. 2016; Kim et al. 2014; Tran et al. 2012; Lasithiotaki et al. 2016; Tate et al. 2016). In that regard, NLRs located in the cytoplasm of cells are known to directly respond to a variety of PAMPs, including the bacterial wall components peptidoglycan, bacterial flagellin, other bacterial toxins and viral proteins (Kersse et al. 2011; Franchi et al. 2006; Pinar et al. 2017). Studies into lung infections have revealed the important role of NLRP3 as an intracellular sensor for bacterial toxins from *Staphylococcus aureus*, *Streptococcus pneumoniae*, *Chlamydia trachomatis* and *Haemophilus influenzae* in lung diseases such as asthma (Kim et al. 2017) and cystic fibrosis (CF) (Yonker et al. 2015). NLRC4 can directly activate caspase-1 via its own CARD domain (unlike NLRP3) and acts as a cytosolic sensor of bacterial flagellin and type II/IV secretion system bacteria such as *Pseudomonas aeruginosa*. NLRC4 is therefore a key modulator of Gram-negative bacterial infection in the lungs (Cai et al. 2012; Yonker et al. 2015). In addition to these NLR-based inflammasomes, AIM2 forms an inflammasome by binding directly to the double-stranded (ds) DNA from numerous cytosolic bacteria and viruses. However, its role in lung disease is not currently understood (Man et al. 2016).

DAMPs are host-derived biomolecules that alert the immune system to a loss of homeostasis by activation of PRRs (Matzinger 1994) (Fig. 4.1). DAMPs/danger molecules can have an endogenous or exogenous origin. Noxious exogenous signals from infectious (bacteria, viruses) and environmental antigens (smoke, silica, asbestos, allergens such as house dust mite (HDM)) can damage resident airway epithelial cells in the lung, which can induce several modes of cell death such as apoptosis (programmed) and necrosis (unprogrammed), resulting in the release of DAMPs into the extracellular space (Messner et al. 2012; Kaczmarek et al. 2013). Programmed cell death, apoptosis, is caspase-dependent, and most of the released DAMPs are retained within apoptotic bodies for phagocytosis by macrophages (Krysko et al. 2010). However, when these apoptotic bodies are not adequately cleared, their presence leads to secondary necrosis resulting from the release of DAMPs (Krysko et al. 2010; Kono and Rock 2008). Necrotic cell death is the most immunogenic form of cell death and leads to a further massive release of DAMPs (Rubartelli and Lotze 2007). In recent years, it has been reported that exogenous stimuli can induce an inflammatory mode of airway epithelial cell death independent of executioner caspase activity in a manner akin to necrosis, and this process has been termed pyroptosis (Dos Santos et al. 2012). Pyroptosis, in contrast to the immunologically silent programmed cell death of apoptosis, is dependent on the inflammatory caspase-1 and is characterised by the rapid loss of plasma membrane integrity, leading to the release of DAMPs. In addition to release of DAMPs, cell death can also lead to release of several cytokines and chemokines such as interleukin (IL)-6 and IL-33, which can also act as DAMPs or danger signals (Hirsiger et al. 2012; Krysko et al. 2012). DAMPs include dsDNA, ROS, heat shock proteins, ATP and extracellular matrix fragments, which can potentiate pro-inflammatory reactions in innate immune (e.g. macrophage) and epithelial cells (Kaczmarek et al. 2013; Kono and Rock 2008; Pouwels et al. 2014). Upon release of ATP into the extracellular space, ATP triggers inflammasome activation by signalling through P2X7 (purinergic receptors) (Lucattelli et al. 2011) or changes in ion influx/efflux from cells (such as K^+) (Latz 2010). Several experimental models have shown that ROS can cause the development of many acute and chronic airway diseases, including fibrosis, asthma, emphysema, ARDS and bronchial carcinogenesis (Birrell and Eltom 2011).

4.1.2 Current Therapies

There are a limited number of specific drugs to block inflammasome activities under development currently. However, there are numerous preclinical inhibitors/antibodies tested in mouse studies, which show promise against up- and downstream key activators of inflammasomes in the lung (Figs. 4.1 and 4.2). Here, we will examine some of these key activators and the efficacy of corresponding preclinical therapeutics that have been tested against them.

The production of ROS has been suggested to act as an upstream modulator of the NLRP3 inflammasome. However, ROS inhibitors block the priming step of NLRP3 inflammasome activation by preventing pro-IL-1β synthesis (Bauernfeind et al. 2011), suggesting ROS inhibitors act at the synthesis level, rather than activation level, of NLRP3. The ROS scavenger *N*-acetyl cysteine (NAC) is one of the widely used antioxidants in vitro (Dostert et al. 2008) or in COPD and pulmonary fibrosis patients (Salve and Atram 2016; Tarrant et al. 2017) to block inflammasome activation. However NAC must be used at high concentrations to be able to block inflammasome activities (Bauernfeind et al. 2011). NecroX-5 is a mitochondrial inhibitor which displays excellent efficacy as an antioxidant focusing on the relationship between mitochondrial ROS and NLRP3 activation in allergic airway diseases such as asthma in mouse models (Kim et al. 2014). NLRP3 inflammasome activation via extracellular ATP acting on the P2X7 receptor signalling has been seen in lung inflammation and lung diseases (Wang et al. 2015). The P2X7 receptor antagonist has successfully blocked P2X7/NLRP3 inflammasome pathway resulting in a significant amelioration of lung injury in mouse models (Wang et al. 2015). Thus far no P2X7R antagonists are used for treating lung diseases in clinic; however, the P2X7 receptor antagonist CE-224-535 is currently in clinical trials to treat arthritis (Arulkumaran et al. 2011). ATP represents a suitable pharmacological target for the development of new effective therapeutic options in the treatment of inflammasome-related lung diseases. In that regard, glyburide, a blocker of K+ channels associated with ATP, has shown to inhibit NLRP3 inflammasome activation and lung inflammation in mouse models of bronchopulmonary dysplasia and cystic fibrosis (Liao et al. 2015; Buchanan et al. 2013). Modulation of ATP levels using the ATP-degrading enzyme apyrase is employed as another method to inhibit ATP-regulated inflammasome activities such as production of IL-1β in pulmonary fibrosis (Riteau et al. 2010).

Caspase-1 activation is important for NLRs and AIM2 inflammasome activities. The administration of caspase-1 inhibitors such as Ac-YVAD-CHO and z-WEHD-fmk has shown to effectively inhibit inflammasome activity by reducing IL-1β in vivo (Churg et al. 2009, Kim et al. 2017). The caspase-1 inhibitor VX-765 is an orally available prodrug, and it reduces the release of IL-1β and IL-18 in patients with cryopyrin-associated periodic syndromes (Wannamaker et al. 2007), but it is not used in lung diseases. IL-1β is one of the main downstream modulators of inflammasome, exerting its inflammatory action by binding to its receptor (IL-1R). The IL-1R antagonist (IL-1RA) prevents IL-1β binding and signalling through IL-1R. Anakinra is the recombinant form of naturally found IL-1RA and is widely used in clinic. However, anakinra is rapidly excreted by the kidney and therefore has a very short half-life, requiring frequent administration by subcutaneous injections (daily) which is associated with hepatotoxicity (Dinarello 2010). This antagonist has been widely used in animal models of asthma, ALI, bronchopulmonary dysplasia and cystic fibrosis (Jones et al. 2014; Kim et al. 2014; Rimessi et al. 2015; Rudloff et al. 2017). The fully humanised monoclonal antibodies against IL-1β (IL-1β mAb) such as canakinumab are also used in an animal model of asthma (Kim et al. 2017) and small clinical trials of asthma and COPD (Rogliani et al. 2015). However, the use of IL-1β mAb in lung diseases needs further investigation. There are also monoclonal antibodies against IL-1R and IL-18 receptor (IL-18R) which block their IL-1-mediated signal transduction. Even though those have exhibited excellent safety for patients (Dinarello 2010), those are not as effective in relieving the symptoms as anakinra in disease settings. However these antibodies have not thus applied for lung diseases.

Despite the emerging role for the inflammasomes in immunity, no drugs directly targeting specific inflammasomes, or with pan-inflammasome activity, have yet been described. MCC950 is a potent selective inhibitor of the NLRP3 inflammasome and has been successfully used in in vivo models of asthma to block NLRP3 activation.

4.2 Role of Inflammasomes in Respiratory Diseases

4.2.1 Role in Acute Lung Diseases

Acute lung injury (ALI) and its most severe form, acute respiratory distress syndrome (ARDS), are major causes of fatal respiratory failure. These diseases often occur as a result of severe viral and bacterial pneumonia, sepsis, burns or even oxygen and mechanical ventilator therapy (reviewed in Umbrello et al. 2016). Increased vascular permeability is a hallmark feature leading to lung oedema and poor arterial oxygenation. These conditions are characterised by the infiltration of neutrophils into the lung and the production of inflammatory mediators including complement activation products, cytokines and chemokines, proteases and oxidants. Currently, the mortality rate for patients who develop ALI is as high as 60%, and current treatments involve mechanical ventilatory support and anti-inflammatory drugs such as corticosteroids.

It is becoming evident that inflammasomes play a role in ALI and ARDS. IL-1β has been shown to be elevated and biologically active in the lungs of patients early after the onset of ALI (Olman et al. 2002; Pugin et al. 1996). Elevated levels of plasma IL-18 protein (Dolinay et al. 2012; Makabe et al. 2012) were shown to be associated with long-term poor prognosis in ALI (Makabe et al. 2012). At the mRNA level, *CASP1* and *IL-1B* and *IL-18* were increased in PBMCs from patients with ARDS (Dolinay et al. 2012). Furthermore, direct effects of IL-1β and IL-18 on lung vascular permeability and fluid transport, which are altered in ALI, have been shown. Treating rats with IL-1β and IL-18 and adenoviral overexpression of IL-1β in mice has been shown to increase vascular permeability in vivo (Ganter et al. 2008; Leff et al. 1994; Jordan et al. 2001). IL-1β has also been shown to inhibit fluid transport across the lung epithelium by decreasing the expression of the epithelial sodium channel alpha subunit (Roux et al. 2005). Collectively, inflammasome activation and the overproduction of IL-1β and IL-18 may play an important role in the pathogenesis of ALI/ARDS.

In the mouse model, instillation of LPS into the lung results in pulmonary oedema, injury, neutrophil infiltration and the production of pro-inflammatory cytokines including IL-1β and IL-18 (Grailer et al. 2014; Jiang et al. 2016). However, NLRP3 and caspase-1 knockout mice are protected from LPS-induced ALI (Grailer et al. 2014; Dolinay et al. 2012; Frank et al. 2008). Grailer et al. also demonstrated that neutrophil and macrophage depletion reduced IL-1β production in the lung following LPS instillation (Grailer et al. 2014), indicating these cells are a major source of this cytokine. Genetic deletion of caspase-1, IL-18 or the IL-1β receptor (IL-1R1) was also shown to reduce ALI in a mechanical ventilation model (Frank et al. 2008). Similar results have been seen with anti-IL-1β, anti-IL-18 antibody treatment or administration of recombinant IL-1R antagonist (IL1-Ra) in an attenuating ventilator model of ALI (Frank et al. 2008; Kuipers et al. 2012; Wu et al. 2013; Jordan et al. 2001). Furthermore, in a two-hit LPS and mechanical ventilation model, NLRP3- and caspase-1-deficient mice or those mice treated with IL-1R antagonist (anakinra) were shown to have diminished IL-1β levels and to be protected from ALI (Jones et al. 2014). Overall, these studies highlight a role for the NLRP3 inflammasome, as well as IL-1β and IL-18 in the pathogenesis of LPS and ventilator-induced ALI mouse models.

The role of inflammasomes in ALI induced by hypoxia or burns is less clear. Hypoxia-induced ALI is a serious complication of prolonged oxygen therapy, and mice lacking NLRP3 have been reported to display decreased (Mizushina et al. 2015) as well as increased (Fukumoto et al. 2013) susceptibility to ALI. The latter study identified that deletion of NLRP3 did not alter IL-1β levels, yet STAT3 responses were abrogated. While increased levels of IL-1β and IL-18 have been observed following burn injury (Ipaktchi et al. 2006; Rana et al. 2005; Han et al. 2015), studies have not directly examined the role of inflammasomes. In one study, the NF-κB inhibitor, BAY11-7082, was shown to dampen NLRP3 activation and to attenuate histological changes and inflammation in burn-induced ALI (Han et al. 2015).

The mechanisms of inflammasome activation during ALI have been examined. In the LPS instillation model, neutrophils were shown to be a source of extracellular histones in vivo, which were shown to activate NLRP3 in a caspase-1- and K+ efflux-dependent manner (Grailer et al. 2014). The P2X7 membrane receptor is activated in response to binding of extracellular ATP, resulting in NLRP3 inflammasome responses (Moncao-Ribeiro et al. 2011; Kolliputi et al. 2010). A role for P2X7 in hypoxia- and LPS-induced ALI was identified by genetic deletion of P2X7 or inhibition of P2X7 with antagonist A438079 treatment, resulting in reduced IL-1β production and inflammation (Galam et al. 2016; Wang et al. 2015). Reactive oxygen metabolites are also thought to play a key role in the pathogenesis of ALI/ARDS. In the presence of ROS, thioredoxin-interacting protein (TXNIP) has been shown to dissociate from thioredoxin (TRX) and bind to NLRP3, leading to its activation (Zhou et al. 2010).

4.2.2 Role in Chronic Lung Diseases

4.2.2.1 Asthma

Asthma is a chronic inflammatory respiratory disease characterised by the infiltration of inflammatory cells (e.g. eosinophils and neutrophils); elevated cytokine levels, including IL-1β and IL-33; production of immunoglobulin E (IgE); airway hyperresponsiveness; and mucus hypersecretion (Madouri et al. 2015; Kim et al. 2014; Tran et al. 2012; Leaker et al. 2017). Mild asthma due to allergic airway inflammation (AAI), typically caused by allergens such as house dust mites (HDM) or grass pollen, involves $CD4^+$ T helper type 2 (T_H2) cells and eosinophils. Severe steroid-resistant asthma due to nonallergic airway inflammation (NAAI), such as microbial (viral or bacterial) invasion, is largely neutrophilic and T_H1/T_H17-dependent (Madouri et al. 2015; Kim et al. 2014, 2017; Tran et al. 2012; McKinley et al. 2008).

IL-1β treatment of mast cells has been shown to increase IgE-mediated T_H2 cytokine secretion, suggesting the NLRP3 may play a role in the manifestation of AAI (Lee et al. 2004; Hultner et al. 2000). In addition, increased levels of NLRP3 and caspase-1 in BAL fluid from asthmatic patients have been observed in comparison with healthy individuals (Kim et al. 2014). These human studies are supported by similar findings in mouse models of neutrophilic asthma (LPS and ovalbumin (OVA) treatment) and AAI (OVA treatment alone) (Kim et al. 2014; Tran et al. 2012). Activation of NLRP3 has been observed in airway epithelial cells and eosinophils in tissue sections from OVA-treated mice, which was shown to be accompanied by the presence of IL-1β and IL-18 (Tran et al. 2012). LPS-/OVA-treated mice deficient in NLRP3 and ASC display reduced levels of pro-inflammatory cytokines IL-1β, IL-5 and IFN-γ, as well as decreased eosinophil numbers in bronchoalveolar lavage (BAL) following OVA challenge (Kim et al. 2014). Furthermore, genetic absence of IL-1R1 or administration of IL-1R antagonist anakinra also reduced eosinophil numbers, suggesting that IL-1β could be the main contributor to the recruitment of tissue eosinophils in asthma (Kim et al. 2014). Of note, treatment of mice with the mitochondrial ROS inhibitor NecroX-5 ablated IL-1β as well as NLRP3 and caspase-1, suggesting that mitochondrial ROS plays a vital role in the activation of NLRP3 and production of functional IL-1β in allergic asthma (Kim et al. 2014). These studies suggest NLRP3 and IL-1β play a multifactorial role in the development of asthma.

On the contrary, in the HDM-induced allergic asthma model, the levels of eosinophils and T_H2 pro-inflammatory cytokines IL-1β, IL-33, IL-4 and IL-5 were increased in BAL fluid of NLRP3-, caspase-1- and ASC-deficient mice following challenge (Madouri et al. 2015). With the use of gene-deficient mice, NLRP4 was shown to play no major role. IL-33 is a strong inducer of T_H2 cytokines, and IL-33 antagonist treatment was found to reverse the enhanced allergic response in caspase-1 knockout mice, suggesting a link between IL-33 and caspase-1. The role of caspase-11 in the HDM model is unclear and warrants further investigation with regard to the possible involvement of a noncanonical inflammasome pathway.

Alarmingly, steroid treatment is largely ineffective in severe asthma, possibly due to an impairment in nuclear translocation of glucocorticoid receptor (GR)α, as evidenced by a reduction of GR staining in the nucleus of PBMCs isolated from steroid-resistant asthmatic patients (Kim et al. 2017; Matthews et al. 2004). Clinically, mRNA expression of *NLRP3*, *ASC*, *CASP1* and *IL-1B* genes in sputum was found to be increased in neutrophilic compared with eosinophilic asthma (Simpson et al. 2014). However, increased levels of IL-1β protein in the sputum were also detected. Recently, Kim et al. demonstrated that increased mRNA expression of *NLRP3* and *IL-1B* is linked to increased neutrophilic inflammation and decreased lung function, as well as severe asthma, despite high-dose steroid treatment (Kim et al. 2017). Using experimental models of *Chlamydia trachomatis* and *Haemophilus influenzae* infection-induced severe steroid-resistant asthma, it was shown that anti-IL-1β antibody, caspase-1 inhibitor (Ac-YVAD-cho) or NLRP3 inhibitor (MCC950) treatment in vivo suppresses IL-1β production and airway hyperresponsiveness. Of note, IL-1β has been shown to promote T_H17 differentiation and IL-17 production (Chung et al. 2009), which is involved in the development of steroid-resistant asthma (McKinley et al. 2008). These studies suggest that regulating NLRP3 responses may be a potential therapy for severe asthma (Kim et al. 2017).

4.2.2.2 Chronic Obstructive Pulmonary Disease (COPD)

COPD is a chronic inflammatory lung disease that encompasses conditions such as chronic bronchitis and emphysema and is the third leading cause of death worldwide (Barnes et al. 2003). The main cause of COPD is cigarette smoke, which triggers potent immune responses leading to chronic inflammation and then to clinically significant COPD in up to 20% of smokers (Lokke et al. 2006). Exposure to cigarette smoke, which contains over 4000 toxins, including LPS, can cause damage to the lung epithelium resulting in recruitment of macrophages and neutrophils and the release of many inflammatory mediators involved in COPD, such as ROS, ATP, chemokines and other growth factors (Valenca et al. 2006; Barbu et al. 2011; Yoshida and Tuder 2007). In this regard, a growing body of evidence implicates inflammasomes and their associated mediators (such as IL-1β, IL-33) in alveolar destruction and small airway obstruction in COPD. Therefore, understanding the pathways and mediators of COPD development will lead to better therapeutic approaches for this debilitating disease, which has no current treatment options.

As discussed above, there is significant crosstalk between the TLR and inflammasome activation pathways, particularly in the first "priming" step required for inflammasome activation. While the role of TLRs in COPD development appear to be protective. This is best evidenced for TLR4, whereby TLR4$^{-/-}$ mice spontaneously develop emphysema as a result of excessive oxidant activity (Zhang et al. 2006; Ruwanpura et al. 2013). Furthermore, TLR4 polymorphisms and downregulated TLR4 expression are observed in lung tissues from emphysematous smokers (Pace et al. 2011; Ruwanpura et al. 2013). It has shown that the bronchial epithelial cells express mRNA of TLRs and release CXCL8 and IL-1β in response to cigarette smoke via TLR9 (Hoppstadter et al. 2010; Akira et al. 2001). Furthermore, cigarette smoke-induced CXCL8 levels are inhibited by TLR4 antibody, and exposure of bronchial epithelial cells to TLR4 or TLR9 ligands resulted in release of CXCL8 and IL-1β through ROS and P2X7 activation (Hoppstadter et al. 2010). Despite these observations, critical functions for TLRs in promoting cigarette smoke-induced lung inflammation leading to COPD have not been investigated.

Even though a role for inflammasome activation in COPD development is not well known, there are emerging lines of evidence that inflammasome-associated mediators (such as ASC, caspase-1, IL-1β and IL-18) are induced in COPD. Extracellular ATP is upregulated in the airways of COPD (Lommatzsch et al. 2010) and correlates with the decline in lung function (Cicko et al. 2010) and increased airway infiltration of inflammatory cells. Expression of P2X7 receptor is also elevated in inflammatory cells (macrophages and neutrophils) in blood from COPD patients (Lommatzsch et al. 2010) and in a cigarette smoke-induced lung inflammation mouse model (Lucattelli et al. 2011). ATP activates the NLRP3 inflammasome through the P2X7 receptor (Lucattelli et al. 2011): however, the role of NLRP3 in the development of COPD is not fully investigated. Despite this observation, caspase-1 is increased in lung tissues of mice following acute cigarette smoke exposure (Churg et al. 2009) and COPD patients who are smokers compared to non-smokers (Eltom et al. 2011). In addition, selective inhibition of caspase-1 using z-WEHD-fmk, a caspase-1 (IL-1-converting enzyme) inhibitor, significantly reduced inflammatory cells and serum IL-1β in an acute cigarette smoke-induced model (Churg et al. 2009). Furthermore, inflammasome-associated cytokines IL-1β and IL-18 are increased in the lungs of COPD patients and cigarette smoke-induced mouse models of COPD (Kang et al. 2007; Hoshino et al. 2007). Cigarette smoke induces caspase-1 activity, as well as IL-1β and IL-18 production, both in vitro (lung epithelial cells) and in vivo (Botelho et al. 2011; Churg et al. 2009; Kang et al. 2007; Hoshino et al. 2007). Moreover, experimentally induced (i.e. cigarette smoke, elastase) emphysema mouse models involving mice either lacking the receptors for IL-1β and IL-18 or treated with an IL-1R antagonist are protected against emphysema (Couillin et al. 2009). Despite these observations, further studies have suggested that even though P2X7 receptor activation, IL-1R signalling and caspase-1 are upregulated in cigarette smoke-induced mouse models of COPD, NLRP3/caspase-1 cleavage of IL-1β is not required for disease phenotype (Eltom et al. 2011; Pauwels et al. 2011). In contrast, a study shows that NLRP3$^{-/-}$ mice are protected against a cigarette smoke-induced COPD-like phenotype in the lung (Yang et al. 2015).

Collectively, these observations highlight the pressing need for informative animal models of COPD, in parallel with complementary human studies to elucidate a causal role for specific inflammasomes in promoting COPD, the identification of which will not only shed new light on the complex molecular and cellular pathogenesis of COPD but also pave the way forward for novel therapeutic approaches.

4.2.2.3 Fibrotic Lung Diseases

Idiopathic Pulmonary Fibrosis (IPF)

Pulmonary fibrosis is involved in a broad range of lung disorders characterised by irreversible destruction of normal lung architecture as well as scarring. This leads to a progressive decline in lung function and impaired gas exchange causing morbidity and mortality. The recruitment of fibroblasts and their activation/proliferation lead to the formation of fibrotic foci and the release of components of the extracellular matrix. The causes of the majority of cases of pulmonary fibrosis are unknown. Innate immune responses are known to be impaired in IPF; however, TLR9 activation is thought to contribute to progression through the differentiation of pulmonary fibroblasts into myofibroblasts (Kirillov et al. 2015).

Increased levels of IL-1β have been observed in the BAL fluid of IPF patients (Kitasato et al. 2004; Pan et al. 1996; Zhang et al. 1993; Lasithiotaki et al. 2016). However, macrophages isolated from the BAL have been shown to have impaired IL-1β production following LPS/ATP treatment ex vivo (Lasithiotaki et al. 2016). In line with these results, transcriptome analysis identified expression of *ASC*, *CASP1*, as well as *IL-1R1* genes as being significantly downregulated in cultured lung fibroblasts from IPF patients, in comparison with healthy controls (Plantier et al. 2016). The impact of IL-1β on fibroblasts in vitro is unclear with both pro- and anti-fibrotic effects being reported (Borthwick 2016; Mia et al. 2014; Furuyama et al. 2008). Alveolar macrophage-derived IL-1β has been shown to play a key role in the initiation of a fibrotic response by upregulating platelet-derived growth factor receptor (PDGF-R) (Lindroos et al. 1997). A number of studies have demonstrated a role for IL-1β in lung fibrosis in vivo in overexpression models or by recombinant delivery of IL-1β (Kolb et al. 2001; Gasse et al. 2007; Lappalainen et al. 2005). The role of IL-1β in IPF may be therefore cell type specific.

In the bleomycin-induced mouse model, lung fibrosis and inflammation were shown to be mediated through IL-1β and IL-1R1/MyD88 signalling (Gasse et al. 2007). It was subsequently identified that this pathway involved IL-17A and IL-23, which are necessary for TGF-β1 production, collagen deposition and evolution to fibrosis (Wilson et al. 2010; Gasse et al. 2011). Currently, the role of NLRP3 or other inflammasome components in IPF is not well understood; however, extracellular ATP activation of NLRP3 may act as a danger signal during IPF. Riteau et al. detected significantly elevated levels of ATP in BAL fluid from patients with both stable and exacerbated IPF (Riteau et al. 2010). Furthermore, instillation of bleomycin was found to rapidly increase levels of ATP in the airways of mice. Local depletion of ATP in wild-type mice with apyrase in treatment or genetic deletion of the ATP receptor P2X7 was shown to reduce neutrophil infiltration, IL-1β production and pulmonary fibrosis in vivo.

Inhibition of NLRP3 and/or IL-1β may therefore provide a therapeutic option for dampening lung fibrosis. Two drugs approved by the United States FDA, pirfenidone (Esbriet by Roche) and nintedanib (Ofev by Boehringer Ingelheim), slow progression of IPF, having been shown to inhibit lung fibrosis in murine models, with these effects being associated with a reduction in IL-1β levels in lung tissue (Oku et al. 2008; Wollin et al. 2014).

Cystic Fibrosis (CF)

CF is an autosomal recessive disorder due to mutations in the CFTR gene leading to an abnormality of chloride channels in mucus and sweat-producing cells. CF patients experience a vicious cycle of infection, inflammation and tissue damage, which progressively impacts on pulmonary function with respiratory failure the primary cause of death (reviewed in Yonker et al. 2015). Pathogens such as *Staphylococcus aureus*, *Pseudomonas aeruginosa*, *Haemophilus influenzae* and *Aspergillus fumigatus* are commonly observed in cystic fibrosis patients. Macrophages and epithelial cells in the lung recognise such pathogens leading to neutrophil recruitment and production of pro-inflammatory cytokines. A number of inflammatory pathways have been shown to be dysregulated in CF (reviewed in Cantin et al. 2015; Yonker et al. 2015). For example, CF bronchial epithelial cells display aberrant PRR signalling and constitutively elevated NF-κB activity (Venkatakrishnan et al. 2000).

Pathogens such as those associated with CF patients can activate NLRP3 and NLRC4 inflammasomes. Mice lacking NLRC4 have been shown to be less susceptible to *P. aeruginosa* infection (Cohen and Prince 2013), suggesting that dysregulation of the response may be detrimental and there is increasing evidence that this occurs in CF. Increased levels of IL-1β in the BAL fluid of CF patients (Bonfield et al. 1995) and polymorphisms in the *IL-1B* gene are reported to be associated with disease severity (Levy et al. 2009). Bronchial epithelial cells and haematopoietic cells have been shown to be a source of IL-1β in CF. Of note, IL-1R signalling activates pathogenic IL-17A-secreting T cells and thereby modulates the T_H17/regulatory T (Treg) cell balance (Basu et al. 2015), important for the control of *Aspergillus fumigatus* colonisation and disease in CF (Iannitti et al. 2013).

In line with the results seen in patients with CF, $CFTR^{-/-}$ mice, which represent a murine model of CF, have been shown to have increased caspase-1 activity and production of IL-1β in the lung following *A. fumigatus* or *P. aeruginosa* infection (Iannitti et al. 2016). NLRP3 expression in lung epithelial cells was found to be higher, while NLRC4 was lower in $CFTR^{-/-}$ mice compared to wild-type controls, suggesting the latter may be defective. These results lead Iannitti et al. to postulate that NLRP3 may play a detrimental role in the absence of NLRC4. NLRC4 was found to induce IL-1RA, which dampens NLRP3 activity and therefore may act as a negative regulator. Lastly, treatment of $CFTR^{-/-}$ mice with IL-1R antagonist anakinra was found to increase survival following *P. aeruginosa* infection and reduce bacterial burden. In an additional study, *P. aeruginosa* infection in CF epithelial cells was found to induce mitochondrial dysfunction and increase mitochondrial Ca^{2+} uptake, leading to increased NLRP3 responses, which could be ameliorated with anakinra treatment (Rimessi et al. 2015). These studies suggest a complex role for NLRP3 and NLRC4 in the pathogenesis of CF that requires further delineation.

Bronchopulmonary Dysplasia (BPD)

BPD is a chronic lung disease of preterm infants with long-term impact (reviewed in Davidson and Berkelhamer 2017). BPD is more common in infants of low birth weight and those who receive mechanical ventilation and oxygen therapy to treat respiratory distress syndrome. There are currently limited therapeutic options available for prevention and treatment of this disease, but one such treatment involves IL-1 receptor antagonist (IL-1RA). The development of BPD is associated with an inflammatory response, including increased numbers of neutrophils and macrophages and elevated levels of IL-1β (Rindfleisch et al. 1996; Watterberg et al. 1994; Kotecha et al. 1996); however, the pathogenic pathways involved are not well defined.

NLRP3 has been implicated in the development of hypoxia-induced lung injury. Using a neonatal model, Liao et al. demonstrated that NLRP3 activation is associated with the development of BPD in murine and primate models (Liao et al. 2015). Neonatal mice exposed to 85% oxygen were shown to have increased caspase-1 activation and apoptotic cells in the lung, IL-1β secretion and airway inflammation. Decreased alveolarisation was also a feature, which was not observed in mice lacking NLRP3. In addition, treatment of hypoxia-exposed neonatal mice with recombinant IL-1RA reduces inflammation and improves alveolarisation, suggesting the NLRP3/IL-1R pathway promotes the disease (Nold et al. 2013; Liao et al. 2015; Rudloff et al. 2017). In line with these results, ventilated preterm baboons were also found to have increased NLRP3 inflammasome activation and IL-1β/IL-1RA ratio in tracheal aspirates. Increased levels of IL-1β have also been observed in amniotic and BAL fluid from infants with BPD (Kotecha et al. 1996; Yoon et al. 1997). Importantly, overexpression of IL-1β in alveolar epithelial cells has been shown to result in respiratory insufficiency and postnatal growth abnormality, associated with increased postnatal mortality of mice (Bry et al. 2007). Overall, the NLRP3/IL-1β pathway appears to promote inflammation in the development of BPD, and inhibition of this pathway has been shown to be of therapeutic benefit in preclinical models.

Inhalation of Pathogenic Pollutants: Asbestosis and Silicosis

Inhalation of pathogenic pollutants such as asbestos and crystalline silica can lead to the development of chronic lung disease (reviewed in Maeda et al. 2010). For example, inhalation of silica and asbestos results in the progressive pulmonary fibrotic disorders silicosis and asbestosis, respectively. Airborne silica particles are commonly encountered occupationally in mining, construction, manufacturing and farming. Asbestosis does not normally manifest for more than 15 years after the initial exposure. There are currently no effective treatments available. Once inhaled, silica particles and asbestos fibres are engulfed by macrophages in the lung, which leads to the induction of inflammation and the development of fibrosis after repeated exposure.

The NLRP3 inflammasome and IL-1β have been implicated in the development of asbestosis and silicosis. Alveolar macrophages from patients with asbestosis have been shown to secrete elevated amounts of IL-1β compared with healthy controls (Kline et al. 1993). Treatment of LPS-primed macrophages with silica or asbestos induces ROS production and K^+ efflux, leading to caspase-1-dependent NLRP3 activation, as well as IL-1β and IL-18 secretion in vitro (Cassel et al. 2008; Hornung et al. 2008; Dostert et al. 2008). Silica has also been reported to activate NLRP3 in human bronchial epithelial cells (Peeters et al. 2013, 2014), and treatment of mice and rats induces caspase-1 activity and IL-1β production in vivo (Cassel et al. 2008; Sarih et al. 1993; Peeters et al. 2014). In the murine model of silicosis, inflammation and collagen deposition was observed in wild-type mice 3 months after treatment: however, this was reduced in mice lacking NLRP3, ASC or IL-1β (Cassel et al. 2008; Sarih et al. 1993). The infiltration of eosinophils and neutrophils into the airways, as well as IL-1β, has also been shown to be NLRP3-dependent 9 days following asbestos treatment (Dostert et al. 2008).

4.2.2.4 Lung Cancer

Lung cancer is strongly associated with chronic lung inflammation triggered by cigarette smoke, and 80% of all lung cancers occur in patients with a history of smoking. Investigations into the role of inflammasomes in lung cancer are in their relatively infancy, with a small volume of literature on the contradictory pro- and anti-tumorigenic roles for inflammasomes largely limited to in vivo studies on other cancers. Therefore, there is a clear and urgent need to elucidate a causal role for inflammasomes in lung cancer development.

While the role of TLRs in lung cancer has been controversial with reports of opposing pro- and anti-tumorigenic functions (e.g. TLR4) (Wang et al. 2017), the role of NLR- and AIM2-containing inflammasomes in lung adenocarcinoma development is ill-defined. The incidence of tumour development is reduced in mice lacking NLRP3; caspase-$1^{-/-}$ and IL-$1R1^{-/-}$ mice were also resistant to tumour development (Chaix et al. 2008). Furthermore, NLRP3 is required for NK-mediated experimentally induced lung cancer and primary tumour growth, with NK cells having an indirect effect as they do not express NLRP3 (Chow et al. 2012). Further indirect evidence for the potential involvement of inflammasomes in lung adenocarcinoma has come from observations that the NLRP3 inflammasome can promote the proliferation and migration of human lung adenocarcinoma cells (e.g. A549) in vitro following ATP and LPS stimulation (Wang et al. 2016; Kong et al. 2015). Also, clinical data have shown that increased mRNA expression of specific inflammasome components (i.e. AIM2) and production of the inflammasome effector cytokines IL-1β and IL-18 are associated with disease grading, invasion and chemoresistance of lung cancer (Kong et al. 2015). However, in light of this paucity of information, the role of inflammasomes in lung tumorigenesis needs comprehensive and urgent study.

4.3 Conclusion/Future Directions

Inflammasomes play an important role in mediating inflammation in the lung and are activated in response to a range of microbial and cellular stress responses. In turn, inflammasome responses need to be tightly regulated, and excessive activation has been implicated in the development of a number of respiratory diseases. Here, we have discussed the recent findings and evidence for a role for inflammasomes, as well as the potent pro-inflammatory cytokines IL-1β and IL-18, in a number of lung pathologies. The activation and regulation of inflammasome responses in the lung is complex. A greater understanding of the molecular subtleties of inflammasome responses in the context of specific lung diseases is imperative to design improved and better-targeted treatments. There is also increasing evidence that inflammasome responses not only occur in myeloid cells, such as alveolar macrophages, but also in epithelial cells and the contribution of each cell type to the pathogenesis of respiratory diseases is not clear. In addition to secretion of IL-1β and IL-18, inflammasome activation results in pyroptosis, an inflammatory form of cell death, and, currently, little is known regarding about the role of pyroptosis in lung diseases. A number of broad-acting antagonists and inhibitors that block different aspects of the inflammasome pathway (Fig. 4.1) or downstream cytokine signalling (Fig. 4.2) have been utilised in small animal models of respiratory disease. The development of cell-targeted therapies that inhibit or blunt pathogenic responses could offer improved efficacy.

References

Akira S, Takeda K, Kaisho T (2001) Toll-like receptors: critical proteins linking innate and acquired immunity. Nat Immunol 2:675–680

Andonegui G, Bonder CS, Green F, Mullaly SC, Zbytnuik L, Raharjo E, Kubes P (2003) Endothelium-derived Toll-like receptor-4 is the key molecule in LPS-induced neutrophil sequestration into lungs. J Clin Invest 111:1011–1020

Arulkumaran N, Unwin RJ, Tam FW (2011) A potential therapeutic role for P2X7 receptor (P2X7R) antagonists in the treatment of inflammatory diseases. Expert Opin Investig Drugs 20:897–915

Bals R, Hiemstra PS (2004) Innate immunity in the lung: how epithelial cells fight against respiratory pathogens. Eur Respir J 23:327–333

Barbu C, Iordache M, Man MG (2011) Inflammation in COPD: pathogenesis, local and systemic effects. Romanian J Morphol Embryol 52:21–27

Barnes PJ, Shapiro SD, Pauwels RA (2003) Chronic obstructive pulmonary disease: molecular and cellular mechanisms. Eur Respir J 22:672–688

Basu R, Whitley SK, Bhaumik S, Zindl CL, Schoeb TR, Benveniste EN, Pear WS, Hatton RD, Weaver CT (2015) IL-1 signaling modulates activation of STAT transcription factors to antagonize retinoic acid signaling and control the TH17 cell-iTreg cell balance. Nat Immunol 16:286–295

Bauernfeind F, Bartok E, Rieger A, Franchi L, Nunez G, Hornung V (2011) Cutting edge: reactive oxygen species inhibitors block priming, but not activation, of the NLRP3 inflammasome. J Immunol 187:613–617

Birrell MA, Eltom S (2011) The role of the NLRP3 inflammasome in the pathogenesis of airway disease. Pharmacol Ther 130:364–370

Bonfield TL, Panuska JR, Konstan MW, Hilliard KA, Hilliard JB, Ghnaim H, Berger M (1995) Inflammatory cytokines in cystic fibrosis lungs. Am J Respir Crit Care Med 152:2111–2118

Borthwick LA (2016) The IL-1 cytokine family and its role in inflammation and fibrosis in the lung. Semin Immunopathol 38:517–534

Botelho FM, Bauer CM, Finch D, Nikota JK, Zavitz CC, Kelly A, Lambert KN, Piper S, Foster ML, Goldring JJ, Wedzicha JA, Bassett J, Bramson J, Iwakura Y, Sleeman M, Kolbeck R, Coyle AJ, Humbles AA, Stampfli MR (2011) IL-1alpha/IL-1R1 expression in chronic obstructive pulmonary disease and mechanistic relevance to smoke-induced neutrophilia in mice. PLoS One 6: e28457

Bry K, Whitsett JA, Lappalainen U (2007) IL-1beta disrupts postnatal lung morphogenesis in the mouse. Am J Respir Cell Mol Biol 36:32–42

Buchanan PJ, Mcnally P, Harvey BJ, Urbach V (2013) Lipoxin A(4)-mediated KATP potassium channel activation results in cystic fibrosis airway epithelial repair. Am J Physiol Lung Cell Mol Physiol 305:L193–L201

Cai S, Batra S, Wakamatsu N, Pacher P, Jeyaseelan S (2012) NLRC4 inflammasome-mediated production of IL-1beta modulates mucosal immunity in the lung against gram-negative bacterial infection. J Immunol 188:5623–5635

Cantin AM, Hartl D, Konstan MW, Chmiel JF (2015) Inflammation in cystic fibrosis lung disease: pathogenesis and therapy. J Cyst Fibros 14:419–430

Cassel SL, Eisenbarth SC, Iyer SS, Sadler JJ, Colegio OR, Tephly LA, Carter AB, Rothman PB, Flavell RA, Sutterwala FS (2008) The Nalp3 inflammasome is essential for the development of silicosis. Proc Natl Acad Sci U S A 105:9035–9040

Chaix J, Tessmer MS, Hoebe K, Fuseri N, Ryffel B, Dalod M, Alexopoulou L, Beutler B, Brossay L, Vivier E, Walzer T (2008) Cutting edge: priming of NK cells by IL-18. J Immunol 181:1627–1631

Chow MT, Tschopp J, Moller A, Smyth MJ (2012) NLRP3 promotes inflammation-induced skin cancer but is dispensable for asbestos-induced mesothelioma. Immunol Cell Biol 90:983–986

Chung Y, Chang SH, Martinez GJ, Yang XO, Nurieva R, Kang HS, Ma L, Watowich SS, Jetten AM, Tian Q, Dong C (2009) Critical regulation of early Th17 cell differentiation by interleukin-1 signaling. Immunity 30:576–587

Churg A, Zhou S, Wang X, Wang R, Wright JL (2009) The role of interleukin-1beta in murine cigarette smoke-induced emphysema and small airway remodeling. Am J Respir Cell Mol Biol 40:482–490

Cicko S, Lucattelli M, Muller T, Lommatzsch M, De Cunto G, Cardini S, Sundas W, Grimm M, Zeiser R, Durk T, Zissel G, Boeynaems JM, Sorichter S, Ferrari D, Di Virgilio F, Virchow JC, Lungarella G, Idzko M (2010) Purinergic receptor inhibition prevents the development of smoke-induced lung injury and emphysema. J Immunol 185:688–697

Cohen TS, Prince AS (2013) Activation of inflammasome signaling mediates pathology of acute *P. aeruginosa* pneumonia. J Clin Invest 123:1630–1637
Couillin I, Vasseur V, Charron S, Gasse P, Tavernier M, Guillet J, Lagente V, Fick L, Jacobs M, Coelho FR, Moser R, Ryffel B (2009) IL-1R1/MyD88 signaling is critical for elastase-induced lung inflammation and emphysema. J Immunol 183:8195–8202
Davidson LM, Berkelhamer SK (2017) Bronchopulmonary dysplasia: chronic lung disease of infancy and long-term pulmonary outcomes. J Clin Med 6:4
De Nardo D, De Nardo CM, Latz E (2014) New insights into mechanisms controlling the NLRP3 inflammasome and its role in lung disease. Am J Pathol 184:42–54
Dinarello CA (2010) Why not treat human cancer with interleukin-1 blockade? Cancer Metastasis Rev 29:317–329
Dolinay T, Kim YS, Howrylak J, Hunninghake GM, An CH, Fredenburgh L, Massaro AF, Rogers A, Gazourian L, Nakahira K, Haspel JA, Landazury R, Eppanapally S, Christie JD, Meyer NJ, Ware LB, Christiani DC, Ryter SW, Baron RM, Choi AM (2012) Inflammasome-regulated cytokines are critical mediators of acute lung injury. Am J Respir Crit Care Med 185:1225–1234
Dos Santos G, Kutuzov MA, Ridge KM (2012) The inflammasome in lung diseases. Am J Physiol Lung Cell Mol Physiol 303:L627–L633
Dostert C, Petrilli V, Van Bruggen R, Steele C, Mossman BT, Tschopp J (2008) Innate immune activation through Nalp3 inflammasome sensing of asbestos and silica. Science 320:674–677
Eltom S, Stevenson CS, Rastrick J, Dale N, Raemdonck K, Wong S, Catley MC, Belvisi MG, Birrell MA (2011) P2X7 receptor and caspase 1 activation are central to airway inflammation observed after exposure to tobacco smoke. PLoS One 6:e24097
Franchi L, Mcdonald C, Kanneganti TD, Amer A, Nunez G (2006) Nucleotide-binding oligomerization domain-like receptors: intracellular pattern recognition molecules for pathogen detection and host defense. J Immunol 177:3507–3513
Frank JA, Pittet JF, Wray C, Matthay MA (2008) Protection from experimental ventilator-induced acute lung injury by IL-1 receptor blockade. Thorax 63:147–153
Franklin BS, Bossaller L, De Nardo D, Ratter JM, Stutz A, Engels G, Brenker C, Nordhoff M, Mirandola SR, Al-Amoudi A, Mangan MS, Zimmer S, Monks BG, Fricke M, Schmidt RE, Espevik T, Jones B, Jarnicki AG, Hansbro PM, Busto P, Marshak-Rothstein A, Hornemann S, Aguzzi A, Kastenmuller W, Latz E (2014) The adaptor ASC has extracellular and 'prionoid' activities that propagate inflammation. Nat Immunol 15:727–737
Fritz JH, Ferrero RL, Philpott DJ, Girardin SE (2006) Nod-like proteins in immunity, inflammation and disease. Nat Immunol 7:1250–1257
Fukumoto J, Fukumoto I, Parthasarathy PT, Cox R, Huynh B, Ramanathan GK, Venugopal RB, Allen-Gipson DS, Lockey RF, Kolliputi N (2013) NLRP3 deletion protects from hyperoxia-induced acute lung injury. Am J Physiol Cell Physiol 305:C182–C189
Furuyama A, Hosokawa T, Mochitate K (2008) Interleukin-1beta and tumor necrosis factor-alpha have opposite effects on fibroblasts and epithelial cells during basement membrane formation. Matrix Biol 27:429–440
Galam L, Rajan A, Failla A, Soundararajan R, Lockey RF, Kolliputi N (2016) Deletion of P2X7 attenuates hyperoxia-induced acute lung injury via inflammasome suppression. Am J Physiol Lung Cell Mol Physiol 310:L572–L581
Ganter MT, Roux J, Miyazawa B, Howard M, Frank JA, Su G, Sheppard D, Violette SM, Weinreb PH, Horan GS, Matthay MA, Pittet JF (2008) Interleukin-1beta causes acute lung injury via alphavbeta5 and alphavbeta6 integrin-dependent mechanisms. Circ Res 102:804–812
Gasse P, Mary C, Guenon I, Noulin N, Charron S, Schnyder-Candrian S, Schnyder B, Akira S, Quesniaux VF, Lagente V, Ryffel B, Couillin I (2007) IL-1R1/MyD88 signaling and the inflammasome are essential in pulmonary inflammation and fibrosis in mice. J Clin Invest 117:3786–3799
Gasse P, Riteau N, Vacher R, Michel ML, Fautrel A, di Padova F, Fick L, Charron S, Lagente V, Eberl G, Le Bert M, Quesniaux VF, Huaux F, Leite-de-Moraes M, Ryffel B, Couillin I (2011)

IL-1 and IL-23 mediate early IL-17A production in pulmonary inflammation leading to late fibrosis. PLoS One 6:e23185

Grailer JJ, Canning BA, Kalbitz M, Haggadone MD, Dhond RM, Andjelkovic AV, Zetoune FS, Ward PA (2014) Critical role for the NLRP3 inflammasome during acute lung injury. J Immunol 192:5974–5983

Han S, Cai W, Yang X, Jia Y, Zheng Z, Wang H, Li J, Li Y, Gao J, Fan L, Hu D (2015) ROS-mediated NLRP3 inflammasome activity is essential for burn-induced acute lung injury. Mediat Inflamm 2015:720457

Hirsiger S, Simmen HP, Werner CM, Wanner GA, Rittirsch D (2012) Danger signals activating the immune response after trauma. Mediat Inflamm 2012:315941

Hoppstadter J, Diesel B, Zarbock R, Breinig T, Monz D, Koch M, Meyerhans A, Gortner L, Lehr CM, Huwer H, Kiemer AK (2010) Differential cell reaction upon Toll-like receptor 4 and 9 activation in human alveolar and lung interstitial macrophages. Respir Res 11:124

Hornung V, Bauernfeind F, Halle A, Samstad EO, Kono H, Rock KL, Fitzgerald KA, Latz E (2008) Silica crystals and aluminum salts activate the NALP3 inflammasome through phagosomal destabilization. Nat Immunol 9:847–856

Hornung V, Ablasser A, Charrel-Dennis M, Bauernfeind F, Horvath G, Caffrey DR, Latz E, Fitzgerald KA (2009) AIM2 recognizes cytosolic dsDNA and forms a caspase-1-activating inflammasome with ASC. Nature 458:514–518

Hoshino T, Kato S, Oka N, Imaoka H, Kinoshita T, Takei S, Kitasato Y, Kawayama T, Imaizumi T, Yamada K, Young HA, Aizawa H (2007) Pulmonary inflammation and emphysema: role of the cytokines IL-18 and IL-13. Am J Respir Crit Care Med 176:49–62

Hultner L, Kolsch S, Stassen M, Kaspers U, Kremer JP, Mailhammer R, Moeller J, Broszeit H, Schmitt E (2000) In activated mast cells, IL-1 up-regulates the production of several Th2-related cytokines including IL-9. J Immunol 164:5556–5563

Iannitti RG, Carvalho A, Cunha C, De Luca A, Giovannini G, Casagrande A, Zelante T, Vacca C, Fallarino F, Puccetti P, Massi-Benedetti C, Defilippi G, Russo M, Porcaro L, Colombo C, Ratclif L, De Benedictis FM, Romani L (2013) Th17/Treg imbalance in murine cystic fibrosis is linked to indoleamine 2,3-dioxygenase deficiency but corrected by kynurenines. Am J Respir Crit Care Med 187:609–620

Iannitti RG, Napolioni V, Oikonomou V, De Luca A, Galosi C, Pariano M, Massi-Benedetti C, Borghi M, Puccetti M, Lucidi V, Colombo C, Fiscarelli E, Lass-Florl C, Majo F, Cariani L, Russo M, Porcaro L, Ricciotti G, Ellemunter H, Ratclif L, De Benedictis FM, Talesa VN, Dinarello CA, van de Veerdonk FL, Romani L (2016) IL-1 receptor antagonist ameliorates inflammasome-dependent inflammation in murine and human cystic fibrosis. Nat Commun 7:10791

Ipaktchi K, Mattar A, Niederbichler AD, Hoesel LM, Vollmannshauser S, Hemmila MR, Su GL, Remick DG, Wang SC, Arbabi S (2006) Attenuating burn wound inflammatory signaling reduces systemic inflammation and acute lung injury. J Immunol 177:8065–8071

Jiang L, Fei D, Gong R, Yang W, Yu W, Pan S, Zhao M, Zhao M (2016) CORM-2 inhibits TXNIP/NLRP3 inflammasome pathway in LPS-induced acute lung injury. Inflamm Res 65:905–915

Jones HD, Crother TR, Gonzalez-Villalobos RA, Jupelli M, Chen S, Dagvadorj J, Arditi M, Shimada K (2014) The NLRP3 inflammasome is required for the development of hypoxemia in LPS/mechanical ventilation acute lung injury. Am J Respir Cell Mol Biol 50:270–280

Jordan JA, Guo RF, Yun EC, Sarma V, Warner RL, Crouch LD, Senaldi G, Ulich TR, Ward PA (2001) Role of IL-18 in acute lung inflammation. J Immunol 167:7060–7068

Kaczmarek A, Vandenabeele P, Krysko DV (2013) Necroptosis: the release of damage-associated molecular patterns and its physiological relevance. Immunity 38:209–223

Kang MJ, Homer RJ, Gallo A, Lee CG, Crothers KA, Cho SJ, Rochester C, Cain H, Chupp G, Yoon HJ, Elias JA (2007) IL-18 is induced and IL-18 receptor alpha plays a critical role in the pathogenesis of cigarette smoke-induced pulmonary emphysema and inflammation. J Immunol 178:1948–1959

Kawai T, Akira S (2010) The role of pattern-recognition receptors in innate immunity: update on Toll-like receptors. Nat Immunol 11:373–384

Kersse K, Bertrand MJ, Lamkanfi M, Vandenabeele P (2011) NOD-like receptors and the innate immune system: coping with danger, damage and death. Cytokine Growth Factor Rev 22:257–276

Kim SR, Kim DI, Kim SH, Lee H, Lee KS, Cho SH, Lee YC (2014) NLRP3 inflammasome activation by mitochondrial ROS in bronchial epithelial cells is required for allergic inflammation. Cell Death Dis 5:e1498

Kim RY, Pinkerton JW, Essilfie AT, Robertson AA, Baines KJ, Brown AC, Mayall JR, Ali MK, Starkey MR, Hansbro NG, Hirota JA, Wood LG, Simpson JL, Knight DA, Wark PA, Gibson PG, O'Neill LA, Cooper MA, Horvat JC, Hansbro PM (2017) Role for NLRP3 inflammasome-mediated, IL-1beta-dependent responses in severe, steroid-resistant asthma. Am J Respir Crit Care Med 196(3):283–297

Kirillov V, Siler JT, Ramadass M, Ge L, Davis J, Grant G, Nathan SD, Jarai G, Trujillo G (2015) Sustained activation of toll-like receptor 9 induces an invasive phenotype in lung fibroblasts: possible implications in idiopathic pulmonary fibrosis. Am J Pathol 185:943–957

Kitasato Y, Hoshino T, Okamoto M, Kato S, Koda Y, Nagata N, Kinoshita M, Koga H, Yoon DY, Asao H, Ohmoto H, Koga T, Rikimaru T, Aizawa H (2004) Enhanced expression of interleukin-18 and its receptor in idiopathic pulmonary fibrosis. Am J Respir Cell Mol Biol 31:619–625

Kline JN, Schwartz DA, Monick MM, Floerchinger CS, Hunninghake GW (1993) Relative release of interleukin-1 beta and interleukin-1 receptor antagonist by alveolar macrophages. A study in asbestos-induced lung disease, sarcoidosis, and idiopathic pulmonary fibrosis. Chest 104:47–53

Kolb M, Margetts PJ, Anthony DC, Pitossi F, Gauldie J (2001) Transient expression of IL-1beta induces acute lung injury and chronic repair leading to pulmonary fibrosis. J Clin Invest 107:1529–1536

Kolliputi N, Shaik RS, Waxman AB (2010) The inflammasome mediates hyperoxia-induced alveolar cell permeability. J Immunol 184:5819–5826

Kong H, Wang Y, Zeng X, Wang Z, Wang H, Xie W (2015) Differential expression of inflammasomes in lung cancer cell lines and tissues. Tumour Biol 36:7501–7513

Kono H, Rock KL (2008) How dying cells alert the immune system to danger. Nat Rev Immunol 8:279–289

Kotecha S, Wilson L, Wangoo A, Silverman M, Shaw RJ (1996) Increase in interleukin (IL)-1 beta and IL-6 in bronchoalveolar lavage fluid obtained from infants with chronic lung disease of prematurity. Pediatr Res 40:250–256

Krysko O, Vandenabeele P, Krysko DV, Bachert C (2010) Impairment of phagocytosis of apoptotic cells and its role in chronic airway diseases. Apoptosis 15:1137–1146

Krysko DV, Garg AD, Kaczmarek A, Krysko O, Agostinis P, Vandenabeele P (2012) Immunogenic cell death and DAMPs in cancer therapy. Nat Rev Cancer 12:860–875

Kuipers MT, Aslami H, Janczy JR, van der Sluijs KF, Vlaar AP, Wolthuis EK, Choi G, Roelofs JJ, Flavell RA, Sutterwala FS, Bresser P, Leemans JC, van der Poll T, Schultz MJ, Wieland CW (2012) Ventilator-induced lung injury is mediated by the NLRP3 inflammasome. Anesthesiology 116:1104–1115

Lappalainen U, Whitsett JA, Wert SE, Tichelaar JW, Bry K (2005) Interleukin-1beta causes pulmonary inflammation, emphysema, and airway remodeling in the adult murine lung. Am J Respir Cell Mol Biol 32:311–318

Lasithiotaki I, Giannarakis I, Tsitoura E, Samara KD, Margaritopoulos GA, Choulaki C, Vasarmidi E, Tzanakis N, Voloudaki A, Sidiropoulos P, Siafakas NM, Antoniou KM (2016) NLRP3 inflammasome expression in idiopathic pulmonary fibrosis and rheumatoid lung. Eur Respir J 47:910–918

Latz E (2010) The inflammasomes: mechanisms of activation and function. Curr Opin Immunol 22:28–33

Latz E, Xiao TS, Stutz A (2013) Activation and regulation of the inflammasomes. Nat Rev Immunol 13:397–411

Leaker BR, Malkov VA, Mogg R, Ruddy MK, Nicholson GC, Tan AJ, Tribouley C, Chen G, de Lepeleire I, Calder NA, Chung H, Lavender P, Carayannopoulos LN, Hansel TT (2017) The nasal mucosal late allergic reaction to grass pollen involves type 2 inflammation (IL-5 and IL-13), the inflammasome (IL-1beta), and complement. Mucosal Immunol 10:408–420

Lee SA, Fitzgerald SM, Huang SK, Li C, Chi DS, Milhorn DM, Krishnaswamy G (2004) Molecular regulation of interleukin-13 and monocyte chemoattractant protein-1 expression in human mast cells by interleukin-1beta. Am J Respir Cell Mol Biol 31:283–291

Leff JA, Baer JW, Bodman ME, Kirkman JM, Shanley PF, Patton LM, Beehler CJ, Mccord JM, Repine JE (1994) Interleukin-1-induced lung neutrophil accumulation and oxygen metabolite-mediated lung leak in rats. Am J Phys 266:L2–L8

Levy H, Murphy A, Zou F, Gerard C, Klanderman B, Schuemann B, Lazarus R, Garcia KC, Celedon JC, Drumm M, Dahmer M, Quasney M, Schneck K, Reske M, Knowles MR, Pier GB, Lange C, Weiss ST (2009) IL1B polymorphisms modulate cystic fibrosis lung disease. Pediatr Pulmonol 44:580–593

Liao J, Kapadia VS, Brown LS, Cheong N, Longoria C, Mija D, Ramgopal M, Mirpuri J, Mccurnin DC, Savani RC (2015) The NLRP3 inflammasome is critically involved in the development of bronchopulmonary dysplasia. Nat Commun 6:8977

Lindroos PM, Coin PG, Badgett A, Morgan DL, Bonner JC (1997) Alveolar macrophages stimulated with titanium dioxide, chrysotile asbestos, and residual oil fly ash upregulate the PDGF receptor-alpha on lung fibroblasts through an IL-1beta-dependent mechanism. Am J Respir Cell Mol Biol 16:283–292

Lokke A, Lange P, Scharling H, Fabricius P, Vestbo J (2006) Developing COPD: a 25 year follow up study of the general population. Thorax 61:935–939

Lommatzsch M, Cicko S, Muller T, Lucattelli M, Bratke K, Stoll P, Grimm M, Durk T, Zissel G, Ferrari D, di Virgilio F, Sorichter S, Lungarella G, Virchow JC, Idzko M (2010) Extracellular adenosine triphosphate and chronic obstructive pulmonary disease. Am J Respir Crit Care Med 181:928–934

Lucattelli M, Cicko S, Muller T, Lommatzsch M, de Cunto G, Cardini S, Sundas W, Grimm M, Zeiser R, Durk T, Zissel G, Sorichter S, Ferrari D, di Virgilio F, Virchow JC, Lungarella G, Idzko M (2011) P2X7 receptor signaling in the pathogenesis of smoke-induced lung inflammation and emphysema. Am J Respir Cell Mol Biol 44:423–429

Madouri F, Guillou N, Fauconnier L, Marchiol T, Rouxel N, Chenuet P, Ledru A, Apetoh L, Ghiringhelli F, Chamaillard M, Zheng SG, Trovero F, Quesniaux VF, Ryffel B, Togbe D (2015) Caspase-1 activation by NLRP3 inflammasome dampens IL-33-dependent house dust mite-induced allergic lung inflammation. J Mol Cell Biol 7:351–365

Maeda M, Nishimura Y, Kumagai N, Hayashi H, Hatayama T, Katoh M, Miyahara N, Yamamoto S, Hirastuka J, Otsuki T (2010) Dysregulation of the immune system caused by silica and asbestos. J Immunotoxicol 7:268–278

Makabe H, Kojika M, Takahashi G, Matsumoto N, Shibata S, Suzuki Y, Inoue Y, Endo S (2012) Interleukin-18 levels reflect the long-term prognosis of acute lung injury and acute respiratory distress syndrome. J Anesth 26:658–663

Man SM, Karki R, Kanneganti TD (2016) AIM2 inflammasome in infection, cancer, and autoimmunity: role in DNA sensing, inflammation, and innate immunity. Eur J Immunol 46:269–280

Matthews JG, Ito K, Barnes PJ, Adcock IM (2004) Defective glucocorticoid receptor nuclear translocation and altered histone acetylation patterns in glucocorticoid-resistant patients. J Allergy Clin Immunol 113:1100–1108

Matzinger P (1994) Tolerance, danger, and the extended family. Annu Rev Immunol 12:991–1045

Mckinley L, Alcorn JF, Peterson A, Dupont RB, Kapadia S, Logar A, Henry A, Irvin CG, Piganelli JD, Ray A, Kolls JK (2008) TH17 cells mediate steroid-resistant airway inflammation and airway hyperresponsiveness in mice. J Immunol 181:4089–4097

Messner B, Frotschnig S, Steinacher-Nigisch A, Winter B, Eichmair E, Gebetsberger J, Schwaiger S, Ploner C, Laufer G, Bernhard D (2012) Apoptosis and necrosis: two different outcomes of cigarette smoke condensate-induced endothelial cell death. Cell Death Dis 3:e424

Mia MM, Boersema M, Bank RA (2014) Interleukin-1beta attenuates myofibroblast formation and extracellular matrix production in dermal and lung fibroblasts exposed to transforming growth factor-beta1. PLoS One 9:e91559

Mizushina Y, Shirasuna K, Usui F, Karasawa T, Kawashima A, Kimura H, Kobayashi M, Komada T, Inoue Y, Mato N, Yamasawa H, Latz E, Iwakura Y, Kasahara T, Bando M, Sugiyama Y, Takahashi M (2015) NLRP3 protein deficiency exacerbates hyperoxia-induced lethality through Stat3 protein signaling independent of interleukin-1beta. J Biol Chem 290:5065–5077

Moncao-Ribeiro LC, Cagido VR, Lima-Murad G, Santana PT, Riva DR, Borojevic R, Zin WA, Cavalcante MC, Rica I, Brando-Lima AC, Takiya CM, Faffe DS, Coutinho-Silva R (2011) Lipopolysaccharide-induced lung injury: role of P2X7 receptor. Respir Physiol Neurobiol 179:314–325

Nold MF, Mangan NE, Rudloff I, Cho SX, Shariatian N, Samarasinghe TD, Skuza EM, Pedersen J, Veldman A, Berger PJ, Nold-Petry CA (2013) Interleukin-1 receptor antagonist prevents murine bronchopulmonary dysplasia induced by perinatal inflammation and hyperoxia. Proc Natl Acad Sci U S A 110:14384–14389

Oku H, Shimizu T, Kawabata T, Nagira M, Hikita I, Ueyama A, Matsushima S, Torii M, Arimura A (2008) Antifibrotic action of pirfenidone and prednisolone: different effects on pulmonary cytokines and growth factors in bleomycin-induced murine pulmonary fibrosis. Eur J Pharmacol 590:400–408

Olman MA, White KE, Ware LB, Cross MT, Zhu S, Matthay MA (2002) Microarray analysis indicates that pulmonary edema fluid from patients with acute lung injury mediates inflammation, mitogen gene expression, and fibroblast proliferation through bioactive interleukin-1. Chest 121:69S–70S

Pace E, Giarratano A, Ferraro M, Bruno A, Siena L, Mangione S, Johnson M, Gjomarkaj M (2011) TLR4 upregulation underpins airway neutrophilia in smokers with chronic obstructive pulmonary disease and acute respiratory failure. Hum Immunol 72:54–62

Pan LH, Ohtani H, Yamauchi K, Nagura H (1996) Co-expression of TNF alpha and IL-1 beta in human acute pulmonary fibrotic diseases: an immunohistochemical analysis. Pathol Int 46:91–99

Pauwels NS, Bracke KR, Dupont LL, Van Pottelberge GR, Provoost S, Vanden Berghe T, Vandenabeele P, Lambrecht BN, Joos GF, Brusselle GG (2011) Role of IL-1alpha and the Nlrp3/caspase-1/IL-1beta axis in cigarette smoke-induced pulmonary inflammation and COPD. Eur Respir J 38:1019–1028

Peeters PM, Perkins TN, Wouters EF, Mossman BT, Reynaert NL (2013) Silica induces NLRP3 inflammasome activation in human lung epithelial cells. Part Fibre Toxicol 10:3

Peeters PM, Eurlings IM, Perkins TN, Wouters EF, Schins RP, Borm PJ, Drommer W, Reynaert NL, Albrecht C (2014) Silica-induced NLRP3 inflammasome activation in vitro and in rat lungs. Part Fibre Toxicol 11:58

Pinar A, Dowling JK, Bitto NJ, Robertson AA, Latz E, Stewart CR, Drummond GR, Cooper MA, Mcauley JL, Tate MD, Mansell A (2017) PB1-F2 peptide derived from avian influenza A virus H7N9 induces inflammation via activation of the NLRP3 inflammasome. J Biol Chem 292:826–836

Plantier L, Renaud H, Respaud R, Marchand-Adam S, Crestani B (2016) Transcriptome of cultured lung fibroblasts in idiopathic pulmonary fibrosis: meta-analysis of publically available microarray datasets reveals repression of inflammation and immunity pathways. Int J Mol Sci 17:2091

Pouwels SD, Heijink IH, ten Hacken NH, Vandenabeele P, Krysko DV, Nawijn MC, van Oosterhout AJ (2014) DAMPs activating innate and adaptive immune responses in COPD. Mucosal Immunol 7:215–226

Pugin J, Ricou B, Steinberg KP, Suter PM, Martin TR (1996) Proinflammatory activity in bronchoalveolar lavage fluids from patients with ARDS, a prominent role for interleukin-1. Am J Respir Crit Care Med 153(6):1850

Rana SN, Li X, Chaudry IH, Bland KI, Choudhry MA (2005) Inhibition of IL-18 reduces myeloperoxidase activity and prevents edema in intestine following alcohol and burn injury. J Leukoc Biol 77:719–728

Ratsimandresy RA, Dorfleutner A, Stehlik C (2013) An update on PYRIN domain-containing pattern recognition receptors: from immunity to pathology. Front Immunol 4:440

Rimessi A, Bezzerri V, Patergnani S, Marchi S, Cabrini G, Pinton P (2015) Mitochondrial Ca2+-dependent NLRP3 activation exacerbates the Pseudomonas aeruginosa-driven inflammatory response in cystic fibrosis. Nat Commun 6:6201

Rindfleisch MS, Hasday JD, Taciak V, Broderick K, Viscardi RM (1996) Potential role of interleukin-1 in the development of bronchopulmonary dysplasia. J Interf Cytokine Res 16:365–373

Riteau N, Gasse P, Fauconnier L, Gombault A, Couegnat M, Fick L, Kanellopoulos J, Quesniaux VF, Marchand-Adam S, Crestani B, Ryffel B, Couillin I (2010) Extracellular ATP is a danger signal activating P2X7 receptor in lung inflammation and fibrosis. Am J Respir Crit Care Med 182:774–783

Rogliani P, Calzetta L, Ora J, Matera MG (2015) Canakinumab for the treatment of chronic obstructive pulmonary disease. Pulm Pharmacol Ther 31:15–27

Roux J, Kawakatsu H, Gartland B, Pespeni M, Sheppard D, Matthay MA, Canessa CM, Pittet JF (2005) Interleukin-1beta decreases expression of the epithelial sodium channel alpha-subunit in alveolar epithelial cells via a p38 MAPK-dependent signaling pathway. J Biol Chem 280:18579–18589

Rubartelli A, Lotze MT (2007) Inside, outside, upside down: damage-associated molecular-pattern molecules (DAMPs) and redox. Trends Immunol 28:429–436

Rudloff I, Cho SX, Bui CB, Mclean C, Veldman A, Berger PJ, Nold MF, Nold-Petry CA (2017) Refining anti-inflammatory therapy strategies for bronchopulmonary dysplasia. J Cell Mol Med 21:1128–1138

Ruwanpura SM, Mcleod L, Lilja AR, Brooks G, Dousha LF, Seow HJ, Bozinovski S, Vlahos R, Hertzog PJ, Anderson GP, Jenkins BJ (2013) Non-essential role for TLR2 and its signaling adaptor Mal/TIRAP in preserving normal lung architecture in mice. PLoS One 8:e78095

Salve VT, Atram JS (2016) N-acetylcysteine combined with home based physical activity: effect on health related quality of life in stable COPD patients - a randomised controlled trial. J Clin Diagn Res 10:OC16–OC19

Sarih M, Souvannavong V, Brown SC, Adam A (1993) Silica induces apoptosis in macrophages and the release of interleukin-1 alpha and interleukin-1 beta. J Leukoc Biol 54:407–413

Schnare M, Holt AC, Takeda K, Akira S, Medzhitov R (2000) Recognition of CpG DNA is mediated by signaling pathways dependent on the adaptor protein MyD88. Curr Biol 10:1139–1142

Simpson JL, Phipps S, Baines KJ, Oreo KM, Gunawardhana L, Gibson PG (2014) Elevated expression of the NLRP3 inflammasome in neutrophilic asthma. Eur Respir J 43:1067–1076

Tarrant BJ, Le Maitre C, Romero L, Steward R, Button BM, Thompson BR, Holland AE (2017) Mucoactive agents for chronic, non-cystic fibrosis lung disease: a systematic review and meta-analysis. Respirology 22(6):1084–1092

Tate MD, Ong JD, Dowling JK, Mcauley JL, Robertson AB, Latz E, Drummond GR, Cooper MA, Hertzog PJ, Mansell A (2016) Reassessing the role of the NLRP3 inflammasome during pathogenic influenza A virus infection via temporal inhibition. Sci Rep 6:27912

Tran HB, Lewis MD, Tan LW, Lester SE, Baker LM, Ng J, Hamilton-Bruce MA, Hill CL, Koblar SA, Rischmueller M, Ruffin RE, Wormald PJ, Zalewski PD, Lang CJ (2012) Immunolocalization of NLRP3 inflammasome in normal murine airway epithelium and changes following induction of ovalbumin-induced airway inflammation. J Allergy (Cairo) 2012:819176

Umbrello M, Formenti P, Bolgiaghi L, Chiumello D (2016) Current concepts of ARDS: a narrative review. Int J Mol Sci 18:E64

Valenca SS, Castro P, Pimenta WA, Lanzetti M, Silva SV, Barja-Fidalgo C, Koatz VL, Porto LC (2006) Light cigarette smoke-induced emphysema and NFkappaB activation in mouse lung. Int J Exp Pathol 87:373–381

Vanaja SK, Rathinam VA, Fitzgerald KA (2015) Mechanisms of inflammasome activation: recent advances and novel insights. Trends Cell Biol 25:308–315

Venkatakrishnan A, Stecenko AA, King G, Blackwell TR, Brigham KL, Christman JW, Blackwell TS (2000) Exaggerated activation of nuclear factor-kappaB and altered IkappaB-beta processing in cystic fibrosis bronchial epithelial cells. Am J Respir Cell Mol Biol 23:396–403

Wang S, Zhao J, Wang H, Liang Y, Yang N, Huang Y (2015) Blockage of P2X7 attenuates acute lung injury in mice by inhibiting NLRP3 inflammasome. Int Immunopharmacol 27:38–45

Wang Y, Kong H, Zeng X, Liu W, Wang Z, Yan X, Wang H, Xie W (2016) Activation of NLRP3 inflammasome enhances the proliferation and migration of A549 lung cancer cells. Oncol Rep 35:2053–2064

Wang K, Wang J, Wei F, Zhao N, Yang F, Ren X (2017) Expression of TLR4 in non-small cell lung cancer is associated with PD-L1 and poor prognosis in patients receiving pulmonectomy. Front Immunol 8:456

Wannamaker W, Davies R, Namchuk M, Pollard J, Ford P, Ku G, Decker C, Charifson P, Weber P, Germann UA, Kuida K, Randle JC (2007) (S)-1-((S)-2-{[1-(4-amino-3-chloro-phenyl)-methanoyl]-amino}-3,3-dimethyl-butanoy l)-pyrrolidine-2-carboxylic acid ((2R,3S)-2-ethoxy-5-oxo-tetrahydro-furan-3-yl)-amide (VX-765), an orally available selective interleukin (IL)-converting enzyme/caspase-1 inhibitor, exhibits potent anti-inflammatory activities by inhibiting the release of IL-1beta and IL-18. J Pharmacol Exp Ther 321:509–516

Watterberg KL, Carmichael DF, Gerdes JS, Werner S, Backstrom C, Murphy S (1994) Secretory leukocyte protease inhibitor and lung inflammation in developing bronchopulmonary dysplasia. J Pediatr 125:264–269

Wilson MS, Madala SK, Ramalingam TR, Gochuico BR, Rosas IO, Cheever AW, Wynn TA (2010) Bleomycin and IL-1beta-mediated pulmonary fibrosis is IL-17A dependent. J Exp Med 207:535–552

Wollin L, Maillet I, Quesniaux V, Holweg A, Ryffel B (2014) Antifibrotic and anti-inflammatory activity of the tyrosine kinase inhibitor nintedanib in experimental models of lung fibrosis. J Pharmacol Exp Ther 349:209–220

Wu J, Yan Z, Schwartz DE, Yu J, Malik AB, Hu G (2013) Activation of NLRP3 inflammasome in alveolar macrophages contributes to mechanical stretch-induced lung inflammation and injury. J Immunol 190:3590–3599

Yang W, Ni H, Wang H, Gu H (2015) NLRP3 inflammasome is essential for the development of chronic obstructive pulmonary disease. Int J Clin Exp Pathol 8:13209–13216

Yonker LM, Cigana C, Hurley BP, Bragonzi A (2015) Host-pathogen interplay in the respiratory environment of cystic fibrosis. J Cyst Fibros 14:431–439

Yoon BH, Romero R, Jun JK, Park KH, Park JD, Ghezzi F, Kim BI (1997) Amniotic fluid cytokines (interleukin-6, tumor necrosis factor-alpha, interleukin-1 beta, and interleukin-8) and the risk for the development of bronchopulmonary dysplasia. Am J Obstet Gynecol 177:825–830

Yoshida T, Tuder RM (2007) Pathobiology of cigarette smoke-induced chronic obstructive pulmonary disease. Physiol Rev 87:1047–1082

Zhang Y, Lee TC, Guillemin B, Yu MC, Rom WN (1993) Enhanced IL-1 beta and tumor necrosis factor-alpha release and messenger RNA expression in macrophages from idiopathic pulmonary fibrosis or after asbestos exposure. J Immunol 150:4188–4196

Zhang X, Shan P, Jiang G, Cohn L, Lee PJ (2006) Toll-like receptor 4 deficiency causes pulmonary emphysema. J Clin Invest 116:3050–3059

Zhou R, Tardivel A, Thorens B, Choi I, Tschopp J (2010) Thioredoxin-interacting protein links oxidative stress to inflammasome activation. Nat Immunol 11:136–140

Zuo L, Lucas K, Fortuna CA, Chuang CC, Best TM (2015) Molecular regulation of toll-like receptors in asthma and COPD. Front Physiol 6:312

Chapter 5
Traumatic Injury

Borna Relja and Johann-Philipp Horstmann

Contents

Abstract Traumatic injury as one of the world's most relevant but neglected health concerns results in modulated inflammasome activity, which is closely linked to the development of post-injury complications. Cytokine-producing capacity of cells is important for the appropriate immune response to trauma and requires not only synthesis and transcription of inflammasome components but also their activation. Unfortunately, the precise role of inflammasome in trauma is still largely unknown. However, in the following chapter, we provide an overview on the best described inflammasomes in the various settings of trauma, introducing the recent findings on the *up-to-date* best described NLRP inflammasomes and underlying cytokines in the inflammatory response to trauma.

B. Relja (✉)
Department of Trauma, Hand and Reconstructive Surgery, University Hospital Frankfurt, Goethe University, Frankfurt am Main, Germany
e-mail: info@bornarelja.com

J.-P. Horstmann
Department of Psychiatry and Psychotherapy, University Medicine Mainz, Mainz, Germany

M. D. Cordero, E. Alcocer-Gómez (eds.), *Inflammasomes: Clinical and Therapeutic Implications*, Experientia Supplementum 108,
https://doi.org/10.1007/978-3-319-89390-7_5

Keywords Trauma · Inflammasome · SIRS · CARS · PTI · Immunosuppression · Stress · Outcome

5.1 Introduction

Traumatic injury is one of the world's most relevant but neglected health concerns. Next to severe injury itself resulting in immediate or early death at the scene or within few hours, patients may in the later phase following the insult develop infectious complications (such as sepsis, septic shock, multiple organ dysfunction syndrome, MODS) causing mortality (Wafaisade et al. 2011; Wutzler et al. 2013; The global burden of disease 2013; Osuka et al. 2014). Traumatic injury in those patients, who initially survive beyond the first and early post-injury phase, affects the immune system homeostasis, thereby promoting an increased susceptibility to opportunistic infections and complications. Due to tissue injury, the natural dynamic immune response, which is both pro- and anti-inflammatory, with the aim of reducing damage, is induced. While the pro-inflammatory response is mainly driven by the innate immune system and termed *systemic inflammatory response syndrome* (SIRS), the anti-inflammatory response is mainly orchestrated by the adaptive immune system and is called *compensatory anti-inflammatory response syndrome* (CARS) (Wutzler et al. 2013; Osuka et al. 2014).

Research on post-traumatic complications had long time assumed a biphasic model with an initial SIRS followed by CARS. However, since then, the theory of a simultaneous onset of both SIRS and CARS has gained widespread attention. It interprets the ongoing processes as the organism's effort to strike the delicate balance between a sufficient defense against putative pathogens entering through eventual wounds on the one hand and on the other hand reducing collateral damage by immune cells (Adib-Conquy and Cavaillon 2009; Bhan et al. 2016; Reikeras 2010).

5.2 Post-Traumatic Immune Response

SIRS is a common reaction of the organism toward the tissue injury deriving from an acute trauma (Wutzler et al. 2013; Giannoudis and Pape 2007). It has been characterized as a massive immune reaction, which is extending from the local response to a spreading systemic activation. Several markers such as the clinically used C-reactive protein (CRP) and/or procalcitonin (PCT) but pro-inflammatory cytokines including tumor necrosis factor (TNF)-alpha, interleukin (IL)-1beta, and IL-6 as well are applicable for characterizing SIRS (Wutzler et al. 2013; Meisner et al. 2006; Pape et al. 2007). Systemic levels of circulating inflammatory mediators promote the activation of innate effector cells, e.g., monocytes, macrophages, and granulocytes. Previous attempts to prevent post-traumatic complications in patients who were developing SIRS by administering immunosuppressive agents have not only failed but proved to be detrimental as well (Hotchkiss et al. 2013a, b; Leentjens et al. 2013). Therefore, research outlook in recent years has refocused on the simultaneously beginning CARS (Hotchkiss et al. 2013a, b; Leentjens et al. 2013; Dinarello 2005). CARS is understood to render the patient susceptible to nosocomial infections due to his/her "immunosuppressive" state (Adib-Conquy and Cavaillon 2009; Hotchkiss et al. 2013a; Islam et al. 2016; Pfeifer et al. 2009). Numerous authors assume that CARS represents a shift from a T-helper cell type (Th)1-mediated to a Th2-mediated immune response, thus affecting the adaptive immune system more than the innate system (Bhan et al. 2016; Reikeras 2010; Islam et al. 2016; Xiao et al. 2011). Still, the early peak in anti-inflammatory IL-10 levels and the onset of endotoxin tolerance imply a profound impairment of innate immunity during CARS as well (Bhan et al. 2016; Hotchkiss et al. 2013a; Wutzler et al. 2009).

Summarized, on a biochemical level a parallel increase in pro- and anti-inflammatory mediators and antigens characterizes the post-traumatic immune response. This process is not only activated by foreign nonself material (Janeway Jr. and Medzhitov 2002) but includes endogenous factors or so-called alarmins (e.g., high-mobility group box 1, HMGB1; nucleosomes; histones; adenosine 5′-triphosphate, ATP; reactive oxygen species, ROS; etc.), which are released from necrotic or physiologically "stressed" cells and can activate as well as recruit effector cells of the immune system (Matzinger 2002). Several such endogenous triggers of the systemic post-traumatic inflammation have been described, however, their precise role in humans is still unknown (Bianchi 2007; Manson et al. 2012; Zedler and Faist 2006). Pattern recognition receptors (PRRs), of which notably the membrane-bound Toll-like receptors (TLRs) and cytoplasmic NOD-like receptors (NLRs), are responsible for the detection of (a) alarmins, damage-associated molecular pattern molecules (DAMPs) and (b) exogenous antigens (pathogen-associated molecular pattern molecules, PAMPs) (Matzinger 2002).

Monocytes and macrophages are important effector cells and critical regulators in the post-traumatic response to injury. Human monocytes isolated from trauma patients exert a decreased capability of releasing pro-inflammatory cytokines such as TNF-alpha or IL-1beta after a secondary ex vivo in vitro exposure to endotoxin (lipopolysaccharide, LPS) (Keel et al. 1996; Relja et al. 2015; Spolarics et al. 2003), a phenomenon that has been described as endotoxin tolerance (Cavaillon and Adib-Conquy 2006; Cavaillon et al. 2003). Such paralysis of innate effector cells, e.g., monocytes to produce pro-inflammatory cytokines in response to LPS in vitro, has been also found in sepsis, and has therefore fueled research on the similarities between these two etiologies (Galbraith et al. 2016). Although the exact mechanisms of endotoxin tolerance remain unclarified, downregulation of expression of a multitude of pro-inflammatory genes and/or TLRs has been observed, comprising several interleukins as well as genes involved in their release, such as those of inflammasome components (Bhan et al. 2016; Xiao et al. 2011; Cavaillon and Adib-Conquy 2006; Dobrovolskaia and Vogel 2002; Lendemans et al. 2007; Mendes et al. 2011). Inflammasomes are the multiprotein complexes, which mediate the cleavage of IL-1beta's and IL-18's precursors in their respective bioactive forms and are capable of inducing a specific form of cell death called pyroptosis. Interestingly reduced inflammasome activation in monocytes early after trauma has been described (Relja et al. 2015). Meanwhile the immune system after trauma is so-called primed due to, e.g., alarmin presentation. This aims to protect the organism from tissue injury and spreading infection, but, paradoxically, an excessive inflammatory cascade can develop upon a secondary trigger, e.g., PAMPs ("two-hit" response), and harm the organism and damage tissues by exaggerating the pro-inflammatory response. Furthermore, injury-induced inflammasome activation has been described in several immune cell subsets but primarily macrophages (Osuka et al. 2012). In their work, the authors propose that inflammasome activation plays a protective role in the host response to severe injury; nonetheless the data are conflictive (Osuka et al. 2012).

5.3 Inflammasome Biology in Trauma

Because of its activation via PRRs for pathogens or danger signals, the inflammasome is assumed to represent a common innate immune system recognition pathway as well (Zedler and Faist 2006; Martinon et al. 2009), which is closely linked to trauma- or tissue damage-induced inflammatory response. In contrast to most cytokines, IL-1beta and IL-18 are synthesized in their biologically inactive precursor forms, pro-IL, which accumulate in the cytosol and become cleaved and thereby activated within the inflammasome complex (Martinon et al. 2009; Agostini et al. 2004; Arend et al. 2008; Cerretti et al. 1992; Kwak et al. 2016). The inflammasome is a multiprotein complex, which via limited proteolysis mediates conversion of, e.g., inactive pro-caspase-1 to active caspase-1. Activated caspase-1 cleaves pro-IL and, furthermore, induces pyroptosis (Martinon et al. 2002, 2009). For nearly all inflammasomes, caspase-1 or caspase-5 are obligatory, while apoptosis-associated speck-like protein containing a caspase activation and recruitment domain (CARD) (ASC) is optional (Martinon et al. 2002; Latz et al. 2013). Most described inflammasomes contain a nucleotide oligomerization domain (NOD)-like receptor (NLR) sensor molecule, such as NOD-containing, leucine-rich repeats (LRR)-containing, and pyrin domain-containing (NLRP), e.g., NLRP1 or NLRP3 (Relja et al. 2015; Martinon et al. 2002; Latz et al. 2013; Proell et al. 2013).

The inflammasome activation, and thereby, e.g., IL-1β release, is dependent on intracellularly available components, which are required for its assembly (Latz et al. 2013). Thus, the NLR can be seen as the trigger molecules to set off the inflammasome and the subsequent interleukin secretion, cell lysis, and ultimately inflammation (Martinon et al. 2009). After receiving an activation signal, e.g., ATP-induced activation of P2X7 channels and the efflux of potassium ions, bacterial peptidoglycans, crystalline material, peptide aggregates, bacterial toxins or reactive oxygen species, etc., NLR sensor molecule and ASC aggregate and proteolytically activate pro-caspase-1 to caspase-1 (Martinon et al. 2002; Bauernfeind et al. 2011; Bauernfeind and Hornung 2013). Activated caspase-1 proteolytically cleaves the IL-1beta or IL-18 cytokine precursors to their active forms, thereby initiating the pro-inflammatory response. The NLRs that most commonly constitute inflammasomes are NLRP1, NLRP3, and NLRC4, but AIM2-inflammasomes have been described as well (de Vasconcelos et al. 2016; Schroder and Tschopp 2010). As for the activation, no direct link of the NLRs via binding of either PAMPs or DAMPs could be found; the hypothesis of an indirect activation emerged (Martinon et al. 2009; Muruve et al. 2008). It is commonly assumed that the secretion of IL-1beta must be "prepared" by a priming stimulus that is usually mediated by TLR, which in turn activates the NF-kappaB-pathway upregulating on one hand, the Pro-IL-1beta transcription, as well as, on the other hand, the transcription of inflammasome components (Dinarello 2009; van de Veerdonk et al. 2011). TLR4, for example, can be activated by an array of different PAMPs, such as LPS, as well as DAMPs, such as HMGB1 or HEME (Wegiel et al. 2015). To activate the inflammasome during a second step, the prevailing theory for a long time was that a potassium influx after activation of P2X7 would set off the NLR (Dinarello 2009). P2X7 is commonly activated by binding of ATP (Martinon et al. 2009; Schroder and Tschopp 2010; Dinarello 2011; Rathinam and Fitzgerald 2016; Vladimer et al. 2013). Concerning this, tissue damage after trauma, but also blood transfusion itself, can activate P2X7 and lead to the inflammasome activation with subsequent IL-1beta and IL-18 secretion. Also evidence for activation via reactive oxygen species has been discussed. Recently, the double-stranded RNA-dependent protein kinase (PKR) has been identified as a further player of the inflammasome pathway and has been attributed the capability to activating the inflammasome in its autophosphorylated state (Vladimer et al. 2013; Lu et al. 2012; Yang et al. 2017).

The data supporting this two-step model of activation are conflictive, since numerous studies have illustrated that an activation of macrophages and monocytes can lead to a release of inflammasome-dependent cytokines without a pre-activation (Relja et al. 2015; Chen and Sun 2013; He et al. 2013; Netea et al. 2009). On the other hand, it might be hypothesized that at the patient's admission to the emergency department, a pre-activation in trauma patients might have already taken place, since bacterial endotoxin has already crossed the gut-blood barrier in case of massive blood loss, open wounds, or DAMPs have been released upon tissue damage itself.

Apart from the cleavage of cytokines IL-1beta and IL-18, the inflammasome is capable of inducing cell lysis. In 2001 adding to apoptosis and necrosis, the new concept of pyroptosis has been established (Vande Walle and Lamkanfi 2016). It refers to an orchestrated lysis of the cell, which is initiated by inflammasome activation and consecutive formation of caspase-1-dependent pores of 1–2 nm width (Vande Walle and Lamkanfi 2016). Formation of these pores leads to swelling and lysis of the cell but also to release of intracellular compounds, such as the cleaved interleukins IL-1beta and IL-18 as well as intracellular molecules, which outside of the cell membrane potentially act as DAMPs, e.g., HMGB1 or ATP. Thus, pyroptosis represents a mode of cell death enhancing and spreading with every lysed cell the immune response. It mainly affects the myeloid lineage with monocytes, macrophages, and granulocytes but also occurs in epithelial, endothelial cells and neurons.

5.3.1 Cytokines in Trauma

Cytokines are effectors or rather messenger molecules of the immune system, which is activated upon trauma (van Griensven 2014). They are closely associated with the intensity of the post-traumatic inflammatory response. SIRS and CARS are characterized by the liberation of cytokines, which are recognized as part of the physiologic response to trauma. Tissue injury after trauma results in depressed cell-mediated immunity, such as dysfunctions of circulating monocytes or T cells, leading to an increased risk of infectious complications and unfavorable outcome (Spolarics et al. 2003; Bronkhorst et al. 2015; Kirchhoff et al. 2009; Marik and Flemmer 2012). The suppression of the local inflammation due to trauma-induced SIRS is meant to favor the repair and remodel the damaged tissue. However, recent research suggest that this local immune response counteracts not only with resident cells such as macrophages and dendritic cells but also with circulating cells, such as monocytes, thereby amplifying the post-traumatic immunosuppression (PTI) (Islam et al. 2016). Here, soluble mediators such as cytokines and chemokines (IL-1beta, IL-6, IL-8, IL-10, IL-15, monocyte chemoattractant protein-1 (MCP-1), granulocyte colony-stimulating factor (GCSF), IL-1 receptor antagonist (IL1-RA), eotaxin, IL-4, IL-7, IL-13, and many others) play decisive roles because they recruit and activate neutrophils, monocytes, etc. (Islam et al. 2016; Hazeldine et al. 2016). However, it is important to mention that there are also studies representing controversial findings, i.e., that the trauma patients, who developed sepsis, have enhanced IL-8 levels in endotoxin-stimulated whole blood in comparison to healthy donors (Flach et al. 1999). Summarized, the increase of inflammatory mediators may be useful as an early prognostic or diagnostic marker for post-injury complications.

5.3.1.1 IL-1beta

After severe trauma a shift in the levels of circulating cytokines can be observed. Within the wide array of cytokines affected by this shift, the two inflammasome-dependent ones, IL-1beta and IL-18, can be found (Wutzler et al. 2013; van Griensven 2014). IL-1beta has been described as one of the archetypical pro-inflammatory cytokines. It possesses the capability to activate an array of genes that can be related to and remain silent during the absence of inflammation (Dinarello 2005). It induces, e.g., transcription of inducible nitric oxide synthase (iNOS), cyclooxygenase-2 (COX2), as well as further pro-inflammatory cytokines, such as TNF-alpha or IL-6 (Dinarello 2005, 2009, 2011, 2013). After its expression, the inactive 31 kDa propeptide IL-1beta largely induced via NF-kappaB pathway has to be cleaved into its bioactive 15 kDa form by caspase-1 (Dinarello 2005). Less than 20% of IL-1beta's zymogen get cleaved into the bioactive form, thus making cleavage by caspase-1 the bottleneck of IL-1beta secretion (Dinarello 2009). IL-1beta is mainly produced by monocytes, lymphoid cells, dendritic cells, and natural killer cells. The functional depression of monocytes which is observed in patients after trauma is characterized by a diminished production and release of IL-1beta after exposure to DAMP or PAMP molecules such as LPS (Relja et al. 2015; Kirchhoff et al. 2009). Keeping in mind the PTI, monocytes may play here an important role. During the trauma response, lack of IL-1beta as one of the key players in the post-injury inflammatory reaction may potentiate the immunosuppression and increase the risks for susceptibility to infections and late mortality in critical illness.

Due to the short in vivo half-life of IL-1beta with approximately 10 minutes, there are only few studies elaborating serum levels of IL-1beta after trauma. Neither an increase nor any correlation between post-traumatic IL-1beta levels in serum was found (Frink et al. 2009; Sperry et al. 2008). Serum levels of IL-1beta differed significantly between injured patients and controls, although they were not applicable for determining the severity of injury in trauma patients (Alper et al. 2016). IL-1beta increased in bronchoalveolar lavage fluid (BAL) obtained from trauma patients with ARDS and is assumed to play a major role in its etiology (Bhatia and Moochhala 2004). Also studies on promotor polymorphisms provided data on increased risk for post-traumatic sepsis and multiple organ failure in homozygous carriers of overexpressing IL-1beta promotor genotypes as compared to those with underexpressing promotor genotypes (Wen et al. 2010). In order to mirror the post-traumatic IL-1beta biology, ex vivo in vitro stimulation assays of whole blood or isolated monocytes with endotoxins are applied. The subsequent assessment of LPS-stimulated IL-1beta release can be used as indicator of the monocyte activity (Munoz et al. 1991). Here, the secretion of IL-1beta from isolated monocytes of traumatized patients has been found to be consistently reduced reaching its nadir at 24 h post-trauma and negatively correlating with the probability of post-traumatic complications in humans (Relja et al. 2015; Kirchhoff et al. 2009; Ertel et al. 1995). Regarding the marked pro-inflammatory characteristics of IL-1beta, these findings support the onset of PTI after trauma. In summary, even mild trauma implicated a monocyte suppression, which recovered within 2 days (Wutzler et al. 2009). A more severe injury prolonged this recovery until day 5 after trauma and may contribute to PTI and concomitant complications. However, in another study, inflammasome recovery was not even found during the 10 observatory post-injury days in severely injured trauma patients (Relja et al. 2015). Recently the ability of TLR activation to induce less IL-1beta production in severely injured trauma intensive care unit patients compared with control subjects has been shown reduced, although this difference was not significant there was an obvious trend (Holloway et al. 2016).

Lederer's group has reported that injury activates the inflammasome pathway in injury site draining nodes within 2 h. Interestingly, they have shown that blocking the inflammasome worsened prognosis following injury. While the cytokine profiles of injured mice with blocked inflammasome activation showed decreased interleukin IL-1beta levels, IL-6 was markedly increased (Osuka et al. 2012). With regard to their findings, it is interesting to note that patients who are not able to develop a febrile response after injury, which is IL-1beta dependent, exert worse outcome as compared to those patients who do develop a febrile response after trauma (Mizushima et al. 2009). These findings indicate that an adequate post-injury

IL-1beta response appears beneficial after trauma. A possible mechanistical and therapeutical approach has been provided by our group, showing that the IL-1beta release can be reconstituted in isolated monocytes from trauma patients by recovering the infammasome functionality ex vivo in vitro (Relja et al. 2015).

In an experimental mouse model of spinal cord and brain injury, glial IL-1beta expression was increased (Impellizzeri et al. 2017). Therapeutical reduction of the IL-1beta among other pro-inflammatory mediators such as COX2, iNOS, and TNF-alpha has been associated with significantly decreased glial fibrillary acidic protein hyper-expression, the nuclear translocation and activation of NF-κB, suggesting a novel therapy to control neuroinflammatory conditions associated with spinal cord injury and traumatic brain injury (Impellizzeri et al. 2017). Also in other studies, the processing and release of IL-1beta has been demonstrated following traumatic brain injury and was linked to the NLR inflammasome activation. Yet, similar to other experimental trauma models, there are inconsistencies with regard to the significance of either IL-1beta expression or depletion, because although IL-1beta expression was significantly attenuated in the cortex of Nlrp1$^{(-/-)}$ and Asc$^{(-/-)}$ mice following moderate controlled cortical impact injury, no difference in motor recovery, cell death, or contusion volume has been observed compared to wild type animals (Brickler et al. 2016). In a rat model of intracerebral hemorrhage preventing inflammasome-dependent IL-1beta/IL-18 release using the selective P2X7R antagonist brilliant blue G (BBG) reduced brain edema and neurological deficits (Feng et al. 2015).

Summarized, these findings indicate that the role of IL-1beta released by inflammasome activation is still not fully understood. Interestingly, in the setting of ischemic brain injury, an elevation of IL-1beta levels was paralleled by an elevation of gelatinolytic, but not caspase-1 activity in the injured hemisphere. Moreover, pharmacological inhibition of gelatinases, i.e., matrix metalloproteases (MMP)-2 and MMP-9, prevented cytokine maturation (Amantea et al. 2016). It is important to keep in mind that these findings indicate at other mechanisms than caspase-1, likely involving gelatinases, for the maturation of IL-1beta in that model.

5.3.1.2 IL-18

IL-18 is another inflammasome-regulated pro-inflammatory cytokine, which affects a more specific set of immune cells including NK, B cells, cytotoxic T cells and Th1 cells, by synergistically inducing together with IL-12 the production of interferon (IF)-gamma, which in turn can induce the Th1-carried immune response. In the absence of IL-12, IL-18 induces a Th2-mediated immune response. Thus, it represents an agent of polarizing the spectrum of the adaptive immune system (Dinarello 2013; Lebel-Binay et al. 2000; Wawrocki et al. 2016). Mainly produced by antigen-presenting cells, e.g., macrophages and other cells, IL-18 is expressed as a 24 kDa zymogen, which is processed into its 18 kDa biologically active form. It is assumed that IL-18 gene is expressed constitutively; however, there are conflicting data regarding its induction via TLR-dependent pathways (Horstmann et al. 2016).

Some authors propose HMGB1 to stimulate pro-IL-18 synthesis via NF-κB and p38 MAPK in THP-1 macrophages, while LPS stimulation does not induce the expression of pro-IL-18 mRNA and protein in human primary peripheral blood mononuclear cells (He et al. 2012; Puren et al. 1999). IL-18 has been associated with a number of autoinflammatory diseases, including systemic *lupus erythematodes*, Crohn's disease, psoriasis, and graft-versus-host disease (Dinarello 2009; Lebel-Binay et al. 2000; Boraschi and Dinarello 2006). In a mouse model of sepsis, animals with deletion or blockade of IL-18 had increased survival rates (Dinarello and Fantuzzi 2003). Therefore, with regard to trauma, IL-18 as an important mediator in infectious diseases may be involved in the post-injury inflammatory reaction, causing susceptibility to infections or even mortality in critically ill patients.

The role of IL-18 in the aftermath of trauma has so far found little attention, although the implications of this cytokine for a putative shift in the Th spectrum might be highly relevant for the development of post-injury complications. IL-18 has been elevated in trauma patients, when compared to healthy controls, and was even found to allow prediction to the development of MODS, but not clinical outcome (Heizmann et al. 2008; Mommsen et al. 2009; Roetman et al. 2008). On the other hand, IL-18 was significantly increased in survivors compared with non-survivors in multiply injured trauma patients as discussed before. The secretion of IL-18 after trauma has been shown to be regulated by caspases in healthy subjects, while under the development of sepsis, secretion of IL-18 might circumvent the necessity for caspase activation by cleavage via proteinase-3 (Oberholzer et al. 2000).

Interestingly, mechanical ventilation enhanced IL-18 levels in the lung, serum, and BAL in mice. Similar to abovementioned study, IL-18 neutralization reduced lung injury in response to mechanical ventilation. In samples from clinical centers, IL-18 was found elevated in the plasma of patients with acute respiratory distress syndrome (ARDS) (either sepsis- or trauma-induced ARDS) and was proposed as a novel biomarker of intensive care unit morbidity and mortality (Dolinay et al. 2012).

Among service members who sustained combat-related injuries, IL-18 was determined as potential urinary biomarker (Janak et al. 2017). Here, no differences were found among patients with burns compared to military patients with non-burn trauma (Janak et al. 2017). Gut barrier disruption is frequently implicated in pathogenesis associated with burn and other traumatic injuries. The authors found markedly increased IL-18 (by ~2.5-fold) in the small intestine epithelial cells one day after burn injury in mice (Cannon et al. 2016). In the same study, therapeutical normalization of IL-18 levels has been associated with improved permeability and intestinal transition. In burn-injured mice, IL-1beta and IL-18 were induced by ATP or ATP + LPS stimulation by spleen cells (Osuka et al. 2012). In parallel a significant caspase-1 activation was measured in macrophages and dendritic cells by 4 h after burn injury and peaked by the first post-injury day. Moreover, a significant caspase-1 activation was found in NK cells, CD4 T cells, and B cells; however, CD8 T cells did not demonstrate caspase-1 activation post-injury. In this in vivo study the authors have provided evidence that blocking caspase-1 activation caused significantly higher mortality in burn-injured mice (Osuka et al. 2012). Interestingly, severe injury caused by burn injury induces an early, but not late or sustained inflammasome activation in a variety of immune cell subsets (Osuka et al. 2012). Thus, the inflammasome may not play a significant role in the "two-hit response" to injury, but this remains to be evaluated in further studies. Nonetheless, these data suggest that possible treatment strategies targeting inflammasome activation pathways could be protective for injured patients.

Taken together, these findings document that injury induces inflammasome activation in many immune cell subsets, but primarily in macrophages, and that inflammasome activation plays a protective role in the host response to severe injury. Moreover, in most of these studies, the kinetics of injury-induced inflammasome activation correlated with detectable increases in circulating IL-1β (Osuka et al. 2012). These findings suggest that the inflammasome pathway and its downstream cytokines play critical roles in the recovery from trauma and may constitute a promising therapeutic target in trauma patients.

5.3.2 The NLRP1 Inflammasome

The NLRP1-inflammasome was the first inflammasome to be discovered in 2002 (Martinon et al. 2002). Since then researchers interest has waned in favor of NLRP3, which has been found to be involved in a wide array of inherited pathologies. The peptide chain of NLRP1 is a sequence of PYD-NACHT-LRR-FIIND-CARD. Herein the C-terminal CARD as the death fold region provides the possibility to directly bind and activate caspase-1, while the N-terminal pyrin-rich domain (PYD) allows the optional recruitment of ASC. The central NACHT domain mediates oligomerization after activation due to ligand binding to the LRR region (Martinon et al. 2002, 2009; Schroder and Tschopp 2010; Dagenais et al. 2012). The NLRP1 inflammasome consists of NLRP1 itself and usually caspase-1. Binding of the NLRs PYD region to ASCs PYD region can recruit the adaptor molecule and hence boost enzymatic activity. This however is not necessary for the complexes functioning (Martinon et al. 2002, 2009; Schroder and Tschopp 2010; Dagenais et al. 2012). The modes of activation of the NLRP1 inflammasome have been unclear for a long time. An activation via K^+-influx has been hypothesized, which is mediated by the purinergic receptor P2X7 that is usually activated by extracellular ATP, thus a classical DAMP (Martinon et al. 2002, 2009; Schroder and Tschopp 2010; Dagenais et al. 2012).

Research addressing inflammasomes has made great progress as they play a key role as gatekeeper of IL-1beta secretion and therefore in endotoxin tolerance as well. Alterations in NLRP1 gene expression, which were observed and associated with monocyte deactivation during septic shock, led to the assumption that NLRP1 may play a key role in the post-traumatic inflammatory response as well (Fahy et al. 2008). Thus, it has been shown that NLRP1 expression was upregulated after trauma, but interestingly same monocytes isolated from patients failed to increase NLRP1 expression when confronted with a secondary ex vivo in vitro LPS stimulus (Relja et al. 2015). In parallel the IL-1beta release upon ex vivo in vitro LPS stimulation was reduced as compared to the IL-1beta releasing capacity in both whole blood and isolated monocytes from healthy volunteers (Relja et al. 2015). In more severely injured trauma patients, an "inflammasome recovery," which was evaluated by IL-1beta release, was not observed until day 5 and may explain the susceptibility of the included patients to post-traumatic inflammatory complications, such as sepsis or organ failure. However, the authors did not provide more specific data on the inflammasome itself. In above discussed study, "inflammasome recovery" was not even achieved within the ten observational days after trauma (Relja et al. 2015). In monocytes from healthy individuals, LPS-induced IL-1beta release was associated with enhanced expression of caspase-1 and IL-1beta precursors as well as slightly enhanced NLRP1 and rather continuous expression of PYCARD (Relja et al. 2015). A functional inflammasome activation and thereby IL-1beta release is actually dependent on intracellularly available components, which are required for the inflammasome assembly (Latz et al. 2013). In contrast to the NLRP1 inflammasome complex, which is preassembled in unstimulated neurons or in normal central nervous system, in unstimulated peripheral macrophages, NLRP1 components are not preassembled (de Rivero Vaccari et al. 2014) but rapidly form protein-protein associations upon stimulation by, e.g., muramyl dipeptide (Hsu et al. 2008). Therefore, it is reasonable that in monocytes from healthy volunteers, all components which are required and not constitutively available become expressed upon a PRR stimulus. This mechanism has been confirmed by increased caspase-1 cleavage in parallel to increased gene expression of NLRP1 components as well as increased IL-1beta release (Relja et al. 2015). With regard to trauma patients, reduced IL-1beta secretory capacity of monocytes was primarily associated with the lack of NLRP1 gene expression, while there was no reduced IL-1beta transcription (Relja et al. 2015). In the same study monocytes from trauma patients were isolated and transfected with NLRP1 leading to a restoration of IL-1beta secretion. The abrogated assembly of NLRP1 inflammasome due to missing certain NLRP1 inflammasome components may be the reason for the monocytic deactivation after trauma.

Interestingly, there are significant changes in relative copy numbers for the inflammasome mRNA for caspase-1 regulatory proteins including PYCARD, NLRP1, and caspase-1 in monocytes from critically ill patients (Fahy et al. 2008). Moreover, NLRP1 was higher in patients with septic shock who survived to days 7 and 30 and to hospital discharge as compared with non-survivors (Fahy et al. 2008). The authors propose that sepsis-induced suppression of NLRP1 gene may induce a transient immunosuppression of monocytes, which is actually comparable to the concept of PTI.

With regard to other trauma injury patterns, traumatic brain injury (TBI) activates the NLRP1 inflammasome as an important component of the early innate inflammatory response to injury. It is important to keep in mind that the NLPR1 inflammasome is preassembled in CNS, and it has been suggested that preassembly of inflammasome complexes in CNS cells facilitates a rapid triggering of the innate immune response after CNS trauma. In the CNS, caspase-11 is involved in the processing of IL-1beta and forms protein-protein interactions with other inflammasome proteins as well, but its role remains to be investigated. However, recent findings indicate that caspase-11 has a role in noncanonical inflammasome activation after bacterial infections (Vigano and Mortellaro 2013). In the experimental model of TBI, caspase-1, caspase-11, and expression of the purinergic receptor P2X7 have been increased at 24 h after TBI (Tomura et al. 2012). IL-1beta processing has been associated with the activation of caspase-1 and increased levels of ASC and caspase-11, however, not with increased NLRP1 (de Rivero Vaccari et al. 2009). Nonetheless, the assembly of the NLRP1 inflammasome complex was promoted by TBI (de Rivero Vaccari et al. 2009). Decreasing the expression of these proteins or neutralizing ASC decreased inflammasome signaling in neurons and reduced the innate immune response to TBI after injury (Tomura et al. 2012; de Rivero Vaccari et al. 2009). Opposite to the data obtained from severely injured trauma patients, in TBI patients the data propose that the NLRP1 inflammasome constitutes an important component of the innate CNS inflammatory response after TBI and, therefore, may be a novel therapeutic target for reducing the damaging effects of post-TBI-induced inflammation. Components of the NLRP1 inflammasome in CSF have been found to correlate with clinical outcome after TBI and thus were proposed as clinical predictors (Adamczak et al. 2012).

Similar findings were reported for spinal cord injury as well (de Rivero Vaccari et al. 2012; Lin et al. 2016). Recently, conflictive data have shown that although IL-1beta expression was significantly attenuated in the cortex of $Nlrp1^{(-/-)}$ and $Asc^{(-/-)}$ mice following TBI, no difference in motor recovery, cell death, or contusion volume has been observed compared to wild type (Brickler et al. 2016). The models did partly differ in their severity or methodology, and maybe this caused conflictive suggestions that NLRP1 inflammasome activation does not significantly contribute to acute neural injury in the murine model of TBI. However, the importance of NLRP1 in TBI remains to be elucidated in further studies.

It was shown that stress as conveyed by catecholamines can increase IL-1beta response in monocytes to LPS stimulation (Horstmann et al. 2016; Grisanti et al. 2011). Results regarding changes in the inflammasome pathway give a somewhat intriguing pattern of inflammasome gene expression. Both LPS and adrenergic stimulation led to increases in NLRP1 gene expression, with adrenergic stimulation alone leading to markedly higher NLRP1 gene expression than LPS stimulation alone (Horstmann et al. 2016). Costimulation resulted in NLRP1 expression levels, which were not significantly different from controls (Horstmann et al. 2016). This reflects the observations of NLRP1 expression after trauma with an increase after stressor exposure but failure to further upregulate, when confronted with the PAMP (Relja et al. 2015).

Apart from its role in polytrauma and subsequent development of MOF, the NLRP1 inflammasome was found to cause lung injury mimicking ARDS upon activation in the mouse model. Since the observed effects were independent of IL-1beta and absent in $Casp^{-/-}$ mice, it has been assumed that lung tissue damage was predominantly mediated by pyroptosis of resident lung macrophages, thus spilling DAMPs such as HMGB1, and not by a cascade of interleukins. Interestingly

NLRP1$^{-/-}$ mice were protected from these detrimental effects, highlighting the pivotal role of NLRP1 in lung injury (Kovarova et al. 2012).

5.3.3 The NLRP3 Inflammasome

The NLRP3 monomer consists of a PYD-NACHT-LRR chain. The oligomer complex due to the lack of a CARD region in NLRP3 requires binding of the ASC adaptor molecule in order to activate caspase-1. Binding of both NLRP3s and ASCs PYD regions as well as CARD regions of ASC and caspase-1 then allows for the assembly of the inflammasome (Martinon et al. 2009; Schroder and Tschopp 2010; Dagenais et al. 2012). The activation of NLRP3 has been a subject of a variety of hypotheses in the past. The structural differences of agents triggering NLRP3 led scientists to favor an indirect way of activation against a direct activation. For a long time, the most popular theory was activation by potassium efflux after activation of the purinergic receptor P2X7 (Martinon et al. 2009; Schroder and Tschopp 2010; Dagenais et al. 2012). By now, studies came up with different explanations as how NLRP3 might be activated. One is the concept of activation by phosphorylation via PKR (Lu et al. 2012; Yang et al. 2017). Interestingly, excessive ROS may be crucially involved in the process of regulating the assembly and activation of NLRP3 inflammasomes (Minutoli et al. 2016). The NLRP3 inflammasome can be activated by multiple distinct exogenous and endogenous triggers, including uric acid (Martinon et al. 2006), silica and asbestos (Dostert et al. 2008), bacteria (Muruve et al. 2008), and toxins (Gurcel et al. 2006), which are known for inducing the production of short-lived ROS and ROS scavengers. Moreover, ROS can directly induce NLRP3 activation by promoting the association of thioredoxin-interacting protein (TXNIP) with NLRP3 (Zhou et al. 2010).

Furthermore, histones have been reported to activate the NLRP3 inflammasome and are abundantly present in serum after trauma (Allam et al. 2013, 2014). Activation of the NLRP3 inflammasome with following IL-1beta response due to hyaluronic acid binding to macrophage CD44 receptor marks the inflammasome's role in injury of connective tissue, e.g., musculoskeletal trauma (Yamasaki et al. 2009). Glucocorticoids as neuroendocrine stress signals have been shown in THP macrophages to induce the transcription of NLRP3, supporting the hypothesis of glucocorticoids not only keeping homeostasis in reducing general inflammation but also preparing macrophages for a rapid unfolding of the inflammatory cascade upon DAMP presentation (Busillo et al. 2011). Stress as conveyed by increased serum levels of glucocorticoids has been found to lead to a release of HMGB1 from hippocampal microglia in the rodent model, thus indirectly priming the NLRP3 inflammasome via TLR4 and augmenting the immune response to a physical trauma by its psychological stressor component (Frank et al. 2015). In rodent in vivo experiments of tailshock-induced stress, inhibition of caspase-1 using the caspase-1 inhibitor ac-YVAD-cmk attenuated stress-induced production of IL-1β, IL-18, and IL-6 in both the circulation and peripheral tissues, highlighting the pivotal role for the inflammasome in this process (Maslanik et al. 2013). Genetic variations in NLRP3 gene influence the IL-1beta response of monocytes when stimulated ex vivo with LPS and predict the development of sepsis and MODS in trauma patients (Zhang et al. 2011).

Ischemia is frequently observed in trauma patients suffering from massive blood loss. Experiments using siRNA have demonstrated that the NLRP3 inflammasome mediates pro-inflammatory cytokine release and hepatocellular damage in hepatic ischemia and reperfusion (I/R) (Zhu et al. 2011). Furthermore, ASC/caspase-1/IL-1β complex modulates HGMB1 release and TLR4-dependent hepatic I/R injury (Kamo et al. 2013). There are several studies dealing with the process of inflammasome activation in liver after I/R; however, the cellular processes behind engaging inflammasome-mediated inflammatory responses and injury after I/R are not yet understood. It has been demonstrated that serum IL-1β and protein conversion of the active form increased after reperfusion, peaking at 6 h and then declining again. Protein expression of cleaved caspase-1, ASC, and NLRP3 increased in response to I/R injury (Kim et al. 2015). The authors have shown that activation of the NLRP3 inflammasome plays a role in hepatocellular damage induced by I/R (Kim et al. 2015).

Regarding other tissues such as, e.g., the lung after trauma, the lung's endothelial cells represent a major source of IL-1beta after trauma and constitute the source of pyroptosis leading to acute lung injury (ALI) (Xiang et al. 2011; Yang et al. 2016). Release of IL-1beta, pyroptosis, and subsequent lung injury are mediated by NLRP3, which can be activated by LPS or by the release of DAMPs, such as HMGB1 or ROS, due to tissue stress during hemorrhagic shock or ischemia as described above (Xiang et al. 2011). Another study highlighted the importance of histones for the development of ALI as well (Grailer et al. 2014). In a mouse model of ALI, histones appeared in BAL and activated the NLRP3 inflammasome. Elimination of histones also reduced IL-1beta levels in BAL. Furthermore, histone appearance in BAL seemed to depend on NLRP3 and caspase-1 availability, suggesting a role of the inflammasome for histone release in ALI and therefore a positive feedback loop (Grailer et al. 2014).

Severely injured trauma patients are frequently on mechanical ventilation. Lung damage due the continuous injury caused by mechanical ventilation is assumed to be mediated by an increase in serum levels of DAMPs, such as ATP and ROS, thus activating the NLRP3 inflammasome (Hosseinian et al. 2015; Jones et al. 2014; Kuipers et al. 2012). Interestingly, in one study, NLRP3 activity and IL-beta levels in BAL after LPS administration and mechanical ventilation seemed to influence blood oxygen levels in a mouse model but were independent of alveolar leakage and neutrophil extravasation (Jones et al. 2014). On the other hand, in mice, mechanical ventilation enhanced IL-18 levels in the lung, serum, and BAL, and IL-18-neutralization by antibody treatment, or genetic deletion of IL-18 or caspase-1, reduced markedly lung injury (Dolinay et al. 2012). Similarly, in human patients with ARDS from four clinical centers, inflammasome-related mRNA transcripts (CASP1, IL1B, and IL18) as well as plasma IL-18 were elevated in peripheral blood (Dolinay et al. 2012). The authors suggest the inflammasome pathway, and its downstream cytokines, playing critical roles in ARDS development, and, moreover, IL-18 as a novel biomarker of intensive care unit morbidity and mortality.

Also, iatrogenic activation of the NLRP3 inflammasome due to blood transfusion has been hypothesized (Hosseinian et al. 2015). After the initial traumatic injury to the lungs putatively acting as a first trigger, the efflux of alarmins, such as ATP from transfusion products, sets off the inflammasome and mediates interleukin release. The increased levels of interleukin in the pulmonary milieu then lead to inflammation and cause so-called transfusion-associated lung injury (TRALI) (Land 2013).

The inflammasome-mediated inflammation of pulmonary tissue also marks the result of burn injury. NLRP3 induction and activation has been found in rats' lungs after burn injury, as well as in vitro in alveolar macrophages after they had been stimulated with serum of rats after burn injury. This in vitro NLRP3 activity could be reversed by eliminating ROS from the cell medium (Han et al. 2015). Studies of inflammasome activation upon burn injury have shown an increase in NLRP3 activity in lesioned tissue (Long et al. 2016; Xiao et al. 2016). Rodent in vitro and in vivo studies attained inhibition of the NRLP3 inflammasome with compounds as 3,4-methylenedioxy-β-nitrostyrene (MNS) or the antimalarial drug artemisinin and have shown beneficial effects, such as improved wound healing after burn injury and reduced mortality after burn and sepsis. Insulin resistance because of burns has been hypothesized to be mediated by NLRP3 inflammasome activation and increase in IL-1beta secretion as in *diabetes mellitus* type 2 (Long et al. 2016; Xiao et al. 2016; Stanojcic et al. 2014).

In a rat model of localized trauma, activation of caspase-1 has been found similar in both skin and muscle tissue upon injury alone. When followed by cardiac arrest, however, the activation of caspase-1 and levels of IL-18 were markedly increased in skin and markedly decreased in muscle tissue, suggesting a differential effect of cardiac arrest on inflammasome activation in different tissues (Starzl et al. 2015).

There is emerging evidence that NLRP3 is involved in the process of the CNS disorders development such as ischemic stroke (Yang et al. 2014), Alzheimer's disease (Halle et al. 2008), and pneumococcal meningitis (Hoegen et al. 2011). Liu et al. demonstrated that NLRP3 might play an important role in the inflammatory response in an experimental model of brain trauma (Liu et al. 2013). Ma et al. and others as well have shown before that propofol (2, 6-diisopropylphenol), a well-known lipid-soluble intravenous anesthetic, which is commonly used during surgery, has neuroprotective effects against various forms of brain injury (Ma et al. 2009; Ding et al. 2013). In their recent study, the molecular mechanism, underling the neuroprotective effect of propofol, was further elucidated. Propofol administration substantially suppressed the enhanced expression and activation of NLRP3 inflammasome in the cerebral cortex of TBI rats, findings which were associated with the decreased oxidative stress, cytokine production, and ameliorated cerebral cortex damage (Ma et al. 2016). The authors provide evidence that overactivation of NLRP3 inflammasome in the cerebral cortex may be involved in the process of neuroinflammation during the secondary injury of TBI in rats. The findings of reduced activation in rat's cortices in response to propofol mirror the observation that neuroinflammation after ICH as found in NLRP3 and IL-1beta activation can be attenuated by blocking the glutamatergic NMDA receptor. Although the underlying mechanism for propofol's mode of action is primarily its agonism on GABAergic receptors, it seems that an inhibition of neural activity after brain injury goes together with a reduction in neural tissue inflammation via the NLRP3 inflammasome (Weng et al. 2015).

In a subarachnoidal hemorrhage model, a beneficial role for the antibiotic minocycline has been found in rats. Here the administration of the compound reduced expression and activation of the NLRP3 inflammasome. Consequently, cleavage of caspase-1 as well as pro-IL-1beta was reduced, and brain edema emerging from the disruption of the blood-brain barrier was attenuated. This finding might warrant additional research in minocycline's ability to inhibit the NLRP3 inflammasome (Li et al. 2016).

5.3.4 The AIM2 Inflammasome

AIM2 is a member of the hemopoietic interferon-inducible nuclear 200 (HIN200) family of proteins and can form an inflammasome (Burckstummer et al. 2009; Fernandes-Alnemri et al. 2009; Hornung et al. 2009). It is known to be activated by viral, bacterial, and host ectopic double-stranded DNA, thereby recognizing pathogenic DNA or autoimmune reactions to host nucleic acids, but HMGB1 as well, and subsequently inducing cleavage of pro-caspase-1 and pro-IL-1beta, as well as pyroptosis (Burckstummer et al. 2009; Fernandes-Alnemri et al. 2009; Hornung et al. 2009; Miao et al. 2010). AIM2 can associate with ASC and caspase-1 to form a DNA-responsive inflammasome in, e.g., cortical neurons (Adamczak et al. 2014; Sun et al. 2017). AIM2 inflammasome contributes with ASC to acute brain injury independently of NLRP3 in an experimental stroke model (Denes et al. 2015). Further characterization of the neuronal AIM2 inflammasome will provide important information that will guide us in the development of therapies to treat inflammasomes activated by DNA after injury or viral infections.

As there are no studies regarding AIM2 in experimental trauma or in human scenario, we concentrate here on a hypoxia model to describe the possible relevance of AIM2 in trauma setting. Ischemia/reperfusion (I/R) injury, e.g., to the liver caused by an injurious inflammatory response, occurs after trauma as well. The mechanisms by which organ damage occurs are still unclear, but the involvement of liver-resident Kupffer cells (KCs) with their immune-triggering, ROS-producing, and inflammation-inducing capabilities is indisputable (Wanner et al. 1996). Recently, the activation of the AIM2 inflammasome in liver in I/R has been uncovered (Kim et al. 2015). Consistent with earlier studies, I/R caused an increase in serum but also in cytosolic levels of dsDNA, a ligand of AIM2 (Kim et al. 2015). Both AIM2 protein expression and its interaction with ASC protein were increased after I/R, suggesting that AIM2, a non-NLR protein, was induced by ROS and may also be responsible for caspase-1 activation in hepatic I/R (Kim et al. 2015). The authors have confirmed a crucial role of KCs upon liver damage after I/R, because a depletion of KCs with gadolinium chloride markedly decreased AIM2 inflammasome expression and activation, as well as the level of caspase-1 protein in F4/80-positive cells (Kim et al. 2015). However, because the NLRP3 inflammasome has also affected by the KCs depletion, the sole role of AIM2 after I/R still remains undefined. However, evidence on beneficial effects of caspase-1 activity has been provided in rodent models of I/R. Here higher levels of ROS as well as systemic inflammatory cytokines and liver tissue injury have been observed under caspase-1 knockout. Interestingly a NLRP3 knockout did not alter outcomes, suggesting a crucial role for the AIM2 inflammasome (Menzel et al. 2011; Sun et al. 2013).

5.3.5 The NLRC4 Inflammasome

There is only one study evaluating the importance of the NLRC4 inflammasome for trauma-induced inflammatory response and subsequent tissue damage. Recently, in their PNAS study, Denes et al. have shown that ischemic brain injury was reduced in $ASC^{-/-}$ and $NLRC4^{-/-}$ mice (Denes et al. 2015). The authors did not observe such protective effects in mice deficient for the NLRP3. This was the first evidence for identifying AIM2 and NLRC4 as key drivers of post-traumatic inflammatory responses and, more interestingly, multiple inflammasomes regulating neuronal injury (Denes et al. 2015) (Table 5.1).

5.4 Summary

Though there is clear evidence for an important role of inflammasome activation correlating with the post-traumatic inflammatory response, there is no clear hypothesis if the activation of inflammasomes is rather harmful or beneficial upon trauma. Moreover, it appears that distinct activity of different inflammasomes in different tissues may play a dual role in post-traumatic regeneration. However, it is indisputable that traumatic injury is closely associated with inflammasome activation and/or their inhibition due to, e.g., their deranged assembly. Therefore, identifying not only inflammasomes but also their components, which may be decisive for their functionality, is important therapeutically balancing of the post-injury response. Furthermore, alterations of the inflammasome function appear promising in the setting of post-traumatic therapies and should be further evaluated in future studies. Especially the activation of the inflammasome with subsequent increase in secondary damage in TBI deserves a closer look as evidence suggests that a downregulation on neural activity also decreases inflammatory activity. Given findings on the damaging effect of NMDAR-activity in neuroinflammation, studies using the NMDAR-antagonist and important analgesic ketamine can be potentially promising. Similarly, the inhibition of P2X7 by BBG or NLRP3 inhibition by MNS could provide new

Table 5.1 Overview on some trauma-relevant inflammasomes

Inflammasome/ inflammasome component	Tissue/cell Type	Species	In vivo, in vitro, ex vivo in vitro	Model	Therapy/intervention	Therapy response	Ref.
NLRP1	Monocytes	hn	Ex vivo in vitro	Traumatic injury	NLRP1 transfection	Restauration of IL-1beta response	Relja et al. (2015)
Caspase-1	Lung	ms	In vivo	Mechanical ventilation	Caspase-1 KO	Reduced lung injury	Dolinay et al. (2012)
NLRP3	Lung	ms	In vivo	LPS-induced ALI	LR12	Reduction: lung inflammation, NLRP3 expression, mortality	Liu et al. (2016)
NLRP3	Brain	rt	In vivo	bTBI	Propofol	Reduction: NLRP3 expression + activation, tissue injury	Ma et al. (2016)
NLRP1/ASC	Brain	ms	In vivo	CCl	NLRP1/ASC KO	Attenuation of IL-1beta; no difference in damage	Brickler et al. (2016)
NLRP3	Wounds	rt	In vivo	Burn	3,4-Methylenedioxy-β-nitrostyrene	Reduction: cytokine production, neutrophil infiltration; accelerated wound healing	Xiao et al. (2016)
NLRP3	Lung, heart	ms	In vivo	Burn sepsis	Artemisinin	Reduction: cytokine production, adhesion molecule expression, neutrophil infiltration, mortality	Long et al. (2016)
NLRP3	Brain	ms	In vivo	SAH	Melatonin	Attenuation of NLRP3-associated proteins	Dong et al. (2016)
NLRP3	Microglia	ms	In vivo	ICH	NLRP3 KO	Amelioration of inflammation	Yuan et al. (2015)

promising treatment options in scenarios, where inflammasome suppression proves beneficial. Importantly to notice is that only few studies on inflammasomes with regard to human trauma exist, and, here, a huge gap of knowledge has to be closed in future.

References

Adamczak S et al (2012) Inflammasome proteins in cerebrospinal fluid of brain-injured patients as biomarkers of functional outcome: clinical article. J Neurosurg 117:1119–1125. https://doi.org/10.3171/2012.9.JNS12815
Adamczak SE et al (2014) Pyroptotic neuronal cell death mediated by the AIM2 inflammasome. J Cereb Blood Flow Metab 34:621–629. https://doi.org/10.1038/jcbfm.2013.236
Adib-Conquy M, Cavaillon JM (2009) Compensatory anti-inflammatory response syndrome. Thromb Haemost 101:36–47
Agostini L et al (2004) NALP3 forms an IL-1beta-processing inflammasome with increased activity in Muckle-Wells autoinflammatory disorder. Immunity 20:319–325
Allam R, Darisipudi MN, Tschopp J, Anders HJ (2013) Histones trigger sterile inflammation by activating the NLRP3 inflammasome. Eur J Immunol 43:3336–3342. https://doi.org/10.1002/eji.201243224
Allam R, Kumar SV, Darisipudi MN, Anders HJ (2014) Extracellular histones in tissue injury and inflammation. J Mol Med (Berl) 92:465–472. https://doi.org/10.1007/s00109-014-1148-z
Alper B, Erdogan B, Erdogan MO, Bozan K, Can M (2016) Associations of trauma severity with mean platelet volume and levels of systemic inflammatory markers (IL1beta, IL6, TNFalpha, and CRP). Mediat Inflamm 2016:9894716. https://doi.org/10.1155/2016/9894716
Amantea D et al (2016) Caspase-1-independent maturation of IL-1beta in ischemic brain injury: is there a role for gelatinases? Mini Rev Med Chem 16:729–737
Arend WP, Palmer G, Gabay C (2008) IL-1, IL-18, and IL-33 families of cytokines. Immunol Rev 223:20–38. https://doi.org/10.1111/j.1600-065X.2008.00624.x
Bauernfeind F, Hornung V (2013) Of inflammasomes and pathogens--sensing of microbes by the inflammasome. EMBO Mol Med 5:814–826. https://doi.org/10.1002/emmm.201201771
Bauernfeind F et al (2011) Inflammasomes: current understanding and open questions. Cell Mol Life Sci 68:765–783. https://doi.org/10.1007/s00018-010-0567-4
Bhan C, Dipankar P, Chakraborty P, Sarangi PP (2016) Role of cellular events in the pathophysiology of sepsis. Inflamm Res 65:853–868. https://doi.org/10.1007/s00011-016-0970-x
Bhatia M, Moochhala S (2004) Role of inflammatory mediators in the pathophysiology of acute respiratory distress syndrome. J Pathol 202:145–156. https://doi.org/10.1002/path.1491
Bianchi ME (2007) DAMPs, PAMPs and alarmins: all we need to know about danger. J Leukoc Biol 81:1–5. https://doi.org/10.1189/jlb.0306164
Boraschi D, Dinarello CA (2006) IL-18 in autoimmunity: review. Eur Cytokine Netw 17:224–252
Brickler T et al (2016) Nonessential role for the NLRP1 inflammasome complex in a murine model of traumatic brain injury. Mediat Inflamm 2016:6373506. https://doi.org/10.1155/2016/6373506
Bronkhorst MW, Patka P, Van Lieshout EM (2015) Effects of sequence variations in innate immune response genes on infectious outcome in trauma patients: a comprehensive review. Shock 44:390–396. https://doi.org/10.1097/SHK.0000000000000450
Burckstummer T et al (2009) An orthogonal proteomic-genomic screen identifies AIM2 as a cytoplasmic DNA sensor for the inflammasome. Nat Immunol 10:266–272. https://doi.org/10.1038/ni.1702
Busillo JM, Azzam KM, Cidlowski JA (2011) Glucocorticoids sensitize the innate immune system through regulation of the NLRP3 inflammasome. J Biol Chem 286:38703–38713. https://doi.org/10.1074/jbc.M111.275370

Cannon AR et al (2016) Effects of mesalamine treatment on gut barrier integrity after burn injury. J Burn Care Res 37:283–292. https://doi.org/10.1097/BCR.0000000000000396
Cavaillon JM, Adib-Conquy M (2006) Bench-to-bedside review: endotoxin tolerance as a model of leukocyte reprogramming in sepsis. Crit Care 10:233. https://doi.org/10.1186/cc5055
Cavaillon JM, Adrie C, Fitting C, Adib-Conquy M (2003) Endotoxin tolerance: is there a clinical relevance? J Endotoxin Res 9:101–107. https://doi.org/10.1179/096805103125001487
Cerretti DP et al (1992) Molecular cloning of the interleukin-1 beta converting enzyme. Science 256:97–100
Chen S, Sun B (2013) Negative regulation of NLRP3 inflammasome signaling. Protein Cell 4: 251–258. https://doi.org/10.1007/s13238-013-2128-8
Dagenais M, Skeldon A, Saleh M (2012) The inflammasome: in memory of Dr. Jurg Tschopp. Cell Death Differ 19:5–12. https://doi.org/10.1038/cdd.2011.159
de Rivero Vaccari JP et al (2009) Therapeutic neutralization of the NLRP1 inflammasome reduces the innate immune response and improves histopathology after traumatic brain injury. J Cereb Blood Flow Metab 29:1251–1261. https://doi.org/10.1038/jcbfm.2009.46
de Rivero Vaccari JP et al (2012) P2X4 receptors influence inflammasome activation after spinal cord injury. J Neurosci 32:3058–3066. https://doi.org/10.1523/JNEUROSCI.4930-11.2012
de Rivero Vaccari JP, Dietrich WD, Keane RW (2014) Activation and regulation of cellular inflammasomes: gaps in our knowledge for central nervous system injury. J Cereb Blood Flow Metab 34:369–375. https://doi.org/10.1038/jcbfm.2013.227
de Vasconcelos NM, Van Opdenbosch N, Lamkanfi M (2016) Inflammasomes as polyvalent cell death platforms. Cell Mol Life Sci 73:2335–2347. https://doi.org/10.1007/s00018-016-2204-3
Denes A et al (2015) AIM2 and NLRC4 inflammasomes contribute with ASC to acute brain injury independently of NLRP3. Proc Natl Acad Sci U S A 112:4050–4055. https://doi.org/10.1073/pnas.1419090112
Dinarello CA (2005) Interleukin-1beta. Crit Care Med 33:S460–S462
Dinarello CA (2009) Immunological and inflammatory functions of the interleukin-1 family. Annu Rev Immunol 27:519–550. https://doi.org/10.1146/annurev.immunol.021908.132612
Dinarello CA (2011) A clinical perspective of IL-1beta as the gatekeeper of inflammation. Eur J Immunol 41:1203–1217. https://doi.org/10.1002/eji.201141550
Dinarello CA (2013) Overview of the interleukin-1 family of ligands and receptors. Semin Immunol 25:389–393. https://doi.org/10.1016/j.smim.2013.10.001
Dinarello CA, Fantuzzi G (2003) Interleukin-18 and host defense against infection. J Infect Dis 187(Suppl 2):S370–S384. https://doi.org/10.1086/374751
Ding Z, Zhang J, Xu J, Sheng G, Huang G (2013) Propofol administration modulates AQP-4 expression and brain edema after traumatic brain injury. Cell Biochem Biophys 67:615–622. https://doi.org/10.1007/s12013-013-9549-0
Dobrovolskaia MA, Vogel SN (2002) Toll receptors, CD14, and macrophage activation and deactivation by LPS. Microbes Infect 4:903–914
Dolinay T et al (2012) Inflammasome-regulated cytokines are critical mediators of acute lung injury. Am J Respir Crit Care Med 185:1225–1234. https://doi.org/10.1164/rccm.201201-0003OC
Dong Y et al (2016) Melatonin attenuated early brain injury induced by subarachnoid hemorrhage via regulating NLRP3 inflammasome and apoptosis signaling. J Pineal Res 60:253–262. https://doi.org/10.1111/jpi.12300
Dostert C et al (2008) Innate immune activation through Nalp3 inflammasome sensing of asbestos and silica. Science 320:674–677. https://doi.org/10.1126/science.1156995
Ertel W et al (1995) Downregulation of proinflammatory cytokine release in whole blood from septic patients. Blood 85:1341–1347
Fahy RJ et al (2008) Inflammasome mRNA expression in human monocytes during early septic shock. Am J Respir Crit Care Med 177:983–988. https://doi.org/10.1164/rccm.200703-418OC

Feng L et al (2015) P2X7R blockade prevents NLRP3 inflammasome activation and brain injury in a rat model of intracerebral hemorrhage: involvement of peroxynitrite. J Neuroinflammation 12:190. https://doi.org/10.1186/s12974-015-0409-2

Fernandes-Alnemri T, Yu JW, Datta P, Wu J, Alnemri ES (2009) AIM2 activates the inflammasome and cell death in response to cytoplasmic DNA. Nature 458:509–513. https://doi.org/10.1038/nature07710

Flach R et al (1999) Relation of ex vivo stimulated blood cytokine synthesis to post-traumatic sepsis. Cytokine 11:173–178. https://doi.org/10.1006/cyto.1998.0412

Frank MG, Weber MD, Watkins LR, Maier SF (2015) Stress sounds the alarmin: the role of the danger-associated molecular pattern HMGB1 in stress-induced neuroinflammatory priming. Brain Behav Immun 48:1–7. https://doi.org/10.1016/j.bbi.2015.03.010

Frink M et al (2009) IL-6 predicts organ dysfunction and mortality in patients with multiple injuries. Scand J Trauma Resusc Emerg Med 17:49. https://doi.org/10.1186/1757-7241-17-49

Galbraith N, Walker S, Galandiuk S, Gardner S, Polk HC Jr (2016) The significance and challenges of monocyte impairment: for the ill patient and the surgeon. Surg Infect 17:303–312. https://doi.org/10.1089/sur.2015.245

Giannoudis PV, Pape HC (2007) Trauma and immune reactivity: too much, or too little immune response? Injury 38:1333–1335. https://doi.org/10.1016/j.injury.2007.10.007

Grailer JJ et al (2014) Critical role for the NLRP3 inflammasome during acute lung injury. J Immunol 192:5974–5983. https://doi.org/10.4049/jimmunol.1400368

Grisanti LA et al (2011) Alpha1-adrenergic receptors positively regulate Toll-like receptor cytokine production from human monocytes and macrophages. J Pharmacol Exp Ther 338:648–657. https://doi.org/10.1124/jpet.110.178012

Gurcel L, Abrami L, Girardin S, Tschopp J, van der Goot FG (2006) Caspase-1 activation of lipid metabolic pathways in response to bacterial pore-forming toxins promotes cell survival. Cell 126:1135–1145. https://doi.org/10.1016/j.cell.2006.07.033

Halle A et al (2008) The NALP3 inflammasome is involved in the innate immune response to amyloid-beta. Nat Immunol 9:857–865. https://doi.org/10.1038/ni.1636

Han S et al (2015) ROS-mediated NLRP3 inflammasome activity is essential for burn-induced acute lung injury. Mediat Inflamm:720457. https://doi.org/10.1155/2015/720457(2015).

Hazeldine J, Hampson P, Lord JM (2016) The diagnostic and prognostic value of systems biology research in major traumatic and thermal injury: a review. Burns Trauma 4:33. https://doi.org/10.1186/s41038-016-0059-3

He Q et al (2012) HMGB1 promotes the synthesis of pro-IL-1beta and pro-IL-18 by activation of p38 MAPK and NF-kappaB through receptors for advanced glycation end-products in macrophages. Asian Pac J Cancer Prev 13:1365–1370

He Y, Franchi L, Nunez G (2013) TLR agonists stimulate Nlrp3-dependent IL-1beta production independently of the purinergic P2X7 receptor in dendritic cells and in vivo. J Immunol 190:334–339. https://doi.org/10.4049/jimmunol.1202737

Heizmann O, Koeller M, Muhr G, Oertli D, Schinkel C (2008) Th1- and Th2-type cytokines in plasma after major trauma. J Trauma 65:1374–1378. https://doi.org/10.1097/TA.0b013e31818b257d

Hoegen T et al (2011) The NLRP3 inflammasome contributes to brain injury in pneumococcal meningitis and is activated through ATP-dependent lysosomal cathepsin B release. J Immunol 187:5440–5451. https://doi.org/10.4049/jimmunol.1100790

Holloway TL, Nicholson SE, Rani M, Cap AP, Schwacha MG (2016) Toll-like receptor responses are suppressed in trauma ICU patients. J Surg Res 206:139–145. https://doi.org/10.1016/j.jss.2016.06.056

Hornung V et al (2009) AIM2 recognizes cytosolic dsDNA and forms a caspase-1-activating inflammasome with ASC. Nature 458:514–518. https://doi.org/10.1038/nature07725

Horstmann JP, Marzi I, Relja B (2016) Adrenergic stimulation alters the expression of inflammasome components and interleukins in primary human monocytes. Exp Ther Med 11:297–302. https://doi.org/10.3892/etm.2015.2850

Hosseinian N, Cho Y, Lockey RF, Kolliputi N (2015) The role of the NLRP3 inflammasome in pulmonary diseases. Ther Adv Respir Dis 9:188–197. https://doi.org/10.1177/1753465815586335

Hotchkiss RS, Monneret G, Payen D (2013a) Immunosuppression in sepsis: a novel understanding of the disorder and a new therapeutic approach. Lancet Infect Dis 13:260–268. https://doi.org/10.1016/S1473-3099(13)70001-X

Hotchkiss RS, Monneret G, Payen D (2013b) Sepsis-induced immunosuppression: from cellular dysfunctions to immunotherapy. Nat Rev Immunol 13:862–874. https://doi.org/10.1038/nri3552

Hsu LC et al (2008) A NOD2-NALP1 complex mediates caspase-1-dependent IL-1beta secretion in response to Bacillus anthracis infection and muramyl dipeptide. Proc Natl Acad Sci U S A 105:7803–7808. https://doi.org/10.1073/pnas.0802726105

Impellizzeri D et al (2017) N-Palmitoylethanolamine-oxazoline (PEA-OXA) as a new therapeutic strategy to control neuroinflammation: neuroprotective effects in experimental models of spinal cord and brain injury. J Neurotrauma. https://doi.org/10.1089/neu.2016.4808

Islam MN, Bradley BA, Ceredig R (2016) Sterile post-traumatic immunosuppression. Clin Transl Immunol 5:e77. https://doi.org/10.1038/cti.2016.13

Janak JC et al (2017) Urinary biomarkers are associated with severity and mechanism of injury. Shock 47:593–598. https://doi.org/10.1097/SHK.0000000000000784

Janeway CA Jr, Medzhitov R (2002) Innate immune recognition. Annu Rev Immunol 20:197–216. https://doi.org/10.1146/annurev.immunol.20.083001.084359

Jones HD et al (2014) The NLRP3 inflammasome is required for the development of hypoxemia in LPS/mechanical ventilation acute lung injury. Am J Respir Cell Mol Biol 50:270–280. https://doi.org/10.1165/rcmb.2013-0087OC

Kamo N et al (2013) ASC/caspase-1/IL-1beta signaling triggers inflammatory responses by promoting HMGB1 induction in liver ischemia/reperfusion injury. Hepatology 58:351–362. https://doi.org/10.1002/hep.26320

Keel M et al (1996) Endotoxin tolerance after severe injury and its regulatory mechanisms. J Trauma 41:430–437; discussion 437–438

Kim HY, Kim SJ, Lee SM (2015) Activation of NLRP3 and AIM2 inflammasomes in Kupffer cells in hepatic ischemia/reperfusion. FEBS J 282:259–270. https://doi.org/10.1111/febs.13123

Kirchhoff C et al (2009) Early down-regulation of the pro-inflammatory potential of monocytes is correlated to organ dysfunction in patients after severe multiple injury: a cohort study. Crit Care 13:R88. https://doi.org/10.1186/cc7914

Kovarova M et al (2012) NLRP1-dependent pyroptosis leads to acute lung injury and morbidity in mice. J Immunol 189:2006–2016. https://doi.org/10.4049/jimmunol.1201065

Kuipers MT et al (2012) Ventilator-induced lung injury is mediated by the NLRP3 inflammasome. Anesthesiology 116:1104–1115. https://doi.org/10.1097/ALN.0b013e3182518bc0

Kwak A, Lee Y, Kim H, Kim S (2016) Intracellular interleukin (IL)-1 family cytokine processing enzyme. Arch Pharm Res 39:1556–1564. https://doi.org/10.1007/s12272-016-0855-0

Land WG (2013) Transfusion-related acute lung injury: the work of DAMPs. Transfus Med Hemother 40:3–13. https://doi.org/10.1159/000345688

Latz E, Xiao TS, Stutz A (2013) Activation and regulation of the inflammasomes. Nat Rev Immunol 13:397–411. https://doi.org/10.1038/nri3452

Lebel-Binay S et al (2000) Interleukin-18: biological properties and clinical implications. Eur Cytokine Netw 11:15–26

Leentjens J, Kox M, van der Hoeven JG, Netea MG, Pickkers P (2013) Immunotherapy for the adjunctive treatment of sepsis: from immunosuppression to immunostimulation. Time for a paradigm change? Am J Respir Crit Care Med 187:1287–1293. https://doi.org/10.1164/rccm.201301-0036CP

Lendemans S et al (2007) Toll-like receptor 2 and 4 expression after severe injury is not involved in the dysregulation of the innate immune system. J Trauma 63:740–746. https://doi.org/10.1097/01.ta.0000240451.42238.d1

Li J et al (2016) Minocycline protects against NLRP3 inflammasome-induced inflammation and P53-associated apoptosis in early brain injury after subarachnoid hemorrhage. Mol Neurobiol 53:2668–2678. https://doi.org/10.1007/s12035-015-9318-8

Lin WP et al (2016) Heme oxygenase-1 promotes neuron survival through down-regulation of neuronal NLRP1 expression after spinal cord injury. J Neuroinflammation 13:52. https://doi.org/10.1186/s12974-016-0521-y

Liu HD et al (2013) Expression of the NLRP3 inflammasome in cerebral cortex after traumatic brain injury in a rat model. Neurochem Res 38:2072–2083. https://doi.org/10.1007/s11064-013-1115-z

Liu T et al (2016) Blocking triggering receptor expressed on myeloid cells-1 attenuates lipopolysaccharide-induced acute lung injury via inhibiting NLRP3 inflammasome activation. Sci Rep 6:39473. https://doi.org/10.1038/srep39473

Long H, Xu B, Luo Y, Luo K (2016) Artemisinin protects mice against burn sepsis through inhibiting NLRP3 inflammasome activation. Am J Emerg Med 34:772–777. https://doi.org/10.1016/j.ajem.2015.12.075

Lu B et al (2012) Novel role of PKR in inflammasome activation and HMGB1 release. Nature 488:670–674. https://doi.org/10.1038/nature11290

Ma J, Dong Z, Li QG, Wang JR (2009) Protective effect of propofol against intracerebral hemorrhage injury in rats. Yao Xue Xue Bao 44:344–349

Ma J et al (2016) Propofol inhibits NLRP3 inflammasome and attenuates blast-induced traumatic brain injury in rats. Inflammation 39:2094–2103. https://doi.org/10.1007/s10753-016-0446-8

Manson J, Thiemermann C, Brohi K (2012) Trauma alarmins as activators of damage-induced inflammation. Br J Surg 99(Suppl 1):12–20. https://doi.org/10.1002/bjs.7717

Marik PE, Flemmer M (2012) The immune response to surgery and trauma: implications for treatment. J Trauma Acute Care Surg 73:801–808. https://doi.org/10.1097/TA.0b013e318265cf87

Martinon F, Burns K, Tschopp J (2002) The inflammasome: a molecular platform triggering activation of inflammatory caspases and processing of proIL-beta. Mol Cell 10:417–426

Martinon F, Petrilli V, Mayor A, Tardivel A, Tschopp J (2006) Gout-associated uric acid crystals activate the NALP3 inflammasome. Nature 440:237–241. https://doi.org/10.1038/nature04516

Martinon F, Mayor A, Tschopp J (2009) The inflammasomes: guardians of the body. Annu Rev Immunol 27:229–265. https://doi.org/10.1146/annurev.immunol.021908.132715

Maslanik T et al (2013) The inflammasome and danger associated molecular patterns (DAMPs) are implicated in cytokine and chemokine responses following stressor exposure. Brain Behav Immun 28:54–62. https://doi.org/10.1016/j.bbi.2012.10.014

Matzinger P (2002) The danger model: a renewed sense of self. Science 296:301–305. https://doi.org/10.1126/science.1071059

Meisner M, Adina H, Schmidt J (2006) Correlation of procalcitonin and C-reactive protein to inflammation, complications, and outcome during the intensive care unit course of multiple-trauma patients. Crit Care 10:R1. https://doi.org/10.1186/cc3910

Mendes ME et al (2011) Differential expression of toll-like receptor signaling cascades in LPS-tolerant human peripheral blood mononuclear cells. Immunobiology 216:285–295. https://doi.org/10.1016/j.imbio.2010.07.008

Menzel CL et al (2011) Caspase-1 is hepatoprotective during trauma and hemorrhagic shock by reducing liver injury and inflammation. Mol Med 17:1031–1038. https://doi.org/10.2119/molmed.2011.00015

Miao EA et al (2010) Caspase-1-induced pyroptosis is an innate immune effector mechanism against intracellular bacteria. Nat Immunol 11:1136–1142. https://doi.org/10.1038/ni.1960

Minutoli L et al (2016) ROS-mediated NLRP3 inflammasome activation in brain, heart, kidney, and testis ischemia/reperfusion injury. Oxidative Med Cell Longev 2016:2183026. https://doi.org/10.1155/2016/2183026

Mizushima Y, Ueno M, Idoguchi K, Ishikawa K, Matsuoka T (2009) Fever in trauma patients: friend or foe? J Trauma 67:1062–1065. https://doi.org/10.1097/TA.0b013e3181b848fc

Mommsen P et al (2009) Elevated systemic IL-18 and neopterin levels are associated with post-traumatic complications among patients with multiple injuries: a prospective cohort study. Injury 40:528–534. https://doi.org/10.1016/j.injury.2008.08.007

Munoz C et al (1991) Dysregulation of in vitro cytokine production by monocytes during sepsis. J Clin Invest 88:1747–1754. https://doi.org/10.1172/JCI115493

Muruve DA et al (2008) The inflammasome recognizes cytosolic microbial and host DNA and triggers an innate immune response. Nature 452:103–107. https://doi.org/10.1038/nature06664

Netea MG et al (2009) Differential requirement for the activation of the inflammasome for processing and release of IL-1beta in monocytes and macrophages. Blood 113:2324–2335. https://doi.org/10.1182/blood-2008-03-146720

Oberholzer A et al (2000) Sepsis after severe injury interrupts caspase-dependent processing of interleukin-18. J Trauma 49:11–16; discussion 16–17

Osuka A, Hanschen M, Stoecklein V, Lederer JA (2012) A protective role for inflammasome activation following injury. Shock 37:47–55. https://doi.org/10.1097/SHK.0b013e318234f7ff

Osuka A, Ogura H, Ueyama M, Shimazu T, Lederer JA (2014) Immune response to traumatic injury: harmony and discordance of immune system homeostasis. Acute Med Surg 1(6):63–69

Pape HC et al (2007) Assessment of the clinical course with inflammatory parameters. Injury 38:1358–1364. https://doi.org/10.1016/j.injury.2007.09.026

Pfeifer R, Tarkin IS, Rocos B, Pape HC (2009) Patterns of mortality and causes of death in polytrauma patients--has anything changed? Injury 40:907–911. https://doi.org/10.1016/j.injury.2009.05.006

Proell M, Gerlic M, Mace PD, Reed JC, Riedl SJ (2013) The CARD plays a critical role in ASC foci formation and inflammasome signalling. Biochem J 449:613–621. https://doi.org/10.1042/BJ20121198

Puren AJ, Fantuzzi G, Dinarello CA (1999) Gene expression, synthesis, and secretion of interleukin 18 and interleukin 1beta are differentially regulated in human blood mononuclear cells and mouse spleen cells. Proc Natl Acad Sci U S A 96:2256–2261

Rathinam VA, Fitzgerald KA (2016) Inflammasome complexes: emerging mechanisms and effector functions. Cell 165:792–800. https://doi.org/10.1016/j.cell.2016.03.046

Reikeras O (2010) Immune depression in musculoskeletal trauma. Inflamm Res 59:409–414. https://doi.org/10.1007/s00011-010-0167-7

Relja B et al (2015) Nlrp1 inflammasome is downregulated in trauma patients. J Mol Med (Berl) 93:1391–1400. https://doi.org/10.1007/s00109-015-1320-0

Roetman B et al (2008) Elevated systemic interleukin-18 in multiple injured patients is not related to clinical outcome. J Interf Cytokine Res 28:741–747. https://doi.org/10.1089/jir.2008.0029

Schroder K, Tschopp J (2010) The inflammasomes. Cell 140:821–832. https://doi.org/10.1016/j.cell.2010.01.040

Sperry JL et al (2008) Male gender is associated with excessive IL-6 expression following severe injury. J Trauma 64:572–578; discussion 578–579. doi:https://doi.org/10.1097/TA.0b013e3181650fdf

Spolarics Z et al (2003) Depressed interleukin-12-producing activity by monocytes correlates with adverse clinical course and a shift toward Th2-type lymphocyte pattern in severely injured male trauma patients. Crit Care Med 31:1722–1729. https://doi.org/10.1097/01.CCM.0000063579.43470.AA

Stanojcic M et al (2014) Leukocyte infiltration and activation of the NLRP3 inflammasome in white adipose tissue following thermal injury. Crit Care Med 42:1357–1364. https://doi.org/10.1097/CCM.0000000000000209

Starzl R et al (2015) Cardiac arrest disrupts caspase-1 and patterns of inflammatory mediators differently in skin and muscle following localized tissue injury in rats: insights from data-driven modeling. Front Immunol 6:587. https://doi.org/10.3389/fimmu.2015.00587

Sun Q et al (2013) Caspase 1 activation is protective against hepatocyte cell death by up-regulating beclin 1 protein and mitochondrial autophagy in the setting of redox stress. J Biol Chem 288:15947–15958. https://doi.org/10.1074/jbc.M112.426791

Sun Q et al (2017) Redox-dependent regulation of hepatocyte absent in melanoma 2 inflammasome activation in sterile liver injury in mice. Hepatology 65:253–268. https://doi.org/10.1002/hep.28893

The global burden of disease (2013) World Health Organization. http://www.who.int/healthinfo/global_burden_disease/en/
Tomura S, de Rivero Vaccari JP, Keane RW, Bramlett HM, Dietrich WD (2012) Effects of therapeutic hypothermia on inflammasome signaling after traumatic brain injury. J Cereb Blood Flow Metab 32:1939–1947. https://doi.org/10.1038/jcbfm.2012.99
van de Veerdonk FL, Netea MG, Dinarello CA, Joosten LA (2011) Inflammasome activation and IL-1beta and IL-18 processing during infection. Trends Immunol 32:110–116. https://doi.org/10.1016/j.it.2011.01.003
van Griensven M (2014) Cytokines as biomarkers in polytraumatized patients. Unfallchirurg 117: 699–702. https://doi.org/10.1007/s00113-013-2543-6
Vande Walle L, Lamkanfi M (2016) Pyroptosis. Curr Biol 26:R568–R572. https://doi.org/10.1016/j.cub.2016.02.019
Vigano E, Mortellaro A (2013) Caspase-11: the driving factor for noncanonical inflammasomes. Eur J Immunol 43:2240–2245. https://doi.org/10.1002/eji.201343800
Vladimer GI, Marty-Roix R, Ghosh S, Weng D, Lien E (2013) Inflammasomes and host defenses against bacterial infections. Curr Opin Microbiol 16:23–31. https://doi.org/10.1016/j.mib.2012.11.008
Wafaisade A et al (2011) Epidemiology and risk factors of sepsis after multiple trauma: an analysis of 29,829 patients from the Trauma Registry of the German Society for Trauma Surgery. Crit Care Med 39:621–628. https://doi.org/10.1097/CCM.0b013e318206d3df
Wanner GA et al (1996) Liver ischemia and reperfusion induces a systemic inflammatory response through Kupffer cell activation. Shock 5:34–40
Wawrocki S, Druszczynska M, Kowalewicz-Kulbat M, Rudnicka W (2016) Interleukin 18 (IL-18) as a target for immune intervention. Acta Biochim Pol 63:59–63. https://doi.org/10.18388/abp.2015_1153
Wegiel B, Hauser CJ, Otterbein LE (2015) Heme as a danger molecule in pathogen recognition. Free Radic Biol Med 89:651–661. https://doi.org/10.1016/j.freeradbiomed.2015.08.020
Wen AQ et al (2010) Clinical relevance of IL-1beta promoter polymorphisms (−1470, −511, and −31) in patients with major trauma. Shock 33:576–582. https://doi.org/10.1097/SHK.0b013e3181cc0a8e
Weng X et al (2015) N-methyl-D-aspartic acid receptor 1 (NMDAR1) aggravates secondary inflammatory damage induced by hemin-NLRP3 pathway after intracerebral hemorrhage. Chin J Traumatol 18:254–258
Wutzler S et al (2009) Suppression and recovery of LPS-stimulated monocyte activity after trauma is correlated with increasing injury severity: a prospective clinical study. J Trauma 66: 1273–1280. https://doi.org/10.1097/TA.0b013e3181968054
Wutzler S, Lustenberger T, Relja B, Lehnert M, Marzi I (2013) Pathophysiology of multiple trauma: intensive care medicine and timing of treatment. Chirurg 84:753–758. https://doi.org/10.1007/s00104-013-2477-0
Xiang M et al (2011) Hemorrhagic shock activation of NLRP3 inflammasome in lung endothelial cells. J Immunol 187:4809–4817. https://doi.org/10.4049/jimmunol.1102093
Xiao W et al (2011) A genomic storm in critically injured humans. J Exp Med 208:2581–2590. https://doi.org/10.1084/jem.20111354
Xiao M et al (2016) 3,4-Methylenedioxy-beta-nitrostyrene ameliorates experimental burn wound progression by inhibiting the NLRP3 inflammasome activation. Plast Reconstr Surg 137: 566e–575e. https://doi.org/10.1097/01.prs.0000479972.06934.83
Yamasaki K et al (2009) NLRP3/cryopyrin is necessary for interleukin-1beta (IL-1beta) release in response to hyaluronan, an endogenous trigger of inflammation in response to injury. J Biol Chem 284:12762–12771. https://doi.org/10.1074/jbc.M806084200
Yang F et al (2014) NLRP3 deficiency ameliorates neurovascular damage in experimental ischemic stroke. J Cereb Blood Flow Metab 34:660–667. https://doi.org/10.1038/jcbfm.2013.242
Yang J et al (2016) Hemorrhagic shock primes for lung vascular endothelial cell pyroptosis: role in pulmonary inflammation following LPS. Cell Death Dis 7:e2363. https://doi.org/10.1038/cddis.2016.274
Yang J, Liu Z, Xiao TS (2017) Post-translational regulation of inflammasomes. Cell Mol Immunol 14:65–79. https://doi.org/10.1038/cmi.2016.29

Yuan B et al (2015) Recombinant adenovirus encoding NLRP3 RNAi attenuate inflammation and brain injury after intracerebral hemorrhage. J Neuroimmunol 287:71–75. https://doi.org/10.1016/j.jneuroim.2015.08.002
Zedler S, Faist E (2006) The impact of endogenous triggers on trauma-associated inflammation. Curr Opin Crit Care 12:595–601. https://doi.org/10.1097/MCC.0b013e3280106806
Zhang AQ et al (2011) Clinical relevance of single nucleotide polymorphisms within the entire NLRP3 gene in patients with major blunt trauma. Crit Care 15:R280. https://doi.org/10.1186/cc10564
Zhou R, Tardivel A, Thorens B, Choi I, Tschopp J (2010) Thioredoxin-interacting protein links oxidative stress to inflammasome activation. Nat Immunol 11:136–140. https://doi.org/10.1038/ni.1831
Zhu P et al (2011) Gene silencing of NALP3 protects against liver ischemia-reperfusion injury in mice. Hum Gene Ther 22:853–864. https://doi.org/10.1089/hum.2010.145

Chapter 6
Inflammasome in the Pathogenesis of Pulmonary Diseases

Fengying Xu, Zongmei Wen, Xueying Shi, and Jie Fan

Contents

F. Xu
Department of Anesthesiology, No. 401 Hospital, Qingdao, China

Department of Surgery, University of Pittsburgh School of Medicine, Pittsburgh, USA

Z. Wen
Department of Anesthesiology, Shanghai Pulmonary Hospital, Tongji University School of Medicine, Shanghai, China

X. Shi
Department of Anesthesiology and Critical Care, Shanghai Xinghua Hospital, Jiaotong University School of Medicine, Shanghai, China

J. Fan (✉)
Department of Surgery, University of Pittsburgh School of Medicine, Pittsburgh, USA

Research and Development, Veterans Affairs Pittsburgh Healthcare System, Pittsburgh, USA
e-mail: jif7@pitt.edu

M. D. Cordero, E. Alcocer-Gómez (eds.), *Inflammasomes: Clinical and Therapeutic Implications*, Experientia Supplementum 108,
https://doi.org/10.1007/978-3-319-89390-7_6

Abstract Lung diseases are common and significant causes of illness and death around the world. Inflammasomes have emerged as an important regulator of lung diseases. The important role of IL-1 beta and IL-18 in the inflammatory response of many lung diseases has been elucidated. The cleavage to turn IL-1 beta and IL-18 from their precursors into the active forms is tightly regulated by inflammasomes. In this chapter, we structurally review current evidence of inflammasome-related components in the pathogenesis of acute and chronic lung diseases, focusing on the "inflammasome-caspase-1-IL-1 beta/IL-18" axis.

Keywords Inflammasomes · Infectious pulmonary diseases · Lung injury · Smoking · Chronic obstructive pulmonary disease

6.1 Introduction

In serving its primary function in gas exchange, the lung is constantly exposed to the outside world and is highly susceptible to all kinds of foreign matters. Lung diseases are common and significant causes of illness and death around the world. According to the World Health Organization (WHO) (http://www.who.int/mediacentre/factsheets/fs310/en/), lower respiratory infections, chronic obstructive pulmonary disease (COPD), and tuberculosis are among the top 10 causes of death worldwide. Up to 7.73 million people died of these three diseases, accounting for nearly 14% of all death in 2015, not to mention other pulmonary diseases. In addition, lower respiratory infection is the leading cause of death in low-income economies.

The lack of effective treatment for many pulmonary diseases is at least partly due to our limited understanding of the pathobiology of these diseases. The lung has a defense mechanism consisting of physical barriers and immune cells against infection and injury. Upon insult, such as infection or tissue injury, the innate and adaptive immune system in the lung initiate a series of responses, followed by a period of normalization to restore homeostasis in the lung. Inflammation is one of the immediate responses of the innate immune system, in which cytokines constitute a significant part (Shaikh 2011). Interleukin (IL)-1 beta and its isoform IL-1 alpha are proinflammatory cytokines that exert pleiotropic effects on a variety of cells and play a vital role in acute and chronic inflammatory processes (Ren and Torres 2009). IL-1 signal acting through the type 1 receptor IL-1R1, with the help of IL-1 receptor accessory protein, activates transcription factor nuclear factor-kappa B (NF-kappaB) and activator protein 1 (AP-1). Binding of IL-1 to type 2 receptor IL-1Ra does not lead to downstream signaling, and IL-1Ra is therefore considered a decoy receptor. IL-1 beta has important homeostatic functions under normal circumstances, while its overproduction is implicated in the pathophysiological changes in diverse disease states. IL-18 is another member of the IL-1 family. The IL-18 receptor (IL-18R) is a heterodimer consisting of IL-18R alpha and beta chains. IL-18 also mediates responses and activates NF-kappa B and AP-1, resulting in the production of IFN-gamma, essential for immunity against invading pathogens. In addition, IL-18 can act as a Th2 response inducer in some allergic diseases (Sedimbi et al. 2013).

The important role of IL-1 beta and IL-18 in the inflammatory response of many diseases has been elucidated. The cleavage to turn IL-1 beta and IL-18 from their precursors into the active forms is tightly regulated. Over the past decade, researchers have found that inflammasome is the key component for this process, therefore, critical for the induction of a proper inflammatory response. The core sensor protein of inflammasome comes from the nucleotide-binding oligomerization domain (NOD)-like receptors (NLRs) family. All NLR family members are classified into subfamilies based on the N-terminal domain they contain: NLRAs have transactivator activation domains (ADs); NLRBs have BIR (baculoviral inhibitor of apoptosis repeat) domains; NLRCs have CARD (caspase activation/recruitment domains); and NLRPs have PYD (pyrin domains) (Leissinger et al. 2014). Of these, at least four NLR members (NLRP1, NLRP3, NLRC4, NLRB1) and absent in melanoma 2 (AIM2) can form inflammasome complexes with the adaptor protein apoptosis-associated speck-like protein containing caspase-recruitment domain (ASC) and pro-caspase-1. Inflammasome formation results in the catalysis and activation of caspase-1, which in turn catalyzes the cytokine precursors (Lee et al. 2016).

The detailed description of inflammasomes can be found elsewhere in this book. Normally, inflammasome formation requires a canonical two-step mechanism. Taking the NLRP3 inflammasome as an example: the first signal (e.g., TLR4 or other pattern recognition receptors) stimulates NF-kappa B and enhanced the expression and synthesis of NLRP3. The second signal induced NLRP3 inflammasome assembly. Common signal is provided by P2X purinoreceptor 7 (P2X7R) bond by adenosine triphosphate (ATP), K^+ efflux, lysosome destabilization caused by urate crystals, DNA and reactive oxygen species (ROS) generated in mitochondria, etc.

The discovery of inflammasome has changed our understanding of the pathogenesis of many diseases. Here we structurally summarize current evidence for the involvement of inflammasome in the pathogenesis of acute and chronic lung diseases, focusing on the "inflammasome-caspase-1-IL-1 beta/IL-18" axis.

6.2 Acute Lung Diseases

6.2.1 Infectious Pulmonary Diseases

Despite sophisticated advances in antibiotics, lung infection remains a significant cause of morbidity and mortality. As multidrug-resistant bacteria are emerging, it is a priority to better understand the mechanisms how our immune system combat pathogens. Researches involving the discussion of inflammasome in infectious pulmonary diseases are summarized and listed in Table 6.1 according to publish date. The term "conflicting" in the column of "Contribution" means that not all the element studied was reported to take effect in a particular article. These terms are consistent throughout this chapter.

Table 6.1 Articles containing discussions of inflammasomes in infectious pulmonary disease

ID	Species	Pathogen	Design	Element	Contribution
Bhakdi 1989	*Homo sapiens*	*Staphylococcus aureus*	In vitro, monocyte	IL-1 beta	Yes
Shimokata 1991	*Homo sapiens*	*Mycobacterium tuberculosis*	Ex vivo, pleural fluid	IL-1	Yes
Chensue 1992	*Homo sapiens*	*Mycobacterium tuberculosis*	Ex vivo, granulomas	IL-1 beta	Yes
Hennet 1992	*Mus musculus*	Influenza A virus	In vivo, BALF	IL-1 alpha, IL-1 beta	Yes
Jonas 1994	*Homo sapiens*	*Staphylococcus aureus*	In vitro, T lymphocyte	K^+ efflux	Yes
Law 1996	*Homo sapiens*	*Mycobacterium tuberculosis*	Ex vivo, cells in BALF	IL-1 beta	Yes
Kaukoranta-Tolvanen 1996	*Homo sapiens*	*Chlamydia pneumoniae*	In vitro, macrophage	IL-1 beta	Yes
Kawakami 1997	*Mus musculus*	*Cryptococcus neoformans*	In vivo, lung	IL-18	Yes
Bellamy 1998	*Homo sapiens*	*Mycobacterium tuberculosis*	Ex vivo, blood	IL-1R1, IL-1 alpha, IL-1 beta	Conflicting
Sareneva 1998	*Homo sapiens*	Influenza A virus	In vitro, macrophage	IL-1 beta, IL-18	Yes
Pirhonen 1999	*Homo sapiens*	Influenza A virus	In vitro, monocyte, macrophage	IL-1 beta, IL-18	Yes
Qureshi 1999	*Mus musculus*	*Cryptococcus neoformans*	In vivo, lung	IL-18	Yes
Sugawara 1999	*Mus musculus*	*Mycobacterium tuberculosis*	In vivo, lung	IL-18	Yes
Wilkinson 1999	*Homo sapiens*	*Mycobacterium tuberculosis*	Ex vivo, blood	IL-1Ra, IL-1 beta	No
Juffermans 2000	*Mus musculus*	*Mycobacterium tuberculosis*	In vivo, lung	IL-1R1	Yes
Kawakami 2000a, b	*Mus musculus*	*Cryptococcus neoformans*	In vivo, lung	IL-18	Yes
Netea 2000	*Homo sapiens*	*Chlamydia pneumoniae*	In vitro, macrophage	IL-1 beta	Yes
Tsao 2000	*Homo sapiens*	*Mycobacterium tuberculosis*	Ex vivo, BALF	IL-1 beta	Yes
Yamada 2000	*Mus musculus*	*Mycobacterium tuberculosis*	In vivo, lung	IL-1 alpha, IL-1 beta	Yes
Giacomini 2001	*Homo sapiens*	*Mycobacterium tuberculosis*	In vitro, macrophage, dendritic cell	IL-1	Yes

Sugawara 2001	*Mus musculus*	*Mycobacterium tuberculosis*	In vivo, lung	IL-1R1	Yes
Kinjo 2002	*Mus musculus*	*Mycobacterium tuberculosis*	In vivo, lung	IL-18	Yes
Rupp 2003	*Homo sapiens*	*Chlamydia pneumoniae*	In vitro, macrophage	IL-1 beta, IL-1Ra	Yes
Liu 2004	*Mus musculus*	Influenza A virus	In vivo, lung	IL-18	Yes
Netea 2004	*Homo sapiens* *Mus musculus*	*Chlamydia pneumoniae*	In vitro, macrophage	IL-18	Yes
Schmitz 2005	*Mus musculus*	Influenza virus	In vivo, lung	IL-1R1	Yes
Stasakova 2005	*Homo sapiens*	Influenza A virus	In vitro, macrophage	IL-1 beta, IL-18, caspase-1	Yes
Amer 2006	*Mus musculus*	*Legionella pneumophila*	In vitro, macrophage	NLRC4, caspase-1	Yes
Denton 2007	*Mus musculus*	Influenza A virus	In vivo, lung	IL-18	Yes
Fremond 2007	*Mus musculus*	*Mycobacterium tuberculosis*	In vivo, lung; in vitro, macrophage, dendritic cell	IL-1R, IL-18	Conflicting
Reiniger 2007	*Mus musculus* *Homo sapiens*	*Pseudomonas aeruginosa*	In vivo; in vitro, *epithelial cell*	IL-1 beta, IL-1R1	Yes
Simitsopoulou 2007	*Homo sapiens*	*Aspergillus fumigatus*	In vitro, THP-1	IL-1 beta	Yes
Sutterwala 2007	*Mus musculus*	*Pseudomonas aeruginosa*	In vitro, macrophage	NLRC4, caspase-1, IL-1	Yes
Koo 2008	*Mus musculus*	*Mycobacterium tuberculosis*, *Mycobacterium marinum*	In vitro, macrophage	NLRP3, NLRC4, caspase-1, IL-1 beta, IL-18	Conflicting
Master 2008	*Mus musculus*	*Mycobacterium tuberculosis*	In vivo, lung; in vitro, J774A, RAW 264.7	IL-1 beta	No
Miao 2008	*Mus musculus*	*Pseudomonas aeruginosa*	In vitro, macrophage	NLRC4, caspase-1, IL-1 beta	Yes
Wolk 2008	*Mus musculus*	Influenza A virus	In vivo, BALF	ATP	Yes
Allen 2009	*Mus musculus*	Influenza A virus	In vivo, lung; in vitro, BMDM, EC	NLRP3, NLRC4, ROS, caspase-1	Yes
Case 2009	*Mus musculus*	*Legionella pneumophila*	In vivo, lung; in vitro, macrophage	NLRP3, NLRC4, ASC, IL-1 beta, IL-18	Conflicting

(continued)

Table 6.1 (continued)

ID	Species	Pathogen	Design	Element	Contribution
Craven 2009	*Homo sapiens, Mus musculus*	*Staphylococcus aureus*	In vitro, monocyte	NLRP3, caspase-1, IL-1 beta, IL-18	Yes
Harder 2009	*Mus musculus*	*Streptococcus pyogenes*	In vitro, macrophage	NLRP3, ASC, caspase-1, IL-1 beta, ATP, P2X7R	Conflicting
Ichinohe 2009	*Mus musculus*	Influenza virus	In vivo, lung; in vitro, $CD4^+$ and $CD8^+$ T cells	NLRP3, ASC, caspase-1	Conflicting
Kleinnijenhuis 2009	*Homo sapiens*	*Mycobacterium tuberculosis*	In vitro, macrophage	IL-1 beta, caspase-1, ATP, P2X7R	Conflicting
Kurenuma 2009	*Mus musculus*	*Mycobacterium tuberculosis*	In vitro, macrophage	Caspase-1, IL-1 beta, IL-18, K^+ efflux, P2X7R	Conflicting
Munoz-Planillo 2009	*Mus musculus*	*Staphylococcus aureus*	In vitro, macrophage	NLRP3, caspase-1, IL-1 beta, IL-18, ATP, P2X7R	Conflicting
Thomas 2009	*Mus musculus*	Influenza A virus	In vivo, lung	NLRP3, caspase-1	Yes
Willingham 2009	*Mus musculus, Homo sapiens*	*Klebsiella pneumoniae*	In vitro, THP-1; in vivo, lung	NLRP3, ASC, IL-1 beta, IL-18 NLRC4	Conflicting
Carlsson 2010	*Mus musculus*	*Mycobacterium marinum*	In vivo, lung	NLRP3, ASC, IL-1 beta	Yes
Ichinohe 2010	*Mus musculus*	Influenza A virus	In vitro, macrophage	NLRP3	Yes
Mayer-Barber 2010	*Mus musculus*	*Mycobacterium tuberculosis*	In vivo, lung	IL-1 beta, IL-1R1	Yes
McElvania Tekippe 2010	*Mus musculus, Homo sapiens*	*Mycobacterium tuberculosis*	In vivo, lung; in vitro, macrophage	ASC, caspase-1, NLRP3, NLRC4, IL-1 beta	Conflicting
McNeela 2010	*Mus musculus*	*Streptococcus pneumonia*	In vitro, dendritic cell	NLRP3, IL-1 beta	Yes
Mishra 2010	*Mus musculus*	*Mycobacterium tuberculosis*	In vitro, macrophage	NLRP3, IL-1 beta	Yes
Said-Sadier 2010	*Homo sapiens*	*Aspergillus fumigatus*	In vitro, THP-1	NLRP3, caspase-1, IL-1 beta, K^+ efflux, ROS, Dectin-1, Syk	Yes
Arlehamn 2011	*Mus musculus*	*Pseudomonas aeruginosa*	In vitro, macrophage	NLRC4, caspase-1, IL-1 beta	Yes

Pereira 2011a	*Mus musculus*	*Legionella pneumophila*	In vivo, lung; in vitro, macrophage	NLRC4, caspase-1, IL-1 beta	Yes
Pereira 2011b	*Mus musculus*	*Legionella*	In vivo, lung	NLRC4	Conflicting
Shimada 2011	*Mus musculus*	*Chlamydia pneumoniae*	In vivo, lung	NLRP3, ASC, caspase-1, IL-1 beta, TLR2, MyD88	Yes
Witzenrath 2011	*Mus musculus, Homo sapiens*	*Streptococcus pneumonia*	In vitro, macrophage; in vivo, lung	NLRP3, IL-1 beta	Conflicting
Cai 2012	*Mus musculus, Homo sapiens*	*Klebsiella pneumoniae*	In vitro, macrophage; in vivo, lung	NLRC4, IL-1 beta, IL-1R1	Yes
Chen 2012	*Homo sapiens*	*Mycobacterium kansasii*	In vitro, THP-1	NLRP3, caspase-1, IL-1 beta, K^+ efflux, ROS, lysosome	Yes
Dorhoi 2012	*Mus musculus*	*Mycobacterium tuberculosis*	In vivo, lung; in vitro, macro-phage, dendritic cell, neutrophil	NLRP3, IL-1 beta	Conflicting
Lee 2012	*Homo sapiens*	*Mycobacterium abscessus*	In vitro, macrophage	NLRP3, IL-1 beta, Dectin-1, Syk	Yes
Patankar 2013	*Mus musculus*	*Pseudomonas aeruginosa*	In vitro, macrophage, dendritic cell; in vivo	NLRC4, caspase-1, IL-1 beta	Yes
Rotta Detto Loria 2013	*Homo sapiens*	*Nontypeable Haemophilus influenzae*	In vitro, RAW 264.7, human lung tissue	NLRP3, caspase-1, IL-1 beta, IL-18	Yes
Rimessi 2015	*Homo sapiens*	*Pseudomonas aeruginosa*	In vitro, epithelial cell	NLRP3, caspase-1, mitochondrial Ca^{2+}, IL-1 beta, IL-18,	Yes

Abbreviations in the table are the same as those in the text

6.2.1.1 Influenza Virus

IL-1 Signaling Hennet et al. found an early increase of IL-1 alpha and IL-1 beta peaked between 36 h and 3 days after influenza A virus (IAV) infection in mice (Hennet et al. 1992). Also, IAV-infected human peripheral macrophages secreted IL-1 beta and IL-18 (Sareneva et al. 1998). Similar results were found by Pirhonen et al. with primary human monocytes and differentiated macrophages. Besides, virus-induced IL-1 beta and IL-18 production was significantly blocked by a specific caspase-1 inhibitor (Pirhonen et al. 1999).

The role of IL-1 was further manifested. Schmitz et al. investigated the role of the IL-1R1 signaling during pulmonary antiviral immune responses in $Il1r1^{-/-}$ mice. They demonstrated reduced inflammatory pathology, decreased activation and migration of $CD4^+$ T cells, and greatly diminished immunoglobulin (Ig) M responses in $Il1r1^{-/-}$ mice after influenza virus infection. In contrast, the activation of cytotoxic T lymphocytes and the IgG and IgA antibody responses was intact. Notably, the authors found significantly increased mortality in $Il1r1^{-/-}$ mice after infection (Schmitz et al. 2005).

IL-18 Signaling The role of IL-18 has also been studied in gene knockout mice. $Il18^{-/-}$ mice inoculated with IAV showed increased mortality with the occurrence of pathogenic changes including enhanced virus growth, massive inflammatory infiltration, and elevated nitric oxide production over the first 3 days after respiratory challenge (Liu et al. 2004). Thereafter, Denton et al. found that IL-18 deficiency was associated with delayed virus clearance from the lung and decreased cytokine production by $CD8^+$ T lymphocytes (Denton et al. 2007).

Inflammasome Activation Thomas et al. showed that in vivo activation of the NLRP3 inflammasome by IAV RNA controlled the release of IL-1 beta and IL-18 and modulated the extent of lung pathology. Furthermore $Nlrp3^{-/-}$ and $Casp1^{-/-}$ mice were found more susceptible after IAV infection correlated with decreased cell recruitment and cytokine/chemokine production (Thomas et al. 2009). The in vivo role of NLRP3 inflammasome during influenza virus infection was verified around the same time. Mice lacking Nlrp3, ASC, or caspase-1, but not Nlrc4, exhibited increased mortality and a reduced immune response after influenza virus infection. Using poly I:C and ssRNA40 to analogize virus RNA, the authors concluded that NLRP3 inflammasome could be activated by RNA species dependent on lysosomal maturation and ROS (Allen et al. 2009).

In contrast to studies in certain cell types, like macrophages and epithelial cells, Ichinohe et al. found that ASC and caspase-1, but not NLRP3, were required for $CD4^+$ and $CD8^+$ T cell responses, as well as mucosal IgA secretion and systemic IgG responses to influenza virus infection. This study provided evidence of the requirement for the components of inflammasomes in adaptive immunity to virus infection in vivo (Ichinohe et al. 2009).

The mechanisms by which influenza virus activates the inflammasome have also been studied. Ichinohe et al. showed that the influenza virus M2 protein localized to the Golgi apparatus dependent on the pH gradient to stimulate the NLRP3 inflammasome in primed macrophages and dendritic cells (Ichinohe et al. 2010). Increased ATP has been observed in the bronchoalveolar lavage fluid (BALF) of mice infected with IAV (Wolk et al. 2008), which is usually believed to be a secondary signal in inflammasome activation and assembly.

Interaction Between Host and Microorganism On the other hand, pathogens have also evolved strategies to take advantage of inflammasome-related mechanisms to evade the host immune system. For example, Stasakova et al. reported that several NS1 mutant viruses induced much more biologically active IL-1 beta and IL-18 than wild-type viruses, therefore inducing rapid apoptosis in infected macrophages, which correlated with the enhanced activity of caspase-1 (Stasakova et al. 2005).

6.2.1.2 Mycobacterium

Mycobacterium is a genus of over 150 recognized species, including pathogens known to cause serious diseases in mammals, like tuberculosis (King et al. 2017). Mycobacteria are aerobic and normally known as acid fast. The studies about infection with mycobacteria, in which inflammasome may take a part in were summarized.

IL-1 Signaling Studies in humans have shown that IL-1 was elevated in monocyte-derived macrophages stimulated in vitro (Giacomini et al. 2001), in pleural fluid (Shimokata et al. 1991), in cells obtained from BALF (Law et al. 1996; Tsao et al. 2000), and in granulomas of patients with tuberculosis (Chensue et al. 1992). It was reported that IL-1 beta production was induced by *M. tuberculosis* through pathways involving ERK, p38, and Rip2 after recognition by TLR2/TLR6 and NOD2 receptors (Kleinnijenhuis et al. 2009).

Genetic studies conducted by Bellamy et al. suggested that polymorphisms in IL-1R1 and possibly IL-1 alpha (but not IL-1 beta) significantly associated with tuberculosis (Bellamy et al. 1998). However, Wilkinson et al. reported no allele or genotype in IL-1Ra and IL-1 beta, single or in combination, was associated with an increased risk of tuberculosis (Wilkinson et al. 1999).

In vivo animal studies using Il1r1$^{-/-}$ and IL-1 beta$^{-/-}$ mice displayed acute mortality with increased bacterial burden in the lungs, suggesting an important role for IL-1 beta/IL-1R1 signaling response to *Mycobacterium tuberculosis* (Mayer-Barber et al. 2010). Il1r1$^{-/-}$ mice showed defective granuloma formation containing fewer macrophages and lymphocytes, defective migration of immune cells, and a decrease in IFN-gamma production in the spleen. These changes were associated with increased mortality and an enhanced mycobacterial outgrowth in the lungs and distant organs (Juffermans et al. 2000). IL-1 alpha/beta double knockout mice developed significantly larger granulomas in lungs than wild-type mice after infection with *M. tuberculosis*, suggesting a protective role of IL-1 (Yamada et al. 2000). More precisely, *M. tuberculosis* infection in Il1r1$^{-/-}$ mice led to a profound defect of early control of infection with higher bacterial load in the lung and necrotic pneumonia. While pulmonary CD4^{+} and CD8^{+} T cell responses were unaffected

(Fremond et al. 2007). However, in contrast, Master et al. showed that *M. tuberculosis* prevented inflammasome activation and subsequent IL-1 beta release and zmp1, which encoded a Zn^{2+} metalloprotease, was responsible (Master et al. 2008). Besides, Sugawara reported that $Il1r1^{-/-}$ mice developed significantly larger granulomatous lesions with neutrophil infiltration in the lungs than wild-type mice did, and IFN-gamma production in spleen cells was lower in $Il1r1^{-/-}$ mice (Sugawara et al. 2001).

IL-18 Signaling $Il18^{-/-}$ mice developed marked granulomas compared with wild-type ones after *M. tuberculosis* infection. The granulomatous lesions could be inhibited significantly by exogenous recombinant IL-18. The splenic IFN-gamma levels were also lower in $Il18^{-/-}$ mice (Sugawara et al. 1999). Similarly, $Il18^{-/-}$ mice were more prone to this infection than wild-type mice, and IFN-gamma production was significantly attenuated. Consistently, IL-18 transgenic mice were more resistant to the infection than their littermate mice, and IFN-gamma levels were increased (Kinjo et al. 2002). However, Fremond et al. reported that unlike IL-1 beta, IL-18-dependent pathways seemed to be dispensable in response to *M. tuberculosis* infection (Fremond et al. 2007).

Inflammasome Activation *Mycobacterium tuberculosis* activated the NLRP3 inflammasome and induced a strong IL-1 beta response. The mechanism is not yet fully understood, but it was believed that *M. tuberculosis* induced inflammasome activation involving the export of the 6 kDa early secreted antigenic target (ESAT-6) through a functional protein secretion system ESX-1 (Mishra et al. 2010). The function of ESX-1 in NLRP3 activation was further confirmed by Dorhoi et al. They also concluded that although NLRP3 inflammasome was critical for IL-1 beta secretion in macrophages, $Nlrp3^{-/-}$ mice were not susceptible to *M. tuberculosis* infection, due to NLRP3-independent compensatory IL-1 beta production in lung parenchyma (Dorhoi et al. 2012). Kleinnijenhuis et al. showed that the secretion of IL-1 beta in macrophage depended on the activation of P2X7R by endogenously ATP. However, they also suggested that constitutively expressed caspase-1 in monocyte need not be activated by *M. tuberculosis* (Kleinnijenhuis et al. 2009). Kurenuma et al. found that a genomic locus called "region of difference 1" (RD1) in *Mycobacterium tuberculosis* was essential for the activation of caspase-1 and subsequent secretion of IL-1 beta and IL-18 in macrophages. The activation was induced via RD1-dependent K^+ efflux independent of P2X7R (Kurenuma et al. 2009). While the above experiments were performed in an acute settings, McElvania et al. showed that *M. tuberculosis* induced IL-1 beta secretion in human and mouse macrophages in vitro, depending on ASC, caspase-1, and NLRP3, but not NLCR4. In addition, murine ASC protected the host during chronic *M. tuberculosis* infection, but the effects of caspase-1 and NLRP3 were dispensible (McElvania Tekippe et al. 2010).

Mycobacterium marinum *Mycobacterium marinum* possesses virtually all of the virulence factors associated with *M. tuberculosis*, including the ESX-1 secretion system. Koo et al. identified that NLRP3, caspase-1, ASC, but not NLRC4, were required for the release of IL-1 beta and IL-18 after *M. marinum* or *M. tuberculosis* infection. Mostly important, they showed that mycobacteria-induced ESX-1-dependent lysosome secretion was essential to release, but not to synthesize IL-1 beta and IL-18 in vitro (Koo et al. 2008). In vivo study confirmed the function of ESX-1 secretion system in activating NLRP3 inflammasome. However, the activation of NLRP3 inflammasome did not restrict bacterial growth, indicating a host-detrimental role of this inflammatory pathway in mycobacterial infection (Carlsson et al. 2010).

Mycobacterium abscessus *Mycobacterium abscessus* is one of the common species that causes disseminated infections in patients with cystic fibrosis. It has been reported that NLRP3 inflammasome activation contributed to the antimicrobial responses against *M. abscessus* in human macrophages, and its activation was dependent on dectin-1/Syk signaling (Lee et al. 2012).

Mycobacterium kansasii Live intracellular *Mycobacterium kansasii* has been reported to trigger the activation of the NLRP3 inflammasome, leading to caspase-1 activation and IL-1 beta secretion. Furthermore, K^+ efflux, lysosomal acidification, ROS production, and cathepsin B release played a role in this activation process (Chen et al. 2012).

6.2.1.3 Other Pathogens

Some other pathogens that are common cause of respiratory tract and pulmonary infections are included in this part.

Streptococcus pneumonia *Streptococcus pneumonia* is a frequent colonizer in the upper respiratory tract and a leading cause of infections like pneumonia. McNeela et al. demonstrated that the activation of NLRP3 inflammasome was required for *S. pneumonia* or its virulence factor pneumolysin-mediated enhancement of IL-1 beta secretion in dendritic cells. Furthermore, NLRP3 was required for protective immunity against respiratory infection with *S. pneumonia* (McNeela et al. 2010). Similarly, Witzenrath et al. reported that *S. pneumonia* expressing hemolytic pneumolysin also induced NLRP3-dependent IL-1 beta production in human and murine mononuclear cells. The inflammasome pathway was protective maintaining the pulmonary microvascular barrier. Additionally, the results showed that inflammasome was not activated by bacterial mutants lacking pneumolysin, which could cause invasive disease clinically (Witzenrath et al. 2011).

Staphylococcus aureus Staphylococcal α-hemolysin, an essential virulence factor of *Staphylococcus aureus*, has been shown to be required for the promotion of pneumonia in mouse models. It has long been proven that α-hemolysin could induce IL-1 beta secretion from human monocytes (Bhakdi et al. 1989). Furthermore, α-hemolysin can induce K^+ efflux in host cells (Jonas et al. 1994). Craven et al. demonstrated that α-hemolysin activated the NLRP3 inflammasome resulting in the activation of caspase-1 and secretion of cytokines IL-1 beta and IL-18 in monocyte-derived cells from humans and mice. They also reported that α-hemolysin induced NLRP3-dependent cellular necrosis resulting in the release of endogenous danger-associated molecular patterns (DAMPs) (Craven et al. 2009). Munoz-Planillo et al. further concluded that bacterial lipoproteins released by *S. aureus* were required for NLRP3 and caspase-1 activation triggered by α- and β-hemolysins. Notably, caspase-1 activation was independent of ATP and P2X7R (Munoz-Planillo et al. 2009).

Streptococcus pyogenes Harder et al. found that caspase-1 activation and IL-1 beta secretion were induced by live *Streptococcus pyogenes*. The toxin streptolysin O, NLRP3, and ASC were crucial for the process, while exogenous ATP or the P2X7R was not required (Harder et al. 2009).

Klebsiella pneumoniae NLRP3 inflammasome protected host during infection with *Klebsiella pneumoniae*, as inflammatory response decreased and mortality increased in Nlrp3$^{-/-}$ and Asc$^{-/-}$, but not Nlrc4$^{-/-}$ mice. NLRP3 activated necrosis and triggered HMGB1 release in addition to IL-1 beta as well as IL-18 secretion in macrophages (Willingham et al. 2009). However, NLRC4 has also been found to be of importance for host survival, bacterial clearance, production of IL-1 beta, as well as neutrophil-mediated inflammation following pulmonary *K. pneumoniae* infection. Exogenous IL-1 beta partially rescued survival and restored neutrophil accumulation and cytokine/chemokine expression in the lungs of Nlrc4$^{-/-}$ mice. Furthermore, Il1r1$^{-/-}$ mice displayed a decrease in neutrophilic inflammation after infection (Cai et al. 2012).

Legionella pneumophila NLRC4 is shown to be important in the recognition, response, and resolution of infection with flagellated pathogens. *Legionella pneumophila* is a flagellated, Gram-negative, facultative intracellular pathogen. Amer et al. found that *Legionella*-induced NLRC4-dependent caspase-1 activation to restrict replication in macrophages (Amer et al. 2006). ASC was found to be important for caspase-1 activation during *L. pneumophila* infection. Activation of caspase-1 via ASC did not require sense of flagellin by NLRC4. Besides, activation of caspase-1 in macrophages occurred independently of NLRP3 (Case et al. 2009). Slightly different, Pereira et al. found that NLRC4-dependent growth restriction of *L. pneumophila* was fully due to flagellin. In addition, *L. pneumophila* multiplied better in Nlrc4$^{-/-}$ mice, and macrophages compared with that in caspase-1 deficient ones, suggesting a caspase-1-independent downstream of NLRC4 (Pereira et al. 2011b). The importance of flagellin in activating NLRC4 was further tested, as nonflagellated *Legionella* bypassed the NLRC4 inflammasome-mediated growth restriction (Pereira et al. 2011a). The type 4 secretion system was also suggested to be important for NLRC4- and caspase-1-dependent host response (Silveira and Zamboni 2010).

Chlamydia pneumonia Studies have shown that in vitro *Chlamydia pneumoniae* infection could elicit IL-1 beta and IL-18 secretion (Netea et al. 2000; Kaukoranta-Tolvanen et al. 1996; Netea et al. 2004; Rupp et al. 2003). Shimada et al. demonstrated that *C. pneumoniae* infection in the lung induced NLRP3 inflammasome activation, leading to caspase-1-dependent IL-1 beta secretion. This inflammatory response was critical for host defense against infection, manifested by delayed bacterial clearance and increased mortality in caspase1$^{-/-}$ mice, which could be rescued by recombinant IL-1 beta (Shimada et al. 2011).

Nontypeable Haemophilus influenzae *Nontypeable Haemophilus influenzae* (NTHi) is the most common cause for bacterial exacerbations in COPD. Higher expression of NLRP3 and caspase-1 and a significant induction of IL-1 beta after NTHi stimulation were detected in a murine macrophage cell line. In addition, inhibition of caspase-1 in human lung tissue led to a significant reduction of IL-1 beta and IL-18 (Rotta Detto Loria et al. 2013).

Pseudomonas aeruginosa Reiniger et al. found rapid release of IL-1 beta in response to *Pseudomonas aeruginosa*. And Il1r$^{-/-}$ mice were susceptible to chronic *P. aeruginosa* lung infection (Reiniger et al. 2007). NLRC4 inflammasome was identified critical for optimal bacterial clearance in an in vivo model of lung infection with *P. aeruginosa*. The activation of caspase-1 and secretion of IL-1 beta were triggered by bacterial flagellin and type 3 secretion system (T3SS) (Franchi et al. 2007). The importance of NLRC4 and T3SS was further manifested by Sutterwala et al. and Miao et al. (Miao et al. 2008). Sutterwala et al. also reported that the *P. aeruginosa* strain expressing the exoenzyme U (ExoU, a T3SS effector) phospholipase was able to suppress caspase-1-mediated cytokine production via NLRC4, associated with more severe disease (Sutterwala et al. 2007). Pilin, a major component of the type 4 bacterial pilus, has also been reported to activate NLRC4 inflammasome via the T3SS in *P. aeruginosa* infection (Arlehamn and Evans 2011). Activation of NLRC4 may depend not only on T3SS or flagellin but also on bacterial motility, as caspase-1 activation and IL-1 beta production were reduced when exposed to nonmotile *P. aeruginosa* in macrophages and dendritic cells (Patankar et al. 2013). As an example, the temporal loss of *P. aeruginosa* motility has been described during chronic infections in patients with cystic fibrosis (Luzar et al. 1985; Mahenthiralingam et al. 1994). Recently, Rimessi et al. demonstrated that flagellin of *P. aeruginosa* caused mitochondrial perturbation, which regulated NLRP3 activation and IL-1β and IL-18 processing by mitochondrial Ca2+ in human bronchial epithelial cells (Rimessi et al. 2015).

Cryptococcus neoformans Recombinant IL-18 enhanced the elimination of live *Cryptococcus neoformans* from the lungs, prevented its dissemination to the brain, and increased the survival rate of infected mice. In addition, administration of neutralizing anti-IL-18 antibody exacerbated the infection (Kawakami et al. 1997). They further reported that fungal clearance in the lung was reduced and the levels of IL-12 and IFN-gamma in the sera were significantly lower in Il18$^{-/-}$ mice (Kawakami et al. 2000a). IL-12 and IL-18 have been shown to synergistically increase the fungicidal activity against *C. neoformans*. A single administration of either IL-12 or IL-18 was not effective, while their combination significantly prolonged survival time of infected mice and reduced the fungal growth in lungs (Qureshi et al. 1999). To discriminate the activity of IL-18 from that of counterpart cytokines like IL-12, Kawakami et al. conducted the experiment in IL-12p40$^{-/-}$ mice. Neutralizing anti-IL-18 antibody almost completely abrogated IFN-gamma production, and host response in IL-12p40 and IL-18 double knockout mice was more profoundly impaired than in IL-12p40$^{-/-}$ mice. Moreover, administration of IL-12 as well as IL-18 significantly restored the host resistance (Kawakami et al. 2000b).

Aspergillus fumigatus The release of IL-1 beta was significantly increased from monocytes stimulated with hyphal fragments of *Aspergillus fumigatus* (Simitsopoulou et al. 2007). Further study found that hyphal fragments induced NLRP3 inflammasome assembly, caspase-1 activation, and IL-1 beta release from THP-1 cell line. The activation of NLRP3 required dectin-1/Syk signaling, K^+ efflux, and ROS production (Said-Sadier et al. 2010).

6.2.2 Acute Lung Injury (ALI)

Acute respiratory distress syndrome (ARDS) is the acute onset of hypoxemia with bilateral infiltrates, in the absence of left atrial hypertension. In the 2012 Berlin definition, ALI was reassigned to be a mild type of ARDS. As it is not a single disease, we listed in Table 6.2 the studies on ALI of different causes.

IL-1 Signaling IL-1 beta has been found in BALF from patients with ARDS (Pugin et al. 1996). And it has been previously shown in rats that lung vascular permeability increases after short-term exposure to IL-1 alpha and IL-1 beta (Leff et al. 1994). Ganter et al. demonstrated a role for the alphavbeta5 and alphavbeta6 integrins in mediating IL-1 beta-induced ALI (Ganter et al. 2008).

Inflammasome Activation Our group demonstrated that lipopolysaccharides (LPS) activated NLRP3, enhanced the release of IL-1 beta, and promoted pyroptosis in alveolar macrophages. Meanwhile, IL-1 beta upregulated IL-1R1 through an autocrine mechanism (He et al. 2016). Our group examined the role of the NLRP3 inflammasome in response to hemorrhagic shock in a mouse model of ALI. In our study, pulmonary endothelial cells were the primary source of IL-1 beta secretion after hemorrhagic shock. DAMPs (especially HMGB1) activated NADPH oxidase and caused thioredoxin-interacting protein to associate with NLRP3, leading to inflammasome activation. Notably, endothelial cells were also targets of IL-1 beta, which might cause a range of inflammatory molecules and an amplification of inflammation leading to ALI (Xiang et al. 2011). We further showed that there existed a negative-feedback regulating the activation of inflammasome. While activating NLRP3 inflammasome, LPS also induced pyrin expression, which in turn suppressed the activation of inflammasome in mouse lungs. However, hemorrhagic shock suppressed IL-10 and pyrin expression, therefore significantly enhancing inflammasome activation and IL-1 beta secretion in macrophages and endothelial cells (Xu et al. 2013).

Table 6.2 Articles containing discussion of inflammasomes in ALI

ID	Species	Diseases	Design	Element	Contribution
Leff 1994	*Rattus norvegicus*	ALI	In vivo, lung	IL-1 alpha, IL-1 beta	Yes
Narimanbekov 1995	*Oryctolagus cuniculus*	ALI (mechanical ventilation)	In vivo, lung	IL-1Ra	Yes
Pugin 1996	*Homo sapiens*	ARDS	Ex vivo, BALF	IL-1 beta	Yes
Tremblay 1997	*Rattus norvegicus*	ALI (mechanical ventilation)	Ex vivo, lung	IL-1 beta	Yes
Ranieri 1999	*Homo sapiens*	ARDS (mechanical ventilation)	Ex vivo, BALF	IL-1 beta, IL-1Ra	Yes
Wrigge 2000	*Homo sapiens*	Mechanical ventilation	Ex vivo, blood	IL-1Ra	No
Ricard 2001	*Rattus norvegicus*	ALI (mechanical ventilation)	Ex vivo, lung; in vivo, lung	IL-1 beta	No
Rich 2003	*Rattus norvegicus*	ALI (mechanical ventilation)	Ex vivo, BALF	ATP	Yes
Ma 2005	*Rattus norvegicus* *Mus musculus*	ALI (mechanical ventilation)	Microarray, lung tissue	IL-1 beta	Yes
Lin 2007	*Rattus norvegicus*	LPS + mechanical ventilation	In vivo, lung	IL-1 beta	Yes
Frank 2008	*Rattus norvegicus* *Mus musculus*	ALI (mechanical ventilation)	In vivo, lung	IL-1R1, IL-1Ra	Yes
Ganter 2008	*Mus musculus*	ALI	In vivo, lung	IL-1 beta	Yes
Kolliputi 2010	*Mus musculus*	ALI (hyperoxia)	In vivo, lung, BALF; in vitro, macrophage	NLRP3, K^+ efflux, P2X7R	Yes
Xiang 2011	*Mus musculus*	ALI (hemorrhagic shock)	In vivo, lung; in vitro, endothelial cells	NLRP3, IL-1 beta, ROS	Yes
Dolinay 2012	*Homo sapiens* *Mus musculus*	ALI (mechanical ventilation)	In vivo, lung; in vitro, blood, BALF	IL-1 beta, IL-18, caspase-1	Yes
Kuipers 2012	*Homo sapiens* *Mus musculus*	ALI (mechanical ventilation)	In vivo, lung; in vitro, BALF	NLRP3, ASC, caspase-1, IL-1 beta, uric acid	Yes
Fukumoto 2013	*Mus musculus*	ALI (hyperoxia)	In vivo, lung; in vitro, BALF	NLRP3, IL-1 beta	Yes
Wu 2013	*Mus musculus*	ALI (mechanical ventilation)	In vitro, macrophage	NLRP3, IL-1 beta, IL-18, uric acid, ROS	Yes

(continued)

Table 6.2 (continued)

ID	Species	Diseases	Design	Element	Contribution
Xu 2013	*Mus musculus*	ALI (hemorrhagic shock)	In vivo, lung; in vitro, endothelial cells, macrophage	NLRP3, IL-1 beta	Yes
Mizushina 2015	*Mus musculus*	ALI (hyperoxia)	In vivo, lung, BALF	NLRP3, IL-1 beta	Yes
He 2016	*Mus musculus*	ALI (infection)	In vivo, lung; in vitro, macrophage	NLRP3, IL-1 beta, IL-1R1	Yes

Ventilation-induced lung injury (VILI) is a special type of ALI. Ventilation alone for a short period does not seem sufficient for mediator release and major lung injury in normal lungs (Wrigge et al. 2000). However, mechanical ventilation may augment preexisting lung injury.

IL-1 Signaling A RCT showed that mechanical ventilation caused increased concentrations of IL-1 beta and IL-1Ra in BALF of ARDS patients (Ranieri et al. 1999). In basic research, combined LPS instillation and ventilation synergistically unregulated the production of IL-1 beta in rat lung tissues (Lin et al. 2007). Ventilation with a large tidal volume for 2 h induced the released of IL-1 beta in isolated, unperfused rat lungs with or without LPS injection (Tremblay et al. 1997). However, Ricard et al. reappraised the cytokine production in both in vivo and ex vivo ventilated rat lungs and were unable to detect the release of IL-1 beta (Ricard et al. 2001). In gene expression microarray studies, IL-1 beta has been identified as a candidate gene in rodent (mouse and rat) VILI models (Ma et al. 2005).

Furthermore, recombinant IL-1Ra significantly lowered the concentration of albumin and elastase and decreased neutrophil infiltration in a rabbit model of VILI (Narimanbekov and Rozycki 1995). Similarly, mice deficient in IL1R1 and rats treated with IL-1Ra showed preserved alveolar barrier function, reduced neutrophil recruitment, and decreased epithelial injury and permeability after mechanically ventilation (Frank et al. 2008).

IL-18 Signaling Dolinay et al. reported a critical role of caspase-1 and IL-18 in VILI. A comprehensive gene expression analysis on peripheral blood from patients with ARDS and polymerase chain reaction and ELISA were performed. IL-1 beta and IL-18 transcripts were increased. And human plasma IL-18 levels were correlated with disease severity and mortality in critically ill patients. Besides, mechanical ventilation enhanced IL-18 levels in the lung, serum, and BALF in mice.

Genetic deletion of IL-18 or caspase-1 or treatment with IL-18 neutralizing antibody reduced lung injury and inflammation in response to ventilation (Dolinay et al. 2012).

Inflammasome Activation More directly, Kuipers et al. showed that mRNA levels of ASC were higher in lung brush samples from patients after 5 h of ventilation. Also, ventilation increased relative expression of NLRP3 in alveolar macrophages. Besides, mechanical ventilation increased the expression of NLRP3 and ASC, activated caspase-1, and promoted the release of IL-1 beta in mouse lung. In this process, uric acid was also released and may serve as the ligand for NLRP3. Additionally, mice deficient in NLRP3 or treatment with IL-1 receptor antagonist or glibenclamide displayed less VILI (Kuipers et al. 2012). Using an in vitro model, Wu et al. also demonstrated that alveolar macrophages subjected to cyclic stretch released uric acid, which activated the NLRP3 inflammasome, and induced the release of IL-1 beta and IL-18. They determined that mitochondrial ROS generation was required for NLRP3 activation (Wu et al. 2013). It has long been reported that high-pressure mechanical ventilation significantly increased ATP release in BALF (Rich et al. 2003).

Another form of ALI associated with ventilation is hyperoxic acute lung injury (HALI). Kolliputi et al. reported that hyperoxia-induced K^+ efflux activated the NLRP3 inflammasome via the purinergic P2X7R to cause inflammation and HALI (Kolliputi et al. 2010). Further, they demonstrated that Nlrp3$^{-/-}$ mice had suppressed inflammatory response in BALF and lung tissue and blunted epithelial cell apoptosis to HALI (Fukumoto et al. 2013). Notably, Mizushina et al. found that deficiency in NLRP3 shortened survival under hyperoxic conditions regardless of diminished inflammatory responses. And this lethality was due to Stat3 signaling (Mizushina et al. 2015).

6.3 Chronic Lung Diseases

6.3.1 Smoke and Particles Inhalation

6.3.1.1 Cigarette Smoking (CS)

Table 6.3 lists the studies containing discussion of inflammasomes in smoking.

IL-1 Signaling Cytokine regulation in the lung may be altered by smoke exposure. CS inhalation in smokers (healthy and COPD patients) induced IL-1 beta release in BALF (Kuschner et al. 1996) and in lung tissue and induced sputum (Pauwels et al. 2011). However, there are also studies reporting a lower level of IL-1 beta in macrophage from smokers before and after LPS stimulation (Sauty et al. 1994; Brown et al. 1989). Pauwels et al. demonstrated that pulmonary inflammation after subacute CS exposure could be significantly attenuated by IL-1R1 knockout or neutralizing IL-1 alpha or IL-1 beta (Pauwels et al. 2011). TLR4, MyD88, and IL-1R1 were reported to be involved in the inflammatory response to CS both in vitro and in vivo. Besides, CS-activated macrophages released IL-1 beta only in presence of ATP (Doz et al. 2008).

IL-18 Signaling IL-18 signaling has also been demonstrated to be critical in the response to CS. CS was a potent stimulator of IL-18 and caspases-1. In addition, CS-induced inflammation was significantly decreased in Il18ra$^{-/-}$ mice (Kang et al. 2007).

Table 6.3 Articles containing discussion of inflammasomes in relation to smoking

ID	Species	Design	Element	Contribution
Brown et al. 1989	*Homo sapiens*	In vitro, macrophage	IL-1 beta	No
Sauty 1994	*Homo sapiens*	Ex vivo, BALF	IL-1 beta	No
Kuschner 1996	*Homo sapiens*	Ex vivo, BALF	IL-1 beta	Yes
Kang 2007	*Mus musculus, Homo sapiens*	In vivo, lung; in vitro, macrophage	IL-18, IL-18R	Yes
Doz 2008	*Mus musculus*	In vivo, lung; ex vivo, BALF; in vitro, macrophage	TLR4, MyD88, IL-1R1, IL-1 beta, ATP	Yes
Eltom 2011	*Mus musculus, Homo sapiens*	In vivo, lung; ex vivo, lung	P2X7R, caspase-1, IL-1 beta	Yes
Pauwels 2011	*Mus musculus, Homo sapiens*	In vivo, lung; ex vivo, sputum	IL-1 alpha, IL-1 beta, IL-1R1, NLRP3, caspase-1	Conflicting
Eltom 2014	*Mus musculus*	In vivo, lung	IL-1 beta, IL-18, NLRP3, ASC, NLRC4, AIM2, caspase-1	Yes

CS is closely related to COPD, which will be discussed in Sect. 6.3.2

Inflammasome Activation Direct evidence of inflammasome involvement came from Eltom et al. They demonstrated that NLRP3 and ASC, but not NLRC4 or AIM2, were required for CS-induced IL-1 beta and IL-18 release. Besides, mice deficient in caspase 1/11 had markedly attenuated levels of cytokines and neutrophil infiltration (Eltom et al. 2014). However, CS-induced inflammation and IL-1 alpha production were reported to occur independently of the NLRP3-caspase-1 axis (Pauwels et al. 2011).

Genetic deletion of the P2X7R or using a selective P2X7R inhibitor reduced CS-induced caspase-1 activation and IL-1 beta release during acute CS exposure in vivo. They reported that caspase-1 activity were higher in lung tissue from smokers (Eltom et al. 2011).

6.3.1.2 Inhalation of Particles

Articles containing discussion of inflammasomes in particle inhalation are summarized in Table 6.4.

There are various kinds of particles in industrial and urban life, which can cause injuries to the lungs and pulmonary diseases.

Carbon black nanoparticles have been reported to cause caspase-1 activation, IL-1 beta release, and pyroptosis in alveolar macrophages (Reisetter et al. 2011).

Inorganic materials can trigger NLRP3 response as well. Nano-TiO_2 activated NLRP3 inflammasome and induced IL-1 alpha and beta release in a phagocytosis-independent manner (Yazdi et al. 2010). Nickel nanoparticles induced transient increase of IL-1 beta in rats (Morimoto et al. 2010). Hamilton et al. demonstrated that nickel contamination in multiwalled carbon nanotubules activated NLRP3 via lysosomal disruption in primary macrophages (Hamilton et al. 2012).

Urban particulate matter has been shown to induce IL-1 beta release in human primary bronchial epithelial cells (Fujii et al. 2001). Furthermore, NLRP3 inflammasome was required for the production of IL-1 beta in vivo (Hirota et al. 2012). Diesel exhaust particles (DEP) are a major component of the ambient particulate matter. It was shown that in vitro DEP stimulated IL-1 beta production in monocytes and macrophages (Pacheco et al. 2001; Yang et al. 1997) and in epithelial cells (Boland et al. 1999). $Il1r1^{-/-}$ mice and mice treated with IL-1Ra had reduced inflammation upon DEP exposure. However, the authors concluded that

Table 6.4 Articles containing discussion of inflammasomes in particle inhalation

ID	Species	Particles	Design	Element	Contribution
Yang 1997	*Rattus norvegicus*	Diesel exhaust particles	In vitro, macrophage	IL-1 beta	Yes
Boland 1999	*Homo sapiens*	Diesel exhaust particles	In vitro, epithelial cell line	IL-1 beta	Yes
Fujii 2001	*Homo sapiens*	Particulate matter	In vitro, epithelial cell	IL-1 beta	Yes
Pacheco 2001	*Homo sapiens*	Diesel exhaust particles	In vitro, PBMC	IL-1 beta	Yes
Morimoto 2010	*Rattus norvegicus*	Nickel nanoparticle	In vivo, lung, BALF	IL-1 beta	Yes
Yazdi 2010	*Homo sapiens, Mus musculus*	Nano-titanium dioxide	In vivo, lung; in vitro, macrophage, dendritic cell, keratinocyte	NLRP3, IL-1 alpha, IL-1 beta, IL-1R1	Yes
Provoost 2011	*Mus musculus*	Diesel exhaust particles	In vivo, lung, BALF	IL-1R1, NLRP3, Caspase-1	Conflicting
Reisetter 2011	*Mus musculus, Homo sapiens*	Carbon black nanoparticle	In vitro, macrophage	Caspase-1, IL-1 beta	Yes
Hamilton 2012	*Mus musculus*	Nickel	In vitro, macrophage	NLRP3, lysosome	Yes
Hirota 2012	*Mus musculus*	Particulate matter	In vivo, lung	NLRP3, IL-1 beta	Yes

DEP-initiated inflammation did not depend on NLRP3-caspase-1 pathway (Provoost et al. 2011).

6.3.2 *Chronic Obstructive Pulmonary Disease*

Listed in Table 6.5 are the articles containing discussion on inflammasomes in COPD.

COPD is an important lung and airway disease, and is increasing in incidence, especially in developing countries. COPD may affect over 200 million people worldwide (data from WHO). Long-term cigarette smoking is the most important risk factor that may initiate the disease. The evidence for the involvement of inflammasome in cigarette smoking has been discussed in previous section.

IL-1 Signaling Acute exposure to smoke elevated IL-1 beta, while 6 months of exposure did not. Mice deficient in IL-1R or treatment with pan-caspase or caspase-1 inhibitor were protected from inflammatory cell infiltration and matrix breakdown during acute smoke exposure. After 6 months of exposure, $Il1r^{-/-}$ mice were 65% protected against emphysema and completely protected against small airway remodeling (Churg et al. 2009).

IL-18 Signaling Kang et al. demonstrate that IL-18 is present in exaggerated quantities in the lungs and the serum from patients with COPD (Kang et al. 2007). The levels of IL-18 in induced sputum of patients with COPD were also found to be elevated compared with healthy subjects and were inversely correlated with lung function (% predicted FEV_1 and FEV_1/FVC ratio) (Rovina et al. 2009).

Furthermore, targeted overexpression of IL-18 in murine lungs resulted in widespread pulmonary inflammation, emphysema, mucus metaplasia, and airway remodeling through increased pulmonary $CD4^+$, $CD8^+$, $CD19^+$, and $NK1.1^+$ cells

Table 6.5 Articles containing discussion of inflammasomes in COPD

ID	Species	Design	Element	Contribution
Kang 2007	*Mus musculus, Homo sapiens*	In vivo, lung; in vitro, macrophage	IL-18, IL-18R	Yes
Churg 2009	*Mus musculus*	In vivo, lung	IL-1 beta, IL-1R, IL-18	Yes
Rovina 2009	*Homo sapiens*	Ex vivo, sputum	IL-18	Yes
Cicko 2010	*Mus musculus*	In vivo, lung	ATP	Yes
Kang 2012	*Mus musculus*	In vivo, lung	IL-18	Yes
Bartziokas 2014	*Homo sapiens*	Ex vivo, blood	Uric acid	Yes
Di Stefano 2014	*Homo sapiens*	Ex vivo, bronchial mucosa, BALF	NLRP3, caspase-1, IL-1 beta, IL-18	No

and type 1 cytokine (IFN-gamma), type 2 cytokine (IL-13), and type 17 cytokine (IL-17A) (Kang et al. 2012).

Inflammasome Activation An increased level of ATP has been found in the lungs in a mouse model of smoke-induced acute lung inflammation and emphysema, and the increased ATP level correlated with pulmonary neutrophilia (Cicko et al. 2010).

Serum uric acid levels were higher in patients with more severe airflow limitation and in those having frequent exacerbations. Besides, high uric acid levels correlated with 30-day mortality, prolonged hospitalization, and more aggressive medical care in COPD patients with exacerbations (Bartziokas et al. 2014).

Recently, Di Stefano et al. reported lack of NLRP3 inflammasome activation, with no differences in caspase-1 activation, IL-1 beta, or IL-18 levels in bronchial biopsies or in BALF in patients with stable COPD compared with control subjects (Di Stefano et al. 2014).

6.3.3 Asthma

Asthma is another important lung disease, characterized by allergic reaction. Allergic inflammatory response in asthma is conventionally characterized by the activation of Th2 pathway. The importance of Th17 response has now been recognized (Table 6.6).

IL-1 Signaling Serum IL-1beta levels and expression of IL-1 beta in the bronchial epithelium and submucosal macrophages were higher in patients with asthma compared with control subjects (Thomas and Chhabra 2003; Sousa et al. 1996). In asthmatic patients, IL-1 beta concentrations in the sputum (Konno et al. 1996) and BALF (Broide et al. 1992) of symptomatic patients were significantly higher than that in asymptomatic subjects. BALF from patients with status asthmaticus had an elevated inflammatory activity due to the presence of excessive bioactive IL-1 beta (Tillie-Leblond et al. 1999). Hastie et al. stratified subjects by sputum granulocytes. Those patients with both increased eosinophils and neutrophils had the lowest lung function and increased symptoms. In this subset of patients, IL-1 beta level in the sputum was positively associated with neutrophil counts (Hastie et al. 2010).

In a mouse model, IL-1 beta combined with TNF alpha can contribute to airway hyperresponsiveness and methacholine-induced bronchoconstriction (Horiba et al. 2011). It was also reported that the ovalbumin-induced airway hypersensitivity response was significantly reduced in IL-1 alpha/beta-deficient mice whereas profoundly exacerbated in mice deficient in IL-1Ra, suggesting that IL-1 signaling was required for Th2 response (Nakae et al. 2003). In a model of mild asthma, IL-1R signaling was reported to be required, as eosinophilic inflammation and goblet cell hyperplasia were strongly reduced in $Il1r1^{-/-}$ mice. In contrast, the IL-1R was not required in an allergic model with adjuvant (Schmitz et al. 2003). Wang et al. applied a recombinant adenovirus expressing human IL-1ra in an ovalbumin-sensitized murine model of asthma. Single intranasal delivery before airway antigen challenge

Table 6.6 Articles containing discussion of inflammasomes in asthma

ID	Species	Design	Element	Contribution
Broide 1992	*Homo sapiens*	Ex vivo, BALF	IL-1 beta	Yes
Konno 1996	*Homo sapiens*	Ex vivo, sputum	IL-1 beta	Yes
Sousa 1996	*Homo sapiens*	Ex vivo, bronchial biopsies	IL-1 beta, IL-1Ra	Yes
Tillie-Leblond 1999	*Homo sapiens*	Ex vivo, BALF	IL-1 beta, IL-1Ra	Yes
Schmitz 2003	*Mus musculus*	In vivo, lung	IL-1R1	Conflicting
Thomas 2003	*Homo sapiens*	Ex vivo, blood	IL-1 beta	Yes
Wang 2006	*Mus musculus*	In vivo, lung; ex vivo, BALF	IL-1Ra	Yes
Idzko 2007	*Homo sapiens, Mus musculus*	Ex vivo, BALF	ATP	Yes
Hastie 2010	*Homo sapiens*	Ex vivo, sputum	IL-1 beta	Yes
Ather 2011	*Mus musculus*	In vivo, lung; in vitro, dendritic cell, macrophage	NLRP3, ASC, caspase-1, IL-1 beta	Yes
Besnard 2011	*Mus musculus*	In vivo, lung	NLRP3, ASC, caspase-1, IL-1 alpha, IL-1 beta, IL-1R	Yes
Horiba 2011	*Mus musculus*	In vivo, lung	IL-1 beta	Yes
Kool 2011	*Homo sapiens, Mus musculus*	In vivo, lung	NLRP3, IL-1 beta	No
Allen 2012	*Mus musculus*	In vivo, lung	NLRP3, IL-1 beta, IL-18	No
Martin 2013	*Mus musculus*	In vivo, lung	NLRP3, IL-1R	Conflicting
Kim 2014	*Homo sapiens, Mus musculus*	In vivo, lung, ex vivo, BALF; in vitro, epithelial cell	NLRP3, caspase-1, IL-1 beta, ROS	Yes

significantly decreased the severity of airway hyperresponsiveness, reduced pulmonary infiltration, and decreased peribronchial inflammation (Wang et al. 2006).

Inflammasome Activation Direct evidence showed that allergic airway inflammation depended on NLRP3 inflammasome activation, as Th2 lymphocyte activation and cytokine production were reduced in mice deficient in NLRP3, ASC, or caspase-1. The critical role of IL-1R1 signaling was also confirmed in mice deficient in IL-1R1, IL-1 beta, and IL-1 alpha (Besnard et al. 2011). Kim et al. recently demonstrated that levels of NLRP3 and caspase-1 in BALF from the patients with asthma were significantly higher than that in healthy subjects. Furthermore, suppression of mitochondrial ROS generation by NecroX-5 attenuated allergic airway inflammation associated with inhibition of NLRP3 inflammasome and caspase-1 activation in primary tracheal epithelial cells and mouse lung tissues. In addition, blockade of IL-1 beta substantially reduced airway inflammation and hyperresponsiveness in asthmatic mice (Kim et al. 2014). It has been shown that gain of function SNPs in human NLRP3 are linked to food-induced anaphylaxis and aspirin-induced asthma (Hitomi et al. 2009). In multiple asthmatic models of mixed Th2/Th17 responses, serum amyloid A activated NLRP3 inflammasome to induce IL-1 beta secretion in dendritic cells and macrophages and promote $CD4^+$ T cells to secrete IL-17A in an IL-1-dependent manner (Ather et al. 2011). Similarly but differently, Martin et al. reported the importance of caspase-1 and IL-1R, but not NLRP3, for Th17 development in NO_2-promoted allergic airway disease (Martin et al. 2013). Also, Allen et al. determined that the NLRP3 inflammasome was not required in multiple allergic asthma models in mice. Besides, in all the models, the cytokines IL-1 beta and IL-18 in the lung were below the level of detection (Allen et al. 2012). And Kool et al. suggested that NLRP3 and IL-1 beta did not contribute to the Th2 adjuvant effect of uric acid in mice (Kool et al. 2011).

Elevated ATP was found in the BALF of patients with asthma and ovalbumin-challenged asthmatic mice (Idzko et al. 2007). Consistently, P2X7R was found to be upregulated in acute and chronic asthmatic airway inflammation in mice and humans. Mice deficient in P2X7R or treated with specific P2X7R-antagonist had reduced airway inflammation in asthma models (Muller et al. 2011).

6.3.4 Fibrotic Lung Diseases

6.3.4.1 Idiopathic Pulmonary Fibrosis (IPF)

IPF is a progressive while irreversible disease, with a general poor prognosis. IPF is characterized by a histologic or radiologic pattern of usual interstitial pneumonia and progressive fibrosis of lung parenchyma. Bleomycin is a chemotherapeutic drug used clinically for a variety of human malignancies. However, a high dose of bleomycin can lead to lethal lung injury and pulmonary fibrosis in human patients, as well as in rodent models. Therefore rodent models of bleomycin-induced lung fibrosis have been widely used for the investigation of human IPF. Bleomycin-induced fibrosis is also discussed in this section. Table 6.7 listed the articles containing discussion of inflammasomes in IPF.

IL-1 Signaling Pan et al. observed that in IPF patients, cytokine IL-1 beta was positive in alveolar macrophages and type 2 pneumocytes in acute pulmonary fibrotic changes, but not in areas of old fibrosis, suggesting that IL-1 beta may play a role in the initial pulmonary fibrotic responses (Pan et al. 1996). In vitro study found that alveolar macrophages from healthy human subjects released IL-1 beta after bleomycin challenge (Scheule et al. 1992).

With a single base variation at position +2018 of the IL-1Ra gene, there is an increased risk of developing cryptogenic fibrosing alveolitis (Whyte et al. 2000). Two other studies examining another polymorphism at intron 2 of the IL-1Ra gene found no association with increased susceptibility to IPF (Hutyrova et al. 2002; Riha et al. 2004).

Gasse et al. reported that in bleomycin-induced lung inflammation fibrosis depended on IL-1R1 signaling, as neutralization of IL-1 beta or specific blockage of IL-1R1 by antibody reduced bleomycin-induced pathology (Gasse et al. 2007). Overexpression of IL-1 beta for 7–10 days in rats was reported to induce an increase of TGF-beta in BALF and progressive interstitial fibrogenesis for the next 60 days, resembling human pulmonary fibrosis (Kolb et al. 2001). IL-1 beta was further reported sufficient to induce IL-17 production, required for inflammatory response to bleomycin (Wilson et al. 2010; Gasse et al. 2011).

Table 6.7 Articles containing discussion of inflammasomes in IPF

ID	Species	Design	Element	Contribution
Scheule 1992	*Homo sapiens*	In vitro, macrophage	IL-1 beta	Yes
Pan 1996	*Homo sapiens*	Ex vivo, lung tissue specimens	IL-1 beta	Conflicting
Whyte 2000	*Homo sapiens*	Ex vivo, lung biopsy	IL-1Ra	Yes
Kolb 2001	*Rattus norvegicus*	In vivo, lung; ex vivo, BALF	IL-1 beta	Yes
Kuwano 2001	*Mus musculus*	In vivo, lung	Caspase-1	Yes
Hutyrova 2002	*Homo sapiens*	Ex vivo, blood	IL-1Ra	No
Kitasato 2004	*Homo sapiens*	Ex vivo, lung tissue, serum, BALF	IL-18, IL-18R alpha	Yes
Riha 2004	*Homo sapiens*	Ex vivo, blood	IL-1Ra	No
Nakatani-Okuda 2005	*Mus musculus*	In vivo, lung	IL-18	Yes
Gasse 2007	*Mus musculus*	In vivo, lung	IL-1R1, IL-1 beta, ASC, IL-18	Conflicting
Gasse 2009	*Mus musculus*	In vivo, lung	NLRP3, caspase-1, IL-1 beta, IL-18, uric acid	Conflicting
Hoshino 2009	*Homo sapiens, Mus musculus*	Ex vivo, lung tissue; in vivo, lung, serum	IL-18, IL-18R alpha, IL-1 beta, caspase-1	Yes
Riteau et al. 2010	*Homo sapiens, Mus musculus*	In vivo, lung; ex vivo, BALF	ATP, P2X7R, IL-1 beta	Yes
Liu 2011	*Homo sapiens*	Ex vivo, serum, BALF	IL-18	No
Xu 2012	*Homo sapiens, Mus musculus*	In vivo, lung, in vitro, macrophage	NLRP3, caspase-1,	Yes

IL-18 Signaling Kitasato et al. reported elevated levels of IL-18 in the serum and BALF of patients with IPF and strongly expressed IL-18 and IL-18R alpha in the fibroblastic foci (Kitasato et al. 2004). Hoshino et al. reported excessive IL-18 and IL-18R alpha expression in the lungs of patients with bleomycin-induced lung injury. They also found that intravenous administration of bleomycin induced the expression of IL-1 beta and IL-18 in the serum and lungs of mice. Moreover, lung injury, assessed by fibrosis score, hydroxyproline levels, and wet lung weight, was significantly attenuated in mice deficient in caspase-1, IL-18, or IL-18R alpha (Hoshino et al. 2009).

However, Liu et al. failed to find an increase in serum and BALF levels of IL-18 in IPF patients (Liu et al. 2011). Il18−/− mice showed much worse lung injuries after treatment with bleomycin, as assessed by survival rate, histological images, and leukocyte infiltration. Besides, pretreatment with IL-18 before bleomycin instillation appeared to be protective in lung injuries (Nakatani-Okuda et al. 2005).

Inflammasome Activation Bleomycin-induced lung injury depended on NLRP3 inflammasome, as mice deficient in NLRP3 or caspase-1 displayed reduced neutrophil influx and IL-1 beta production in the lung. It was found that bleomycin-induced inflammasome activation is mediated by uric acid. Reduction of uric acid levels with inhibitor or uricase led to a decrease in IL-1 beta production, lung inflammation, and fibrosis. In addition, bleomycin-induced inflammation was IL-18-independent (Gasse et al. 2009). It has also been reported that mice lacking ASC had reduced neutrophil recruitment and a reduction in IL-1 beta production following bleomycin challenge (Gasse et al. 2007). Another example is from the research of statin. Numerous case reports suggested that statins could cause various types of interstitial lung diseases. Statin pretreatment enhances caspase-1-mediated responses in vivo and in vitro, which could be abolished in macrophages from mice deficient in NLRP3 (Xu et al. 2012).

The role of caspase-1 in bleomycin-induced lung injury has also been investigated. Kuwano et al. reported that bleomycin enhanced caspase-1 activity in addition to elevated expression in inflammatory cells. They also demonstrated that a pan-caspase inhibitor zVAD-FMK was able to attenuate bleomycin-induced lung injuries (Kuwano et al. 2001)

ATP levels were elevated in BALF from patients with IPF and from mice treated with bleomycin. Mice deficient in P2X7R or neutralized against ATP in the airways potently inhibit bleomycin-induced lung inflammation and remodeling (Riteau et al. 2010).

6.3.4.2 Cystic Fibrosis

Cystic fibrosis is caused by mutations of the cystic fibrosis transmembrane conductance regulator and is the most common autosomal recessive disorder in western countries. Patients with cystic fibrosis often experience recurrent and chronic infections with *Pseudomonas aeruginosa*, as well as *Staphylococcus aureus* and *Haemophilus influenzae* (discussed in Sect. 1.1.3).

Grassme et al. demonstrated the activation of caspase-1 and upregulation and membrane recruitment of ASC in the lungs of CF mice. These activations were associated with elevated levels of the signaling lipid-derived mediator, ceramide. Consistently, they also observed a normalization of IL-1 beta in the lungs after treatment with caspase-1 inhibitors (Grassme et al. 2014).

6.3.4.3 Silicosis

Crystalline silica is very common in occupational and environmental settings. Prolonged exposure in the workplace may lead to the development of silicosis, which is irreversible, progressive pulmonary fibrosis. Silica exposure is a high-priority public health concern. Alveolar macrophages, and their production of IL-1 beta, have been suggested to play a crucial role during the early inflammatory response after exposure to silica. Table 6.8 lists the articles containing discussion of inflammasomes in silicosis.

IL-1 Signaling Silica induced a release of IL-1 beta in human alveolar macrophages in a caspase-1-dependent manner (Iyer et al. 1996) and in the lungs of silica-exposed mice (Davis et al. 1998).

A polymorphism in IL-1Ra (+2018), but not IL-1 beta (+3953), was increased in a population of Caucasian coal miners with silicosis, indicating that this variant may confer susceptibility to developing silicosis (Yucesoy et al. 2001).

In addition, neutralizing IL-1 beta with monoclonal antibody reduced silica-induced inflammation and fibrosis by inhibiting mRNA expression of inflammatory and fibrogenic mediators (TGF beta, collagen I, and fibronectin) and modulating the Th1/Th2 balance toward a Th2-dominant response (Guo et al. 2013). The anti-fibrotic effect of inhibiting IL-1 beta was also reported by Piguet et al. where the administration of recombinant IL-1Ra reduced collagen deposition and the formation of fibrotic nodules in mice (Piguet et al. 1993). More directly, exposure of mice deficient in IL-1 beta to silica resulted in reduced lung inflammation, apoptosis, and significantly smaller silicotic lesions than in wild-type mice over a 12 weeks course (Srivastava et al. 2002).

Inflammasome Activation Stimulation of macrophages with silica resulted in the secretion of IL-1 beta and IL-18 in an inflammasome-dependent manner, as macrophages deficient in NLRP3, ASC, or caspase-1 all displayed a marked defect in their ability to secrete cytokines. They also found that activation of the NLRP3 inflammasome by silica required both a K^+ efflux and the generation of ROS (Cassel et al. 2008). Similarly, NLRP3 inflammasome activation was triggered by ROS generated by NADPH oxidase. In a model of asbestos inhalation, Nalp3$^{-/-}$ mice showed diminished recruitment of inflammatory cells to the lungs, paralleled by lower cytokine production (Dostert et al. 2008). Hornung et al. demonstrated that silica activated caspase-1 and induced the release of mature IL-1 beta in human PBMCs. IL-1 mediated the neutrophil influx after exposure to silica crystals. The phagocytosis of silica by macrophages resulted in lysosomal destabilization and

Table 6.8 Articles containing discussion of inflammasomes in silicosis

ID	Species	Design	Element	Contribution
Piguet 1993	*Mus musculus*	In vivo, lung	IL-1 beta	Yes
Iyer 1996	*Homo sapiens*	In vitro, macrophage	IL-1 beta	Yes
Davis 1998	*Mus musculus*	In vivo, lung	IL-1 beta	Yes
Yucesoy 2001	*Homo sapiens*	Ex vivo, lung tissue	IL-1Ra, IL-1 beta	Conflicting
Srivastava 2002	*Mus musculus*	In vivo, lung	IL-1 beta	Yes
Cassel 2008	*Mus musculus*	In vitro, macrophage	NLRP3, ASC, caspase-1, IL-1 beta, IL-18, K^+ efflux, ROS	Yes
Dostert 2008	*Homo sapiens, Mus musculus*	In vitro, macrophage	NLRP3, ASC, caspase-1, IL-1 beta, ROS, NADPH	Yes
Hornung 2008	*Homo sapiens, Mus musculus*	In vitro, macrophage	NLRP3, caspase-1, IL-1 beta, lysosomal destabilization	Yes
Ji 2012	*Homo sapiens*	Ex vivo, blood	NLRP3	Yes
Riteau 2012	*Homo sapiens, Mus musculus*	In vitro, macrophage	NLRP3, IL-1 beta, ATP, P2X7R	Yes
Guo 2013	*Mus musculus*	In vivo, lung	IL-1 beta	Yes
Peeters 2013	*Homo sapiens*	In vitro, epithelial cell	NLRP3, caspase-1, IL-1 beta	Yes
Moncao-Ribeiro 2014	*Mus musculus*	In vitro, macrophage	IL-1 beta, P2X7R	Yes

subsequent rupture releasing proteolytic enzymes, such as cathepsin B into the cytosol, and the activation of the NLRP3 inflammasome (Hornung et al. 2008). In a case-control study, Ji et al. found that an SNP in the NLRP3 gene (rs1539019) was associated with a significant increase in coal workers pneumoconiosis in a Chinese population. This association was more pronounced in patients with stage I disease suggesting a potential role for the NLRP3 inflammasome in the development of silicosis (Ji et al. 2012). NLPR3 activation has also been reported in nonmyeloid cells. NLRP3 activation, as well as activation of caspase-1, led to maturation and secretion of IL-1 beta in human bronchial epithelial cell lines and primary human bronchial epithelial cells (Peeters et al. 2013).

ATP was released by macrophages after exposure to silica. The activation of the NLRP3 inflammasome relied on purinergic receptors and pannexin/connexin hemichannels. The use of specific P2X7 receptor inhibitors, or abrogation of ATP in primed human monocytic cell lines, was able to prevent silica-induced IL-1 beta production (Riteau et al. 2012). This was further manifested in P2X7R knockout mice. Inflammatory cell infiltration and collagen deposition, cell apoptosis, and

NF-κB activation as well as TGF-beta, nitric oxide, ROS, and IL-1 beta secretion were reduced in knockout mice (Moncao-Ribeiro et al. 2014).

6.3.4.4 Asbestosis

Similar to silicosis, asbestosis often occurs as an occupational disease, particularly in developing countries. The inhalation of asbestos can also lead to lung cancer, mesothelioma, and pleural diseases. The articles containing discussion of inflammasomes in asbestosis are summarized in Table 6.9.

IL-1 Signaling Cells recovered in BALF or alveolar macrophages from patients with asbestosis were reported to release higher levels of IL-1 beta in comparison with control groups (Zhang et al. 1993; Perkins et al. 1993). In vivo models also demonstrated that asbestos exposure can result in enhanced IL-1 beta secretion in BALF (Haegens et al. 2007).

Inflammasome Activation Hillegass et al. reported that asbestos exposure was associated with an increase in NLRP3 expression and caspase-1 activation in mesothelial cells, leading to secreted IL-1 beta and IL-18, which could be attenuated by downregulation of NLRP3. They also reported that asbestos challenge had no significant effect on the NLRP1 or AIM2 inflammasomes (Hillegass et al. 2013). Girardelli et al. reported that in a cohort of Italian patients with asbestos-induced mesothelioma, SNPs in the NLRP1, but not NLRP3 gene, may be associated with the disease (Girardelli et al. 2012). Furthermore, Nlrp3$^{-/-}$ mice were reported to have defects in IL-1 beta secretion and immune cell recruitment following asbestos exposure. However, NLRP3 was not critical in the chronic development of asbestos-induced mesothelioma, as a similar incidence of malignant mesothelioma in knockout mice (Chow et al. 2012).

Table 6.9 Articles containing discussion of inflammasome in asbestosis

ID	Species	Design	Element	Contribution
Perkins 1993	*Homo sapiens*	In vitro, alveolar macrophage	IL-1 beta	Yes
Zhang 1993	*Homo sapiens*	In vitro, BALF cells	IL-1 beta	Yes
Haegens 2007	*Mus musculus*	In vivo, lung, BALF	IL-1 beta	Yes
Chow 2012	*Mus musculus*	In vivo, lung	NLRP3, IL-1 beta	Conflicting
Girardelli 2012	*Mus musculus*	In vitro, macrophage	IL-1 beta, P2X7R	Yes
Hillegass 2013	*Homo sapiens, Mus musculus*	In vitro, mesothelial cells	NLRP3, caspase-1, IL-1 beta, IL-18, NLRP1, AIM2	Conflicting

Table 6.10 Articles containing discussion of inflammasomes in pulmonary hypertension

ID	Species	Design	Element	Contribution
Villegas 2013	*Mus musculus*	In vivo, lung	NLRP3, caspase-1, IL-1 beta, IL-18	Yes
Cero 2015	*Mus musculus*	In vivo, lung	NLRP3, ASC, IL-1 beta, IL-18	Conflicting

6.3.5 *Pulmonary Hypertension*

Pulmonary hypertension is characterized by sustained elevation of the pulmonary arterial pressure (>25 mm Hg). Prolonged high pressure in pulmonary artery system may lead to right ventricular failure. Table 6.10 lists the studies on inflammasomes in pulmonary hypertension.

In a mice model, hypoxia exposure caused pulmonary hypertension, including increased right ventricular systolic pressure and pulmonary vascular remodeling, along with activation of the NLRP3 inflammasome and caspase-1, as well as IL-1 beta and IL-18 production. These effects could be reversed with a superoxide dismutase mimetic (Villegas et al. 2013). In another study, Asc$^{-/-}$ mice, but not Nlrp3$^{-/-}$, mice were resistant to hypoxia-induced pulmonary hypertension, as evidenced by no significant changes in levels of caspase-1, IL-18, or IL-1 beta, reduced right ventricular systolic pressure and reduced pulmonary vascular remodeling, indicating the possible involvement of alternate inflammasome complexes involving ASC (Cero et al. 2015).

6.4 Conclusion

Inflammasomes have emerged as an important regulator of the innate immune system and have significantly affected the understanding of the pathogenesis of many diseases. In this chapter, we reviewed the evidence of inflammasome-related components in the progression of pulmonary diseases. We can easily appreciate how the discovery of the inflammasome affects our understanding of the role of IL-1 and IL-18 signaling in lung disease. Still, in some diseases, the importance of inflammasomes has not been fully investigated. Besides, NLRP3 inflammasome in macrophage is currently the most clearly defined type. The potential of other inflammasomes in nonmyeloid cells needs to be further studied in the process of injury and recovery in lung diseases.

References

Allen IC, Scull MA, Moore CB, Holl EK, McElvania-TeKippe E, Taxman DJ, Guthrie EH, Pickles RJ, Ting JP (2009) The NLRP3 inflammasome mediates in vivo innate immunity to influenza A virus through recognition of viral RNA. Immunity 30(4):556–565. https://doi.org/10.1016/j.immuni.2009.02.005

Allen IC, Jania CM, Wilson JE, Tekeppe EM, Hua X, Brickey WJ, Kwan M, Koller BH, Tilley SL, Ting JP (2012) Analysis of NLRP3 in the development of allergic airway disease in mice. J Immunol (Baltimore, MD: 1950) 188(6):2884–2893. https://doi.org/10.4049/jimmunol.1102488

Amer A, Franchi L, Kanneganti TD, Body-Malapel M, Ozoren N, Brady G, Meshinchi S, Jagirdar R, Gewirtz A, Akira S, Nunez G (2006) Regulation of Legionella phagosome maturation and infection through flagellin and host Ipaf. J Biol Chem 281(46):35217–35223. https://doi.org/10.1074/jbc.M604933200

Arlehamn CS, Evans TJ (2011) *Pseudomonas aeruginosa* pilin activates the inflammasome. Cell Microbiol 13(3):388–401. https://doi.org/10.1111/j.1462-5822.2010.01541.x

Ather JL, Ckless K, Martin R, Foley KL, Suratt BT, Boyson JE, Fitzgerald KA, Flavell RA, Eisenbarth SC, Poynter ME (2011) Serum amyloid A activates the NLRP3 inflammasome and promotes Th17 allergic asthma in mice. J Immunol (Baltimore, MD: 1950) 187(1):64–73. https://doi.org/10.4049/jimmunol.1100500

Bartziokas K, Papaioannou AI, Loukides S, Papadopoulos A, Haniotou A, Papiris S, Kostikas K (2014) Serum uric acid as a predictor of mortality and future exacerbations of COPD. Eur Respir J 43(1):43–53. https://doi.org/10.1183/09031936.00209212

Bellamy R, Ruwende C, Corrah T, McAdam KP, Whittle HC, Hill AV (1998) Assessment of the interleukin 1 gene cluster and other candidate gene polymorphisms in host susceptibility to tuberculosis. Tuber Lung Dis 79(2):83–89. https://doi.org/10.1054/tuld.1998.0009

Besnard AG, Guillou N, Tschopp J, Erard F, Couillin I, Iwakura Y, Quesniaux V, Ryffel B, Togbe D (2011) NLRP3 inflammasome is required in murine asthma in the absence of aluminum adjuvant. Allergy 66(8):1047–1057. https://doi.org/10.1111/j.1398-9995.2011.02586.x

Bhakdi S, Muhly M, Korom S, Hugo F (1989) Release of interleukin-1 beta associated with potent cytocidal action of staphylococcal alpha-toxin on human monocytes. Infect Immun 57 (11):3512–3519

Boland S, Baeza-Squiban A, Fournier T, Houcine O, Gendron MC, Chevrier M, Jouvenot G, Coste A, Aubier M, Marano F (1999) Diesel exhaust particles are taken up by human airway epithelial cells in vitro and alter cytokine production. Am J Phys 276(4 Pt 1):L604–L613

Broide DH, Lotz M, Cuomo AJ, Coburn DA, Federman EC, Wasserman SI (1992) Cytokines in symptomatic asthma airways. J Allergy Clin Immunol 89(5):958–967

Brown GP, Iwamoto GK, Monick MM, Hunninghake GW (1989) Cigarette smoking decreases interleukin 1 release by human alveolar macrophages. Am J Phys 256(2 Pt 1):C260–C264

Cai S, Batra S, Wakamatsu N, Pacher P, Jeyaseelan S (2012) NLRC4 inflammasome-mediated production of IL-1beta modulates mucosal immunity in the lung against gram-negative bacterial infection. J Immunol 188(11):5623–5635. https://doi.org/10.4049/jimmunol.1200195

Carlsson F, Kim J, Dumitru C, Barck KH, Carano RA, Sun M, Diehl L, Brown EJ (2010) Host-detrimental role of Esx-1-mediated inflammasome activation in mycobacterial infection. PLoS Pathog 6(5):e1000895. https://doi.org/10.1371/journal.ppat.1000895

Case CL, Shin S, Roy CR (2009) Asc and Ipaf inflammasomes direct distinct pathways for caspase-1 activation in response to *Legionella pneumophila*. Infect Immun 77(5):1981–1991. https://doi.org/10.1128/iai.01382-08

Cassel SL, Eisenbarth SC, Iyer SS, Sadler JJ, Colegio OR, Tephly LA, Carter AB, Rothman PB, Flavell RA, Sutterwala FS (2008) The Nalp3 inflammasome is essential for the development of silicosis. Proc Natl Acad Sci U S A 105(26):9035–9040. https://doi.org/10.1073/pnas.0803933105

Cero FT, Hillestad V, Sjaastad I, Yndestad A, Aukrust P, Ranheim T, Lunde IG, Olsen MB, Lien E, Zhang L, Haugstad SB, Loberg EM, Christensen G, Larsen KO, Skjonsberg OH (2015) Absence of the inflammasome adaptor ASC reduces hypoxia-induced pulmonary hypertension in mice. Am J Physiol Lung Cell Mol Physiol 309(4):L378–L387. https://doi.org/10.1152/ajplung.00342.2014

Chen CC, Tsai SH, Lu CC, Hu ST, Wu TS, Huang TT, Said-Sadier N, Ojcius DM, Lai HC (2012) Activation of an NLRP3 inflammasome restricts *Mycobacterium kansasii* infection. PLoS One 7 (4):e36292. https://doi.org/10.1371/journal.pone.0036292

Chensue SW, Warmington KS, Berger AE, Tracey DE (1992) Immunohistochemical demonstration of interleukin-1 receptor antagonist protein and interleukin-1 in human lymphoid tissue and granulomas. Am J Pathol 140(2):269–275

Chow MT, Tschopp J, Moller A, Smyth MJ (2012) NLRP3 promotes inflammation-induced skin cancer but is dispensable for asbestos-induced mesothelioma. Immunol Cell Biol 90 (10):983–986. https://doi.org/10.1038/icb.2012.46

Churg A, Zhou S, Wang X, Wang R, Wright JL (2009) The role of interleukin-1beta in murine cigarette smoke-induced emphysema and small airway remodeling. Am J Respir Cell Mol Biol 40(4):482–490. https://doi.org/10.1165/rcmb.2008-0038OC

Cicko S, Lucattelli M, Muller T, Lommatzsch M, De Cunto G, Cardini S, Sundas W, Grimm M, Zeiser R, Durk T, Zissel G, Boeynaems JM, Sorichter S, Ferrari D, Di Virgilio F, Virchow JC, Lungarella G, Idzko M (2010) Purinergic receptor inhibition prevents the development of smoke-induced lung injury and emphysema. J Immunol 185(1):688–697. https://doi.org/10.4049/jimmunol.0904042

Craven RR, Gao X, Allen IC, Gris D, Bubeck Wardenburg J, McElvania-Tekippe E, Ting JP, Duncan JA (2009) *Staphylococcus aureus* alpha-hemolysin activates the NLRP3-inflammasome in human and mouse monocytic cells. PLoS One 4(10):e7446. https://doi.org/10.1371/journal.pone.0007446

Davis GS, Pfeiffer LM, Hemenway DR (1998) Persistent overexpression of interleukin-1beta and tumor necrosis factor-alpha in murine silicosis. J Environ Pathol Toxicol Oncol 17(2):99–114

Denton AE, Doherty PC, Turner SJ, La Gruta NL (2007) IL-18, but not IL-12, is required for optimal cytokine production by influenza virus-specific CD8+ T cells. Eur J Immunol 37 (2):368–375. https://doi.org/10.1002/eji.200636766

Di Stefano A, Caramori G, Barczyk A, Vicari C, Brun P, Zanini A, Cappello F, Garofano E, Padovani A, Contoli M, Casolari P, Durham AL, Chung KF, Barnes PJ, Papi A, Adcock I, Balbi B (2014) Innate immunity but not NLRP3 inflammasome activation correlates with severity of stable COPD. Thorax 69(6):516–524. https://doi.org/10.1136/thoraxjnl-2012-203062

Dolinay T, Kim YS, Howrylak J, Hunninghake GM, An CH, Fredenburgh L, Massaro AF, Rogers A, Gazourian L, Nakahira K, Haspel JA, Landazury R, Eppanapally S, Christie JD, Meyer NJ, Ware LB, Christiani DC, Ryter SW, Baron RM, Choi AM (2012) Inflammasome-regulated cytokines are critical mediators of acute lung injury. Am J Respir Crit Care Med 185 (11):1225–1234. https://doi.org/10.1164/rccm.201201-0003OC

Dorhoi A, Nouailles G, Jorg S, Hagens K, Heinemann E, Pradl L, Oberbeck-Muller D, Duque-Correa MA, Reece ST, Ruland J, Brosch R, Tschopp J, Gross O, Kaufmann SH (2012) Activation of the NLRP3 inflammasome by *Mycobacterium tuberculosis* is uncoupled from susceptibility to active tuberculosis. Eur J Immunol 42(2):374–384. https://doi.org/10.1002/eji.201141548

Dostert C, Petrilli V, Van Bruggen R, Steele C, Mossman BT, Tschopp J (2008) Innate immune activation through Nalp3 inflammasome sensing of asbestos and silica. Science (New York, NY) 320(5876):674–677. https://doi.org/10.1126/science.1156995

Doz E, Noulin N, Boichot E, Guénon I, Fick L, Le BM, Lagente V, Ryffel B, Schnyder B, Quesniaux VF (2008) Cigarette smoke-induced pulmonary inflammation is TLR4/MyD88 and IL-1R1/MyD88 signaling dependent. J Immunol 180(9):1169–1178

Eltom S, Stevenson CS, Rastrick J, Dale N, Raemdonck K, Wong S, Catley MC, Belvisi MG, Birrell MA (2011) P2X7 receptor and caspase 1 activation are central to airway inflammation observed after exposure to tobacco smoke. PLoS One 6(9):e24097

Eltom S, Belvisi MG, Stevenson CS, Maher SA, Dubuis E, Fitzgerald KA, Birrell MA (2014) Role of the inflammasome-caspase1/11-IL-1/18 axis in cigarette smoke driven airway inflammation: an insight into the pathogenesis of COPD. PLoS One 9(11):e112829

Franchi L, Stoolman J, Kanneganti TD, Verma A, Ramphal R, Nunez G (2007) Critical role for Ipaf in *Pseudomonas aeruginosa*-induced caspase-1 activation. Eur J Immunol 37(11):3030–3039. https://doi.org/10.1002/eji.200737532

Frank JA, Pittet JF, Wray C, Matthay MA (2008) Protection from experimental ventilator-induced acute lung injury by IL-1 receptor blockade. Thorax 63(2):147–153. https://doi.org/10.1136/thx.2007.079608

Fremond CM, Togbe D, Doz E, Rose S, Vasseur V, Maillet I, Jacobs M, Ryffel B, Quesniaux VF (2007) IL-1 receptor-mediated signal is an essential component of MyD88-dependent innate response to *Mycobacterium tuberculosis* infection. J Immunol 179(2):1178–1189

Fujii T, Hayashi S, Hogg JC, Vincent R, Van Eeden SF (2001) Particulate matter induces cytokine expression in human bronchial epithelial cells. Am J Respir Cell Mol Biol 25(3):265–271

Fukumoto J, Fukumoto I, Parthasarathy PT, Cox R, Huynh B, Ramanathan GK, Venugopal RB, Allen-Gipson DS, Lockey RF, Kolliputi N (2013) NLRP3 deletion protects from hyperoxia-induced acute lung injury. Am J Physiol Cell Physiol 305(2):C182–C189. https://doi.org/10.1152/ajpcell.00086.2013

Ganter MT, Roux J, Miyazawa B, Howard M, Frank JA, Su G, Sheppard D, Violette SM, Weinreb PH, Horan GS, Matthay MA, Pittet JF (2008) Interleukin-1beta causes acute lung injury via alphavbeta5 and alphavbeta6 integrin-dependent mechanisms. Circ Res 102(7):804–812. https://doi.org/10.1161/circresaha.107.161067

Gasse P, Mary C, Guenon I, Noulin N, Charron S, Schnyder-Candrian S, Schnyder B, Akira S, Quesniaux VF, Lagente V, Ryffel B, Couillin I (2007) IL-1R1/MyD88 signaling and the inflammasome are essential in pulmonary inflammation and fibrosis in mice. J Clin Invest 117 (12):3786–3799. https://doi.org/10.1172/jci32285

Gasse P, Riteau N, Charron S, Girre S, Fick L, Petrilli V, Tschopp J, Lagente V, Quesniaux VF, Ryffel B, Couillin I (2009) Uric acid is a danger signal activating NALP3 inflammasome in lung injury inflammation and fibrosis. Am J Respir Crit Care Med 179(10):903–913. https://doi.org/10.1164/rccm.200808-1274OC

Gasse P, Riteau N, Vacher R, Michel ML, Fautrel A, di Padova F, Fick L, Charron S, Lagente V, Eberl G, Le Bert M, Quesniaux VF, Huaux F, Leite-de-Moraes M, Ryffel B, Couillin I (2011) IL-1 and IL-23 mediate early IL-17A production in pulmonary inflammation leading to late fibrosis. PLoS One 6(8):e23185. https://doi.org/10.1371/journal.pone.0023185

Giacomini E, Iona E, Ferroni L, Miettinen M, Fattorini L, Orefici G, Julkunen I, Coccia EM (2001) Infection of human macrophages and dendritic cells with *Mycobacterium tuberculosis* induces a differential cytokine gene expression that modulates T cell response. J Immunol 166 (12):7033–7041

Girardelli M, Maestri I, Rinaldi RR, Tognon M, Boldorini R, Bovenzi M, Crovella S, Comar M (2012) NLRP1 polymorphisms in patients with asbestos-associated mesothelioma. Infect Agent Cancer 7(1):25. https://doi.org/10.1186/1750-9378-7-25

Grassme H, Carpinteiro A, Edwards MJ, Gulbins E, Becker KA (2014) Regulation of the inflammasome by ceramide in cystic fibrosis lungs. Cell Physiol Biochem 34(1):45–55. https://doi.org/10.1159/000362983

Guo J, Gu N, Chen J, Shi T, Zhou Y, Rong Y, Zhou T, Yang W, Cui X, Chen W (2013) Neutralization of interleukin-1 beta attenuates silica-induced lung inflammation and fibrosis in C57BL/6 mice. Arch Toxicol 87(11):1963–1973. https://doi.org/10.1007/s00204-013-1063-z

Haegens A, Barrett TF, Gell J, Shukla A, Macpherson M, Vacek P, Poynter ME, Butnor KJ, Janssen-Heininger YM, Steele C, Mossman BT (2007) Airway epithelial NF-kappaB activation modulates asbestos-induced inflammation and mucin production in vivo. J Immunol 178 (3):1800–1808

Hamilton RF Jr, Buford M, Xiang C, Wu N, Holian A (2012) NLRP3 inflammasome activation in murine alveolar macrophages and related lung pathology is associated with MWCNT nickel contamination. Inhal Toxicol 24(14):995–1008. https://doi.org/10.3109/08958378.2012.745633

Harder J, Franchi L, Munoz-Planillo R, Park JH, Reimer T, Nunez G (2009) Activation of the Nlrp3 inflammasome by *Streptococcus pyogenes* requires streptolysin O and NF-kappa B activation but proceeds independently of TLR signaling and P2X7 receptor. J Immunol 183 (9):5823–5829. https://doi.org/10.4049/jimmunol.0900444

Hastie AT, Moore WC, Meyers DA, Vestal PL, Li H, Peters SP, Bleecker ER (2010) Analyses of asthma severity phenotypes and inflammatory proteins in subjects stratified by sputum granulocytes. J Allergy Clin Immunol 125(5):1028–1036.e1013. https://doi.org/10.1016/j.jaci.2010.02.008

He X, Qian Y, Li Z, Fan EK, Li Y, Wu L, Billiar TR, Wilson MA, Shi X, Fan J (2016) TLR4-upregulated IL-1beta and IL-1RI promote alveolar macrophage pyroptosis and lung inflammation through an autocrine mechanism. Sci Rep 6:31663. https://doi.org/10.1038/srep31663

Hennet T, Ziltener HJ, Frei K, Peterhans E (1992) A kinetic study of immune mediators in the lungs of mice infected with influenza A virus. J Immunol 149(3):932–939

Hillegass JM, Miller JM, MacPherson MB, Westbom CM, Sayan M, Thompson JK, Macura SL, Perkins TN, Beuschel SL, Alexeeva V, Pass HI, Steele C, Mossman BT, Shukla A (2013) Asbestos and erionite prime and activate the NLRP3 inflammasome that stimulates autocrine cytokine release in human mesothelial cells. Part Fibre Toxicol 10:39. https://doi.org/10.1186/1743-8977-10-39

Hirota JA, Hirota SA, Warner SM, Stefanowicz D, Shaheen F, Beck PL, Macdonald JA, Hackett TL, Sin DD, Van Eeden S, Knight DA (2012) The airway epithelium nucleotide-binding domain and leucine-rich repeat protein 3 inflammasome is activated by urban particulate matter. J Allergy Clin Immunol 129(4):1116–1125.e1116. https://doi.org/10.1016/j.jaci.2011.11.033

Hitomi Y, Ebisawa M, Tomikawa M, Imai T, Komata T, Hirota T, Harada M, Sakashita M, Suzuki Y, Shimojo N, Kohno Y, Fujita K, Miyatake A, Doi S, Enomoto T, Taniguchi M, Higashi N, Nakamura Y, Tamari M (2009) Associations of functional NLRP3 polymorphisms with susceptibility to food-induced anaphylaxis and aspirin-induced asthma. J Allergy Clin Immunol 124(4):779–785.e776. https://doi.org/10.1016/j.jaci.2009.07.044

Horiba M, Qutna N, Gendapodi P, Agrawal S, Sapkota K, Abel P, Townley RG (2011) Effect of IL-1beta and TNF-alpha vs IL-13 on bronchial hyperresponsiveness, beta2-adrenergic responses and cellularity of bronchial alveolar lavage fluid. Auton Autacoid Pharmacol 31 (3–4):37–49. https://doi.org/10.1111/j.1474-8673.2011.00465.x

Hornung V, Bauernfeind F, Halle A, Samstad EO, Kono H, Rock KL, Fitzgerald KA, Latz E (2008) Silica crystals and aluminum salts activate the NALP3 inflammasome through phagosomal destabilization. Nat Immunol 9(8):847–856. https://doi.org/10.1038/ni.1631

Hoshino T, Okamoto M, Sakazaki Y, Kato S, Young HA, Aizawa H (2009) Role of proinflammatory cytokines IL-18 and IL-1beta in bleomycin-induced lung injury in humans and mice. Am J Respir Cell Mol Biol 41(6):661–670. https://doi.org/10.1165/rcmb.2008-0182OC

Hutyrova B, Pantelidis P, Drabek J, Zurkova M, Kolek V, Lenhart K, Welsh KI, Du Bois RM, Petrek M (2002) Interleukin-1 gene cluster polymorphisms in sarcoidosis and idiopathic pulmonary fibrosis. Am J Respir Crit Care Med 165(2):148–151. https://doi.org/10.1164/ajrccm.165.2.2106004

Ichinohe T, Lee HK, Ogura Y, Flavell R, Iwasaki A (2009) Inflammasome recognition of influenza virus is essential for adaptive immune responses. J Exp Med 206(1):79–87. https://doi.org/10.1084/jem.20081667

Ichinohe T, Pang IK, Iwasaki A (2010) Influenza virus activates inflammasomes via its intracellular M2 ion channel. Nat Immunol 11(5):404–410. https://doi.org/10.1038/ni.1861

Idzko M, Hammad H, van Nimwegen M, Kool M, Willart MA, Muskens F, Hoogsteden HC, Luttmann W, Ferrari D, Di Virgilio F, Virchow JC Jr, Lambrecht BN (2007) Extracellular ATP triggers and maintains asthmatic airway inflammation by activating dendritic cells. Nat Med 13 (8):913–919. https://doi.org/10.1038/nm1617

Iyer R, Hamilton RF, Li L, Holian A (1996) Silica-induced apoptosis mediated via scavenger receptor in human alveolar macrophages. Toxicol Appl Pharmacol 141(1):84–92. https://doi.org/10.1006/taap.1996.0263

Ji X, Hou Z, Wang T, Jin K, Fan J, Luo C, Chen M, Han R, Ni C (2012) Polymorphisms in inflammasome genes and risk of coal workers' pneumoconiosis in a Chinese population. PLoS One 7(10):e47949. https://doi.org/10.1371/journal.pone.0047949

Jonas D, Walev I, Berger T, Liebetrau M, Palmer M, Bhakdi S (1994) Novel path to apoptosis: small transmembrane pores created by staphylococcal alpha-toxin in T lymphocytes evoke internucleosomal DNA degradation. Infect Immun 62(4):1304–1312

Juffermans NP, Florquin S, Camoglio L, Verbon A, Kolk AH, Speelman P, van Deventer SJ, van Der Poll T (2000) Interleukin-1 signaling is essential for host defense during murine pulmonary tuberculosis. J Infect Dis 182(3):902–908. https://doi.org/10.1086/315771

Kang MJ, Homer RJ, Gallo A, Lee CG, Crothers KA, Cho SJ, Rochester C, Cain H, Chupp G, Yoon HJ, Elias JA (2007) IL-18 is induced and IL-18 receptor alpha plays a critical role in the pathogenesis of cigarette smoke-induced pulmonary emphysema and inflammation. J Immunol 178(3):1948–1959

Kang MJ, Choi JM, Kim BH, Lee CM, Cho WK, Choe G, Kim DH, Lee CG, Elias JA (2012) IL-18 induces emphysema and airway and vascular remodeling via IFN-gamma, IL-17A, and IL-13. Am J Respir Crit Care Med 185(11):1205–1217. https://doi.org/10.1164/rccm.201108-1545OC

Kaukoranta-Tolvanen SS, Teppo AM, Laitinen K, Saikku P, Linnavuori K, Leinonen M (1996) Growth of *Chlamydia pneumoniae* in cultured human peripheral blood mononuclear cells and induction of a cytokine response. Microb Pathog 21(3):215–221. https://doi.org/10.1006/mpat.1996.0056

Kawakami K, Qureshi MH, Zhang T, Okamura H, Kurimoto M, Saito A (1997) IL-18 protects mice against pulmonary and disseminated infection with *Cryptococcus neoformans* by inducing IFN-gamma production. J Immunol 159(11):5528–5534

Kawakami K, Koguchi Y, Qureshi MH, Kinjo Y, Yara S, Miyazato A, Kurimoto M, Takeda K, Akira S, Saito A (2000a) Reduced host resistance and Th1 response to *Cryptococcus neoformans* in interleukin-18 deficient mice. FEMS Microbiol Lett 186(1):121–126

Kawakami K, Koguchi Y, Qureshi MH, Miyazato A, Yara S, Kinjo Y, Iwakura Y, Takeda K, Akira S, Kurimoto M, Saito A (2000b) IL-18 contributes to host resistance against infection with *Cryptococcus neoformans* in mice with defective IL-12 synthesis through induction of IFN-gamma production by NK cells. J Immunol 165(2):941–947

Kim SR, Kim DI, Kim SH, Lee H, Lee KS, Cho SH, Lee YC (2014) NLRP3 inflammasome activation by mitochondrial ROS in bronchial epithelial cells is required for allergic inflammation. Cell Death Dis 5:e1498. https://doi.org/10.1038/cddis.2014.460

King HC, Khera-Butler T, James P, Oakley BB, Erenso G, Aseffa A, Knight R, Wellington EM, Courtenay O (2017) Environmental reservoirs of pathogenic mycobacteria across the Ethiopian biogeographical landscape. PLoS One 12(3):e0173811

Kinjo Y, Kawakami K, Uezu K, Yara S, Miyagi K, Koguchi Y, Hoshino T, Okamoto M, Kawase Y, Yokota K, Yoshino K, Takeda K, Akira S, Saito A (2002) Contribution of IL-18 to Th1 response and host defense against infection by *Mycobacterium tuberculosis*: a comparative study with IL-12p40. J Immunol 169(1):323–329

Kitasato Y, Hoshino T, Okamoto M, Kato S, Koda Y, Nagata N, Kinoshita M, Koga H, Yoon DY, Asao H, Ohmoto H, Koga T, Rikimaru T, Aizawa H (2004) Enhanced expression of interleukin-18 and its receptor in idiopathic pulmonary fibrosis. Am J Respir Cell Mol Biol 31(6):619–625. https://doi.org/10.1165/rcmb.2003-0306OC

Kleinnijenhuis J, Joosten LA, van de Veerdonk FL, Savage N, van Crevel R, Kullberg BJ, van der Ven A, Ottenhoff TH, Dinarello CA, van der Meer JW, Netea MG (2009) Transcriptional and inflammasome-mediated pathways for the induction of IL-1beta production by *Mycobacterium tuberculosis*. Eur J Immunol 39(7):1914–1922. https://doi.org/10.1002/eji.200839115

Kolb M, Margetts PJ, Anthony DC, Pitossi F, Gauldie J (2001) Transient expression of IL-1beta induces acute lung injury and chronic repair leading to pulmonary fibrosis. J Clin Invest 107 (12):1529–1536. https://doi.org/10.1172/jci12568

Kolliputi N, Shaik RS, Waxman AB (2010) The inflammasome mediates hyperoxia-induced alveolar cell permeability. J Immunol 184(10):5819–5826. https://doi.org/10.4049/jimmunol.0902766

Konno S, Gonokami Y, Kurokawa M, Kawazu K, Asano K, Okamoto K, Adachi M (1996) Cytokine concentrations in sputum of asthmatic patients. Int Arch Allergy Immunol 109 (1):73–78

Koo IC, Wang C, Raghavan S, Morisaki JH, Cox JS, Brown EJ (2008) ESX-1-dependent cytolysis in lysosome secretion and inflammasome activation during mycobacterial infection. Cell Microbiol 10(9):1866–1878. https://doi.org/10.1111/j.1462-5822.2008.01177.x

Kool M, Willart MA, van Nimwegen M, Bergen I, Pouliot P, Virchow JC, Rogers N, Osorio F, Reis e Sousa C, Hammad H, Lambrecht BN (2011) An unexpected role for uric acid as an inducer of T helper 2 cell immunity to inhaled antigens and inflammatory mediator of allergic asthma. Immunity 34(4):527–540. https://doi.org/10.1016/j.immuni.2011.03.015

Kuipers MT, Aslami H, Janczy JR, van der Sluijs KF, Vlaar AP, Wolthuis EK, Choi G, Roelofs JJ, Flavell RA, Sutterwala FS, Bresser P, Leemans JC, van der Poll T, Schultz MJ, Wieland CW (2012) Ventilator-induced lung injury is mediated by the NLRP3 inflammasome. Anesthesiology 116(5):1104–1115. https://doi.org/10.1097/ALN.0b013e3182518bc0

Kurenuma T, Kawamura I, Hara H, Uchiyama R, Daim S, Dewamitta SR, Sakai S, Tsuchiya K, Nomura T, Mitsuyama M (2009) The RD1 locus in the *Mycobacterium tuberculosis* genome contributes to activation of caspase-1 via induction of potassium ion efflux in infected macrophages. Infect Immun 77(9):3992–4001. https://doi.org/10.1128/iai.00015-09

Kuschner WG, D'Alessandro A, Wong H, Blanc PD (1996) Dose-dependent cigarette smoking-related inflammatory responses in healthy adults. Eur Respir J 9(10):1989–1994

Kuwano K, Kunitake R, Maeyama T, Hagimoto N, Kawasaki M, Matsuba T, Yoshimi M, Inoshima I, Yoshida K, Hara N (2001) Attenuation of bleomycin-induced pneumopathy in mice by a caspase inhibitor. Am J Physiol Lung Cell Mol Physiol 280(2):L316–L325

Law K, Weiden M, Harkin T, Tchou-Wong K, Chi C, Rom WN (1996) Increased release of interleukin-1 beta, interleukin-6, and tumor necrosis factor-alpha by bronchoalveolar cells lavaged from involved sites in pulmonary tuberculosis. Am J Respir Crit Care Med 153 (2):799–804. https://doi.org/10.1164/ajrccm.153.2.8564135

Lee HM, Yuk JM, Kim KH, Jang J, Kang G, Park JB, Son JW, Jo EK (2012) *Mycobacterium abscessus* activates the NLRP3 inflammasome via Dectin-1-Syk and p62/SQSTM1. Immunol Cell Biol 90(6):601–610. https://doi.org/10.1038/icb.2011.72

Lee S, Suh GY, Ryter SW, Choi AM (2016) Regulation and function of the nucleotide binding domain leucine-rich repeat-containing receptor, pyrin domain-containing-3 inflammasome in lung disease. Am J Respir Cell Mol Biol 54(2):151–160. https://doi.org/10.1165/rcmb.2015-0231TR

Leff JA, Baer JW, Bodman ME, Kirkman JM, Shanley PF, Patton LM, Beehler CJ, McCord JM, Repine JE (1994) Interleukin-1-induced lung neutrophil accumulation and oxygen metabolite-mediated lung leak in rats. Am J Phys 266(1 Pt 1):L2–L8

Leissinger M, Kulkarni R, Zemans RL, Downey GP, Jeyaseelan S (2014) Investigating the role of nucleotide-binding oligomerization domain-like receptors in bacterial lung infection. Am J Respir Crit Care Med 189(12):1461–1468. https://doi.org/10.1164/rccm.201311-2103PP

Lin SM, Lin HC, Lee KY, Huang CD, Liu CY, Wang CH, Kuo HP (2007) Ventilator-induced injury augments interleukin-1beta production and neutrophil sequestration in lipopolysaccharide-treated lungs. Shock (Augusta, Ga) 28(4):453–460. https://doi.org/10.1097/shk.0b013e3180487fb5

Liu B, Mori I, Hossain MJ, Dong L, Takeda K, Kimura Y (2004) Interleukin-18 improves the early defence system against influenza virus infection by augmenting natural killer cell-mediated cytotoxicity. J Gen Virol 85(Pt 2):423–428. https://doi.org/10.1099/vir.0.19596-0

Liu DH, Cui W, Chen Q, Huang CM (2011) Can circulating interleukin-18 differentiate between sarcoidosis and idiopathic pulmonary fibrosis? Scand J Clin Lab Invest 71(7):593–597. https://doi.org/10.3109/00365513.2011.597871

Luzar MA, Thomassen MJ, Montie TC (1985) Flagella and motility alterations in *Pseudomonas aeruginosa* strains from patients with cystic fibrosis: relationship to patient clinical condition. Infect Immun 50(2):577–582

Ma SF, Grigoryev DN, Taylor AD, Nonas S, Sammani S, Ye SQ, Garcia JG (2005) Bioinformatic identification of novel early stress response genes in rodent models of lung injury. Am J Physiol Lung Cell Mol Physiol 289(3):L468–L477. https://doi.org/10.1152/ajplung.00109.2005

Mahenthiralingam E, Campbell ME, Speert DP (1994) Nonmotility and phagocytic resistance of *Pseudomonas aeruginosa* isolates from chronically colonized patients with cystic fibrosis. Infect Immun 62(2):596–605

Martin RA, Ather JL, Lundblad LK, Suratt BT, Boyson JE, Budd RC, Alcorn JF, Flavell RA, Eisenbarth SC, Poynter ME (2013) Interleukin-1 receptor and caspase-1 are required for the Th17 response in nitrogen dioxide-promoted allergic airway disease. Am J Respir Cell Mol Biol 48(5):655–664. https://doi.org/10.1165/rcmb.2012-0423OC

Master SS, Rampini SK, Davis AS, Keller C, Ehlers S, Springer B, Timmins GS, Sander P, Deretic V (2008) *Mycobacterium tuberculosis* prevents inflammasome activation. Cell Host Microbe 3 (4):224–232. https://doi.org/10.1016/j.chom.2008.03.003

Mayer-Barber KD, Barber DL, Shenderov K, White SD, Wilson MS, Cheever A, Kugler D, Hieny S, Caspar P, Nunez G, Schlueter D, Flavell RA, Sutterwala FS, Sher A (2010) Caspase-1 independent IL-1beta production is critical for host resistance to *Mycobacterium tuberculosis* and does not require TLR signaling in vivo. J Immunol 184(7):3326–3330. https://doi.org/10.4049/jimmunol.0904189

McElvania Tekippe E, Allen IC, Hulseberg PD, Sullivan JT, McCann JR, Sandor M, Braunstein M, Ting JP (2010) Granuloma formation and host defense in chronic *Mycobacterium tuberculosis* infection requires PYCARD/ASC but not NLRP3 or caspase-1. PLoS One 5(8):e12320. https://doi.org/10.1371/journal.pone.0012320

McNeela EA, Burke A, Neill DR, Baxter C, Fernandes VE, Ferreira D, Smeaton S, El-Rachkidy R, McLoughlin RM, Mori A, Moran B, Fitzgerald KA, Tschopp J, Petrilli V, Andrew PW, Kadioglu A, Lavelle EC (2010) Pneumolysin activates the NLRP3 inflammasome and promotes proinflammatory cytokines independently of TLR4. PLoS Pathog 6(11):e1001191. https://doi.org/10.1371/journal.ppat.1001191

Miao EA, Ernst RK, Dors M, Mao DP, Aderem A (2008) *Pseudomonas aeruginosa* activates caspase 1 through Ipaf. Proc Natl Acad Sci U S A 105(7):2562–2567. https://doi.org/10.1073/pnas.0712183105

Mishra BB, Moura-Alves P, Sonawane A, Hacohen N, Griffiths G, Moita LF, Anes E (2010) *Mycobacterium tuberculosis* protein ESAT-6 is a potent activator of the NLRP3/ASC inflammasome. Cell Microbiol 12(8):1046–1063. https://doi.org/10.1111/j.1462-5822.2010.01450.x

Mizushina Y, Shirasuna K, Usui F, Karasawa T, Kawashima A, Kimura H, Kobayashi M, Komada T, Inoue Y, Mato N, Yamasawa H, Latz E, Iwakura Y, Kasahara T, Bando M, Sugiyama Y, Takahashi M (2015) NLRP3 protein deficiency exacerbates hyperoxia-induced lethality through Stat3 protein signaling independent of interleukin-1beta. J Biol Chem 290 (8):5065–5077. https://doi.org/10.1074/jbc.M114.603217

Moncao-Ribeiro LC, Faffe DS, Santana PT, Vieira FS, da Graca CL, Marques-da-Silva C, Machado MN, Caruso-Neves C, Zin WA, Borojevic R, Takiya CM, Coutinho-Silva R (2014) P2X7 receptor modulates inflammatory and functional pulmonary changes induced by silica. PLoS One 9(10):e110185. https://doi.org/10.1371/journal.pone.0110185

Morimoto Y, Ogami A, Todoroki M, Yamamoto M, Murakami M, Hirohashi M, Oyabu T, Myojo T, Nishi K, Kadoya C, Yamasaki S, Nagatomo H, Fujita K, Endoh S, Uchida K, Yamamoto K, Kobayashi N, Nakanishi J, Tanaka I (2010) Expression of inflammation-related cytokines following intratracheal instillation of nickel oxide nanoparticles. Nanotoxicology 4 (2):161–176. https://doi.org/10.3109/17435390903518479

Muller T, Vieira RP, Grimm M, Durk T, Cicko S, Zeiser R, Jakob T, Martin SF, Blumenthal B, Sorichter S, Ferrari D, Di Virgillio F, Idzko M (2011) A potential role for P2X7R in allergic airway inflammation in mice and humans. Am J Respir Cell Mol Biol 44(4):456–464. https://doi.org/10.1165/rcmb.2010-0129OC

Munoz-Planillo R, Franchi L, Miller LS, Nunez G (2009) A critical role for hemolysins and bacterial lipoproteins in *Staphylococcus aureus*-induced activation of the Nlrp3 inflammasome. J Immunol 183(6):3942–3948. https://doi.org/10.4049/jimmunol.0900729

Nakae S, Komiyama Y, Yokoyama H, Nambu A, Umeda M, Iwase M, Homma I, Sudo K, Horai R, Asano M, Iwakura Y (2003) IL-1 is required for allergen-specific Th2 cell activation and the development of airway hypersensitivity response. Int Immunol 15(4):483–490

Nakatani-Okuda A, Ueda H, Kashiwamura S, Sekiyama A, Kubota A, Fujita Y, Adachi S, Tsuji Y, Tanizawa T, Okamura H (2005) Protection against bleomycin-induced lung injury by IL-18 in mice. Am J Physiol Lung Cell Mol Physiol 289(2):L280–L287. https://doi.org/10.1152/ajplung.00380.2004

Narimanbekov IO, Rozycki HJ (1995) Effect of IL-1 blockade on inflammatory manifestations of acute ventilator-induced lung injury in a rabbit model. Exp Lung Res 21(2):239–254

Netea MG, Selzman CH, Kullberg BJ, Galama JM, Weinberg A, Stalenhoef AF, Van der Meer JW, Dinarello CA (2000) Acellular components of *Chlamydia pneumoniae* stimulate cytokine production in human blood mononuclear cells. Eur J Immunol 30(2):541–549. https://doi.org/10.1002/1521-4141(200002)30:2<541::aid-immu541>3.0.co;2-x

Netea MG, Kullberg BJ, Jacobs LE, Verver-Jansen TJ, van der Ven-Jongekrijg J, Galama JM, Stalenhoef AF, Dinarello CA, Van der Meer JW (2004) *Chlamydia pneumoniae* stimulates IFN-gamma synthesis through MyD88-dependent, TLR2- and TLR4-independent induction of IL-18 release. J Immunol 173(2):1477–1482

Pacheco KA, Tarkowski M, Sterritt C, Negri J, Rosenwasser LJ, Borish L (2001) The influence of diesel exhaust particles on mononuclear phagocytic cell-derived cytokines: IL-10, TGF-beta and IL-1 beta. Clin Exp Immunol 126(3):374–383

Pan LH, Ohtani H, Yamauchi K, Nagura H (1996) Co-expression of TNF alpha and IL-1 beta in human acute pulmonary fibrotic diseases: an immunohistochemical analysis. Pathol Int 46 (2):91–99

Patankar YR, Lovewell RR, Poynter ME, Jyot J, Kazmierczak BI, Berwin B (2013) Flagellar motility is a key determinant of the magnitude of the inflammasome response to *Pseudomonas aeruginosa*. Infect Immun 81(6):2043–2052. https://doi.org/10.1128/iai.00054-13

Pauwels NS, Bracke KR, Dupont LL, Van Pottelberge GR, Provoost S, Vanden Berghe T, Vandenabeele P, Lambrecht BN, Joos GF, Brusselle GG (2011) Role of IL-1alpha and the Nlrp3/caspase-1/IL-1beta axis in cigarette smoke-induced pulmonary inflammation and COPD. Eur Respir J 38(5):1019–1028. https://doi.org/10.1183/09031936.00158110

Peeters PM, Perkins TN, Wouters EF, Mossman BT, Reynaert NL (2013) Silica induces NLRP3 inflammasome activation in human lung epithelial cells. Part Fibre Toxicol 10:3. https://doi.org/10.1186/1743-8977-10-3

Pereira MS, Marques GG, Dellama JE, Zamboni DS (2011a) The Nlrc4 inflammasome contributes to restriction of pulmonary infection by flagellated *Legionella* spp. that trigger pyroptosis. Front Microbiol 2:33. https://doi.org/10.3389/fmicb.2011.00033

Pereira MS, Morgantetti GF, Massis LM, Horta CV, Hori JI, Zamboni DS (2011b) Activation of NLRC4 by flagellated bacteria triggers caspase-1-dependent and -independent responses to restrict *Legionella pneumophila* replication in macrophages and in vivo. J Immunol 187 (12):6447–6455. https://doi.org/10.4049/jimmunol.1003784

Perkins RC, Scheule RK, Hamilton R, Gomes G, Freidman G, Holian A (1993) Human alveolar macrophage cytokine release in response to in vitro and in vivo asbestos exposure. Exp Lung Res 19(1):55–65

Piguet PF, Vesin C, Grau GE, Thompson RC (1993) Interleukin 1 receptor antagonist (IL-1ra) prevents or cures pulmonary fibrosis elicited in mice by bleomycin or silica. Cytokine 5 (1):57–61

Pirhonen J, Sareneva T, Kurimoto M, Julkunen I, Matikainen S (1999) Virus infection activates IL-1 beta and IL-18 production in human macrophages by a caspase-1-dependent pathway. J Immunol 162(12):7322–7329

Provoost S, Maes T, Pauwels NS, Vanden Berghe T, Vandenabeele P, Lambrecht BN, Joos GF, Tournoy KG (2011) NLRP3/caspase-1-independent IL-1beta production mediates diesel exhaust particle-induced pulmonary inflammation. J Immunol 187(6):3331–3337. https://doi.org/10.4049/jimmunol.1004062

Pugin J, Ricou B, Steinberg KP, Suter PM, Martin TR (1996) Proinflammatory activity in bronchoalveolar lavage fluids from patients with ARDS, a prominent role for interleukin-1. Am J Respir Crit Care Med 153(6 Pt 1):1850–1856. https://doi.org/10.1164/ajrccm.153.6.8665045

Qureshi MH, Zhang T, Koguchi Y, Nakashima K, Okamura H, Kurimoto M, Kawakami K (1999) Combined effects of IL-12 and IL-18 on the clinical course and local cytokine production in murine pulmonary infection with *Cryptococcus neoformans*. Eur J Immunol 29(2):643–649. https://doi.org/10.1002/(sici)1521-4141(199902)29:02<643::aid-immu643>3.0.co;2-e

Ranieri VM, Suter PM, Tortorella C, De Tullio R, Dayer JM, Brienza A, Bruno F, Slutsky AS (1999) Effect of mechanical ventilation on inflammatory mediators in patients with acute respiratory distress syndrome: a randomized controlled trial. JAMA 282(1):54–61

Reiniger N, Lee MM, Coleman FT, Ray C, Golan DE, Pier GB (2007) Resistance to *Pseudomonas aeruginosa* chronic lung infection requires cystic fibrosis transmembrane conductance regulator-modulated interleukin-1 (IL-1) release and signaling through the IL-1 receptor. Infect Immun 75(4):1598–1608. https://doi.org/10.1128/iai.01980-06

Reisetter AC, Stebounova LV, Baltrusaitis J, Powers L, Gupta A, Grassian VH, Monick MM (2011) Induction of inflammasome-dependent pyroptosis by carbon black nanoparticles. J Biol Chem 286(24):21844–21852. https://doi.org/10.1074/jbc.M111.238519

Ren K, Torres R (2009) Role of interleukin-1beta during pain and inflammation. Brain Res Rev 60 (1):57–64. https://doi.org/10.1016/j.brainresrev.2008.12.020

Ricard JD, Dreyfuss D, Saumon G (2001) Production of inflammatory cytokines in ventilator-induced lung injury: a reappraisal. Am J Respir Crit Care Med 163(5):1176–1180. https://doi.org/10.1164/ajrccm.163.5.2006053

Rich PB, Douillet CD, Mahler SA, Husain SA, Boucher RC (2003) Adenosine triphosphate is released during injurious mechanical ventilation and contributes to lung edema. J Trauma 55 (2):290–297. https://doi.org/10.1097/01.ta.0000078882.11919.af

Riha RL, Yang IA, Rabnott GC, Tunnicliffe AM, Fong KM, Zimmerman PV (2004) Cytokine gene polymorphisms in idiopathic pulmonary fibrosis. Intern Med J 34(3):126–129. https://doi.org/10.1111/j.1444-0903.2004.00503.x

Rimessi A, Bezzerri V, Patergnani S, Marchi S, Cabrini G, Pinton P (2015) Mitochondrial Ca2+-dependent NLRP3 activation exacerbates the *Pseudomonas aeruginosa*-driven inflammatory response in cystic fibrosis. Nat Commun 6:6201. https://doi.org/10.1038/ncomms7201

Riteau N, Gasse P, Fauconnier L, Gombault A, Couegnat M, Fick L, Kanellopoulos J, Quesniaux VF, Marchand-Adam S, Crestani B, Ryffel B, Couillin I (2010) Extracellular ATP is a danger signal activating P2X7 receptor in lung inflammation and fibrosis. Am J Respir Crit Care Med 182(6):774–783. https://doi.org/10.1164/rccm.201003-0359OC

Riteau N, Baron L, Villeret B, Guillou N, Savigny F, Ryffel B, Rassendren F, Le Bert M, Gombault A, Couillin I (2012) ATP release and purinergic signaling: a common pathway for particle-mediated inflammasome activation. Cell Death Dis 3:e403. https://doi.org/10.1038/cddis.2012.144

Rotta Detto Loria J, Rohmann K, Droemann D, Kujath P, Rupp J, Goldmann T, Dalhoff K (2013) Nontypeable *Haemophilus influenzae* infection upregulates the NLRP3 inflammasome and leads to caspase-1-dependent secretion of interleukin-1beta – a possible pathway of exacerbations in COPD. PLoS One 8(6):e66818. https://doi.org/10.1371/journal.pone.0066818

Rovina N, Dima E, Gerassimou C, Kollintza A, Gratziou C, Roussos C (2009) Interleukin-18 in induced sputum: association with lung function in chronic obstructive pulmonary disease. Respir Med 103(7):1056–1062. https://doi.org/10.1016/j.rmed.2009.01.011

Rupp J, Kothe H, Mueller A, Maass M, Dalhoff K (2003) Imbalanced secretion of IL-1beta and IL-1RA in *Chlamydia pneumoniae*-infected mononuclear cells from COPD patients. Eur Respir J 22(2):274–279

Said-Sadier N, Padilla E, Langsley G, Ojcius DM (2010) *Aspergillus fumigatus* stimulates the NLRP3 inflammasome through a pathway requiring ROS production and the Syk tyrosine kinase. PLoS One 5(4):e10008. https://doi.org/10.1371/journal.pone.0010008

Sareneva T, Matikainen S, Kurimoto M, Julkunen I (1998) Influenza A virus-induced IFN-alpha/beta and IL-18 synergistically enhance IFN-gamma gene expression in human T cells. J Immunol 160(12):6032–6038

Sauty A, Mauel J, Philippeaux MM, Leuenberger P (1994) Cytostatic activity of alveolar macrophages from smokers and nonsmokers: role of interleukin-1 beta, interleukin-6, and tumor necrosis factor-alpha. Am J Respir Cell Mol Biol 11(5):631–637. https://doi.org/10.1165/ajrcmb.11.5.7946392

Scheule RK, Perkins RC, Hamilton R, Holian A (1992) Bleomycin stimulation of cytokine secretion by the human alveolar macrophage. Am J Phys 262(4 Pt 1):L386–L391

Schmitz N, Kurrer M, Kopf M (2003) The IL-1 receptor 1 is critical for Th2 cell type airway immune responses in a mild but not in a more severe asthma model. Eur J Immunol 33 (4):991–1000. https://doi.org/10.1002/eji.200323801

Schmitz N, Kurrer M, Bachmann MF, Kopf M (2005) Interleukin-1 is responsible for acute lung immunopathology but increases survival of respiratory influenza virus infection. J Virol 79 (10):6441–6448. https://doi.org/10.1128/jvi.79.10.6441-6448.2005

Sedimbi SK, Hagglof T, Karlsson MC (2013) IL-18 in inflammatory and autoimmune disease. Cell Mol Life Sci CMLS 70(24):4795–4808. https://doi.org/10.1007/s00018-013-1425-y

Shaikh PZ (2011) Cytokines & their physiologic and pharmacologic functions in inflammation: a review. Int J Pharm Life Sci 2(10):1247–1263

Shimada K, Crother TR, Karlin J, Chen S, Chiba N, Ramanujan VK, Vergnes L, Ojcius DM, Arditi M (2011) Caspase-1 dependent IL-1beta secretion is critical for host defense in a mouse model of *Chlamydia pneumoniae* lung infection. PLoS One 6(6):e21477. https://doi.org/10.1371/journal.pone.0021477

Shimokata K, Saka H, Murate T, Hasegawa Y, Hasegawa T (1991) Cytokine content in pleural effusion. Comparison between tuberculous and carcinomatous pleurisy. Chest 99(5):1103–1107

Silveira TN, Zamboni DS (2010) Pore formation triggered by *Legionella* spp. is an Nlrc4 inflammasome-dependent host cell response that precedes pyroptosis. Infect Immun 78 (3):1403–1413. https://doi.org/10.1128/iai.00905-09

Simitsopoulou M, Roilides E, Likartsis C, Ioannidis J, Orfanou A, Paliogianni F, Walsh TJ (2007) Expression of immunomodulatory genes in human monocytes induced by voriconazole in the presence of *Aspergillus fumigatus*. Antimicrob Agents Chemother 51(3):1048–1054. https://doi.org/10.1128/aac.01095-06

Sousa AR, Lane SJ, Nakhosteen JA, Lee TH, Poston RN (1996) Expression of interleukin-1 beta (IL-1beta) and interleukin-1 receptor antagonist (IL-1ra) on asthmatic bronchial epithelium. Am J Respir Crit Care Med 154(4 Pt 1):1061–1066. https://doi.org/10.1164/ajrccm.154.4.8887608

Srivastava KD, Rom WN, Jagirdar J, Yie TA, Gordon T, Tchou-Wong KM (2002) Crucial role of interleukin-1beta and nitric oxide synthase in silica-induced inflammation and apoptosis in mice. Am J Respir Crit Care Med 165(4):527–533. https://doi.org/10.1164/ajrccm.165.4.2106009

Stasakova J, Ferko B, Kittel C, Sereinig S, Romanova J, Katinger H, Egorov A (2005) Influenza A mutant viruses with altered NS1 protein function provoke caspase-1 activation in primary human macrophages, resulting in fast apoptosis and release of high levels of interleukins 1beta and 18. J Gen Virol 86(Pt 1):185–195. https://doi.org/10.1099/vir.0.80422-0

Sugawara I, Yamada H, Kaneko H, Mizuno S, Takeda K, Akira S (1999) Role of interleukin-18 (IL-18) in mycobacterial infection in IL-18-gene-disrupted mice. Infect Immun 67 (5):2585–2589

Sugawara I, Yamada H, Hua S, Mizuno S (2001) Role of interleukin (IL)-1 type 1 receptor in mycobacterial infection. Microbiol Immunol 45(11):743–750

Sutterwala FS, Mijares LA, Li L, Ogura Y, Kazmierczak BI, Flavell RA (2007) Immune recognition of *Pseudomonas aeruginosa* mediated by the IPAF/NLRC4 inflammasome. J Exp Med 204 (13):3235–3245. https://doi.org/10.1084/jem.20071239

Thomas SS, Chhabra SK (2003) A study on the serum levels of interleukin-1beta in bronchial asthma. J Indian Med Assoc 101(5):282, 284, 286 passim

Thomas PG, Dash P, Aldridge JR Jr, Ellebedy AH, Reynolds C, Funk AJ, Martin WJ, Lamkanfi M, Webby RJ, Boyd KL, Doherty PC, Kanneganti TD (2009) The intracellular sensor NLRP3 mediates key innate and healing responses to influenza A virus via the regulation of caspase-1. Immunity 30(4):566–575. https://doi.org/10.1016/j.immuni.2009.02.006

Tillie-Leblond I, Pugin J, Marquette CH, Lamblin C, Saulnier F, Brichet A, Wallaert B, Tonnel AB, Gosset P (1999) Balance between proinflammatory cytokines and their inhibitors in bronchial lavage from patients with status asthmaticus. Am J Respir Crit Care Med 159(2):487–494. https://doi.org/10.1164/ajrccm.159.2.9805115

Tremblay L, Valenza F, Ribeiro SP, Li J, Slutsky AS (1997) Injurious ventilatory strategies increase cytokines and c-fos m-RNA expression in an isolated rat lung model. J Clin Invest 99 (5):944–952. https://doi.org/10.1172/jci119259

Tsao TC, Hong J, Li LF, Hsieh MJ, Liao SK, Chang KS (2000) Imbalances between tumor necrosis factor-alpha and its soluble receptor forms, and interleukin-1beta and interleukin-1 receptor antagonist in BAL fluid of cavitary pulmonary tuberculosis. Chest 117(1):103–109

Villegas LR, Kluck D, Field C, Oberley-Deegan RE, Woods C, Yeager ME, El Kasmi KC, Savani RC, Bowler RP, Nozik-Grayck E (2013) Superoxide dismutase mimetic, MnTE-2-PyP, attenuates chronic hypoxia-induced pulmonary hypertension, pulmonary vascular remodeling, and activation of the NALP3 inflammasome. Antioxid Redox Signal 18(14):1753–1764. https://doi.org/10.1089/ars.2012.4799

Wang CC, Fu CL, Yang YH, Lo YC, Wang LC, Chuang YH, Chang DM, Chiang BL (2006) Adenovirus expressing interleukin-1 receptor antagonist alleviates allergic airway inflammation in a murine model of asthma. Gene Ther 13(19):1414–1421. https://doi.org/10.1038/sj.gt.3302798

Whyte M, Hubbard R, Meliconi R, Whidborne M, Eaton V, Bingle C, Timms J, Duff G, Facchini A, Pacilli A, Fabbri M, Hall I, Britton J, Johnston I, Di Giovine F (2000) Increased risk of fibrosing alveolitis associated with interleukin-1 receptor antagonist and tumor necrosis factor-alpha gene polymorphisms. Am J Respir Crit Care Med 162(2 Pt 1):755–758. https://doi.org/10.1164/ajrccm.162.2.9909053

Wilkinson RJ, Patel P, Llewelyn M, Hirsch CS, Pasvol G, Snounou G, Davidson RN, Toossi Z (1999) Influence of polymorphism in the genes for the interleukin (IL)-1 receptor antagonist and IL-1beta on tuberculosis. J Exp Med 189(12):1863–1874

Willingham SB, Allen IC, Bergstralh DT, Brickey WJ, Huang MT, Taxman DJ, Duncan JA, Ting JP (2009) NLRP3 (NALP3, Cryopyrin) facilitates in vivo caspase-1 activation, necrosis, and HMGB1 release via inflammasome-dependent and -independent pathways. J Immunol 183 (3):2008–2015. https://doi.org/10.4049/jimmunol.0900138

Wilson MS, Madala SK, Ramalingam TR, Gochuico BR, Rosas IO, Cheever AW, Wynn TA (2010) Bleomycin and IL-1beta-mediated pulmonary fibrosis is IL-17A dependent. J Exp Med 207 (3):535–552. https://doi.org/10.1084/jem.20092121

Witzenrath M, Pache F, Lorenz D, Koppe U, Gutbier B, Tabeling C, Reppe K, Meixenberger K, Dorhoi A, Ma J, Holmes A, Trendelenburg G, Heimesaat MM, Bereswill S, van der Linden M, Tschopp J, Mitchell TJ, Suttorp N, Opitz B (2011) The NLRP3 inflammasome is differentially activated by pneumolysin variants and contributes to host defense in pneumococcal pneumonia. J Immunol 187(1):434–440. https://doi.org/10.4049/jimmunol.1003143

Wolk KE, Lazarowski ER, Traylor ZP, Yu EN, Jewell NA, Durbin RK, Durbin JE, Davis IC (2008) Influenza A virus inhibits alveolar fluid clearance in BALB/c mice. Am J Respir Crit Care Med 178(9):969–976. https://doi.org/10.1164/rccm.200803-455OC

Wrigge H, Zinserling J, Stuber F, von Spiegel T, Hering R, Wetegrove S, Hoeft A, Putensen C (2000) Effects of mechanical ventilation on release of cytokines into systemic circulation in patients with normal pulmonary function. Anesthesiology 93(6):1413–1417

Wu J, Yan Z, Schwartz DE, Yu J, Malik AB, Hu G (2013) Activation of NLRP3 inflammasome in alveolar macrophages contributes to mechanical stretch-induced lung inflammation and injury. J Immunol 190(7):3590–3599. https://doi.org/10.4049/jimmunol.1200860

Xiang M, Shi X, Li Y, Xu J, Yin L, Xiao G, Scott MJ, Billiar TR, Wilson MA, Fan J (2011) Hemorrhagic shock activation of NLRP3 inflammasome in lung endothelial cells. J Immunol 187(9):4809–4817. https://doi.org/10.4049/jimmunol.1102093

Xu JF, Washko GR, Nakahira K, Hatabu H, Patel AS, Fernandez IE, Nishino M, Okajima Y, Yamashiro T, Ross JC, Estepar RS, Diaz AA, Li HP, Qu JM, Himes BE, Come CE, D'Aco K, Martinez FJ, Han MK, Lynch DA, Crapo JD, Morse D, Ryter SW, Silverman EK, Rosas IO, Choi AM, Hunninghake GM (2012) Statins and pulmonary fibrosis: the potential role of NLRP3 inflammasome activation. Am J Respir Crit Care Med 185(5):547–556. https://doi.org/10.1164/rccm.201108-1574OC

Xu P, Wen Z, Shi X, Li Y, Fan L, Xiang M, Li A, Scott MJ, Xiao G, Li S, Billiar TR, Wilson MA, Fan J (2013) Hemorrhagic shock augments Nlrp3 inflammasome activation in the lung through impaired pyrin induction. J Immunol 190(10):5247–5255. https://doi.org/10.4049/jimmunol.1203182

Yamada H, Mizumo S, Horai R, Iwakura Y, Sugawara I (2000) Protective role of interleukin-1 in mycobacterial infection in IL-1 alpha/beta double-knockout mice. Lab Investig 80(5):759–767

Yang HM, Ma JY, Castranova V, Ma JK (1997) Effects of diesel exhaust particles on the release of interleukin-1 and tumor necrosis factor-alpha from rat alveolar macrophages. Exp Lung Res 23 (3):269–284. https://doi.org/10.3109/01902149709087372

Yazdi AS, Guarda G, Riteau N, Drexler SK, Tardivel A, Couillin I, Tschopp J (2010) Nanoparticles activate the NLR pyrin domain containing 3 (Nlrp3) inflammasome and cause pulmonary inflammation through release of IL-1alpha and IL-1beta. Proc Natl Acad Sci U S A 107 (45):19449–19454. https://doi.org/10.1073/pnas.1008155107

Yucesoy B, Vallyathan V, Landsittel DP, Sharp DS, Weston A, Burleson GR, Simeonova P, McKinstry M, Luster MI (2001) Association of tumor necrosis factor-alpha and interleukin-1 gene polymorphisms with silicosis. Toxicol Appl Pharmacol 172(1):75–82. https://doi.org/10.1006/taap.2001.9124

Zhang Y, Lee TC, Guillemin B, Yu MC, Rom WN (1993) Enhanced IL-1 beta and tumor necrosis factor-alpha release and messenger RNA expression in macrophages from idiopathic pulmonary fibrosis or after asbestos exposure. J Immunol 150(9):4188–4196

Chapter 7
Inflammasome and Oral Diseases

Pedro Bullon, Luis E. Pavillard, and Rafael de la Torre-Torres

Contents

Abstract One of the main steps in the development of the life in the earth is multicellularity. It enables cell differentiation and the development of morphological structures within an organism and is an essential factor in how to recognize friendly cells that are part of the multicellular organism and which foreign organisms can be harmful. Recognition includes devices such as the major histocompatibility complex (MHC), and the pattern recognition receptors (PRRs). PRRs are a group of proteins expressed by cells of the innate immune system that identify two classes of products: pathogen-associated molecular patterns (PAMPs), related to microbial pathogens, and damage-associated molecular patterns (DAMPs), associated with cell components that are released during cell damage or death. All these activate the inflammasome, which is a multiprotein oligomer that includes caspase 1, PYCARD, NALP, and caspase 5 (also known as caspase 11 or ICH-3). It is responsible for

P. Bullon (✉) · L. E. Pavillard · R. de la Torre-Torres
Departament Periodontology, Facultad de Odontología, Universidad de Sevilla, Sevilla, Spain
e-mail: pbullon@us.es

M. D. Cordero, E. Alcocer-Gómez (eds.), *Inflammasomes: Clinical and Therapeutic Implications*, Experientia Supplementum 108,
https://doi.org/10.1007/978-3-319-89390-7_7

activation of inflammatory processes and has been shown to induce cell pyroptosis, a programmed cell death distinct from apoptosis, and promotes the maturation of the inflammatory cytokines interleukin 1β (IL-1β) and interleukin 18 (IL-18). We review whether inflammasome is related to diseases that can occur in the oral cavity. The mouth is always a possible environment for the development of pathological conditions because of the wide variety of microorganisms. Small variations in the equilibrium of the oral flora can cause disorders that could affect the organism in a systemic form. We provide data on periodontal disease, candidiasis, herpes virus, oral cancer, caries, and other oral diseases. There are very few papers that study this issue; therefore, we need more investigation and publications about inflammatory molecular processes, and more specifically, related to the inflammasome complex.

Keywords Periodontal diseases · Inflammasome · Oral diseases · Cancer · Caries

7.1 Introduction

Life is the ability to utilize and control energy derived from the sun that distinguishes every unicellular or multicellular organism. One of the main steps in the development of lives on Earth is the multicellularity by which the concept of what an individual organism is has been redefined and involves the transition from the microscopic to the macroscopic domain. Multicellularity enables cell differentiation and development of morphological structures within an organism that require cell–cell adhesion and intercellular communications to coordinate various activities (Lyons and Kolter 2015). It is an essential factor in how to recognize friendly cells that are part of the multicellular organism and which foreign organisms can be harmful. It is named allorecognition, which is defined as the ability of an individual organism to distinguish its tissues from those of another. Also, the different levels of cell development in an organism assume various types of energy consumption, the production of molecules, and waste disposal. All of them involve the production of molecules or damaged cellular structures that need to be identified and eliminated. When the external or internal aggression has been identified, the immune system starts to work and the host defense system comprising many biological structures and processes within an organism is activated. Therefore, it is essential for a cell that is part of a superior multicellular organism to have a device to detect foreign aggression that eliminates all the harmful molecules or structures. These devices include the major histocompatibility complex (MHC) and the pattern recognition receptors (PRRs). The MHC is a type of cell surface protein indispensable for the acquired immune system to identify foreign molecules in vertebrates that determines histocompatibility. PRRs are a group of proteins expressed by cells of the innate immune system that identify two classes of products: pathogen-associated molecular patterns (PAMPs), related to microbial pathogens, and damage-associated molecular patterns (DAMPs), associated with cell components that are released during cell damage or death. These may be on the membrane surface, e.g., Toll-like receptors (TLRs) and C-type lectin receptors (CLRs), or within the cytoplasm, e.g., NOD-like receptors (NLRs) and RIG-I-like receptors. All these activate the inflammasome, which is a multiprotein oligomer that includes caspase 1, PYCARD, NALP, and caspase 5 (also known as caspase 11 or ICH-3). It is responsible for activation of inflammatory processes and has been shown to induce cell pyroptosis, which is a programmed cell death distinct from apoptosis, and promotes the maturation of the inflammatory cytokines interleukin 1β (IL-1β) and interleukin 18 (IL-18) (Patel et al. 2017).

If life is energy, the main issue for all organisms is to obtain nutrients to keep the metabolic process active. In prokaryotic cells, the first step of nutrient intake consists of a break through the cell membrane with the mechanism of endocytosis. The main characteristic in the eukaryotic cells is the presence of membrane-bounded organelles that allow the energetic and metabolic advantage offered by the intracellular digestion of substrates (de Duve 2007). The multicellular organisms need more energy and have evolved a cell specialization by developing a gastrointestinal tract. It started as a simple blind sac-like structure and then became more heterogeneous, regionalized, and acquired a second opening, with a mouth and an anus (Soukup et al. 2013). The gastrointestinal tract takes in food, then digests it with the decomposition of highly insoluble food ingredients into small water-soluble food molecules, so that they can be absorbed into the blood plasma without any immunity reaction, and finally expels the remaining waste as feces. The oral cavity is the place in which the digestive process starts with chewing, salivation, and swallowing. Therefore, it is essential to keep it healthy and avoid diseases. However, the oral microbiota is abundant, diverse, and has high a capacity for producing pathological conditions. Oral diseases are among the most common human diseases. For instance, periodontal disease can affect up to 90% of the worldwide population and is an infectious disease related to some systemic conditions, such as cardiovascular disease and diabetes (Pihlstrom et al. 2005).

Our goal is to review whether the inflammasome is related to diseases that can occur in the oral cavity.

7.2 Periodontal Disease

Periodontal disease is a chronic inflammatory illness that affects many adults, and it is characterized by a chronic infection related to Gram-negative anaerobic bacteria in the dental biofilm. It leads to the irreversible destruction of tissues supporting the teeth and is clinically detectable by periodontal pockets and alveolar bone loss (Pihlstrom et al. 2005). Studies report how severe periodontitis affects 5–20% of adult populations worldwide, and it is one of the preeminent causes of tooth loss in both developed and developing countries (Pihlstrom et al. 2005; Petersen et al. 2005; Jin et al. 2011). It has been suggested that over 50% of the European population suffer from some form of periodontitis and over 10% have severe disease, with prevalence increasing to 70–85% of the population aged 60–65 years (König et al. 2010).

In nature, the progression of periodontitis is inflammatory, with the main triggers of oral inflammation usually residing in the oral microbes and the balance of its components (Yilmaz and Lee 2015). Within the host's inflammatory response, various biochemicals are strongly associated with the severity and progression of periodontal disease, interleukins (IL-1, IL-18), prostaglandins (PGE2), and matrix metalloproteinases (MMPs) (Orozco et al. 2006). IL-1β has been found to be significantly increased in the periodontal tissues and gingival fluid from diseased sites, compared with healthy sites (Stashenko et al. 1991; Masada et al. 1990). IL-1β up-regulates MMPs and down-regulates tissue inhibitors of metalloproteinase production (Ohshima et al. 1994), and is also a powerful and potent bone-resorbing cytokine (Schwartz et al. 1997), which suggests that it plays a role in degrading the extracellular matrix in periodontitis (Bascones-Martínez et al. 2012).

Although inflammasome signaling is becoming well established in the progression of various diseases, there is intriguingly mounting evidence supporting the association of the oral microbiome with the same array of conditions (Han and Wang 2013). The oral cavity is essentially a diverse ecosystem, harboring vast numbers of oral microorganisms, and can serve as a reservoir for possible systemic dissemination of microorganisms or their components and the release of inflammatory signals, possibly leading to inflammation at distant body sites (Amodini Rajakaruna et al. 2012). With advances in technologies for microbial detection, a diverse group of oral species has additionally been directly detected in several systemic chronic diseases (Detert et al. 2010).

Inflammasomes are emerging as chief regulators of the host innate immune defense system in chronic inflammatory diseases, and their role against microbial pathogens is becoming critical in controlling and limiting invading microbes. On the other hand, increasing numbers of microorganisms and their virulence factors are found to function by targeting inflammasomes and modulating IL-1β and IL-18 processing, which, taken together, could be involved in the development and/or progression of various inflammatory diseases, including periodontal disease (Davis et al. 2011; Kim and Jo 2013).

7.2.1 NLRP3 in Periodontitis

Inflammasome complexes appear to assume a pivotal role in periodontal disease and the inflammasome-associated inflammatory mediators involved in the progression of the disease have been highlighted by several clinical studies. The relationship between the interleukin-1 cytokine family and the NLRP3 inflammasome complex has been revealed by Bostanci et al. (2009). The findings indicated that higher expression levels of NLRP3 and NLRP2, but not of apoptosis-associated speck-like protein containing a carboxyl-terminal caspase recruitment domain (ASC), were detected in gingival tissue samples from patients with three forms of the periodontal disease (gingivitis, chronic periodontitis [CP], and generalized aggressive periodontitis [G-AgP]) compared with healthy subjects. The mRNA expression of NLRP3, its putative antagonist NLRP2, its effector molecule ASC, IL-1β, and IL-18 are detected by quantitative real-time polymerase chain reaction (qPCR) in gingival tissues from patients with gingivitis, CP, G-AgP, and healthy subjects. The data indicated that NLRP3 and NLRP2, but not ASC, were significantly expressed at higher levels in the three forms of the disease compared with healthy subjects. Park et al. (2014) had similar results in gingival crevicular fluid and gingival tissue samples from peritonitis sites compared with healthy sites. Examining the distribution and intensity of NLRP3, NLRP1, and AIM2 expression in gingival tissues with different types of periodontitis (patients with chronic and G-AgP compared with healthy subjects) by qPCR and immunohistochemistry, Xue et al. (2015) found that overall intensity of NLRP3 expression was significantly higher in CP or G-AgP than in healthy tissue, more considerably in the periodontal epithelium. NLRP1 was barely expressed in the samples, whereas absent in melanoma 2 (AIM2) was better represented in the CP group. In another study, gingival samples from patients with CP, and with CP and type 2 diabetes mellitus compared with healthy subjects, were analyzed using immunohistochemistry (Huang et al. 2015). Compared with control subjects, NLRP3 and IL-1β were significantly up-regulated in the gingival epithelium of patients with CP and/or type 2 diabetes mellitus, but the authors observed no differences between groups with periodontitis with or without type 2 diabetes mellitus. Recently, a study in NLRP3-deficient mice infected with *Porphyromonas gingivalis* showed that this bacterial challenge significantly increased the loss of alveolar bone; gingival expression of pro-IL-1β, pro-IL-18, and receptor activator of nuclear factor kappa-B ligand; production of IL-1β, IL-6, and IL-18; and caspase-1 activity in peritoneal macrophages of wild-type mice. In contrast, it did not affect NLRP3-deficient mice. Meanwhile, mRNA expression of OPG in gingival tissue and peritoneal IL-6 production were significantly higher in NLRP3 knock-out (KO) mice (Yamaguchi et al. 2017). This evidence suggests that different inflammasomes, such as NLRP1, NLRP2, NLRP3, and AIM2, and their products IL-1β and IL-18 are over-expressed in gingival tissues from patients with periodontal diseases compared with healthy controls, confirming that these complexes are involved in the pathogenesis of periodontitis to different degrees.

Inflammatory and immune mechanisms turned on by infectious agents are crucial in the development of atherosclerosis. Numerous epidemiological studies have demonstrated that host immune reactions against persistent infectious pathogens, including *Porphyromonas gingivalis*, may promote the development of atherosclerosis (Kurita-Ochiai et al. 2015). In 2015, Yamaguchi et al. challenged orally spontaneously hyperlipidemic mice with wild-type *P. gingivalis* significantly increasing the area of aorta covered with atherosclerotic plaque and alveolar bone loss, compared with gingipain-null mutant or FimA-deficient mutant strains (Yamaguchi et al. 2015). The challenge also increased IL-1β, IL-18 and TNF-α production in peritoneal macrophages, and gingival or aortic gene expression of NLRP3, pro-IL-1β, pro-IL-18, and pro-caspase-1. In another study performed by Velsko et al. (2015), the ability of a polymicrobial consortium of *P. gingivalis*, *Treponema denticola*, *Tannerella forsythia*, and *Fusibacterium nucleatum* to colonize the periodontium and induce local and systemic inflammatory responses was investigated. The author's observations suggested that polybacterial infection with periodontal pathogens triggers aortic TLR and inflammasome signaling and increases aortic oxidative stress.

7.2.2 Porphyromonas gingivalis

Porphyromonas gingivalis is the bacterium most frequently associated with CP and can be detected in up to 85% of the disease sites (Yang et al. 2004), whereas in healthy sites, it is rarely found or only in small numbers (Bostanci and Belibasakis 2012). The presence of *P. gingivalis* in periodontal pockets may be a predictor of disease progression (van Winkelhoff et al. 2002) and a positive correlation is found between the quantity of *P. gingivalis* and pocket depth (Kawada et al. 2004). *P. gingivalis*, like all Gram-negative bacterial species, is covered by a lipopolysaccharide (LPS), which is a component of the outer membrane recognized by the host that can trigger intracellular signaling events. The affinity of LPS to its PRRs, such as the TLRs and CD14, enables discernment between commensal and pathogenic species. The *P. gingivalis* LPS is a stimulator of proinflammatory responses and bone resorption, as demonstrated in experimental animal models. Owing to the importance of this bacterium in periodontal diseases, several in vitro studies have investigated its effects on different cell populations of the periodontium.

One of the strategies for observing the effects of *P. gingivalis* on the tissues is to challenge a certain cell line with live cultures of this bacterium, an approach that several studies have followed. Bostanci et al. (2009) challenged the human myelomonocytic cell line, Mono-Mac-6, with *P. gingivalis*, and observed that although the untreated cells showed low levels of expression of NLRP3, the infected cells showed a high concentration of expression of NLRP3, IL-1β, and IL-18 (Yamaguchi et al. 2015). Park et al. in a recent in vitro study, examined the mechanisms of activation of NLRP3 and IL-1β secretion in a human acute monocytic leukemia cell type (THP-1) differentiated to macrophages. It has been discovered that activation of both NLRP3 and AIM2 is necessary for the secretion of *P. gingivalis*-induced caspase-1-dependent IL-1β via TLRs 2 and 4. Some studies have added DAMPs to their methodology. Yilmaz et al. found that *P. gingivalis* down-regulated NLRP3 expression and induced production of pro-IL-1β, but only promoted the secretion of mature IL-1β upon stimulation with danger signal extracellular ATP in a primary gingival epithelial cell model (Yilmaz et al. 2010). Moreover, *P. gingivalis* may induce the production of DAMPs itself. Jun et al. observed that *P. gingivalis* induced activation of caspase-1, caspase-4, and induced pyroptotic cell death in THP-1-derived macrophages, but only at low concentrations. These results suggest that *P. gingivalis* might modulate the host immune responses, in favor of pathogen survival and persistence. *P. gingivalis* induced the release of ATP, too, which ultimately leads to caspase-1 activation (Jun et al. 2017).

Another strategy for studying *P. gingivalis* in vitro has been to use culture supernatants or the bacterial LPS as PAMPs. Hamedi et al. reported that

P. gingivalis supernatants differentially regulated IL-1β and IL-18 from human monocytes. *P. gingivalis* enhanced IL-1β and IL-18 mRNA expression, the former being induced earlier, but transiently. IL-18 up-regulation was not affected by heat inactivation or chemical inhibition of gingipains, whereas both treatments resulted in a 50% reduction of IL-1β expression. Purified *P. gingivalis* LPS enhanced both IL-1β and IL-18 expression. However, only IL-1β, but not IL-18, secretion was detected and was up-regulated by *P. gingivalis*. Therefore, cytokines of the IL-1 family may participate via different pathways in the pathogenesis of periodontitis (Hamedi et al. 2009). Champaiboon et al. (2014) challenged primary human monocyte-derived macrophages (M1 and M2 macrophages) and human coronary artery endothelial cells (HCAECs) with *P. gingivalis* LPS as PAMPs and cholesterol crystals (CCs) as DAMPs. The authors found a marked release of IL-1β from LPS-primed M1 and M2 macrophages treated with CCs. On the other hand, HCAECs showed no release of IL-1β in response to *P. gingivalis* LPS priming and treatment with either CCs or extracellular danger molecule ATP. The authors conclude that the mechanistic role of periodontal infection in inflammasome activation as a cause of atherosclerotic vascular disease requires further investigation (Champaiboon et al. 2014).

Some authors have studied strategies to reduce the expression of NLRP3 and its products. Li et al. induced the production of heme oxygenase-1 (HO-1), a ubiquitous inducible cellular stress protein and an endogenous cytoprotective enzyme, by hemin on gingival epithelial cells (GECs) and compared the results when cells were challenged with LPS, with or without hemin. The cells cultivated with LPS + hemin demonstrated less NLRP3 formation and overexpression and lower production of IL-1β, leading to the conclusion that the activation of HO-1 protects LPS-induced inflammatory damage in GECs and that it may be used as a target for the prevention and treatment of CP (Li et al. 2014).

Some researchers suggest that *P. gingivalis* might modulate specific inflammasome components and successfully colonize and persist in host cells. Belisakis et al. challenged gingival fibroblasts with a 10-species and a 9-species biofilm model, the second without *P. gingivalis*. The authors observed that the exclusion of *P. gingivalis* from the biofilm partially rescued NLRP3 and IL-1β expression, concluding that subgingival biofilms down-regulate NLRP3 and IL-1β expression, partly because of *P. gingivalis* (Belibasakis et al. 2013). Another study by Taxman et al. (2012) performed in an in vitro mouse macrophage model, demonstrated that *P. gingivalis* could synergistically regulate the invasion of host cells by *F. nucleatum* by inhibiting both *F. nucleatum*-induced IL-1β and IL-18 processing and *F. nucleatum*-promoted cell death (Taxman et al. 2012).

P. gingivalis is not an aggressor of the inflammatory response, but rather an opportunist that can cross-talk with the host and subvert its defense mechanisms. Using this strategy, *P. gingivalis* prolongs its survival and becomes established in the periodontal pocket (Hajishengallis et al. 2011). Preferably, it deregulates the innate immunity, which may, in turn, impair adaptive immunity (Pathirana et al. 2010). Major representative examples of these abilities are its capacity to degrade human defensins (Carlisle et al. 2009), its resistance to oxidative burst killing by polymorphonuclear neutrophils (PMNs) (Mydel et al. 2006) and its ability to inhibit "at will" the production of critical proinflammatory cytokines (Bostanci et al. 2007).

7.2.3 Aggregatibacter actinomycetemcomitans

Aggregatibacter actinomycetemcomitans is a gram-negative capnophilic coccobacillus, which is commonly isolated from the oral cavity of adolescents and young adults afflicted by aggressive periodontal disease states (Slots et al. 1980). In localized juvenile periodontitis, it was described as affecting first molars and then incisors preferentially (Armitage 1999). *A. actinomycetemcomitans* possesses different well-studied virulence factors, among which leukotoxin is suggested to play a significant role in the pathogenicity. Leukotoxin belongs to the repeats-in-toxin (RTX) family, which is produced by many other Gram-negative pathogens such as *Actinobacillus pleuropneumoniae*, *Escherichia coli*, *Bordetella pertussis*, and *Mannheimia haemolytica*. The toxins of the RTX family selectively kill human leukocytes by inducing apoptosis and lysis (Kelk et al. 2011).

As with *P. gingivalis*, some studies have researched the effects of *A. actinomycetemcomitans*, using live strains of the bacteria. Zhao et al. infected human osteoblastic cells with *A. actinomycetemcomitans* and observed that the apoptosis of the cells was significantly enhanced. Expression levels of both NLRP3 and ASC were increased dramatically after exposure to the bacteria. The secretion of mature IL-1β and IL-18 was extensively induced in infected cells compared with the non-invasion group (Zhao et al. 2014). In another in vitro study, leukotoxin and cytolethal distending toxin gene KO mutant strains of *A. actinomycetemcomitans* were used to challenge human mononuclear leukocytes. Only up-regulation of NLRP3, IL-1β, IL-18, and reduction of NLRP6 were observed, but no other inflammasome components, such as ASC, were affected (Belibasakis and Johansson 2012).

As for studies using leukotoxin to challenge different cell strains in vitro, Kelk et al. (2005) tested human macrophages with leukotoxin or LPS from *A. actinomycetemcomitans* or LPS from *Escherichia coli*. Leukotoxin induced abundant production of IL-1β and caspase 1 compared with controls, proving that leukotoxin from *A. actinomycetemcomitans* can trigger inflammatory reactions on other cells, not only on leucocytes (Kelk et al. 2005).

The common pathway usually described for the induction of IL-1β is through the NLRP3/caspase-1 pathway, but Okinaga et al. (2015), in an in vitro study on mouse macrophages, described periodontopathic invasion with *A. actinomycetemcomitans* inducing the production of ROS and the release of cathepsin B. Moreover, IL-1β processing was down-regulated by inhibition of these molecules, but not caspase-1 or NLRP3, suggesting that *A. actinomycetemcomitans* invasion in mouse macrophages might induce IL-1β production, which is dependent upon ROS and cathepsin B, but not NLRP3/caspase-1 activity (Okinaga et al. 2015).

Xylitol is a well-known anti-caries and anti-inflammatory agent, but its effect on the inflammasome had not been researched. Kim et al. (2016) studied the effects of inflammasome activation, by supplementing xylitol to macrophages infected with *A. actinomycetemcomitans*. The authors observed that xylitol inhibited the production of IL-1β and AIM2 inflammasome, seen in the control group, by suppressing the internalization of *A. actinomycetemcomitans* into cells (Kim et al. 2016). Xylitol may be used as a therapeutic weapon for the prevention of periodontal inflammation caused by *A. actinomycetemcomitans*.

7.2.4 Treponema denticola, Tannerella forsythia, *and* Mycoplasma salivarium

Treponema denticola is a spirochete that is identified in several gingivitis cases, especially its presence in necrotizing ulcer gingivitis, root canal infection, and acute apical abscesses. *T. denticola* is a relevant pathogen in periodontal and pulpal processes, its aggressiveness is due to a diversity of virulence factors, emphasizing its dentilisin, mobility and its ability to modulate the defensive response of the host (Dashper et al. 2011). *T. forsythia* is an anaerobic Gram-negative member of the *Cytophaga-Bacteroides* family, which was initially described as *Bacteroides forsythus* by Tanner and Stillman (1993) and later reclassified as *Tannerella forsythia* by Sakamoto et al. (2002) based on 16S rRNA phylogenetic analysis. *T. forsythia* is associated more frequently and/or at higher levels with various forms of the disease, including gingivitis and chronic and aggressive periodontitis, than with health. Several studies have also implicated *T. forsythia* in the progression of clinical attachment loss associated with periodontitis (Sharma 2010).

Jun et al. (2008, 2012) in two in vitro studies, described a pathway of how *T. denticola* activates the NLRP3 inflammasome through Td92, a protein present on its surface. The direct interaction of Td92 with the cell membrane integrin α5β1 resulted in ATP release and K+ efflux, which are the main events in NLRP3 activation (Jun et al. 2008, 2012). Jun et al. also studied how macrophages reacted to the infection with *T. denticola* and *T. forsythia*. Both *T. denticola* and *T. forsythia*, induced pyroptotic cell death and the activation of caspase-1 and caspase-4 in macrophages (Jun et al. 2017).

Mycoplasmas, the smallest self-replicating microorganisms without cell walls, cause various infectious diseases in humans and animals, such as atypical pneumonia, nongonococcal urethritis, and arthritis. *Mycoplasma salivarium* is a non-fermenting species and is part of the human oral microbial flora and inhabits the level of gums and dental plaque. This microorganism is isolated more frequently from the periodontal cavity of the subjects with this disease (Engel and Kenny 1970). The antibody response to this mycoplasma is significantly high in patients compared with healthy subjects (Watanabe et al. 1986). *Mycoplasma salivarium* induces the production of IL-6 and IL-8 in gingival fibroblasts (Shibata et al. 1997).

Sugiyama et al. (2015) studied the association between *M. salivarium* and periodontitis and elucidated the etiological roles of *M. salivarium* in periodontal diseases. Their study determined whether *M. salivarium* can activate the inflammasome to induce IL-1β production by innate immune cells such as dendritic cells or macrophages and, if so, what kinds of inflammasomes are activated by *M. salivarium*, *M. pneumoniae*, and their heat-killed cells. The authors observed that live and heat-killed *M. salivarium* and *M. pneumoniae* cells induced the production of IL-1β by dendritic cells and pyroptosis. Live *M. salivarium* and *M. pneumoniae* lost the ability to induce IL-1β production by macrophages from ASC- and caspase-1-deficient mice almost completely, but not entirely on macrophages from NLRP3-deficient mice. These results suggest that live *M. salivarium* and *M. pneumoniae* might be capable of activating several types of inflammasomes including the NLRP3 inflammasome (Sugiyama et al. 2015).

7.3 Other Oral Infectious Diseases

7.3.1 Candidiasis

Humans need to be protected from the damage a huge variety of microorganisms can cause. When we talk about the fungi kingdom we need to highlight the *Candida* family and especially the *Candida albicans* species. *Candida* is an ascomycete (Arendorf and Walker 1979), opportunistic (Repetto et al. 2012), polymorphic fungi. Fungal infections are becoming increasingly prevalent (Richardson and Moyes 2015). *Candida* species are the fourth most common pathogens in nosocomial bloodstream infections in the USA and Europe (Chen et al. 2013). *Candida albicans* colonizes in an asymptomatic way 65% of healthy people (Joly and Sutterwala 2010). Its overgrowth is limited by competing commensal bacteria and host defense (Tomalka et al. 2011). Alterations in the normal flora cause *Candida* overgrowth. It is usually produced by antibiotic treatment or immunocompromised states such as AIDS, during chemotherapy or following allogenic transplantation (Joly and Sutterwala 2010). *Candida* overgrowth results on oropharyngeal candidiasis (OPC, also recognized as thrush) or denture stomatitis (Abu-Elteen and Abu-Alteen 1998).

Candida albicans can grow in several forms: unicellular yeast, pseudo hyphae and hyphae. *Candida* is capable of changing the morphological and physical structure during growth and this development is reversible, which helps to potentiate its pathogenicity. Multiple forms are often found simultaneously. On the one hand, *Candida* as unicellular yeasts is typically associated with widespread dissemination (commensalism), controlled by neutrophils and macrophages. On the other hand, growth of pseudohyphae and hyphae is commonly shown in infections of the mucosal surfaces, controlled by T-cells (Repetto et al. 2012).

There are very few papers that directly relate inflammasome to *Candida albicans* and oral diseases, none of them in humans, some in vivo using murine models, and others in vitro. The experiments carried out in mice used the same strain. All were female C57BL/6 mice. The method used to infect the mice was very similar in all the studies presented. After a brief period with antibiotic coverage, mice were infected with the fungus. Small scratches were made on the dorsal surface of the tongue, limited to the stratum corneum (Hise et al. 2009).

In one of the publications the authors compared wild-type (WT) mice with interleukin 1 (IL-1) receptor-deficient mice (IL-1r1−/−). They studied the effect of *Candida albicans* infection depending on the time since inoculation. Mice were divided into groups according to the moment of the sacrifice. Three, 7, 14, and 21 days after infection the animals were euthanized and scored clinically. Samples from tongue and kidney were removed to determine local and systemic grades of infection. Results show how IL-1r1−/− mice had higher levels of local colonization and systemic dissemination. This group also showed lower survival rates compared with WT mice. The same findings were observed comparing WT mice with caspase 1 (casp-1)-, NLRP3-, and ASC-deficient mice (Hise et al. 2009).

In another study, the periods of time when the mice were euthanized were the same as in the study explained above. In this case, WT mice were compared with NLRP3-, NLRC4-, and ASC-deficient mice. Tongues were removed after sacrifice and evaluated with a microscope. Comparing two inflammasome complexes of the NLR family, it is shown how genetic knock-down of a single NLR inflammasome could have an important effect on the expression of other NLR proteins. NLRC4-deficient mice responded worst to Candida infection than NLRP3 KO mice (NLRP3−/−) and WT mice measuring the same parameters described above. These data reveal that NLRP3 and NLRC4 play different roles in immunity against *Candida*, NLRC4 being more important in the last period of the fungal infection (Tomalka et al. 2011).

It is also important to mention that we have seen a relationship between the results found from studies that discuss oral and vulvovaginal candidiasis. Similar environmental conditions and characteristics are described. There is a paper where the authors used mice of the same strain, but in this case, the inoculation was intravaginal, concentrating their efforts on studying the role of IL-22. It has been demonstrated that IL-22 controls the process by which NLRP3 recruits neutrophils and promotes inflammation when it is activated. IL-22 also activates NLRC4 so that it can produce IL-1 receptor antagonist (IL-1Ra). WT mice were compared with NLRP3- and NLRC4-deficient mice. Interestingly, NLRP3 inhibition matches NLRC4 activation. NLRP3 is associated with casp-1 activation, polymorphonuclear activation, and inflammatory damage in vulvovaginal candidiasis. NLRC4 is suggested to be part of a process that limits inflammation. Continuing with the study, they used human samples of vulvovaginal candidiasis to examine results in vitro. The results found were the same as in the murine model. They conclude that a lower production of IL-1Ra and IL-22 might be a risk factor for recurrent vulvovaginal candidiasis (Borghi et al. 2015).

Macrophages from WT in addition to NLRP3- and ASC-deficient mice were evaluated in the presence of *Candida albicans* to investigate the NLRP3–ASC–caspase-1 axis and IL-1β production. An active form of IL-1β was only located in supernatants of WT macrophages. The induction and processing of IL-1β are controlled and mediated by the NLRP3–ASC–caspase1 axis (Hise et al. 2009).

7.3.2 Herpes Virus

Many species of virus can affect the oral tissues. We are going to look for information about the herpes virus (HV) family. We can find the herpes simplex virus (HSV), human cytomegalovirus (HCMV), varicella zoster virus (VZV), and Epstein–Barr virus (EBV). HSV is the most frequently studied, particularly type 1 (HSV-1). HSV-1 is frequently associated with facial and oral lesions. HSV-2 is related to genital and neonatal infections. We briefly explained some of the characteristics of HSV-1.

HSV-1 is a ubiquitous (Xu et al. 2006), common and highly contagious pathogen that infects most people (90%) (Smith and Robinson 2002). It is an icosahedral, enveloped, nuclear-replicating, double-stranded DNA virus belonging to the neurotropic *Alphaherpesvirinae* subfamily. Primary infection occurs during childhood and usually affects the oral mucosa. In most cases, with very few symptoms, making it very difficult to diagnose correctly at an early stage. A lifelong infection takes place in the trigeminal ganglia, keeping the virus latent. It periodically activates to a lytic state producing recurrent lesions at the site of primary infection. HSV can cause fatal infections such as encephalitis, blindness, or even the death in immunocompromised patients.

Breast cancer tumor suppressor protein (BRCA1) and interferon inducible protein 16 (IFI16) were studied in human microvascular dermal endothelial cells (HMVECs) and human fibroblasts. BCRA1 is a DNA damage repair sensor and transcription regulator. IFI16 is a restriction factor for HCMV and HSV-1 (Johnson et al. 2014). IFI16 is a sequence-independent nuclear innate sensor recognizing nuclear replication of herpes virus such as Kaposi's sarcoma-associated herpes virus (KSHV), EBV or HSV-1. The recognition takes place in the infected cell nucleus, then forming an inflammasome complex with ASC and pro-caspase-1 (Johnson et al. 2013). It finally produces IL-1β. BCRA1 works together with the IFI16 inflammasome, increasing expression levels during de novo KSHV, EBV, and HSV-1 infection, and in latent KSHV or EBV infection (Dutta et al. 2015a). In a similar paper, Johnson et al. (2013) demonstrate that IFI16 and NLRP3 inflammasomes were activated by HSV-1 infection, promoting IL-1β maturation. Although the host immune system responds to the virus, the authors explained that HSV-1 has evolved defense mechanisms to evade host reaction (Johnson et al. 2013).

The interaction of HVS1 infection with TLR2 is critical, modulating the production of proinflammatory cytokines, for example, IL-6. Excessive signaling can cause too much inflammation and tissue damage (Wang et al. 2012). The balance between proinflammation and down-regulation mechanisms is important. CD200R1 is a protein expressed on myeloid and glial cells, which interacts with CD200 expressed on neurons, epithelial cells, endothelial cells, and lymphocytes. Their junction occurs to initiate inhibitory signaling (Mihrshahi and Brown 2010). CD200R1-deficient mice generated lower levels of IL-1β and IL-6. They are also protected from intracranial HSV-1 infection, increasing the survival rates (Soberman et al. 2012).

Herpes virus infection causes NLRP3 redistribution to the nucleus. Mice and human corneal tissues were used to study keratitis and how NLRP3 was expressed with HSV-1 infection. NLRP3, Casp-1, and IL-1β levels were higher in mice infected with HSV-1, presenting a partial redistribution of NLRP3 to the nucleus. The same results were found in human tissues (Wang et al. 2015). In contrast, Miettinen et al. (2012) published a paper experimenting with human macrophages in which they affirmed that the inflammasome is not activated in response to HSV-1 infection. HSV-1 infection can activate a wide variety of proteins in the absence of inflammasome activation (Miettinen et al. 2012).

HSV-1 infection was studied in KO and normal NLRP3 mice in facial infection and keratitis. Deficient mice had more severe and earlier keratitis, and higher levels of IL-1β and IL-18 in the early stages of infection. Elevated recruitment of neutrophils and elevated levels of CD4+ T cells occurred at advanced stages of infection. To conclude, the NLRP3 inflammasome plays a specific role against keratitis pathogenesis and acts as a regulator and a beneficial support (Gimenez et al. 2016).

We finish describing the HV family by speaking of VZV. VZV induces formation of the NLRP3 inflammasome and consequently the production of IL-1β. This was shown in human T helper-1 (TH-1) cells, fibroblasts, and melanoma cells. NLRP3 recruitment is independent of AIM2, revealing different pathways. VZV triggers formation of the NLRP3 inflammasome complex with activated caspase-1 in the absence of AIM2 (Nour et al. 2011).

7.4 Cancer

Oral cancer is the most common cancer of epithelial origin in the head and neck and the sixth most common cancer overall. Nowadays, oral cancer represents 3% of all new cancers diagnosed and this is rapidly increasing (Parkin et al. 2005; Reid et al. 2000). Most oral cancers are oral cavity squamous cell carcinomas (OSCCs). They are usually locally aggressive with moderate recurrence (Funk et al. 2002). The 5-year survival after treatment (surgery, radiotherapy, and chemotherapy) is only 50% because the diagnosis is often made in the late stages (Siegel et al. 2016; Ferlay et al. 2015). The most frequent locations are the tongue, buccal area, gingiva, lips, floor of the mouth, and hard palate. It is important to clarify that most of the cases start from a potentially malignant disorder that is important to detect and treat as soon as possible. The continuous exposure of this lesion to carcinogens such as tobacco, alcohol or chronic inflammation (25% of malignancies are associated with chronic inflammation and/or infection) (Mantovani et al. 2008) may force it evolve to malignization by a process called carcinogenesis (Lippman et al. 2005).

NLRP3 inflammasome was studied in 20 biopsied cases of OSCCs including the malignant tumor and the adjacent nonpathological tissues. It was revealed that the expression levels of NLRP3 inflammasome-associated genes (ASC, casp-1, Il-1β , and NLRP3) were higher in the tumor tissues, in addition to the protein expression levels, using immunohistochemical (IHC) techniques. These levels are related to clinical and pathological characteristics of OSCCs. The authors asked themselves how these levels could influence the overall survival (OS), disease-specific survival (DSS), and disease-free survival (DFS), showing that the up-regulation of ASC was the only independent predictor. Finally, they also added to their paper an experiment in vitro where they concluded that ASC facilitates migration and invasion of OSCC cell lines, promoting metastasis (Wu et al. 2016).

There are many publications related to cancer from South Asia, more precisely Taiwan. One of these papers shows the consequences of a high prevalence of a traditional custom, such as betel nut chewing. The total number of patients (all of them men) were divided into four groups: young control, middle-aged control, betel nut chewing, and oral cancer. An ELISA technique was used to measure cytokines and hormones: IL-1β, IL-6, IL-15, tumor necrosis factor α (TNF-α), cortisol, and testosterone in plasma. The results published show elevation of cortisol, lower levels of IL-1β, Il-15, and TNF-α, and lower testosterone concentrations in the betel nut chewing group (Hu et al. 2016).

Seven OSCC-related salivary biomarkers of periodontitis were studied in several groups: smokers, nonsmoker patients, patients without periodontal disease but with OSCC, and control patients (a total of 105 patients). The only biomarker that showed significant differences between the OSCC patient group and the rest was S100 calcium binding protein P (S100P) being much higher in the OSCC group. Cytokines that promote inflammation such as IL-1β and Il-8 did not show statistically differences between the groups (Cheng et al. 2017).

There are several papers that focus on IL-18, particularly the Il-18 polymorphism −607 A/C. Nilkaeo and Bhuvanath (2006) suggested that IL-18 might be a proinflammatory cytokine produced by cancer cells. On the one hand, Tsai et al. (2013) concluded in their paper that the IL-18 −137 G/C gene polymorphism could be considered an important factor that increases the susceptibility to OSCC, although they also suggested that it could be proposed as a protective factor for OSCC progression (Tsai et al. 2013). On the other hand, Eleftherios et al. tell us that IL-18 −607 A/C polymorphism gene expression is not a contributing factor in oral carcinogenesis or risk of cancer (Vairaktaris et al. 2007).

Interleukin-18 plays an important role in the regulation of OSCCs, specifically when it is localized in the tongue. It promotes the growth inhibition of OSCCs and helps in the regulation of cell apoptosis and gene expression. It could be used in the future after further investigations as a potential clinical and therapeutic target (Liu et al. 2012).

Interleukin-1β is also a very well recognized proinflammatory cytokine that can regulate cancer growth and metastasis in a secondary position. This is possible because it interacts by controlling the expression of CXCR4, a specific chemokine that regulates processes in cancer development. It is also suggested that CXCR4 may be a link between inflammation and cancer (Sun et al. 2015). In addition, this cytokine may be used as a marker of malignant transformation in oral leukoplakia, one of the most frequently studied potentially malignant disorders (Dutta et al. 2015b).

In brief, we have just discussed IL-1β and IL-18, suggesting that an important relationship might exist among inflammation, the inflammasome, and cancer. High levels of IL-1β and IL-18 are associated with cancer. Likewise, activation of the autophagic process or mitochondrial oxidative stress are related to cancer. All this information is important in continuing to investigate cancer genesis and prognosis (Zhiyu et al. 2016).

7.5 Caries

Pulp lesions can be caused by many factors. Dental caries is the most prevalent cause. It is a chronic infectious disease. It provokes the demineralization of the most superficial tissue the teeth have, the enamel, and the subsequent interior tissue, called dentin, causing pulp injury (Akira et al. 2006; Bolwell 1999). We are not aware of the prevalence of dental pulp lesions. Clinically, we can classify them as pulp (dentin hypersensitivity, pulpitis, and necrosis) and periapical diseases (apical periodontitis and others with no endodontic etiology) (Moller et al. 1981). Experimental studies demonstrate that the presence of bacteria is essential for the development and progression of pulp inflammation (Kakehashi et al. 1965).

Dental pulp reacts to caries infection with inflammation. It has some peculiarities, such as its location in a hard chamber or the exclusive blood and lymphatic circulation. These facts make pulp inflammation processes difficult to control and resolve. Pulpitis is the most common inflammatory disease in mammals. We describe pulpitis as a nonspecific inflammation of the dental pulp. If it remains untreated, it can lead to the patient's death, but it is commonly solved with soft dental tissue removal and root canal treatment (Akira et al. 2006). Pulpitis occurs when caries infection has contact with dentine tubules, which are connected to the pulp tissues (Bortoluci and Medzhitov 2010). The first line of defense the teeth have against caries progression is the activation of the innate immune system components located in the pulp (Chang et al. 1998).

Apical periodontitis is also an inflammatory lesion that occurs in the most apical zone of the teeth because of bacterial infection of endodontic–periodontic origin. It is characterized by bone resorption in this location (Hong et al. 2004).

The relationship between pulpitis and the presence of the AIM2 inflammasome was an unknown fact. The authors exposed only the dental pulp of the right maxillary molars of male Sprague–Dawley rats and compared the levels of expression of AIM2 in both sides. In the in vitro study they cultivate the cells of the mandibular incisors of those rats. AIM2 inflammasome expression was higher in damaged tissues. IL-1β and caspase-1 levels increased whereas AIM2 levels were higher. A direct relationship between AIM2 inflammasome and inflammatory processes such as pulpitis was affirmed (Wang et al. 2013).

The NLRP3 inflammasome complex has been studied related to pulpitis in humans. The third molars of 27 people were extracted: 9 of them had no pathological condition, 9 had reversible pulpitis, and the remaining 9 had irreversible pulpitis. Dental pulp fibroblasts were cultivated for experimentation. Depending on the grade of pulpitis, NLRP3 expression varies. Wisdom teeth with irreversible pulpitis showed significant differences in the levels of IL-1β, caspase-1, and of course, NLRP3. Only in irreversible pulpitis was caspase-1 present in an active form. In this study and the one before the analysis of the cells, the same results are revealed as those seen in the in vivo studies (Jiang et al. 2015).

A similar experiment was reported by Oliveira et al. (2009). Ten healthy wisdom teeth and another ten with pulpectomies. Pulp fibroblasts were analyzed with or without stimulation of *Escherichia coli* LPS. The purpose was to study the role that IL-1β and IL-8 play in a healthy state and in dental infections. As we assumed before reading, damaged pulp tissues presented higher levels of IL-1β and IL-8 (Oliveira et al. 2009).

A very important microorganism is found in infected dental pulp, even after canal treatment. *Enterococcus faecalis* plays an important role in inflammasome activation. Macrophages were infected with this bacterium. The results were that *Enterococcus faecalis* released ATP as a danger signal, making it higher in the extracellular space, which promoted the activation of the NLRP3 inflammasome, and consequently caspase-1 activation and IL-1β secretion (Yang et al. 2014). Wang et al. (2016) also studied the role of *Enterococcus faecalis* in inducing apical periodontitis in rats and the relationship it has with the NLRP3 inflammasome. Both papers reached the same results, but this latter experiment also studied the effect of the lipoteichoic acid (derived from *Enterococcus faecalis*) on the expression of the NLRP3 inflammasome, suggesting that it might act as a directly stimulating factor of this inflammasome complex (Wang et al. 2016).

There are higher levels of proinflammatory cytokines in dental periapical lesions (granulomas and cysts). The cytokine levels vary depending on the apical lesion size (larger or smaller than 5 mm) and the symptomatology. Symptomatic lesions produce higher levels of IL-1β, IL-6, and IL-8. On the one hand, the smaller apical lesions were related to higher numbers of mononuclear phagocytes; on the other hand, higher levels of TNF-α, IL-6, and IL-10, in addition to a higher concentration of CD8+ T cells, were found in large apical lesions (Gazivoda et al. 2009).

On the whole, these papers demonstrate that dental tissue damage is related to inflammation, the activation of some inflammasome complexes, and the secretion of proinflammatory cytokines.

7.6 Other Oral Diseases

Idiopathic burning mouth syndrome (iBMS) is considered by the International Association for the Study of Pain to be a chronic distinctive nosological entity in which patients perceive spontaneous burning sensations and/or other dysesthesias (tingling or itching) (Braud et al. 2013). Normally associated with xerostomia and dysgeusia (de Moraes et al. 2012). It commonly appears in the lips, hard palate, tongue and/or other oral mucosal surfaces with no clinical or laboratory signs that could demonstrate abnormalities (Suh et al. 2009; Kim et al. 2012). Mostly post-menopausal, stressed women are affected by iBMS, where it can rise up to 12% (Bergdahl and Bergdahl 1999). The origin and the way in which the pain is produced in the patient are still unknown (Forssell et al. 2015).

Liquen planus is a chronic inflammatory disease that can affect the mucous membranes of squamous cell origin and the skin (Qiu et al. 2017). Oral liquen planus (OLP) affects the oral mucosa, but skin alterations may not always be found. It appears more frequently in women than in men (Mozaffari et al. 2016). Comparing clinical characteristics, we can distinguish six types: reticular, papular, plaque-like, atrophic, erosive, and bullous (Shen et al. 2012). We can also classify them as non-erosive OLP and erosive OLP subtypes. This subdivision is important to differentiate the potential risks of malignization (Jun et al. 2008). The prevalence of OLP lesions is 0.5–2% (Setterfield et al. 2000), increasing if the patient smokes or has a high consumption of alcohol (Torrente-Castells et al. 2010).

Rare levels of expression of a variety of proinflammatory cytokines have been found in OLP lesions, and in the serum and saliva of those patients (Lu et al. 2015), suggesting a direct relationship between inflammation and OLP. The etiology of OLP is still unknown, but it is clear that immune dysregulation plays a superlative role (Lodi et al. 2005). Several treatments using different methods have been proposed. Topical and systemic products such as corticosteroids or immunosuppressants are the first ones most frequently used (Canter et al. 2007).

Kho et al. (2013) studied two different markers associated with the mucosal defense system, mucin 1 (MUC1) and TLR-2, in patients with iBMS or OLP, compared with control subjects . MUC1 is the membrane-bound mucin expressed by the salivary glands and oral epithelial cells (Hori et al. 2007). TLR-2 is an associated protein expressed in cell membranes that belongs to a family of proteins called PRRs, which recognize PAMPs. NLRP3 inflammasome needs two signals for its activation. MUC1 and TLR-2 play important roles during the priming or first signal. Levels of MUC1 were increased in patients with iBMS in comparison with patients with OLP or controls. These data may be important for confirming in further studies that an increase in the levels of subclinical inflammation could be involved in the intensity and perception of burning symptoms (Lopez-Jornet et al. 2011).

7.7 Conclusions

The mouth is always a possible environment for developing pathological conditions owing to the wide variety of microorganisms that exist there. Small variations in the equilibrium of the oral flora cause disorders that could affect the organism in a systemic form. Many publications have been written about oral diseases, but very few of them try to relate this information to the inflammasome complex. Therefore, further investigation and publications on inflammatory molecular processes are needed, in particular, those more specifically related to the inflammasome complex.

References

Abu-Elteen KH, Abu-Alteen RM (1998) The prevalence of *Candida albicans* populations in the mouths of complete denture wearers. New Microbiol 21:41–48

Akira S, Uematsu S, Takeuchi O (2006) Pathogen recognition and innate immunity. Cell 124:783–801

Amodini Rajakaruna G, Umeda M, Uchida K, Furukawa A, Yuan B, Suzuki Y, Noriko E, Izumi Y, Eishi Y (2012) Possible translocation of periodontal pathogens into the lymph nodes draining the oral cavity. J Microbiol (Seoul, Korea) 50:827–836

Arendorf TM, Walker DM (1979) Oral candidal populations in health and disease. Br Dent J 147:267–272

Armitage GC (1999) Development of a classification system for periodontal diseases and conditions. Ann Periodontol 4:1–6

Bascones-Martínez A, Valderrama G, Vijande F, Puyet-Catalina A, Bascones-Ilundain J, Arias-Herrera S, Garrido-Pertierra A (2012) Interleukin-1 gene polymorphisms and periodontal status in a Spanish population. Mol Med Rep 5:1335–1339

Belibasakis GN, Johansson A (2012) *Aggregatibacter actinomycetemcomitans* targets nLRP3 and nLRP6 inflammasome expression in human mononuclear leukocytes. Cytokine 59:124–130

Belibasakis GN, Guggenheim B, Bostanci N (2013) Down-regulation of nLRP3 inflammasome in gingival fibroblasts by subgingival biofilms: involvement of porphyromonas gingivalis. Innate Immun 19:3–9

Bergdahl M, Bergdahl J (1999) Burning mouth syndrome: prevalence and associated factors. J Oral Pathol Med 28:350–354

Bolwell GP (1999) Role of active oxygen species and NO in plant defence responses. Curr Opin Plant Biol 2:287–294

Borghi M, De Luca A, Puccetti M, Jaeger M, Mencacci A, Oikonomou V, Pariano M, Garlanda C, Moretti S, Bartoli A, Sobel J, van de Veerdonk FL, Dinarello CA, Netea MG, Romani L (2015) Pathogenic NLRP3 inflammasome activity during candida infection is negatively regulated by IL-22 via activation of NLRC4 and IL-1Ra. Cell Host Microbe 18:198–209

Bortoluci KR, Medzhitov R (2010) Control of infection by pyroptosis and autophagy: role of TLR and NLR. Cell Mol Life Sci 67:1643–1651

Bostanci N, Belibasakis GN (2012) *Porphyromonas gingivalis*: an invasive and evasive opportunistic oral pathogen. FEMS Microbiol Lett 333:1–9

Bostanci N, Allaker R, Johansson U, Rangarajan M, Curtis MA, Hughes FJ, McKay IJ (2007) Interleukin-1alpha stimulation in monocytes by periodontal bacteria: antagonistic effects of porphyromonas gingivalis. Oral Microbiol Immunol 22:52–60

Bostanci N, Emingil G, Saygan B, Turkoglu O, Atilla G, Curtis MA, Belibasakis GN (2009) Expression and regulation of the nALP3 inflammasome complex in periodontal diseases. Clin Exp Immunol 157:415–422

Braud A, Touré B, Agbo-Godeau S, Descroix V, Boucher Y (2013) Characteristics of pain assessed with visual analog scale and questionnaire in burning mouth syndrome patients: a pilot study. J Orofac Pain 27:235–242

Canter PH, Wider B, Ernst E (2007) The antioxidant vitamins A, C, E and selenium in the treatment of arthritis: a systematic review of randomized clinical trials. Rheumatology (Oxford) 46:1223–1233

Carlisle MD, Srikantha RN, Brogden KA (2009) Degradation of human alpha- and beta-defensins by culture supernatants of *Porphyromonas gingivalis* strain 381. J Innate Immun 1:118–122

Champaiboon C, Poolgesorn M, Wisitrasameewong W, Sa-Ard-Iam N, Rerkyen P, Mahanonda R (2014) Differential inflammasome activation by *Porphyromonas gingivalis* and cholesterol crystals in human macrophages and coronary artery endothelial cells. Atherosclerosis 235:38–44

Chang YC, Huang FM, Cheng MH, Chou LS, Chou MY (1998) In vitro evaluation of the cytotoxicity and genotoxicity of root canal medicines on human pulp fibroblasts. J Endod 24:604–606

Chen LY, Kuo SC, Wu HS, Yang SP, Chan YJ, Chen LK, Wang FD (2013) Associated clinical characteristics of patients with candidemia among different Candida species. J Microbiol Immunol Infect 46:463–468

Cheng YL, Jordan L, Chen HS, Kang D, Oxford L, Plemons J, Parks H, Rees T (2017) Chronic periodontitis can affect the levels of potential oral cancer salivary mRNA biomarkers. J Periodontal Res 52:428–437

Dashper SG, Seers CA, Tan KH, Reynolds EC (2011) Virulence factors of the oral spirochete *Treponema denticola*. J Dent Res 90:691–703

Davis BK, Wen H, Ting JP-Y (2011) The inflammasome nLRs in immunity, inflammation, and associated diseases. Annu Rev Immunol 29:707–735

De Duve C (2007) The origin of eukaryotes: a reappraisal. Nat Rev Genet 8:395–403

de Moraes M, do Amaral Bezerra BA, da Rocha Neto PC et al (2012) Randomized trials for the treatment of burning mouth syndrome: an evidence-based review of the literature. J Oral Pathol Med 41:281–287

Detert J, Pischon N, Burmester GR, Buttgereit F (2010) The association between rheumatoid arthritis and periodontal disease. Arthritis Res Ther 12:218

Dutta D, Dutta S, Veettil MV, Roy A, Ansari MA, Iqbal J, Chikoti L, Kumar B, Johnson KE, Chandran B (2015a) BRCA1 regulates IFI16 mediated nuclear innate sensing of herpes viral DNA and subsequent induction of the innate inflammasome and interferon-β responses. PLoS Pathog 11:e1005030

Dutta A, Banerjee A, Saikia N, Phookan J, Baruah MN, Baruah S (2015b) Negative regulation of natural killer cell in tumor tissue and peripheral blood of oral squamous cell carcinoma. Cytokine 76:123–130

Engel LD, Kenny GE (1970) *Mycoplasma salivarium* in human gingival sulci. J Periodontal Res 5:163–171

Ferlay J, Soerjomataram I, Dikshit R et al (2015) Cancer incidence and mortality worldwide: sources, methods and major patterns in GLOBOCAN 2012. Int J Cancer 136:359–386

Forssell H, Jääskeläinen S, List T et al (2015) An update on pathophysiological mechanisms related to idiopathic oro-facial pain conditions with implications for management. J Oral Rehabil 42:300–342

Funk GF, Karnell LH, Robinson RA, Zhen WK, Trask DK, Hoffman HT (2002) Presentation, treatment, and outcome of oral cavity cancer: a National Cancer Data Base report. Head Neck 24:165–180

Gazivoda D, Dzopalic T, Bozic B, Tatomirovic Z, Brkic Z, Colic M (2009) Production of proinflammatory and immunoregulatory cytokines by inflammatory cells from periapical lesions in culture. J Oral Pathol Med 38:605–111

Gimenez F, Bhela S, Dogra P, Harvey L, Varanasi SK, Jaggi U, Rouse BT (2016) The inflammasome NLRP3 plays a protective role against a viral immunopathological lesion. J Leukoc Biol 99:647–657

Hajishengallis G, Liang S, Payne MA, Hashim A, Jotwani R, Eskan MA, McIntosh ML, Alsam A, Kirkwood KL, Lambris JD, Darveau RP, Curtis MA (2011) Low-abundance biofilm species orchestrates inflammatory periodontal disease through the commensal microbiota and complement. Cell Host Microbe 10:497–506

Hamedi M, Belibasakis GN, Cruchley AT, Rangarajan M, Curtis MA, Bostanci N (2009) *Porphyromonas gingivalis* culture supernatants differentially regulate interleukin-1beta and interleukin-18 in human monocytic cells. Cytokine 45:99–104

Han YW, Wang X (2013) Mobile microbiome: oral bacteria in extra-oral infections and inflammation. J Dent Res 92:485–491

Hise AG, Tomalka J, Ganesan S, Patel K, Hall BA, Brown GD, Fitzgerald KA (2009) An essential role for the NLRP3 inflammasome in host defense against the human fungal pathogen Candida albicans. Cell Host Microbe 5:487–497

Hong CY, Lin SK, Kok SH, Cheng SJ, Lee MS, Wang TM et al (2004) The role of lipopolysaccharide in infectious bone resorption of periapical lesion. J Oral Pathol Med 33:162–169

Hori Y, Sugiyama H, Soma T, Nishida K (2007) Expression of membrane-associated mucins in cultivated human oral mucosal epithelial cells. Cornea 26:65–69

Hu S, Chen WC, Hwang GS, Chen ST, Kuo SB, Chen Y, Idova G, Wang SW (2016) Changes in plasma steroids and cytokines levels in betel chewing patients in Taiwan. Steroids 111:134–138

Huang X, Yang X, Ni J, Xie B, Liu Y, Xuan D, Zhang J (2015) Hyperglucose contributes to periodontitis: involvement of the nLRP3 pathway by engaging the innate immunity of oral gingival epithelium. J Periodontol 86:327–335

Jiang W, Lv H, Wang H, Wang D, Sun S, Jia Q, Wang P, Song B, Ni L (2015) Activation of the NLRP3/caspase-1 inflammasome in human dental pulp tissue and human dental pulp fibroblasts. Cell Tissue Res 36:541–555

Jin LJ, Armitage GC, Klinge B, Lang NP, Tonetti M, Williams RC (2011) Global oral health inequalities: task group–periodontal disease. Adv Dent Res 23:221–226

Johnson KE, Chikoti L, Chandran B (2013) Herpes simplex virus 1 infection induces activation and subsequent inhibition of the IFI16 and NLRP3 inflammasomes. J Virol 87:5005–5018

Johnson KE, Bottero V, Flaherty S, Dutta S, Singh VV, Chandran B (2014) IFI16 restricts HSV-1 replication by accumulating on the hsv-1 genome, repressing HSV-1 gene expression, and directly or indirectly modulating histone modifications. PLoS Pathog 10:e1004503

Joly S, Sutterwala FS (2010) Fungal pathogen recognition by the NLRP3 inflammasome. Virulence 1:276–280

Jun H-K, Kang Y-M, Lee H-R, Lee S-H, Choi B-K (2008) Highly conserved surface proteins of oral spirochetes as adhesins and potent inducers of proinflammatory and osteoclastogenic factors. Infect Immun 76:2428–2438

Jun H-K, Lee S-H, Lee H-R, Choi B-K (2012) Integrin $\alpha 5\beta 1$ activates the NLRP3 inflammasome by direct interaction with a bacterial surface protein. Immunity 36:755–768

Jun H-K, Jung Y-J, Choi B-K (2017) *Treponema denticola*, *Porphyromonas gingivalis*, and *Tannerella forsythia* induce cell death and release of endogenous danger signals. Arch Oral Biol 73:72–78

Kakehashi S, Stanley HR, Fitzgerald RJ (1965) The effects of surgical exposures of dental pulps in germ-free and conventional laboratory rats. Oral Surg Oral Med Oral Pathol 20:340–349

Kawada M, Yoshida A, Suzuki N, Nakano Y, Saito T, Oho T, Koga T (2004) Prevalence of *Porphyromonas gingivalis* in relation to periodontal status assessed by real-time pCR. Oral Microbiol Immunol 19:289–292

Kelk P, Claesson R, Hänström L, Lerner UH, Kalfas S, Johansson A (2005) Abundant secretion of bioactive interleukin-1beta by human macrophages induced by *Actinobacillus actinomycetemcomitans* leukotoxin. Infect Immun 73:453–458

Kelk P, Abd H, Claesson R, Sandström G, Sjöstedt A, Johansson A (2011) Cellular and molecular response of human macrophages exposed to *Aggregatibacter actinomycetemcomitans* leukotoxin. Cell Death Dis 2:e126

Kho HS, Chang JY, Kim YY, Kim Y (2013) MUC1 and toll-like receptor-2 expression in burning mouth syndrome and oral lichen planus. Arch Oral Biol 58:837–842

Kim J-J, Jo E-K (2013) NLRP3 inflammasome and host protection against bacterial infection. J Korean Med Sci 28:1415–1423

Kim HI, Kim YY, Chang JY, Ko JY, Kho HS (2012) Salivary cortisol, 17b-estradiol, progesterone, dehydroepiandrosterone, and a-amylase in patients with burning mouth syndrome. Oral Dis 18:613–620

Kim S, Park MH, Song Y-R, Na HS, Chung J (2016) *Aggregatibacter actinomycetemcomitans*-induced aIM2 inflammasome activation is suppressed by xylitol in differentiated tHP-1 macrophages. J Periodontol 87:e116–e126

König J, Holtfreter B, Kocher T (2010) Periodontal health in Europe: future trends based on treatment needs and the provision of periodontal services–position paper 1. Eur J Dent Educ 14(Suppl 1):4–24

Kurita-Ochiai T, Jia R, Cai Y, Yamaguchi Y, Yamamoto M (2015) Periodontal disease-induced atherosclerosis and oxidative stress. Antioxidants (Basel) 4:577–590

Li H, Zhou X, Zhang J (2014) Induction of heme oxygenase-1 attenuates lipopolysaccharide-induced inflammasome activation in human gingival epithelial cells. Int J Mol Med 34:1039–1044

Lippman SM, Sudbo J, Hong WK (2005) Oral cancer prevention and the evolution of molecular-targeted drug development. J Clin Oncol 23:346–356

Liu W, Han B, Sun B, Gao Y, Huang Y, Hu M (2012) Overexpression of interleukin-18 induces growth inhibition, apoptosis and gene expression changes in a human tongue squamous cell carcinoma cell line. J Int Med Res 40:537–544

Lodi G, Scully C, Carrozzo M, Griffiths M, Sugerman PB, Thongprasom K (2005) Current controversies in oral lichen planus: report of an international consensus meeting. I. Viral infections and etiopathogenesis. Oral Surg Oral Med Oral Pathol Oral Radiol Endod 100:40–51

Lopez-Jornet P, Camacho-Alonso F, Andujar-Mateos P (2011) A prospective, randomized study on the efficacy of tongue protector in patients with burning mouth syndrome. Oral Dis 17:277–282

Lu R, Zhang J, Sun W, Du G, Zhou G (2015) Inflammation-related cytokines in oral lichen planus: an overview. J Oral Pathol Med 44:1–14

Lyons NA, Kolter R (2015) On the evolution of bacterial multicellularity. Curr Opin Microbiol 24:21–28

Mantovani A, Allavena P, Sica A, Balkwill F (2008) Cancer related inflammation. Nature 454:436–444

Masada MP, Persson R, Kenney JS, Lee SW, Page RC, Allison AC (1990) Measurement of interleukin-1 alpha and -1 beta in gingival crevicular fluid: implications for the pathogenesis of periodontal disease. J Periodontal Res 25:156–163

Miettinen JJ, Matikainen S, Nyman TA (2012) Global secretome characterization of herpes simplex virus 1-infected human primary macrophages. J Virol 86:12770–12778

Mihrshahi R, Brown MH (2010) Downstream of tyrosine kinase 1 and 2 play opposing roles in CD200 receptor signaling. J Immunol 185:7216–7222

Moller AJ, Fabricius L, Dahlen G et al (1981) Influence on periapical tissues of indigenous oral bacteria and necrotic pulp tissue in monkeys. Scand J Dent Res 89:475–484

Mozaffari HR, Rahmani M, Rezaei F, Sadeghi M, Sharifi R, Ejtehadi A (2016) Evaluation of oral lichen planus frequency in patients referred to pathology centers of Kermanshah city, during 2008 to 2011. Sch J App Med Sci 4:2200–2202

Mydel P, Takahashi Y, Yumoto H, Sztukowska M, Kubica M, Gibson FC, Kurtz DM, Travis J, Collins LV, Nguyen K-A, Genco CA, Potempa J (2006) Roles of the host oxidative immune response and bacterial antioxidant rubrerythrin during porphyromonas gingivalis infection. PLoS Pathog 2:e76
Nilkaeo A, Bhuvanath S (2006) Role of interleukin-18 in modulation of oral carcinoma cell proliferation. Mediat Inflamm 2006:67120
Nour AM, Reichelt M, Ku CC, Ho MY, Heineman TC, Arvin AM (2011) Varicella-zoster virus infection triggers formation of an interleukin-1β (IL-1β)-processing inflammasome complex. J Biol Chem 286(20):17921–17933
Ohshima M, Otsuka K, Suzuki K (1994) Interleukin-1 beta stimulates collagenase production by cultured human periodontal ligament fibroblasts. J Periodontal Res 29:421–429
Okinaga T, Ariyoshi W, Nishihara T (2015) *Aggregatibacter actinomycetemcomitans* invasion induces interleukin-1β production through reactive oxygen species and cathepsin B. J Interf Cytokine Res 35:431–440
Oliveira AC, Rubio M, Fontes A, Sampaio M, Neves B (2009) Interleukin-1 beta and interleukin-8 in healthy and inflamed dental pulps. J Appl Oral Sci 171:527–532
Orozco A, Gemmell E, Bickel M, Seymour GJ (2006) Interleukin-1beta, interleukin-12 and interleukin-18 levels in gingival fluid and serum of patients with gingivitis and periodontitis. Oral Microbiol Immunol 21:256–260
Park E, Na HS, Song Y-R, Shin SY, Kim Y-M, Chung J (2014) Activation of nLRP3 and aIM2 inflammasomes by porphyromonas gingivalis infection. Infect Immun 82:112–123
Parkin DM, Bray F, Ferlay J, Pisani P (2005) Global cancer statistics. CA Cancer J Clin 55:74–108
Patel MN, Carroll RG, Galván-Peña S, Mills EL, Olden R, Triantafilou M, Wolf AI, Bryant CE, Triantafilou K, Masters SL (2017) Inflammasome priming in sterile inflammatory disease. Trends Mol Med 23:165–180
Pathirana RD, O'Brien-Simpson NM, Reynolds EC (2010) Host immune responses to porphyromonas gingivalis antigens. Periodontol 52:218–237
Petersen PE, Bourgeois D, Ogawa H, Estupinan-Day S, Ndiaye C (2005) The global burden of oral diseases and risks to oral health. Bull World Health Organ 83:661–669
Pihlstrom BL, Michalowicz BS, Johnson NW (2005) Periodontal diseases. Lancet 366:1809–1820
Qiu M, Wang J, He P, Xv H, Wei J, Gong W, Jiang L, Chen Q (2017) Association between −1082 A/G polymorphism in IL-10 and oral lichen planus: a meta-analysis. J Dermatol Sci 85:252–253
Reid BC, Winn DM, Morse DE, Pendrys DG (2000) Head and neck in situ carcinoma: incidence, trends, and survival. Oral Oncol 36:414–420
Repetto EC, Giacomazzi CG, Castelli F (2012) Hospital-related outbreaks due to rare fungal pathogens: a review of the literature from 1990 to June 2011. Eur J Clin Microbiol Infect Dis 31:2897–2904
Richardson JP, Moyes DL (2015) Adaptive immune responses to *Candida albicans* infection. Virulence 6:327–337
Sakamoto M, Suzuki M, Umeda M, Ishikawa I, Benno Y (2002) Reclassification of *Bacteroides forsythus* (Tanner et al. 1986) as *Tannerella forsythensis* corrig., gen. nov., comb. nov. Int J Syst Evol Microbiol 52:841–849
Schwartz Z, Goultschin J, Dean DD, Boyan BD (1997) Mechanisms of alveolar bone destruction in periodontitis. Periodontology 14:158–172
Setterfield JF, Black MM, Challacombe SJ (2000) The management of oral lichen planus. Clin Exp Dermatol 25:176–182
Sharma A (2010) Virulence mechanisms of *Tannerella forsythia*. Periodontology 54:106–116
Shen ZY, Liu W, Zhu LK, Feng JQ, Tang GY, Zhou ZT (2012) A retrospective clinicopathological study on oral lichen planus and malignant transformation: analysis of 518 cases. Med Oral Patol Oral Cir Bucal 17:943–947
Shibata K, Hasebe A, Sasaki T, Watanabe T (1997) *Mycoplasma salivarium* induces interleukin-6 and interleukin-8 in human gingival fibroblasts. FEMS Immunol Med Microbiol 19:275–283

Siegel RL, Miller KD, Jemal A (2016) Cancer statistics. CA Cancer J Clin 66:7–30
Slots J, Reynolds HS, Genco RJ (1980) *Actinobacillus actinomycetemcomitans* in human periodontal disease: a cross-sectional microbiological investigation. Infect Immun 29:1013–1120
Smith JS, Robinson NJ (2002) Age-specific prevalence of infection with herpes simplex virus types 2 and 1: a global review. J Infect Dis 186:3–28
Soberman RJ, MacKay CR, Vaine CA, Ryan GB, Cerny AM, Thompson MR, Nikolic B, Primo V, Christmas P, Sheiffele P, Aronov L, Knipe DM, Kurt-Jones EA (2012) CD200R1 supports HSV-1 viral replication and licenses pro-inflammatory signaling functions of TLR2. PLoS One 7:e47740
Soukup V, Horácek I, Cerny R (2013) Development and evolution of the vertebrate primary mouth. J Anat 222:79–99
Stashenko P, Fujiyoshi P, Obernesser MS, Prostak L, Haffajee AD, Socransky SS (1991) Levels of interleukin 1 beta in tissue from sites of active periodontal disease. J Clin Periodontol 18:548–554
Sugiyama M, Saeki A, Hasebe A, Kamesaki R, Yoshida Y, Kitagawa Y, Suzuki T, Shibata K (2015) Activation of inflammasomes in dendritic cells and macrophages by *Mycoplasma salivarium*. Mol Oral Microbiol 31:259–269
Suh KI, Kim YK, Kho HS (2009) Salivary levels of IL-1beta, IL-6, IL-8, and TNF-alpha in patients with burning mouth syndrome. Arch Oral Biol 54:797–802
Sun Y, Zhu D, Wang G, Wang D, Zhou H, Liu X, Jiang M, Liao L, Zhou Z, Hu J (2015) Pro-inflammatory cytokine IL-1β up-regulates CXC chemokine receptor 4 via Notch and ERK signaling pathways in tongue squamous cell carcinoma. PLoS One 10:e0132677
Tanner A, Stillman N (1993) Oral and dental infections with anaerobic bacteria: clinical features, predominant pathogens, and treatment. Clin Infect Dis 16(Suppl 4):S304–S309
Taxman DJ, Lei Y, Zhang S, Holley-Guthrie E, Offenbacher S, Ting JPY (2012) ASC-dependent rIP2 kinase regulates reduced pGE2 production in chronic periodontitis. J Dent Res 91:877–882
Tomalka J, Ganesan S, Azodi E, Patel K, Majmudar P, Hall BA, Fitzgerald KA, Hise AG (2011) A novel role for the NLRC4 inflammasome in mucosal defenses against the fungal pathogen *Candida albicans*. PLoS Pathog 7:e10002379
Torrente-Castells E, Figueiredo R, Berini-Aytés L, Gay-Escoda C (2010) Clinical features of oral lichen planus. A retrospective study of 65 cases. Med Oral Patol Oral Cir Bucal 15:685–690
Tsai HT, Hsin CH, Hsieh YH, Tang CH, Yang SF, Lin CW, Chen MK (2013) Impact of interleukin-18 polymorphisms-607A/C and -137G/C on oral cancer occurrence and clinical progression. PLoS One 8:e83572
Vairaktaris E, Serefoglou ZC, Yapijakis C, Agapi C, Vassiliou S, Nkenke E, Antonis V, Sofia S, Neukam FW, Patsouris E (2007) The interleukin-18 -607A/C polymorphism is not associated with risk for oral cancer. Anticancer Res 27:4011–4014
van Winkelhoff AJ, Loos BG, van der Reijden WA, van der Velden U (2002) *Porphyromonas gingivalis*, bacteroides forsythus and other putative periodontal pathogens in subjects with and without periodontal destruction. J Clin Periodontol 29:1023–1028
Velsko IM, Chukkapalli SS, Rivera-Kweh MF, Zheng D, Aukhil I, Lucas AR, Larjava H, Kesavalu L (2015) Periodontal pathogens invade gingiva and aortic adventitia and elicit inflammasome activation in αvβ6 integrin-deficient mice. Infect Immun 83:4582–4593
Wang JP, Bowen GN, Zhou S, Cerny A, Zacharia A et al (2012) Role of specific innate immune responses in herpes simplex virus infection of the central nervous system. J Virol 86:2273–2281
Wang Y, Zhai S, Wang H, Jia Q, Jiang W, Zhang X, Zhang A, Liu J, Ni L (2013) Absent in melanoma 2 (AIM2) in rat dental pulp mediates the inflammatory response during pulpitis. J Endod 39:1390–1394
Wang SL, Zhao G, Zhu W, Dong XM, Liu T, Li YY, Song WG, Wang YQ (2015) Herpes simplex virus-1 infection or Simian virus 40-mediated immortalization of corneal cells causes permanent translocation of NLRP3 to the nuclei. Int J Ophthalmol 8:46–51

Wang L, Jin H, Ye D, Wang J, Ao X, Dong M, Niu W (2016) *Enterococcus faecalis* lipoteichoic acid-induced NLRP3 inflammasome via the activation of the nuclear factor kappa B pathway. J Endod 42:1093–1100

Watanabe T, Matsuura M, Seto K (1986) Enumeration, isolation, and species identification of mycoplasmas in saliva sampled from the normal and pathological human oral cavity and antibody response to an oral mycoplasma. J Clin Microbiol 23:1034–1038

Wu CS, Chang KP, OuYang CN, Kao HK, Hsueh C, Chen LC, Cheng HY, Liang Y, Liou W, Liang CL, Chang YS (2016) ASC contributes to metastasis of oral cavity squamous cell carcinoma. Oncotarget 7:50074–50085

Xu F, Sternberg MR, Kottiri BJ, McQuillan GM, Lee FK, Nahmias AJ, Berman SM, Markowitz LE (2006) Trends in herpes simplex virus type 1 and type 2 seroprevalence in the United States. JAMA 296:964–973

Xue F, Shu R, Xie Y (2015) The expression of nLRP3, nLRP1 and aIM2 in the gingival tissue of periodontitis patients: RT-pCR study and immunohistochemistry. Arch Oral Biol 60:948–958

Yamaguchi Y, Kurita-Ochiai T, Kobayashi R, Suzuki T, Ando T (2015) Activation of the nLRP3 inflammasome in *Porphyromonas gingivalis*-accelerated atherosclerosis. Pathog Dis 73:ftv011

Yamaguchi Y, Kurita-Ochiai T, Kobayashi R, Suzuki T, Ando T (2017) Regulation of the nLRP3 inflammasome in *Porphyromonas gingivalis*-accelerated periodontal disease. Inflamm Res 66:59–65

Yang H-W, Huang Y-F, Chou M-Y (2004) Occurrence of *Porphyromonas gingivalis* and *Tannerella forsythensis* in periodontally diseased and healthy subjects. J Periodontol 75:1077–1083

Yang HH, Jun HK, Jung YJ, Choi BK (2014) *Enterococcus faecalis* activates caspase-1 leading to increased interleukin-1 beta secretion in macrophages. J Endod 40:1587–1592

Yilmaz Ö, Lee KL (2015) The inflammasome and danger molecule signaling: at the crossroads of inflammation and pathogen persistence in the oral cavity. Periodontology 69:83–95

Yilmaz Ö, Sater AA, Yao L, Koutouzis T, Pettengill M, Ojcius DM (2010) ATP-dependent activation of an inflammasome in primary gingival epithelial cells infected by *Porphyromonas gingivalis*. Cell Microbiol 12:188–198

Zhao P, Liu J, Pan C, Pan Y (2014) NLRP3 inflammasome is required for apoptosis of *Aggregatibacter actinomycetemcomitans*-infected human osteoblastic mG63 cells. Acta Histochem 116:1119–1124

Zhiyu W, Wang N, Wang Q, Peng C, Zhang J, Liu P, Ou A, Zhong S, Cordero MD, Lin Y (2016) The inflammasome: an emerging therapeutic oncotarget for cancer prevention. Oncotarget 7:50766–50780

Chapter 8
Inflammasomes in the Kidney

Holly L. Hutton, Maliha A. Alikhan, and A. Richard Kitching

Contents

H. L. Hutton · M. A. Alikhan
Centre for Inflammatory Diseases, Monash University Department of Medicine, Monash Medical Centre, Clayton, VIC, Australia

A. R. Kitching (✉)
Centre for Inflammatory Diseases, Monash University Department of Medicine, Monash Medical Centre, Clayton, VIC, Australia

Department of Nephrology, Monash Health, Clayton, VIC, Australia

Department of Paediatric Nephrology, Monash Health, Clayton, VIC, Australia
e-mail: richard.kitching@monash.edu

M. D. Cordero, E. Alcocer-Gómez (eds.), *Inflammasomes: Clinical and Therapeutic Implications*, Experientia Supplementum 108,
https://doi.org/10.1007/978-3-319-89390-7_8

Abstract Inflammasomes influence a diverse range of kidney disease, including acute and chronic kidney diseases, and those mediated by innate and adaptive immunity. Both IL-18 and in particular IL-1β are validated therapeutic targets in several kidney diseases. In addition to leukocyte-derived inflammasomes, renal tissue cells express functional inflammasome components. Furthermore, a range of endogenous substances that directly activate inflammasomes also mediate kidney injury. Many of the functional studies have focussed on the NLRP3 inflammasome, and there is also evidence for the involvement of other inflammasomes in some conditions. While, at least in some disease, the mechanistic details of the involvement of the inflammasome remain to be elucidated, therapies focussed on inflammasomes and their products have potential in treating kidney disease in the future.

Keywords Inflammasome · Glomerulonephritis · Acute kidney injury · Diabetic nephropathy · Interleukin-1β

The kidney is prone to injury from a range of metabolic, toxic and immunologic insults. The NLRP3 inflammasome is present in kidney tissue cells as well as immune cells within the kidney (such as infiltrating and resident mononuclear phagocytes) and has been implicated in the pathogenesis of a broad spectrum of kidney diseases. Cellular stress caused by an ischaemic, septic or nephrotoxic insult, or by albuminuria itself, provokes the release of NLRP3 inflammasome-activating 'danger signals', or damage-associated molecular proteins (DAMPs) from cells (Akcay et al. 2009; Liu et al. 2014). Other substances associated with kidney injury, such as cholesterol emboli (Duewell et al. 2010) and monosodium urate (Martinon et al. 2006), can themselves act as a signal for NLRP3-ASC oligomerisation and inflammasome formation. After activation, the NLRP3 inflammasome promotes kidney injury via pro-inflammatory cytokines and may also switch on fibrotic (Vilaysane et al. 2010) and cell death pathways (Shen et al. 2016). Thus, the NLRP3 inflammasome may be a mechanistic link between processes that have long been known to cause kidney injury and the injurious inflammatory response that follows. Inhibition of the NLRP3 inflammasome is protective in several models of glomerular and tubulointerstitial kidney disease, indicating its potential for treatment of human disease.

The role of inflammasomes other than NLRP3 in the kidney is less well defined. The AIM2 inflammasome, activated by double-stranded DNA (Fernandes-Alnemri et al. 2009), has been shown to have a potential role in hepatitis B-related glomerulonephritis (Du et al. 2013; Zhen et al. 2014) and in experimental lupus (Choubey and

Panchanathan 2017), although it has not been directly linked to lupus nephritis. The NLRC4 inflammasome is important in pathogenic IL-1β production in a mouse model of diabetes mellitus (Yuan et al. 2016). Although bacterial infection has been linked to some types of glomerulonephritis and is the trigger for the NLRP1 and NLRC4 inflammasomes, few studies have focussed on a role for these inflammasomes in autoimmune kidney disease (Man and Kanneganti 2015).

8.1 Basic Renal Structure and Function

The primary role of the kidney is to filter blood, maintaining levels of electrolytes such as sodium and potassium and excreting nitrogenous wastes such as urea, excess body water and electrolytes into the urine. The functional unit of a kidney a nephron, comprised of a glomerulus, a specialised filter through which blood passes from an afferent arteriole into capillaries that have evolved to generate an ultrafiltrate that accumulates in the urinary (Bowman's) space, and a tubule, through which the glomerular filtrate passes from the urinary space, with reabsorption of most of the water and solutes, before being excreted as urine. In healthy humans, kidneys contain on average approximately 900,000 nephrons, although this number varies depending on the ethnicity of the population studied and factors associated with the fetomaternal environment, which influences kidney development (Puelles et al. 2016). Kidneys consist of an outer cortex, inner medulla and renal pelvis. Nephrons traverse the kidney so that the cortex contains the glomeruli and the medulla contains tubules formed into pyramids, which facilitate the reabsorption and excretion of water and electrolytes and the formation of concentrated urine. The apex of pyramids is comprised of collecting tubules, which drain into calyces that in turn drain into the renal pelvis and ureter for excretion.

The structure of the glomerulus allows fluid and small solutes to pass from the bloodstream through the filtration barrier into the urinary space, but large molecules, including plasma proteins, are largely retained within the microvasculature. The glomerular tuft is a network of glomerular capillaries, lying on a scaffold of mesangial cells and extracellular matrix. The glomerular filtration barrier, functioning as a charge and size selective molecular sieve, is composed of three layers: endothelial cells, glomerular basement membrane (GBM) and podocytes (Haraldsson and Jeansson 2009). Glomerular capillaries are on one side of the glomerular filtration barrier with the urinary space (Bowman's space) on the adjacent side. Glomerular endothelial cells line the capillary lumen and contain fenestrae, small transcellular 'holes' covered with endothelial glycocalyx, which aid filtration. The glomerular basement membrane is composed of extracellular matrix macromolecules including type IV collagen, laminin and proteoglycans. Podocytes line the GBM on the side of the urinary space. They are highly specialised epithelial cells with long, interdigitating foot processes that wrap around the glomerular capillaries (Brinkkoetter et al. 2013). Damage to any layer of the glomerular filtration barrier results in a leak of proteins into the urinary space, and proteinuria is thus regarded as one of the hallmarks of glomerular disease. Kidney diseases may be seen as primarily due to pathology of an intrinsic kidney cell, i.e. mesangial cells as the target of immunologic injury in IgA nephropathy (Tsai et al. 2017), or primarily due to activation or infiltration of immune cells, such as neutrophils in anti-neutrophil cytoplasmic autoantibody (ANCA)-associated vasculitis (Falk et al. 1990). However, the extensive cross talk between different types of cells in the kidney, including intrinsic kidney cells and leukocytes, means that damage to one cell type usually has effects on multiple different cells (Kitching and Hutton 2016). A summary of the function of glomerular cells and some common kidney diseases linked to dysfunction of each cell type is shown in Table 8.1.

Table 8.1 Key functions and responses of intrinsic glomerular cells

Renal cell type	Normal function and features	Responses to injury	Examples of relevant diseases	NLRP3 inflammasome present
Mesangial cells	Maintain structural architecture of glomerulus Mesangial matrix homeostasis Regulate filtration surface area Phagocytose apoptotic cells	Lysis with healthy remodelling Apoptosis Hypertrophy Proliferation and matrix expansion leading to glomerulosclerosis	IgA nephropathy Diabetic nephropathy	Likely
Glomerular endothelial cells	Fenestrations and glycocalyx facilitate selective permeability and filtration	Apoptosis Loss of fenestrations Widening of cell-cell junctions, transcellular holes Glycocalyx damage; loss of GAG synthesis	ANCA-associated glomerulonephritis Lupus nephritis (classes III and IV) Haemolytic-uremic syndrome Diabetic nephropathy	Likely
Podocytes	Foot processes wrap around capillaries Adherence to GBM	Apoptosis Foot process effacement Detachment from GBM; podocyte loss	Minimal change disease Focal and segmental glomerulosclerosis Diabetic nephropathy	Likely
Parietal epithelial cells	Line Bowman's capsule Several subsets of cells likely with different functions	Apoptosis Migration to glomerular tuft, production ECM proteins leading to glomerulosclerosis Proliferation leading to crescent and pseudocrescent formation	Crescentic glomerulonephritis Focal and segmental glomerulosclerosis	Unknown

Adapted from Kitching and Hutton (2016)
ANCA anti-neutrophil cytoplasmic antibody, *GAG* glycosaminoglycan, *GBM* glomerular basement membrane

8.2 NLRP3 Inflammasome Activation in the Kidney

The NLRP3 inflammasome has primarily been described in myeloid cells, particularly macrophages and dendritic cells (Martinon et al. 2009) but also neutrophils (Chen et al. 2016). In inflammatory kidney diseases, leukocyte retention within glomeruli occurs via specialised adhesion molecules. Immune cells may be involved in local tissue inflammation or repair or may influence systemic adaptive immunity which is a key driver of many glomerular diseases (Kitching and Hutton 2016). The presence of the NLRP3 inflammasome in murine kidney tissue cells has been demonstrated in several different studies, generally by showing that these cells can produce IL-1β and IL-18 or by demonstrating co-localisation of inflammasome components such as NLRP3 with ASC and/or caspase-1 in kidney biopsies. Tubular epithelial cells, in both mouse and human, express components for NLRP3 inflammasome complex and produce active IL-1β and/or IL-18 (Faust et al. 2002; Homsi et al. 2006; Lichtnekert et al. 2011; Wang et al. 2015a). While an early murine in vitro study cast doubt on the presence of the inflammasome in glomerular endothelial cells, mesangial cells and podocytes (Lichtnekert et al. 2011), subsequent studies have shown varying levels of evidence of inflammasome activation in these cells (Abais et al. 2013; Shahzad et al. 2015; Zhang et al. 2012; Zhou et al. 2010). Human kidney biopsy studies have documented the presence of the inflammasome in podocytes, mesangial cells, tubular epithelium and intercalated cells (Chun et al. 2016; Gauer et al. 2007; Shahzad et al. 2015). Selected studies documenting the presence of the NLRP3 inflammasome in mouse and human intrinsic glomerular cells are presented in Table 8.2. However, these data should be interpreted with caution, as the presence of ASC and NLRP3 mRNA or co-localisation of NLRP3 with ASC/caspase-1 is not in itself robust evidence for inflammasome activation. Furthermore, ELISAs fail to differentiate between active, cleaved IL-1β or IL-18 and their pro-cytokine forms. Currently, the 'gold standard' readout for inflammasome activation, Western blotting of activated caspase-1, IL-1β or IL-18, can be technically difficult and may not reflect organ specificity (Ludwig-Portugall et al. 2016). The presence of the NLRP3 inflammasome in intrinsic kidney cells is thus still an area of some debate. It may be that NLRP3 inflammasome expression occurs constitutively in tubular cells, but not in other kidney cells, unless switched on during specific diseases in a particular cell type. A study that compared expression of NLRP3 in normal human kidneys, obtained from nephrectomies, and biopsy samples of patients with IgA nephropathy supports this theory, showing the presence of inflammasome components in tubular cells of both groups and in glomerular mesangial cells in patients with IgA nephropathy but not in healthy kidneys (Chun et al. 2016).

Some authors have viewed the data showing the presence of the NLRP3 inflammasome components in renal cells as being evidence that renal cell-derived, rather than immune cell-derived, NLRP3 inflammasome activation is the key driver of pathology in kidney diseases (Chun et al. 2016). Although this may be the case, it can be difficult to interpret the relative contributions of inflammasome activation in

Table 8.2 Selected studies showing evidence for inflammasome activation in intrinsic kidney cells

Disease	Cells involved	Evidence of inflammasome activation	References
Hyperhomocysteinaemia	Podocytes (cultured)	RT-PCR showed NLRP3, ASC and caspase-1 mRNA in cultured podocytes Size exclusion chromatography determined the presence of ASC-NLRP3 complexes in podocyte cultures Co-localisation of NLRP3 with ASC or caspase-1 in podocytes using confocal microscopy and immunofluorescence Small amounts of IL-1β from podocyte cultures by ELISA (not differentiating active from pro-IL-1β)	Zhang et al. (2012)
Primary glomerular diseases	Podocytes (human)	NLRP3 and caspase-1 co-localisation in kidney biopsy of subjects with glomerular disease significantly increased compared to controls. Caspase-1 co-localised with podocyte marker synaptopodin	Xiong et al. (2015)
Lupus nephritis (MRL-Fas^{lpr} mice)	TEC (mouse)	IL-18 detected in sera and kidney tissues by ELISA, RT-PCR and Western blotting. IL-18 production by primary TECs, detected by RT-PCR, ELISA and Western blotting	Faust et al. (2002)
Glycerol-induced AKI	TEC (mouse)	IHC for IL-1β and IL-18 on kidney biopsies localised these to tubules, Western blot on kidney homogenate to determine active IL-1β and IL-18	Homsi et al. (2006)
IgA nephropathy	TEC (but not glomerular cells) in healthy human kidneys. TEC and mesangial cells in IgA nephropathy	Healthy human kidneys stained with immunoperoxidase or processed for indirect IF and confocal microscopy. NLRP3 localised primarily to tubules with absent glomerular staining In vitro NLRP3 present in human proximal tubular cell but not podocyte culture	Chun et al. (2016)
Healthy human kidney	Tubular epithelium	IL-18 mRNA and protein detected by PCR, in situ hybridization and Western blotting in normal human kidneys. IHC located IL-18 to nephron segments of the distal convoluted tubule and to parts of the collecting duct. Confocal microscopy for IL-18 was expressed in intercalated cells	Gauer et al. (2007)

(continued)

Table 8.2 (continued)

Disease	Cells involved	Evidence of inflammasome activation	References
IgA nephropathy	Mouse TEC and mesangial cell cultures	Immune complexes induced secretion of IL-1β in cultured TEC and mesangial cells; IL-1β produced was much reduced in NLRP3-deficient TEC and mesangial cells	Tsai et al. (2017)
Diabetic nephropathy	Mesangial cells (subject to high glucose or LPS)	IL-1β, caspase-1 and NLRP3 mRNA and protein detected in cultured mesangial cells by RT-PCR and immunoblot	Feng et al. (2016)
Diabetes	Endothelial cells and podocytes (mouse and human)	Active IL-1β from murine primary podocyte cultures detected using ELISA. Partial co-localisation of NLRP3 or cleaved caspase-1 with podocytes and glomerular endothelial cells in histological sections of diabetic humans or mice	Shahzad et al. (2015)

ELISA enzyme-linked immunosorbent assay, *RT-PCR* reverse transcription–polymerase chain reaction, *IHC* immunohistochemistry, *TEC* tubular epithelial cells

intrinsic renal cells and immune cells in disease models. For example, one study inhibited ASC in a kidney-restricted fashion, delivering siRNA via the renal artery, and showed improvements in proteinuria and glomerular sclerosis (Zhang et al. 2012). However, the conclusion that the inflammasome in intrinsic renal cells was important in this case is, strictly speaking, not necessarily accurate, as the siRNA was also delivered to intrarenal immune cells such as the network of intrarenal mononuclear phagocytes (macrophages and dendritic cells). Bone marrow chimeric mice, as used in studies of the NLRP3 inflammasome in diabetic nephropathy (Shahzad et al. 2015) may be useful in this context. NLRP3 inflammasome activation may have tissue-specific effects in certain conditions. For example, Bakker et al. used bone marrow chimeric mice to study tubular repair after ischaemia-reperfusion injury (IRI) and found that intrinsic kidney-derived NLRP3 impaired tubular regeneration, whereas leukocyte-associated NLRP3 was associated with tubular apoptosis (Bakker et al. 2014). To add further complexity, it is likely that different types of myeloid cells require different triggers for NLRP3 inflammasome activation. For example, monocytes may respond to purified LPS alone (Netea et al. 2010), and murine neutrophil inflammasomes are not activated by particulate matter as those of macrophages are (Chen et al. 2016). It remains to be seen whether there are specific triggers that cause inflammasome activation in intrinsic kidney cells that differ from the triggers required by immune cells.

8.2.1 Other Inflammasomes in the Kidney

Although a role for AIM2 and NLRC4 inflammasomes in the pathogenesis of several renal diseases has been postulated (Du et al. 2013; Yuan et al. 2016; Zhen et al. 2014), as yet there is little information on the presence of these inflammasomes in intrinsic kidney cells.

8.3 How Does Inflammasome Activation Promote Kidney Injury?

8.3.1 Effects of IL-1β on Leukocytes and Kidney Tissue Cells

The IL-1 receptor (IL-1R) is present on a variety of cell types, including leukocytes and intrinsic kidney cells. It can be bound by either of the isoforms of IL-1, IL-1α or IL-1β, both pro-inflammatory cytokines with similar but slightly distinct biological actions. Pro-fibrotic and inflammatory mediators induced by IL-1/IL-1R interactions include IL-6, tumour necrosis factor (TNF), prostaglandins, TGF-β and tissue matrix metalloproteinases (MMPs), highly relevant to a range of kidney diseases (Gabay et al. 2010). Experiments using mice deficient for the IL-1R highlight the important role of IL-1 in inflammatory cell recruitment to the kidney (although not distinguishing between the actions of IL-1α and IL-1β); IL-1R-deficient mice are protected from acute severe glomerulonephritis, IRI and experimental renal fibrosis (Furuichi et al. 2006; Jones et al. 2009; Timoshanko et al. 2004a). IL-1 (which here refers to both IL-1α and IL-1β) promotes inflammatory cell recruitment both by inducing the expression of adhesion molecules on endothelial cells and promoting the production of chemokines by stromal cells (Gabay et al. 2010). IL-1 also plays a role in adaptive immunity, being important in the differentiation of Th17 cells (Chung et al. 2009).

IL-1 has directly detrimental effects on intrinsic kidney cells. In experimental rapidly progressive glomerulonephritis, bone marrow chimeric studies showed that leukocyte-derived IL-1β mediated its injurious effects via IL-1R present on intrinsic kidney cells (Timoshanko et al. 2004b). IL-1β may play a role in the biological effects of neutrophil gelatinase-associated lipocalin (NGAL). NGAL, though best known as a biomarker of kidney injury, has a number of potential biological roles depending on its iron chelation status. IL-1β is a potent stimulus for NGAL release from proximal tubular and collecting duct cells (Bonnemaison et al. 2017; Konno et al. 2016). Tissue matrix metalloproteinases (MMPs) disrupt the extracellular matrix such as the GBM (but could also limit excessive matrix deposition in fibrosis). Several MMPs and tissue inhibitors of matrix metalloproteinases (TIMPs) are secreted by renal cells in response to cytokines, and IL-1β may contribute to MMP induced tissue injury by augmenting inhibition of endogenous MMP inhibitor TIMP-1 in the presence of TNF in mesangial cells (Nee et al. 2007). One study examining the effect of inflammation on lipid metabolism found that

IL-1β treatment caused cultured human mesangial cells to accumulate cholesterol esters, which lead to production of reactive oxygen species (ROS) and endoplasmic reticulum stress (Zhong et al. 2015).

Two mechanisms for inflammasome-mediated podocyte injury have been postulated. The formation of the large intracellular inflammasome complex in podocytes may lead to podocyte dysfunction by interfering with intracellular signalling (Xiong et al. 2015; Zhang et al. 2012). IL-1β (either produced by leukocytes or acting in an autocrine manner) can adversely affect the production of important podocyte proteins such as nephrin (Takano et al. 2007), compromising podocyte structural integrity and function. Human glomerular endothelial cells use all three forms of intercellular junctions, tight junctions, adherens junctions and gap junctions, to assist in the maintenance of glomerular filtration barrier. IL-1β is known to disrupt tight and adherens junctions: in a study where human glomerular endothelial cells were treated with IL-1β, cells showed increased permeability and an increase in expression of VE-cadherin, which may represent a compensatory mechanism against the disruption of the other inter-endothelial junctions by IL-1β (Du et al. 2015).

8.3.2 *Effects of IL-18 on Leukocytes and Kidney Tissue Cells*

IL-18 was first characterised as a promoter of IFNγ release and in the presence of IL-12 or IL-15 enhances development of Th1 cells. However, in the absence of IL-12 or IL-15, IL-18 has pro-inflammatory actions similar to IL-1 family cytokines (Novick et al. 2013). IL-18 binds to a specific receptor, IL-18R, resulting in a signalling cascade leading to the activation of NF-κB and p38 mitogen-activated protein kinases and the production of downstream pro-inflammatory cytokines (Bombardieri et al. 2007; Yamamura et al. 2001). IL-18 seems to be particularly important in neutrophil activation and recruitment to the kidney, where these cells can then go on to release injurious chemokines, cytokines and ROS (Futosi et al. 2013). IL-18 can also promote the production of pro-inflammatory cytokines by mesangial cells (Schrijvers et al. 2004). In a murine IRI, bone marrow-derived IL-18 was seen to mediate renal injury and was associated with tubular cell damage and increased intrarenal neutrophil and macrophage accumulation (Wu et al. 2008). IL-18 has a local pro-inflammatory and pro-fibrotic role in experimental Th1-dependent GN (Kitching et al. 2005) and unilateral ureteral obstruction (UUO), a model of CKD (Bani-Hani et al. 2009).

Although there is little data on the specific effects of IL-18 on glomerular cells, several studies have focussed on the effect of IL-18 on renal tubular epithelial cells. Both the STAT-3 (Matsui et al. 2013) and TLR4 (Meldrum et al. 2012) pathways in tubular epithelial cells are likely to be important in mediating IL-18's pro-fibrotic effects in unilateral ureteric obstruction mouse models and in vitro in human tubular epithelial cells. IL-18 also induces pro-apoptotic signalling via a FasL-dependent mechanism and may be a significant mediator of tubular cell apoptosis (Zhang et al. 2011).

8.4 Inflammasomes in Acute Kidney Injury

Acute kidney injury (AKI) is an abrupt decline in renal function resulting in the retention of nitrogenous waste, commonly associated by oligoanuria and assessed clinically by serum creatinine and urea measurements. While AKI is often reversible once the underlying insult is treated, it is a global public health concern associated with high morbidity, mortality and healthcare costs (Mehta et al. 2015). AKI has multiple aetiologies, with ischaemic injury to the kidney due to major surgery, sepsis or exposure to nephrotoxins accounting for many cases. While the aetiologies vary between patients, acute tubular necrosis (ATN) is a common histological feature of several forms of AKI, characterised by widespread tubular cell necrosis and the formation of intratubular casts derived from sloughed cells and cellular debris. Damaged tubular and endothelial cells release endogenous DAMPs that activate pattern recognition receptors (PPRs) and initiate innate immune responses. In animal models of AKI, specific components of necrotic cellular debris such as histones, heat-shock proteins, biglycan, HMGB1 and hyaluronan are capable of inducing IL-1β in an NLRP3-dependent manner (Iyer et al. 2009). There is substantial data supporting the role of the inflammasome in numerous models of AKI including IRI, cisplatin-induced nephrotoxicity as well as AKI induced by sepsis, contrast medium and rhabdomyolysis, as summarised in Table 8.3.

8.4.1 Ischaemia-Reperfusion Injury

Renal ischaemia with subsequent reperfusion is a common cause of AKI. In addition, it is an obligatory component of renal transplantation that when severe, leads to delayed allograft function. In murine bilateral renal IRI, NLRP3-deficient mice are functionally protected 24 hours post ischaemia (Iyer et al. 2009), with less intrarenal IL-1β. This protection is associated with fewer infiltrating neutrophils and less CXCL1 (a key neutrophil chemoattractant) in the renal interstitium. However, ASC deficiency shows no protection against early renal dysfunction, although it did provide partial protection after 5 days, with reduced neutrophil recruitment and reduced levels of intrarenal CXCL1 and IL-1β. The pathogenicity of NLRP3 in IRI was confirmed in subsequent studies, but different mechanisms of protection were reported (Kim et al. 2013; Shigeoka et al. 2010). Shigeoka et al. suggested that NLRP3 promoted injury in an inflammasome-independent manner: mice deficient in ASC, caspase-1, IL-1R and IL-18 showed no difference in histological and functional injury 24 hours after ischemia (Shigeoka et al. 2010). However, they observed significantly less renal tubular apoptosis in absence of NLRP3, suggesting that NLRP3 mediated renal IRI by promoting tubular apoptosis. Kim et al. again confirmed a protective effect from IRI in *Nlrp3*$^{-/-}$ mice but, paradoxically, with increased caspase-1 activity (Kim et al. 2013), having previously demonstrated that caspase-1-deficient mice are resistant to IRI. While collectively these studies implicate the NLRP3 inflammasome in pathological acute inflammatory response during

Table 8.3 The function of inflammasome components in experimental acute kidney injury

Acute kidney injury model	Intervention	Renal effect	Canonical or noncanonical signalling	References
Renal ischaemia-reperfusion injury				
	Nlrp3$^{-/-}$	Protective	Both	Iyer et al. (2009), Kim et al. (2013), Shigeoka et al. (2010)
	Asc$^{-/-}$	Equivocal	Canonical	Iyer et al. (2009), Shigeoka et al. (2010)
	Casp1$^{-/-}$	Equivocal	ND	Melnikov et al. (2001), Shigeoka et al. (2010)
	Il1r$^{-/-}$	Equivocal	ND	Haq et al. (1998), Shigeoka et al. (2010)
	IL-1RA	Protective	ND	Rusai et al. (2008)
	Il18$^{-/-}$	Equivocal	ND	Shigeoka et al. (2010), Wu et al. (2008)
	IL-18BP-Tg	Protective	ND	He et al. (2008), Wu et al. (2008)
	Anti-IL-18 Ab	Protective	ND	Melnikov et al. (2001)
Cisplatin nephrotoxicity				
	Nlrp3$^{-/-}$	No difference	–	Kim et al. (2013)
	Asc$^{-/-}$	Protective	Canonical	Chan et al. (2014)
	Casp1$^{-/-}$	Protective	ND	Faubel et al. (2004)
	Il18$^{-/-}$	Protective	ND	Okui et al. (2012)
	Il18ra$^{-/-}$	Worse	ND	Nozaki et al. (2012)
	IL-18BP-Tg	No difference	–	Faubel et al. (2007)
	Anti-IL-18 Ab	No difference	–	Faubel et al. (2007)
	IL-1RA	No difference	–	Faubel et al. (2007)
Sepsis-induced AKI				
	Nlrp3$^{-/-}$	Protective	Canonical	Cao et al. (2015)
	Caspase-1 inhibitor	Protective	Canonical	Cao et al. (2015)
Rhabdomyolysis-induced AKI				
	Nlrp3$^{-/-}$	Protective	Noncanonical	Komada et al. (2015)
	Asc$^{-/-}$	Protective	Noncanonical	Komada et al. (2015)
	Casp1$^{-/-}$	Protective	Noncanonical	Komada et al. (2015)
	Il1b$^{-/-}$	Protective	Noncanonical	Komada et al. (2015)
	Caspase-1 inhibitor	Protective	Canonical	Homsi et al. (2006)
Contrast medium-induced AKI				
	Nlrp3$^{-/-}$	Protective	ND	Shen et al. (2016)

Ab antibody, *IL-1RA* IL-1 receptor antagonist, *IL-18BP-Tg* IL-18 binding protein transgenic mice, *ND* not determined renal ischemia, the contribution of other inflammasome members in renal IRI remains to be established. As AIM2 binds to cytosolic DNA (Fernandes-Alnemri et al. 2009; Hornung et al. 2009), and necrotic renal tubular cells have been reported to release extracellular DNA following renal ischemia (Jansen et al. 2017), the AIM2 inflammasome may be pathogenic in AKI due to IRI.

Both IL-1 and IL-18 have been implicated in the pathogenesis of AKI due to IRI. The lack of a functional IL-1R limits renal dysfunction in IRI (Haq et al. 1998), and treatment with anakinra, an IL-1R antagonist (IL-1RA), impairs the inflammatory response and accelerates renal repair processes (Rusai et al. 2008). In addition to urinary IL-18 being a biomarker for tubular inflammation and predicting mortality risk after severe AKI (Coca et al. 2008), IL-18 promotes renal macrophage recruitment, with deficiency or neutralisation (via antisera or IL-18 binding protein transgenic mice) protecting mice from IRI (He et al. 2008; Melnikov et al. 2001; Wu et al. 2008).

8.4.2 Cisplatin Nephrotoxicity

Cisplatin is an inorganic platinum-based chemotherapeutic agent widely used in the treatment of many solid-organ malignancies. However, cisplatin nephrotoxicity is a common dose-dependent complication with 25–35% of patients being affected after a single dose of cisplatin treatment (dos Santos et al. 2012). Cisplatin concentrates in the S3 segment of the proximal tubule, where it induces both necrotic and apoptotic cell death, as well as pro-inflammatory responses (Peres and da Cunha 2013). There are conflicting data on the role of inflammasome components and products in cisplatin-induced AKI in mice, caspase-1 activity increases prior to the development of severe renal failure and *Casp1*$^{-/-}$ mice are protected (Faubel et al. 2004), with caspase-1 deficiency attenuating the increased intrarenal IL-1β and IL-18 found in this model (Faubel et al. 2007). However, although *Il18*$^{-/-}$ mice are protected from cisplatin nephrotoxicity with reduced renal dysfunction and accelerated clearance of cisplatin (Okui et al. 2012), IL-18Rα-deficient mice have increased cisplatin nephrotoxicity (Nozaki et al. 2012). Furthermore, inhibition of IL-1β with IL-1RA and IL-18 with the use of IL-18 antiserum and IL-18BP-Tg mice or combination therapy with IL-1RA and IL-18 antiserum seems not to be sufficient to prevent cisplatin-induced renal damage in mice (Faubel et al. 2007).

ASC is increased in the kidneys of mice with cisplatin-induced AKI (Kim et al. 2013), and ASC deficiency is protective against cisplatin nephrotoxicity with reduced renal dysfunction, ATN, and renal IL-1β and IL-17A levels (Chan et al. 2014). NLRP3 is abundantly present in macrophages and renal proximal tubules of normal mice, there are conflicting reports as to whether its expression is increased after exposure to cisplatin in vitro or in vivo and *Nlrp3*$^{-/-}$ mice are not protected against cisplatin-induced AKI (Kim et al. 2013; Lee et al. 2015; Zhang et al. 2014). Zhang et al. explored the role of the purinergic $2X_7$ receptor ($P2X_7R$) in cisplatin-induced AKI (Zhang et al. 2014). $P2X_7R$, a ligand-gated ion channel activated by high concentrations of extracellular ATP, triggers a strong potassium efflux and subsequent NLRP3 activation (Franceschini et al. 2015). Although $P2X_7R$ is not expressed in renal tissues of control mice, it is upregulated in renal tubules after cisplatin administration (Zhang et al. 2014). A selective $P2X_7R$ antagonist attenuated cisplatin-induced AKI (Zhang et al. 2014). While $P2X_7R$ may have other pro-inflammatory roles, blocking $P2X_7R$ also decreased the expression of the NLRP3 inflammasome components and downstream inflammatory cytokines in the kidney, implicating the $P2X_7R$-NLRP3 axis in cisplatin nephrotoxicity (Zhang et al. 2014).

In addition to mediating NLRP3 inflammasome function, ASC is an adapter protein for several other inflammasome components including NLRP1, NLRC4 and AIM2. NLRP1 is reportedly increased in the kidney after cisplatin administration (Kim et al. 2013). In addition to sensing microbial stimuli, NLRP1 has been reported to detect reductions in cellular ATP. Given that cisplatin alters intracellular ATP levels in proximal tubules, it is plausible that NLRP1 participates in cisplatin-induced AKI (Liao and Mogridge 2013; Miller et al. 2010; Peres and da Cunha 2013). Similarly, AIM2 may also play an important role in cisplatin-induced AKI, as cisplatin binds to DNA causing DNA strand breaks in mitochondrial DNA (Miller et al. 2010).

8.4.3 Sepsis-Induced AKI

Sepsis, a systemic inflammatory response to infection, is the most common cause of AKI (Rossaint and Zarbock 2016). A growing body of evidence suggests that inflammation, oxidative stress, microvascular dysfunction and tubular epithelial responses are involved in the pathogenesis of this complex condition (Zarbock et al. 2014). Several studies have reported on the participation of NLRP3 and its inflammasome components in sepsis-induced AKI using a cecal ligation and puncture (CLP) model (Cao et al. 2015; Wang et al. 2015b; Zhao et al. 2016). Sepsis-induced kidney damage is accompanied by an upregulation of intrarenal NLRP3, ASC and caspase-1 expression and IL-1β and IL-18 in the serum and kidney (Cao et al. 2015). Inhibiting the inflammasome using *Nlrp3*$^{-/-}$ mice and a caspase-1 inhibitor attenuated CLP-induced renal dysfunction and limited renal neutrophil infiltration, ASC and caspase-1 expression and IL-1β and IL-18 level in the serum and kidney (Cao et al. 2015). Other strategies, including low-dose carbon monoxide (Wang et al. 2015b) and sirtuin 3 (a NAD^+-dependent deacetylase that regulates mitochondrial function by limiting oxidative stress) (Zhao et al. 2016), also limit sepsis-induced AKI and downregulates inflammasome components. Collectively these studies provide insight into the role of the NLRP3 inflammasome in sepsis-induced AKI, but further research is required to determine the underlying mechanisms that link NLRP3 activation and sepsis-induced kidney damage.

8.4.4 Rhabdomyolysis-Induced AKI

Rhabdomyolysis is caused by muscle damage due to a range of insults, leading to the release of myoglobin and other intracellular contents. AKI is a common complication of rhabdomyolysis with up to 50% of patients developing some degree of kidney injury

(Gois et al. 2016). The released myoglobin is deposited in renal proximal tubular cells causing inflammation, necrosis and oxidative damage. Inflammasomes have been implicated in glycerol-induced rodent models of rhabdomyolysis-induced AKI (RI-AKI) (Komada et al. 2015). Caspase inhibition limits rat RI-AKI (Homsi et al. 2006), and mice deficient in NLRP3, ASC, caspase-1 or IL-1β were protected from RI-AKI by preventing the initial inflammatory response that mediates renal tubular damage. Increased renal tubular cell NLRP3 expression initiated inflammatory responses and apoptotic cell death, independent of IL-1β processing suggesting a noncanonical role of NLRP3 during this early phase of RI-AKI (Komada et al. 2015). The endogenous danger signal responsible for NLRP3 activation was not identified in the study, although the heme protein hemin, ferrous and ferric myoglobin, released into circulation following muscle damage, did not induce NLRP3 inflammasome activation in primary tubular and collecting duct epithelial cells in vitro (Komada et al. 2015). Uric acid might be the endogenous danger signal activating the inflammasome cascade in RI-AKI (Gois et al. 2016). Allopurinol, used clinically to lower serum uric acid levels, attenuated renal dysfunction in rat RI-AKI and reduced oxidative stress, tubular apoptosis and renal inflammation by inhibiting the inflammasome cascade with reduced active caspase-1 levels (Gois et al. 2016).

8.4.5 Contrast Medium-Induced AKI

With the wide use of iodinated contrast media in radiological procedures for medical diagnosis and treatment of disease, contrast medium-induced AKI (CI-AKI) has become the third leading cause of hospital-acquired AKI (Shen et al. 2016). While the pathogenesis of CI-AKI is not entirely clear, contrast medium seems to have direct cytotoxic effects on renal tubular cells by inducing apoptosis, the generation of ROS, and indirectly by hemodynamic effects (Sadat et al. 2015). Shen et al. showed that the NLRP3 inflammasome mediates CI-AKI through modulating tubular apoptosis both in vitro using a human renal proximal tubular cell line (HK-2 cells) and experimentally in vivo by administering contrast to mice with a single kidney unilateral nephrectomy model with the administration of media (Shen et al. 2016).

8.5 The Inflammasome in Crystal Nephropathies

Crystal nephropathies are a number of acute and chronic kidney disorders related to crystal deposition or formation inside the kidney, most frequently involving the tubulointerstitium. The kidney is highly susceptible to intrarenal crystal formation or deposition because of the high concentration of ions and molecules reached in the tubulointerstitium as a result of glomerular filtration. It is well established that the NLRP3 inflammasome can be activated by crystalline material via potassium efflux secondary to lysosomal rupture in phagocytic innate immune cells (Hornung et al.

2008). Several crystalline substances implicated in kidney disease, including calcium oxalate, monosodium urate, calcium phosphate and cholesterol embolism, are well-characterised activators of the NLRP3 inflammasome (Hutton et al. 2016).

Calcium oxalate is responsible for kidney stones in approximately 70–80% of kidney stone patients (Darisipudi and Knauf 2016). Two key studies have shown that calcium oxalate crystals activate the NLRP3 inflammasome both in AKI (Mulay et al. 2013) and in progressive renal failure (Knauf et al. 2013). Mulay et al. using a mouse model of crystal nephropathy induced by a high oxalate diet comprehensively showed that intrarenal inflammation, tubular damage and renal dysfunction were attenuated in mice lacking ASC and NLRP3 and their downstream mediators MyD88, caspase-1, IL-1R and IL-18 (Mulay et al. 2013). Calcium oxalate crystals activated renal dendritic cells to secrete IL-1β in an inflammasome-dependent manner; ATP released due to calcium oxalate-mediated tubular damage was potentially activating the inflammasome in this setting (Mulay et al. 2013). *Nlrp3*$^{-/-}$ mice are also protected in chronic calcium oxalate-induced renal disease, seen in primary hyperoxaluria and other crystallopathies (Knauf et al. 2013). A recent microarray study also revealed a potential role for ROS in activating the NLRP3 inflammasome via thioredoxin-interacting protein (TXNIP), a crucial protein that plays a role in regulating ROS production in cells, leading to a robust inflammatory response in the kidneys of rats with hyperoxaluria and calcium oxalate nephrolithiasis (Joshi et al. 2015).

Adenine overload also induces intrarenal crystal precipitation resulting in tubular atrophy and renal fibrosis, with a role for inflammasomes. ASC and caspase-1 are pathogenic in adenine-induced renal inflammation and fibrosis in mice (Correa-Costa et al. 2011), and the NLRP3-specific inhibitor CP-456773 blocked NLRP3 activation in dendritic cells and downstream IL-1β and IL-18 production, attenuating renal inflammation and fibrosis in murine crystal nephropathy induced by diets rich in adenine or oxalate (Ludwig-Portugall et al. 2016). However, delayed treatment, although reducing intrarenal inflammasome activation and inflammation, did not reverse renal fibrosis once it was established.

Hyperuricaemia is epidemiologically associated with an increased risk of AKI and progressive CKD (Iseki et al. 2004; Kaushik and Choo 2016; Obermayr et al. 2008). Experimental studies implicate a variety of mechanisms by which hyperuricaemia causes renal disease, including inflammation provoked by monosodium urate crystals (Isaka et al. 2016). Within the kidney, uric acid preferentially precipitates and forms monosodium urate crystals in collecting ducts and even tophi in the surrounding interstitium of the renal medulla (Mulay et al. 2014). In gout, the potential for monosodium urate crystals to induce the NLRP3 inflammasome is well established (Martinon et al. 2006). Similar mechanisms are likely to operate in the kidney, where monosodium urate induces lysosomal rupture when phagocytosed with mitochondrial damage and ROS production (Emmerson et al. 1990). In vitro studies demonstrate that soluble uric acid or monosodium urate crystals induce NLRP3 activation via a TLR4-dependent pathway in both human primary renal proximal tubular epithelial cells and rat mesangial cells (Hong et al. 2015; Xiao et al. 2015a, b). However, at this stage functional data linking monosodium urate crystals with NLRP3 inflammasome activation in animal models of renal disease is lacking.

Uromodulin is a sticky particle-forming protein secreted exclusively by the thick ascending limb of the distal tubule (Anders and Schaefer 2014; Leemans et al. 2014). Due to its adhesive nature, uromodulin coats all particles in the distal tubule including renal crystals. Distal tubular injury facilitates uromodulin leakage into the interstitial compartment, where it acts as a DAMP and activates interstitial dendritic cells in a TLR4- and NLRP3-dependent manner (Darisipudi et al. 2012; Saemann et al. 2005). The particulate nature of uromodulin favours phagocytosis and endosomal destabilisation in dendritic cells, activating the NLRP3 inflammasome (Darisipudi et al. 2012).

Cystinosis, a rare autosomal recessive disease caused by mutations in the *CTNS* gene, is characterised by the lysosomal accumulation of cystine, leading to the formation of cystine crystals in multiple organs, including the kidney. Infantile cystinosis represents the most severe phenotype with progressive renal impairment and end-stage renal disease (ESRD) that may occur before 10 years of age (Darisipudi and Knauf 2016). Cystine crystals are endogenous inflammasome-activating stimuli, implying that the inflammasome plays a role in the pathogenesis of this disease (Prencipe et al. 2014).

8.6 The Inflammasome in Chronic Kidney Injury and Disease

Unilateral ureteric ligation (UUO) in rodents is often used to study mechanisms of renal fibrosis and progressive chronic kidney disease (CKD). Several studies have employed this model to investigate the role of the inflammasome in the development of CKD. Biglycan, an extracellular matrix component and an endogenous ligand for TLR4 and TLR2, acts as a DAMP for NLRP3 inflammasome activation in UUO, initiating caspase-1-mediated maturation and secretion of IL-1β (Babelova et al. 2009). Vilaysane et al. demonstrated that *Nlrp3*$^{-/-}$ mice developed less tubular injury, inflammation and fibrosis after UUO (Vilaysane et al. 2010), a phenotype associated with reduced caspase-1 activation and IL-1β and IL-18 maturation. Bone marrow chimeras revealed these effects were mediated by NLRP3 in both haematopoietic and non-haematopoietic cells, but IL-1β/IL-18 processing in renal tubular epithelial cells could not be detected, suggesting a noncanonical role for NLRP3 in tubular cells (Vilaysane et al. 2010). The same group revealed that NLRP3 promotes pro-fibrotic TGF-β-mediated signalling and Smad2 and Smad3 phosphorylation in renal epithelium independently of forming a caspase-1-activating inflammasome (Wang et al. 2013). However, the role of the NLRP3 inflammasome during UUO remains controversial. ASC-deficient mice have significantly reduced renal inflammatory responses and improved renal injury and fibrosis after UUO (Komada et al. 2014). Mechanistically, this study found that ATP induces inflammasome activation in ASC, expressing collecting duct epithelial cells via P2X-potassium efflux and ROS-dependent pathways in vitro, but this was not examined in vivo (Komada et al. 2014). Further studies have suggested roles of NLRP3 via promoting mitochondrial dysfunction and the subsequent release of mature IL-1β and IL-18 (Guo et al. 2017), while in UUO, IL-36 signalling may also activate the NLRP3 inflammasome in both immune cells and renal epithelial cells (Chi et al. 2017). Despite the discrepancies in mechanism, most studies have found that NLRP3 signalling promotes injury in UUO. In contrast, at least one study does not support a pathogenic role for NLRP3 with NLRP3 deficiency resulting in increased interstitial oedema and vascular leakage, potentially due to reduced expression of intercellular junction components (Pulskens et al. 2014). Experimental administration of a number of anti-inflammatory mediators such as milk fat globule-epidermal growth factor 8 (Brissette et al. 2016), Danggui Buxue Tang (Wang et al. 2016), aliskiren (Wang et al. 2015c) and fluorofenidone (Zheng et al. 2017) all attenuate experimental renal fibrosis and inflammatory responses after UUO, potentially by inhibiting NLRP3 inflammasome activation. In a cohort of renal biopsies from patients with nondiabetic kidney disease, renal mRNA levels of NLRP3 correlated with renal functional impairment, further supporting that NLRP3 contributes to the pathogenesis of CKD (Vilaysane et al. 2010).

8.6.1 Diabetic Nephropathy

Diabetic nephropathy (DN) is the leading cause of ESKD worldwide and is increasing in prevalence (Molitch et al. 2015). Albuminuria is an early sign of DN, with in many, an irreversible decline in renal function subsequently occurring over years. Unlike Type 1 diabetes mellitus (T1DM), which is an autoimmune disease with destruction of pancreatic islet cells causing insulinopenia, Type 2 diabetes mellitus (T2DM) is characterised by insulin resistance, often seen in concert with other aspects of the metabolic syndrome including obesity, hypertension, hyperlipidaemia and hyperuricaemia. Obesity and hypertension in themselves cause proteinuria and hasten progression of diabetic kidney disease and also contribute to an inflammatory state in which there is aberrant NLRP3 inflammasome activation (Mastrocola et al. 2018). While some processes common to diabetic nephropathy implicate the NLRP3 inflammasome in both Type 1 and Type 2 diabetes, the role of the inflammasome in T2DM bears special consideration, because of the range of substances that may function as inflammasome-activating DAMPs in this condition. This section will focus on the inflammasome in diabetic kidney disease, rather than in diabetes itself or in other aspects of the metabolic syndrome, which is discussed in this chapter (metabolic disease).

Histologically, DN is characterised by glomerular hypertrophy, followed by accumulation of extracellular matrix proteins. Tubular hypertrophy also occurs early and may progress to interstitial fibrosis and tubular atrophy over time (Pourghasem et al. 2015). Macrophages are present in the kidneys of diabetic humans and in rodent models of DN (Chow et al. 2004; Nguyen et al. 2006); the extent of inflammatory cell infiltrate correlates with the decline in renal function, and

CCL2 deficiency in mice limited intrarenal macrophage infiltrates and diabetic nephropathy, suggesting a causative link (Awad et al. 2011; Lim and Tesch 2012). Thus, although it has been known for some time that sterile inflammation is associated with progression of DN, the NLRP3 inflammasome as a link between these two entities has only been explored more recently. Supporting the experimental data detailed below is human data showing that urinary IL-18 and serum IL-1β levels are elevated in patients with diabetic nephropathy compared to diabetic patients without albuminuria, and levels correlate closely with the degree of albuminuria (Nakamura et al. 2005; Shahzad et al. 2015). Additionally, polymorphisms in the promoter region of the IL-18 gene may be associated with the development of diabetic nephropathy in diabetic patients (Elneam et al. 2016).

Various studies have implicated the NLRP3 inflammasome in diabetic nephropathy using mouse models. One of the most comprehensive used both db/db mice, which develop diabetes in the setting of insulin resistance and obesity (modelling T2DM), and uninephrectomised mice treated with the pancreatic β-cell toxin streptozotocin (STZ, modelling T1DM) (Shahzad et al. 2015). While 8-week-old db/db mice do not have signs of renal disease, the development of albuminuria and renal histological changes occurs by 12 weeks. Caspase-1 and NLRP3 deficiency were both protective in the db/db and insulinopenic STZ models. IL-1RA (anakinra) administered to 8-week-old db/db mice for 12 weeks limited albuminuria and extracellular matrix accumulation. IL-1RA was also administered to 12-week-old db/db mice with established renal disease for 8 weeks, resulting in a normalisation of albuminuria and renal histology. Interestingly, though IL-1RA resulted in weight loss and improved blood glucose levels only in the db/db mice, it did not affect metabolic parameters in STZ mice, indicating the renoprotective effect was, at least partially, independent of the observed metabolic improvements. The lack of protection seen in irradiated wild-type mice transplanted with *Nlrp3*$^{-/-}$ or *Casp1*$^{-/-}$ bone marrow and the fact that *Nlrp3*$^{-/-}$ mice transplanted with wild-type bone marrow remained protected indicate that intrinsic renal cell-derived, rather than immune cell-derived, NLRP3 inflammasome activation drives disease. Supporting this, NLRP3 inflammasome activation (in the form of co-localisation of cleaved caspase-1 and NLRP3) was seen both in glomerular endothelial cells and in podocytes of db/db mice and of diabetic humans. Cleaved IL-1β was upregulated in glucose-stressed human podocytes and mouse glomerular endothelial cells compared to cells treated with control substances. This IL-1β production was caspase-1 dependent, indicating canonical NLRP3 inflammasome activation occurred in these intrinsic renal cells.

8.6.2 NLRP3 Inflammasome-Activating Substances in T2DN

There are a number of exogenous and endogenous substances that are present or increased in DN that can also activate the NLRP3 inflammasome; multiple studies have linked a particular inflammasome-activating DAMP to renal disease in T2DM. Several studies point to mitochondrial ROS as being a likely activator of NLRP3-

ASC in diabetic nephropathy, including ex vivo studies in human monocytes and macrophages and db/db mice treated with mitochondrial ROS inhibitor (Mirza et al. 2014; Shahzad et al. 2015). High glucose itself may stimulate the expression of NLRP3 (Chen et al. 2013). A study by Gao et al. found this occurs due to hyperglycaemia-induced expression of thioredoxin-interacting protein, which causes activation of the gp91 (phox) subunit of NADPH oxidase, which then activates NLRP3 (Gao et al. 2015). Islet amyloid polypeptide, produced in response to elevated blood glucose by pancreatic islet cells and secreted with insulin, can activate the NLRP3 inflammasome via disruption of phagolysosomes as well as cathepsin-B and cathepsin-L (Masters et al. 2010). Hyperuricaemia commonly occurs in metabolic syndrome, often found in T2DM. Monosodium urate crystals may act as an inflammasome activator in this situation (Kim et al. 2015). ROS, which are produced in greater amounts in the adipose tissue of obese compared to nonobese individuals (Furukawa et al. 2004), and certain fatty acids (L'Homme et al. 2013; Wen et al. 2011) can also act as an NLRP3 inflammasome-activating signal. In practice, it may be a combination of more than one of these substances that causes NLRP3 inflammasome activation and inflammation in DN.

8.6.3 NLRC4 in Diabetic Nephropathy

Although most studies have focussed on the role of the NLRP3 inflammasome in this disease, NLRC4 inflammasome expression was found to be increased in the kidneys of diabetic humans, and mice deficient in NLRC4 were protected in a STZ model of diabetes, with decreased intrarenal macrophage accumulation (Yuan et al. 2016).

8.6.4 Inflammasome Blocking Treatments for T2DM

Although IL-1 blockade has shown promise in mouse models of DN (Shahzad et al. 2015), trials in humans have focussed on glycaemic control and other metabolic effects rather than renal disease (Larsen et al. 2007; Mandrup-Poulsen et al. 2010). A number of other treatments have been used in mouse models. For example, in an STZ-induced murine model of diabetic nephropathy, with hyperlipidaemia and hyperuricaemia, mice were found to express high levels of NLRP3 and downstream cytokines IL-1β and IL-18. Treatment with allopurinol and quercetin (which have uric acid and lipid-lowering effects) limited NLRP3-ASC inflammasome activation and improved renal histology. Extracellular ATP, an endogenous DAMP activating the inflammasome via the $P2X_7R$, promotes renal interstitial inflammation. Both apyrase, which consumes extracellular ATP, and compounds targeting the $P2X_7R$ limit glucose stimulated NLRP3 expression and IL-1β and IL-18 release (Chen et al. 2013). Although approaches targeting a specific inflammasome activator are useful in mouse models, their effects may be limited in real-world human disease, unless a unifying inflammasome activator is found.

8.7 Inflammasomes and Glomerulonephritis

Autoimmunity is the predominant pathogenic process underlying most forms of glomerulonephritis (Holdsworth et al. 2016). Autoantibodies may be directed against components of the glomerulus, as occurs in membranous nephropathy and anti-GBM disease. Circulating autoantibodies that target antigens which are not specific to the kidney, such as double-stranded DNA (dsDNA) in lupus, and neutrophil components in anti-neutrophil cytoplasmic antibody (ANCA)-associated vasculitis (AAV), can also cause kidney injury (Suarez-Fueyo et al. 2017). In some diseases, such as anti-GBM disease, the pathogenic antigen and autoantibody have been well characterised (Ooi et al. 2017), whereas in others, such as idiopathic minimal change disease, a circulating factor causing immune-mediated damage is likely to be present, but is not yet identified (Bierzynska and Saleem 2017). Although part of the innate immune system, inflammasomes can modulate adaptive immune responses, contributing to loss of tolerance and autoimmunity, via effects on T-cell differentiation. Autoreactive $CD4^+$ T cells not only promote autoantibody production (as T follicular helper cells) but also act as local effectors and are key players in a number of renal autoimmune diseases, including lupus (Okamoto et al. 2012), anti-GBM disease (Ooi et al. 2013; Salama et al. 2001) and AAV (Ooi et al. 2012). On activation, naïve $CD4^+$ T cells differentiate into functionally distinct subsets, with characteristic patterns of cytokine secretion. The inflammasome has effects on $CD4^+$ T-cell fate determination via IL-1β and IL-18. IL-18 is important in Th1 responses (Novick et al. 2013), whereas IL-1β is essential for Th17 cell differentiation from naïve T cells (Joosten et al. 2013). As Th1 and Th17 cells actively participate in experimental models of glomerulonephritis (Summers et al. 2011, 2009), the modulation of adaptive immune responses by the inflammasome may be important in a number of autoimmune kidney conditions.

8.7.1 Models of Anti-GBM Disease and Immune Complex Glomerulonephritis

Anti-GBM disease is a rare condition characterised by the deposition of antibodies targeting the non-collagenous domain of type IV collagen (α3(IV)NC1) within the GBM, with glomerular linear IgG deposition seen in kidney biopsies. Disease is mediated by autoreactive T and B cells (Holdsworth et al. 2016) resulting in rapidly progressive glomerulonephritis and often pulmonary haemorrhage, with ESRD and often death if left untreated. Mouse models of 'anti-GBM disease' usually involve injection of heterologous anti-basement membrane globulin raised in another species (sometimes called nephrotoxic serum nephritis). These models, which are not autoimmune, have two phases: the initial (heterologous) phase is the direct effect of the antibodies binding to the GBM; the second (autologous) phase occurs when antibodies and T cells are produced that target the heterologous globulin bound to glomerular capillary walls. While having some value in defining effector responses in severe glomerular disease, autologous phase 'anti-GBM disease' is not autoimmune and should not be confused with true autoimmune models of this disease (Odobasic et al. 2014; Ooi et al. 2014, 2017; Wu et al. 2002).

Endogenous IL-1 and IL-18 are pathogenic in autologous phase 'anti-GBM' glomerulonephritis (Kitching et al. 2005; Lan et al. 1993). $PX2_7R$ is increased in mesangial cells and glomerular macrophages in murine autologous phase 'anti-GBM' glomerulonephritis and in humans with autoimmune GN (Turner et al. 2007). $PX2_7$-deficient mice and mice treated with a $PX2_7$ inhibitor were protected from glomerular injury in autologous phase 'anti-GBM' GN (Taylor et al. 2009). Lichtnekert et al. studied the role of inflammasomes in heterologous phase 'anti-GBM GN' where passive antibody transfer induces leukocyte-mediated injury. Pro-IL-1β, caspase-1, NLRP3 but not ASC mRNA were induced in kidneys of mice injected with anti-GBM. NLRC1, NLRP4 and AIM2 mRNA were undetectable. As in previous studies (Kitching et al. 2005; Lan et al. 1993; Tang et al. 1994; Timoshanko et al. 2004a), IL-1 and IL-18 were both pathogenic, but ASC, NLRP3 and caspase-1 deficiency did not protect against disease. While this study reported that mesangial cells, glomerular endothelial cells and podocytes did not secrete IL-1β (Lichtnekert et al. 2011), several other studies indicate that glomerular cells can produce IL-1β (Table 8.2).

8.7.2 Lupus Nephritis

Renal injury in lupus nephritis is mediated by immune complex deposition as well as other effectors, with the role of innate immune pathways being increasingly recognised (Bagavant and Fu 2009). Type I interferon is a central mediator in the pathogenesis of systemic lupus erythematosus (SLE) (Crow 2014), and IL-18, a strong interferon inducer, is increased in the serum of people with SLE, with IL-18 levels correlating with the presence of lupus nephritis and proteinuria (Calvani et al. 2004). IL-18 deficiency in autoimmune-prone Fas-deficient (MRL-Fas^{lpr}) mice prolonged survival and attenuated renal disease (Kinoshita et al. 2004). Several inflammasomes have been implicated in the pathogenesis of SLE. Mouse and human data suggest a role for NLRP3 and AIM2 inflammasomes, and a study of single-nucleotide polymorphisms in seven inflammasome-related genes found that polymorphisms in the NLRP1 inflammasome (but not in AIM2 or NLRP3) were associated with both lupus and the development of lupus nephritis (Pontillo et al. 2012). Intrarenal caspase-1 and NLRP3 are increased in human lupus nephritis (Kahlenberg et al. 2011). The AIM2 inflammasome, activated by cytosolic DNA, is implicated in people with SLE with AIM2 expression correlating with clinical disease severity (Zhang et al. 2013). In murine lupus induced by apoptotic DNA immunisation, AIM2 expression was increased in renal macrophages and correlated with dsDNA levels. Silencing of AIM2 expression limited autoantibody production and renal disease (Zhang et al. 2013).

In vivo evidence supporting the role of the NLRP3 inflammasome in lupus is found in pristine-induced murine lupus, in which blockade of caspase-1 was protective, resulting in a reduction in autoantibodies, glomerulonephritis and inhibition of the development of the type I IFN response (Kahlenberg and Kaplan 2014). Conversely, mice with a gain of function mutation in NLRP3 in the pristine-induced experimental model develop more severe lupus. This phenotype appears to be related to NLRP3 expression in myeloid cells, because Cre recombinase-mediated deletion of this mutant from myeloid cells resulted in significant reduction in disease (Lu et al. 2017). However in the C57BL/6$^{lpr/lpr}$ model of systemic autoimmunity, both ASC and NLRP3 played a regulatory role, with deficiency of either resulting in the development of more severe autoimmunity and diseases, potentially via effects on the SMAD2/SMAD3 signalling pathway (Lech et al. 2015).

Both dsDNA and neutrophil extracellular traps (NETs) can activate the NLRP3 inflammasome in lupus. dsDNA complexes isolated from SLE patients can activate the NLRP3 inflammasome in human monocytes, in a TLR9- and NF-κB-dependent manner. Blocking ROS production and potassium efflux significantly reduced IL-1β production from dsDNA-treated monocytes, indicating the importance of these processes to NLRP3 activation (Shin et al. 2013). NETs, networks of chromatin fibres laced with antimicrobial peptides and enzymes that can be extruded from neutrophils and macrophages, are thought not only to play an important role in host defence but also in the pathogenesis of a variety of autoimmune diseases (Kahlenberg et al. 2013; Kessenbrock et al. 2009). Both NETs and IL-37, an antibacterial protein externalised on NETs, activated the NLRP3 inflammasome in human and murine macrophages; this NET-mediated activation was enhanced in macrophages derived from lupus patients (Kahlenberg et al. 2013). Interestingly, NETosis in macrophages was also promoted by IL-18, potentially leading to a cycle of NET-induced inflammasome activation (Kahlenberg et al. 2013).

8.7.3 ANCA-Associated Vasculitis

The ANCA-associated vasculitides (AAV) are small vessel vasculitides characterised by autoantibodies specific for neutrophil granule components, myeloperoxidase (MPO) and proteinase 3 (PR3), and inflammatory cell infiltration causing damage to the walls of small- and medium-sized blood vessels. The kidney is a commonly affected organ. In these diseases, ANCA bind to activated neutrophils, causing degranulation and injurious ROS production (Jarrot and Kaplanski 2016). Both IL-18 and IL-1β have been shown to be important in the pathogenesis of AAV. Neutrophils are usually primed prior to ANCA-mediated activation.

Traditionally TNF has been used to prime neutrophils ex vivo, but IL-18 primes neutrophils comparably to TNF (Hewins et al. 2006). Patients with active vasculitis have increased serum IL-18 levels, and IL-18 is also upregulated in the kidney in AAV (Hewins et al. 2006; Hultgren et al. 2007). In humans with MPO-AAV intrarenal IL-1β, TLR4 and NLRP3 expression correlated with the severity of tubulointerstitial injury (Tashiro et al. 2016). Experimentally, IL-1RA treatment protected mice from the development of anti-MPO GN in a bone marrow transplant model of AAV. As mice transplanted with bone marrow deficient in dipeptidyl peptidase, required for neutrophil serine protease activation, were also protected, it was proposed that activation of neutrophil IL-1β by serine proteases was more important than inflammasome-mediated IL-1β activation in this disease (Schreiber et al. 2012). In further work, phagocyte NADPH oxidase, thought to be involved in the production of tissue-damaging ROS, was in fact found to downregulate caspase-1, decreasing inflammasome-dependent IL-1β production and protecting against GN (Schreiber et al. 2015). Although this may indicate that there are two, independently acting pathways by which ANCA induces IL-1β production by monocytes, a neutrophil serine protease pathway and an inflammasome-dependent pathway (Schreiber et al. 2015), further work would clarify the roles of IL-1β and of specific inflammasomes in AAV.

8.7.4 IgA Nephropathy

IgA nephropathy is the most common form of primary glomerulonephritis worldwide and can be diagnosed on renal biopsy by the presence of glomerular IgA deposits and mesangial cell proliferation. Kidney damage occurs due to glomerular IgA immune complex deposition promoting innate immune effectors, with subsequent T-cell activation and inflammation (Coppo et al. 2010; Rifai 2007). Predicting the course of IgA nephropathy can be difficult, with severity ranging from it being a mild, non-progressive disease to it progressing to end-stage renal failure. Clinical features, including hypertension, proteinuria and a reduced eGFR on presentation (Barbour and Reich 2012), as well as renal biopsy features including tubulo-interstitial scarring and glomerulosclerosis (Cook 2007), are associated with risk of progression. However NLRP3 expression (primarily seen in renal tubular epithelium) also correlated with progression of kidney disease (Chun et al. 2016). Experimentally IL-1RA-treated mice were protected in experimental IgA nephropathy in which ddY mice spontaneously develop disease (Chen et al. 1997). NLRP3, caspase-1 and IL-1β levels were significantly increased in a passively induced mouse model of IgAN; and NLRP3-deficient mice were also protected in a passive model of IgA nephropathy, with reduced renal macrophage, dendritic cell and T-cell infiltration, reduced T-cell activation and increased T regulatory cells compared to wild-type mice (Tsai et al. 2017). The recent comprehensive study by Tsai et al. demonstrates the role that the NLRP3 inflammasome plays in linking IgA immune complex deposition and T-cell activation. In in vitro studies, IgA immune complexes caused NLRP3-dependent IL-1β production in macrophages and dendritic cells, with both canonical (involving caspase-1) and noncanonical pathways (involving caspase-11 in mice, equivalent to caspase-4/5 in humans) being activated. IgA immune complex-primed bone marrow-derived dendritic cells (BMDCs) from wild-type mice induced proliferation of $CD4^+$ T cells and their production of pro-inflammatory cytokines such as IL-17 and IFNγ. T-cell proliferation and cytokine production were greatly reduced when BMDCs from NLRP3-deficient mice were used. The NLRP3 inflammasome activators in this instance were proposed to be IgA immune complex-induced mitochondrial ROS and mitochondrial DNA release into the cytosol. Supporting this, treatment with a mitochondrial ROS inhibitor reduced NLRP3 expression and IL-1β secretion in IgA immune complex-activated macrophages. Additionally, IgA immune complexes were shown to induce NLRP3-dependent IL-1β from primary mesangial cells and renal tubular endothelial cells, indicating NLRP3 inflammasome activation in these intrinsic kidney cells (Tsai et al. 2017).

8.7.5 Inflammasomes in Other Glomerular Diseases

NLRP3 expression has been assessed in kidney biopsies of patients with a variety of nondiabetic kidney diseases with increased NLRP3 and caspase-1 mRNA expression found in all subgroups of kidney disease (Vilaysane et al. 2010; Xiong et al. 2015). In one of these studies, NLRP3 expression correlated with renal impairment and glomerular sclerosis on biopsy, a marker of advanced kidney damage (Xiong et al. 2015), suggesting that NLRP3 might be a common pathway in the progression of disease. Focal and segmental glomerulosclerosis (FSGS) is a kidney disease that can occur in a primary, autoimmune form, or in a secondary form as a result of another insult to the kidney, such as hypertension, obesity or diabetes. Astaxanthin, a compound which exerts suppressive effects on NLRP3 inflammasome activation as well as antioxidant effects, was renoprotective in a mouse model of FSGS induced by Adriamycin (Liu et al. 2015). However, the potential roles of inflammasomes in FSGS and other glomerular diseases such as membranous nephropathy and minimal change disease have not yet been defined.

8.8 Conclusions and Future Directions

While there is much still to learn, there is clear evidence that inflammasomes are relevant to a variety of renal diseases and represent potential therapeutic targets. Ideally a 'common pathway' involved in multiple renal diseases may be identified, disruption of which can treat a number of human kidney diseases. However, the wide variety of causes (and therefore mechanisms of disease) imply that detailed studies of multiple types of kidney disease are likely to be required, with clinical observations being backed by mechanistic and functional studies in relevant in vitro and in vivo models of disease. The advent of more specific inflammasome inhibitors emphasises the need to understand how different inflammasomes contribute to disease in preclinical models.

References

Abais JM, Zhang C, Xia M, Liu Q, Gehr TW, Boini KM, Li PL (2013) NADPH oxidase-mediated triggering of inflammasome activation in mouse podocytes and glomeruli during hyper-homocysteinemia. Antioxid Redox Signal 18:1537–1548

Akcay A, Nguyen Q, Edelstein CL (2009) Mediators of inflammation in acute kidney injury. Mediat Inflamm 2009:137072

Anders HJ, Schaefer L (2014) Beyond tissue injury-damage-associated molecular patterns, toll-like receptors, and inflammasomes also drive regeneration and fibrosis. J Am Soc Nephrol 25: 1387–1400

Awad AS, Kinsey GR, Khutsishvili K, Gao T, Bolton WK, Okusa MD (2011) Monocyte/macrophage chemokine receptor CCR2 mediates diabetic renal injury. Am J Physiol Ren Physiol 301: F1358–F1366

Babelova A, Moreth K, Tsalastra-Greul W, Zeng-Brouwers J, Eickelberg O, Young MF, Bruckner P, Pfeilschifter J, Schaefer RM, Grone HJ, Schaefer L (2009) Biglycan, a danger signal that activates the NLRP3 inflammasome via toll-like and P2X receptors. J Biol Chem 284:24035–24048

Bagavant H, Fu SM (2009) Pathogenesis of kidney disease in systemic lupus erythematosus. Curr Opin Rheumatol 21:489–494

Bakker PJ, Butter LM, Claessen N, Teske GJ, Sutterwala FS, Florquin S, Leemans JC (2014) A tissue-specific role for Nlrp3 in tubular epithelial repair after renal ischemia/reperfusion. Am J Pathol 184:2013–2022

Bani-Hani AH, Leslie JA, Asanuma H, Dinarello CA, Campbell MT, Meldrum DR, Zhang H, Hile K, Meldrum KK (2009) IL-18 neutralization ameliorates obstruction-induced epithelial-mesenchymal transition and renal fibrosis. Kidney Int 76:500–511

Barbour SJ, Reich HN (2012) Risk stratification of patients with IgA nephropathy. Am J Kidney Dis 59:865–873

Bierzynska A, Saleem M (2017) Recent advances in understanding and treating nephrotic syndrome. F1000Res 6:121

Bombardieri M, McInnes IB, Pitzalis C (2007) Interleukin-18 as a potential therapeutic target in chronic autoimmune/inflammatory conditions. Expert Opin Biol Ther 7:31–40

Bonnemaison ML, Marks ES, Boesen EI (2017) Interleukin-1beta as a driver of renal NGAL production. Cytokine 91:38–43

Brinkkoetter PT, Ising C, Benzing T (2013) The role of the podocyte in albumin filtration. Nat Rev Nephrol 9:328–336

Brissette MJ, Laplante P, Qi S, Latour M, Cailhier JF (2016) Milk fat globule epidermal growth factor-8 limits tissue damage through inflammasome modulation during renal injury. J Leukoc Biol 100:1135–1146

Calvani N, Richards HB, Tucci M, Pannarale G, Silvestris F (2004) Up-regulation of IL-18 and predominance of a Th1 immune response is a hallmark of lupus nephritis. Clin Exp Immunol 138:171–178

Cao Y, Fei D, Chen M, Sun M, Xu J, Kang K, Jiang L, Zhao M (2015) Role of the nucleotide-binding domain-like receptor protein 3 inflammasome in acute kidney injury. FEBS J 282: 3799–3807

Chan AJ, Alikhan MA, Odobasic D, Gan PY, Khouri MB, Steinmetz OM, Mansell AS, Kitching AR, Holdsworth SR, Summers SA (2014) Innate IL-17A-producing leukocytes promote acute kidney injury via inflammasome and Toll-like receptor activation. Am J Pathol 184:1411–1418

Chen A, Sheu LF, Chou WY, Tsai SC, Chang DM, Liang SC, Lin FG, Lee WH (1997) Interleukin-1 receptor antagonist modulates the progression of a spontaneously occurring IgA nephropathy in mice. American J Kidney Dis 30:693–702

Chen K, Zhang J, Zhang W, Zhang J, Yang J, Li K, He Y (2013) ATP-P2X4 signaling mediates NLRP3 inflammasome activation: a novel pathway of diabetic nephropathy. Int J Biochem Cell Biol 45:932–943

Chen KW, Bezbradica JS, Gross CJ, Wall AA, Sweet MJ, Stow JL, Schroder K (2016) The murine neutrophil NLRP3 inflammasome is activated by soluble but not particulate or crystalline agonists. Eur J Immunol 46:1004–1010

Chi HH, Hua KF, Lin YC, Chu CL, Hsieh CY, Hsu YJ, Ka SM, Tsai YL, Liu FC, Chen A (2017) IL-36 signaling facilitates activation of the NLRP3 inflammasome and IL-23/IL-17 axis in renal inflammation and fibrosis. J Am Soc Nephrol 28:2022–2037

Choubey D, Panchanathan R (2017) Absent in melanoma 2 proteins in SLE. Clin Immunol 176: 42–48

Chow F, Ozols E, Nikolic-Paterson DJ, Atkins RC, Tesch GH (2004) Macrophages in mouse type 2 diabetic nephropathy: correlation with diabetic state and progressive renal injury. Kidney Int 65:116–128

Chun J, Chung H, Wang X, Barry R, Taheri ZM, Platnich JM, Ahmed SB, Trpkov K, Hemmelgarn B, Benediktsson H, James MT, Muruve DA (2016) NLRP3 localizes to the tubular epithelium in human kidney and correlates with outcome in IgA nephropathy. Sci Rep 6:24667

Chung Y, Chang SH, Martinez GJ, Yang XO, Nurieva R, Kang HS, Ma L, Watowich SS, Jetten AM, Tian Q, Dong C (2009) Critical regulation of early Th17 cell differentiation by interleukin-1 signaling. Immunity 30:576–587

Coca SG, Yalavarthy R, Concato J, Parikh CR (2008) Biomarkers for the diagnosis and risk stratification of acute kidney injury: a systematic review. Kidney Int 73:1008–1016

Cook HT (2007) Interpretation of renal biopsies in IgA nephropathy. Contrib Nephrol 157:44–49

Coppo R, Amore A, Peruzzi L, Vergano L, Camilla R (2010) Innate immunity and IgA nephropathy. J Nephrol 23:626–632

Correa-Costa M, Braga TT, Semedo P, Hayashida CY, Bechara LR, Elias RM, Barreto CR, Silva-Cunha C, Hyane MI, Goncalves GM, Brum PC, Fujihara C, Zatz R, Pacheco-Silva A, Zamboni DS, Camara NO (2011) Pivotal role of Toll-like receptors 2 and 4, its adaptor molecule MyD88, and inflammasome complex in experimental tubule-interstitial nephritis. PLoS One 6:e29004

Crow MK (2014) Type I interferon in the pathogenesis of lupus. J Immunol 192:5459–5468

Darisipudi MN, Knauf F (2016) An update on the role of the inflammasomes in the pathogenesis of kidney diseases. Pediatr Nephrol 31:535–544

Darisipudi MN, Thomasova D, Mulay SR, Brech D, Noessner E, Liapis H, Anders HJ (2012) Uromodulin triggers IL-1beta-dependent innate immunity via the NLRP3 inflammasome. J Am Soc Nephrol 23:1783–1789

dos Santos NA, Carvalho Rodrigues MA, Martins NM, dos Santos AC (2012) Cisplatin-induced nephrotoxicity and targets of nephroprotection: an update. Arch Toxicol 86:1233–1250

Du W, Zhen J, Zheng Z, Ma S, Chen S (2013) Expression of AIM2 is high and correlated with inflammation in hepatitis B virus associated glomerulonephritis. J Inflamm 10:37

Du L, Dong F, Guo L, Hou Y, Yi F, Liu J, Xu D (2015) Interleukin-1beta increases permeability and upregulates the expression of vascular endothelial-cadherin in human renal glomerular endothelial cells. Mol Med Rep 11:3708–3714

Duewell P, Kono H, Rayner KJ, Sirois CM, Vladimer G, Bauernfeind FG, Abela GS, Franchi L, Nunez G, Schnurr M, Espevik T, Lien E, Fitzgerald KA, Rock KL, Moore KJ, Wright SD, Hornung V, Latz E (2010) NLRP3 inflammasomes are required for atherogenesis and activated by cholesterol crystals. Nature 464:1357–1361

Elneam AI, Mansour NM, Zaki NA, Taher MA (2016) Serum Interleukin-18 and Its Gene Haplotypes Profile as Predictors in Patients with Diabetic Nephropathy. Open Access Maced J Med Sci 4:324–328
Emmerson BT, Cross M, Osborne JM, Axelsen RA (1990) Reaction of MDCK cells to crystals of monosodium urate monohydrate and uric acid. Kidney Int 37:36–43
Falk RJ, Terrell RS, Charles LA, Jennette JC (1990) Anti-neutrophil cytoplasmic autoantibodies induce neutrophils to degranulate and produce oxygen radicals in vitro. Proc Natl Acad Sci U S A 87:4115–4119
Faubel S, Ljubanovic D, Reznikov L, Somerset H, Dinarello CA, Edelstein CL (2004) Caspase-1-deficient mice are protected against cisplatin-induced apoptosis and acute tubular necrosis. Kidney Int 66:2202–2213
Faubel S, Lewis EC, Reznikov L, Ljubanovic D, Hoke TS, Somerset H, Oh DJ, Lu L, Klein CL, Dinarello CA, Edelstein CL (2007) Cisplatin-induced acute renal failure is associated with an increase in the cytokines interleukin (IL)-1beta, IL-18, IL-6, and neutrophil infiltration in the kidney. J Pharmacol Exp Ther 322:8–15
Faust J, Menke J, Kriegsmann J, Kelley VR, Mayet WJ, Galle PR, Schwarting A (2002) Correlation of renal tubular epithelial cell-derived interleukin-18 up-regulation with disease activity in MRL-Faslpr mice with autoimmune lupus nephritis. Arthritis Rheum 46:3083–3095
Feng H, Gu J, Gou F, Huang W, Gao C, Chen G, Long Y, Zhou X, Yang M, Liu S, Lu S, Luo Q, Xu Y (2016) High glucose and lipopolysaccharide prime NLRP3 inflammasome via ROS/TXNIP pathway in mesangial cells. J Diabetes Res 2016:6973175
Fernandes-Alnemri T, Yu JW, Datta P, Wu J, Alnemri ES (2009) AIM2 activates the inflammasome and cell death in response to cytoplasmic DNA. Nature 458:509–513
Franceschini A, Capece M, Chiozzi P, Falzoni S, Sanz JM, Sarti AC, Bonora M, Pinton P, Di Virgilio F (2015) The P2X7 receptor directly interacts with the NLRP3 inflammasome scaffold protein. FASEB J 29:2450–2461
Furuichi K, Wada T, Iwata Y, Kokubo S, Hara A, Yamahana J, Sugaya T, Iwakura Y, Matsushima K, Asano M, Yokoyama H, Kaneko S (2006) Interleukin-1-dependent sequential chemokine expression and inflammatory cell infiltration in ischemia-reperfusion injury. Crit Care Med 34:2447–2455
Furukawa S, Fujita T, Shimabukuro M, Iwaki M, Yamada Y, Nakajima Y, Nakayama O, Makishima M, Matsuda M, Shimomura I (2004) Increased oxidative stress in obesity and its impact on metabolic syndrome. J Clin Invest 114:1752–1761
Futosi K, Fodor S, Mocsai A (2013) Neutrophil cell surface receptors and their intracellular signal transduction pathways. Int Immunopharmacol 17:638–650
Gabay C, Lamacchia C, Palmer G (2010) IL-1 pathways in inflammation and human diseases. Nat Rev Rheumatol 6:232–241
Gao P, He FF, Tang H, Lei CT, Chen S, Meng XF, Su H, Zhang C (2015) NADPH oxidase-induced NALP3 inflammasome activation is driven by thioredoxin-interacting protein which contributes to podocyte injury in hyperglycemia. J Diabetes Res 2015:504761
Gauer S, Sichler O, Obermuller N, Holzmann Y, Kiss E, Sobkowiak E, Pfeilschifter J, Geiger H, Muhl H, Hauser IA (2007) IL-18 is expressed in the intercalated cell of human kidney. Kidney Int 72:1081–1087
Gois PH, Canale D, Volpini RA, Ferreira D, Veras MM, Andrade-Oliveira V, Camara NO, Shimizu MH, Seguro AC (2016) Allopurinol attenuates rhabdomyolysis-associated acute kidney injury: renal and muscular protection. Free Radic Biol Med 101:176–189
Guo H, Bi X, Zhou P, Zhu S, Ding W (2017) NLRP3 deficiency attenuates renal fibrosis and ameliorates mitochondrial dysfunction in a mouse unilateral ureteral obstruction model of chronic kidney disease. Mediat Inflamm 2017:8316560
Haq M, Norman J, Saba SR, Ramirez G, Rabb H (1998) Role of IL-1 in renal ischemic reperfusion injury. J Am Soc Nephrol 9:614–619
Haraldsson B, Jeansson M (2009) Glomerular filtration barrier. Curr Opin Nephrol Hypertens 18: 331–335

He Z, Lu L, Altmann C, Hoke TS, Ljubanovic D, Jani A, Dinarello CA, Faubel S, Edelstein CL (2008) Interleukin-18 binding protein transgenic mice are protected against ischemic acute kidney injury. Am J Physiol Ren Physiol 295:F1414–F1421

Hewins P, Morgan MD, Holden N, Neil D, Williams JM, Savage CO, Harper L (2006) IL-18 is upregulated in the kidney and primes neutrophil responsiveness in ANCA-associated vasculitis. Kidney Int 69:605–615

Holdsworth SR, Gan PY, Kitching AR (2016) Biologics for the treatment of autoimmune renal diseases. Nat Rev Nephrol 12:217–231

Homsi E, Janino P, de Faria JB (2006) Role of caspases on cell death, inflammation, and cell cycle in glycerol-induced acute renal failure. Kidney Int 69:1385–1392

Hong W, Hu S, Zou J, Xiao J, Zhang X, Fu C, Feng X, Ye Z (2015) Peroxisome proliferator-activated receptor gamma prevents the production of NOD-like receptor family, pyrin domain containing 3 inflammasome and interleukin 1beta in HK-2 renal tubular epithelial cells stimulated by monosodium urate crystals. Mol Med Rep 12:6221–6226

Hornung V, Bauernfeind F, Halle A, Samstad EO, Kono H, Rock KL, Fitzgerald KA, Latz E (2008) Silica crystals and aluminum salts activate the NALP3 inflammasome through phagosomal destabilization. Nat Immunol 9:847–856

Hornung V, Ablasser A, Charrel-Dennis M, Bauernfeind F, Horvath G, Caffrey DR, Latz E, Fitzgerald KA (2009) AIM2 recognizes cytosolic dsDNA and forms a caspase-1-activating inflammasome with ASC. Nature 458:514–518

Hultgren O, Andersson B, Hahn-Zoric M, Almroth G (2007) Serum concentration of interleukin-18 is up-regulated in patients with ANCA-associated vasculitis. Autoimmunity 40:529–531

Hutton HL, Ooi JD, Holdsworth SR, Kitching AR (2016) The NLRP3 inflammasome in kidney disease and autoimmunity. Nephrology 21:736–744

Isaka Y, Takabatake Y, Takahashi A, Saitoh T, Yoshimori T (2016) Hyperuricemia-induced inflammasome and kidney diseases. Nephrol Dial Transplant 31:890–896

Iseki K, Ikemiya Y, Inoue T, Iseki C, Kinjo K, Takishita S (2004) Significance of hyperuricemia as a risk factor for developing ESRD in a screened cohort. Am J Kidney Dis 44:642–650

Iyer SS, Pulskens WP, Sadler JJ, Butter LM, Teske GJ, Ulland TK, Eisenbarth SC, Florquin S, Flavell RA, Leemans JC, Sutterwala FS (2009) Necrotic cells trigger a sterile inflammatory response through the Nlrp3 inflammasome. Proc Natl Acad Sci U S A 106:20388–20393

Jansen MP, Emal D, Teske GJ, Dessing MC, Florquin S, Roelofs JJ (2017) Release of extracellular DNA influences renal ischemia reperfusion injury by platelet activation and formation of neutrophil extracellular traps. Kidney Int 91:352–364

Jarrot PA, Kaplanski G (2016) Pathogenesis of ANCA-associated vasculitis: an update. Autoimmun Rev 15:704–713

Jones LK, O'Sullivan KM, Semple T, Kuligowski MP, Fukami K, Ma FY, Nikolic-Paterson DJ, Holdsworth SR, Kitching AR (2009) IL-1RI deficiency ameliorates early experimental renal interstitial fibrosis. Nephrol Dial Transplant 24:3024–3032

Joosten LA, Netea MG, Dinarello CA (2013) Interleukin-1beta in innate inflammation, autophagy and immunity. Semin Immunol 25:416–424

Joshi S, Wang W, Peck AB, Khan SR (2015) Activation of the NLRP3 inflammasome in association with calcium oxalate crystal induced reactive oxygen species in kidneys. J Urol 193:1684–1691

Kahlenberg JM, Kaplan MJ (2014) The inflammasome and lupus: another innate immune mechanism contributing to disease pathogenesis? Curr Opin Rheumatol 26:475–481

Kahlenberg JM, Thacker SG, Berthier CC, Cohen CD, Kretzler M, Kaplan MJ (2011) Inflammasome activation of IL-18 results in endothelial progenitor cell dysfunction in systemic lupus erythematosus. J Immunol 187:6143–6156

Kahlenberg JM, Carmona-Rivera C, Smith CK, Kaplan MJ (2013) Neutrophil extracellular trap-associated protein activation of the NLRP3 inflammasome is enhanced in lupus macrophages. J Immunol 190:1217–1226

Kaushik M, Choo JC (2016) Serum uric acid and AKI: is it time? Clin Kidney J 9:48–50

Kessenbrock K, Krumbholz M, Schonermarck U, Back W, Gross WL, Werb Z, Grone HJ, Brinkmann V, Jenne DE (2009) Netting neutrophils in autoimmune small-vessel vasculitis. Nat Med 15:623–625

Kim HJ, Lee DW, Ravichandran K, D OK, Akcay A, Nguyen Q, He Z, Jani A, Ljubanovic D, Edelstein CL (2013) NLRP3 inflammasome knockout mice are protected against ischemic but not cisplatin-induced acute kidney injury. J Pharmacol Exp Ther 346:465–472

Kim SM, Lee SH, Kim YG, Kim SY, Seo JW, Choi YW, Kim DJ, Jeong KH, Lee TW, Ihm CG, Won KY, Moon JY (2015) Hyperuricemia-induced NLRP3 activation of macrophages contributes to the progression of diabetic nephropathy. Am J Physiol Ren Physiol 308:F993–f1003

Kinoshita K, Yamagata T, Nozaki Y, Sugiyama M, Ikoma S, Funauchi M, Kanamaru A (2004) Blockade of IL-18 receptor signaling delays the onset of autoimmune disease in MRL-Faslpr mice. J Immunol 173:5312–5318

Kitching AR, Hutton HL (2016) The players: cells involved in glomerular disease. Clin J Am Soc Nephrol 11:1664–1674

Kitching AR, Turner AL, Wilson GR, Semple T, Odobasic D, Timoshanko JR, O'Sullivan KM, Tipping PG, Takeda K, Akira S, Holdsworth SR (2005) IL-12p40 and IL-18 in crescentic glomerulonephritis: IL-12p40 is the key Th1-defining cytokine chain, whereas IL-18 promotes local inflammation and leukocyte recruitment. J Am Soc Nephrol 16:2023–2033

Knauf F, Asplin JR, Granja I, Schmidt IM, Moeckel GW, David RJ, Flavell RA, Aronson PS (2013) NALP3-mediated inflammation is a principal cause of progressive renal failure in oxalate nephropathy. Kidney Int 84:895–901

Komada T, Usui F, Shirasuna K, Kawashima A, Kimura H, Karasawa T, Nishimura S, Sagara J, Noda T, Taniguchi S, Muto S, Nagata D, Kusano E, Takahashi M (2014) ASC in renal collecting duct epithelial cells contributes to inflammation and injury after unilateral ureteral obstruction. Am J Pathol 184:1287–1298

Komada T, Usui F, Kawashima A, Kimura H, Karasawa T, Inoue Y, Kobayashi M, Mizushina Y, Kasahara T, Taniguchi S, Muto S, Nagata D, Takahashi M (2015) Role of NLRP3 inflammasomes for rhabdomyolysis-induced acute kidney injury. Sci Rep 5:10901

Konno T, Nakano R, Mamiya R, Tsuchiya H, Kitanaka T, Namba S, Kitanaka N, Okabayashi K, Narita T, Sugiya H (2016) Expression and function of interleukin-1beta-induced neutrophil gelatinase-associated lipocalin in renal tubular cells. PLoS One 11:e0166707

L'Homme L, Esser N, Riva L, Scheen A, Paquot N, Piette J, Legrand-Poels S (2013) Unsaturated fatty acids prevent activation of NLRP3 inflammasome in human monocytes/macrophages. J Lipid Res 54:2998–3008

Lan HY, Nikolic-Paterson DJ, Zarama M, Vannice JL, Atkins RC (1993) Suppression of experimental crescentic glomerulonephritis by the interleukin-1 receptor antagonist. Kidney Int 43: 479–485

Larsen CM, Faulenbach M, Vaag A, Volund A, Ehses JA, Seifert B, Mandrup-Poulsen T, Donath MY (2007) Interleukin-1-receptor antagonist in type 2 diabetes mellitus. N Engl J Med 356: 1517–1526

Lech M, Lorenz G, Kulkarni OP, Grosser MO, Stigrot N, Darisipudi MN, Gunthner R, Wintergerst MW, Anz D, Susanti HE, Anders HJ (2015) NLRP3 and ASC suppress lupus-like autoimmunity by driving the immunosuppressive effects of TGF-beta receptor signalling. Ann Rheum Dis 74:2224–2235

Lee DW, Faubel S, Edelstein CL (2015) A pan caspase inhibitor decreases caspase-1, IL-1alpha and IL-1beta, and protects against necrosis of cisplatin-treated freshly isolated proximal tubules. Ren Fail 37:144–150

Leemans JC, Kors L, Anders HJ, Florquin S (2014) Pattern recognition receptors and the inflammasome in kidney disease. Nat Rev Nephrol 10:398–414

Liao KC, Mogridge J (2013) Activation of the Nlrp1b inflammasome by reduction of cytosolic ATP. Infect Immun 81:570–579

Lichtnekert J, Kulkarni OP, Mulay SR, Rupanagudi KV, Ryu M, Allam R, Vielhauer V, Muruve D, Lindenmeyer MT, Cohen CD, Anders HJ (2011) Anti-GBM glomerulonephritis involves IL-1 but is independent of NLRP3/ASC inflammasome-mediated activation of caspase-1. PLoS One 6:e26778

Lim AK, Tesch GH (2012) Inflammation in diabetic nephropathy. Mediat Inflamm 2012:146154
Liu D, Xu M, Ding LH, Lv LL, Liu H, Ma KL, Zhang AH, Crowley SD, Liu BC (2014) Activation of the Nlrp3 inflammasome by mitochondrial reactive oxygen species: a novel mechanism of albumin-induced tubulointerstitial inflammation. Int J Biochem Cell Biol 57:7–19
Liu G, Shi Y, Peng X, Liu H, Peng Y, He L (2015) Astaxanthin attenuates adriamycin-induced focal segmental glomerulosclerosis. Pharmacology 95:193–200
Lu A, Li H, Niu J, Wu S, Xue G, Yao X, Guo Q, Wan N, Abliz P, Yang G, An L, Meng G (2017) Hyperactivation of the NLRP3 inflammasome in myeloid cells leads to severe organ damage in experimental lupus. J Immunol 198:1119–1129
Ludwig-Portugall I, Bartok E, Dhana E, Evers BD, Primiano MJ, Hall JP, Franklin BS, Knolle PA, Hornung V, Hartmann G, Boor P, Latz E, Kurts C (2016) An NLRP3-specific inflammasome inhibitor attenuates crystal-induced kidney fibrosis in mice. Kidney Int 90:525–539
Man SM, Kanneganti TD (2015) Regulation of inflammasome activation. Immunol Rev 265:6–21
Mandrup-Poulsen T, Pickersgill L, Donath MY (2010) Blockade of interleukin 1 in type 1 diabetes mellitus. Nat Rev Endocrinol 6:158–166
Martinon F, Petrilli V, Mayor A, Tardivel A, Tschopp J (2006) Gout-associated uric acid crystals activate the NALP3 inflammasome. Nature 440:237–241
Martinon F, Mayor A, Tschopp J (2009) The inflammasomes: guardians of the body. Annu Rev Immunol 27:229–265
Masters SL, Dunne A, Subramanian SL, Hull RL, Tannahill GM, Sharp FA, Becker C, Franchi L, Yoshihara E, Chen Z, Mullooly N, Mielke LA, Harris J, Coll RC, Mills KH, Mok KH, Newsholme P, Nunez G, Yodoi J, Kahn SE, Lavelle EC, O'Neill LA (2010) Activation of the NLRP3 inflammasome by islet amyloid polypeptide provides a mechanism for enhanced IL-1beta in type 2 diabetes. Nat Immunol 11:897–904
Mastrocola R, Aragno M, Alloatti G, Collino M, Penna C, Pagliaro P (2018) Metaflammation: tissue-specific alterations of the NLRP3 inflammasome platform. Curr Med Chem 25: 1294–1310
Matsui F, Rhee A, Hile KL, Zhang H, Meldrum KK (2013) IL-18 induces profibrotic renal tubular cell injury via STAT3 activation. Am J Physiol Ren Physiol 305:F1014–F1021
Mehta RL, Cerda J, Burdmann EA, Tonelli M, Garcia-Garcia G, Jha V, Susantitaphong P, Rocco M, Vanholder R, Sever MS, Cruz D, Jaber B, Lameire NH, Lombardi R, Lewington A, Feehally J, Finkelstein F, Levin N, Pannu N, Thomas B, Aronoff-Spencer E, Remuzzi G (2015) International Society of Nephrology's 0by25 initiative for acute kidney injury (zero preventable deaths by 2025): a human rights case for nephrology. Lancet 385: 2616–2643
Meldrum KK, Zhang H, Hile KL, Moldower LL, Dong Z, Meldrum DR (2012) Profibrotic effect of interleukin-18 in HK-2 cells is dependent on stimulation of the Toll-like receptor 4 (TLR4) promoter and increased TLR4 expression. J Biol Chem 287:40391–40399
Melnikov VY, Ecder T, Fantuzzi G, Siegmund B, Lucia MS, Dinarello CA, Schrier RW, Edelstein CL (2001) Impaired IL-18 processing protects caspase-1-deficient mice from ischemic acute renal failure. J Clin Invest 107:1145–1152
Miller RP, Tadagavadi RK, Ramesh G, Reeves WB (2010) Mechanisms of cisplatin nephrotoxicity. Toxins 2:2490–2518
Mirza RE, Fang MM, Weinheimer-Haus EM, Ennis WJ, Koh TJ (2014) Sustained inflammasome activity in macrophages impairs wound healing in type 2 diabetic humans and mice. Diabetes 63:1103–1114
Molitch ME, Adler AI, Flyvbjerg A, Nelson RG, So WY, Wanner C, Kasiske BL, Wheeler DC, de Zeeuw D, Mogensen CE (2015) Diabetic kidney disease: a clinical update from kidney disease: improving global outcomes. Kidney Int 87:20–30
Mulay SR, Kulkarni OP, Rupanagudi KV, Migliorini A, Darisipudi MN, Vilaysane A, Muruve D, Shi Y, Munro F, Liapis H, Anders HJ (2013) Calcium oxalate crystals induce renal inflammation by NLRP3-mediated IL-1beta secretion. J Clin Invest 123:236–246
Mulay SR, Evan A, Anders HJ (2014) Molecular mechanisms of crystal-related kidney inflammation and injury. Implications for cholesterol embolism, crystalline nephropathies and kidney stone disease. Nephrol Dial Transplant 29:507–514

Nakamura A, Shikata K, Hiramatsu M, Nakatou T, Kitamura T, Wada J, Itoshima T, Makino H (2005) Serum interleukin-18 levels are associated with nephropathy and atherosclerosis in Japanese patients with type 2 diabetes. Diabetes Care 28:2890–2895
Nee L, Tuite N, Ryan MP, McMorrow T (2007) TNF-alpha and IL-1 beta-mediated regulation of MMP-9 and TIMP-1 in human glomerular mesangial cells. Nephron Exp Nephrol 107:e73–e86
Netea MG, Simon A, van de Veerdonk F, Kullberg BJ, Van der Meer JW, Joosten LA (2010) IL-1beta processing in host defense: beyond the inflammasomes. PLoS Pathog 6:e1000661
Nguyen D, Ping F, Mu W, Hill P, Atkins RC, Chadban SJ (2006) Macrophage accumulation in human progressive diabetic nephropathy. Nephrology 11:226–231
Novick D, Kim S, Kaplanski G, Dinarello CA (2013) Interleukin-18, more than a Th1 cytokine. Semin Immunol 25:439–448
Nozaki Y, Kinoshita K, Yano T, Asato K, Shiga T, Hino S, Niki K, Nagare Y, Kishimoto K, Shimazu H, Funauchi M, Matsumura I (2012) Signaling through the interleukin-18 receptor alpha attenuates inflammation in cisplatin-induced acute kidney injury. Kidney Int 82:892–902
Obermayr RP, Temml C, Gutjahr G, Knechtelsdorfer M, Oberbauer R, Klauser-Braun R (2008) Elevated uric acid increases the risk for kidney disease. J Am Soc Nephrol 19:2407–2413
Odobasic D, Ghali JR, O'Sullivan KM, Holdsworth SR, Kitching AR (2014) Glomerulonephritis Induced by Heterologous Anti-GBM Globulin as a Planted Foreign Antigen. Curr Protoc Immunol 106:15.26.11–15.26.20
Okamoto A, Fujio K, Tsuno NH, Takahashi K, Yamamoto K (2012) Kidney-infiltrating CD4+ T-cell clones promote nephritis in lupus-prone mice. Kidney Int 82:969–979
Okui S, Yamamoto H, Li W, Gamachi N, Fujita Y, Kashiwamura S, Miura D, Takai S, Miyazaki M, Urade M, Okamura H, Ueda H (2012) Cisplatin-induced acute renal failure in mice is mediated by chymase-activated angiotensin-aldosterone system and interleukin-18. Eur J Pharmacol 685:149–155
Ooi JD, Chang J, Hickey MJ, Borza DB, Fugger L, Holdsworth SR, Kitching AR (2012) The immunodominant myeloperoxidase T-cell epitope induces local cell-mediated injury in anti-myeloperoxidase glomerulonephritis. Proc Natl Acad Sci U S A 109:E2615–E2624
Ooi JD, Chang J, O'Sullivan KM, Pedchenko V, Hudson BG, Vandenbark AA, Fugger L, Holdsworth SR, Kitching AR (2013) The HLA-DRB1*15:01-restricted Goodpasture's T cell epitope induces GN. J Am Soc Nephrol 24:419–431
Ooi JD, Gan PY, Odobasic D, Holdsworth SR, Kitching AR (2014) T cell mediated autoimmune glomerular disease in mice. Curr Protoc Immunol 107:15.27.11–15.27.19
Ooi JD, Petersen J, Tan YH, Huynh M, Willett ZJ, Ramarathinam SH, Eggenhuizen PJ, Loh KL, Watson KA, Gan PY, Alikhan MA, Dudek NL, Handel A, Hudson BG, Fugger L, Power DA, Holt SG, Coates PT, Gregersen JW, Purcell AW, Holdsworth SR, La Gruta NL, Reid HH, Rossjohn J, Kitching AR (2017) Dominant protection from HLA-linked autoimmunity by antigen-specific regulatory T cells. Nature 545:243–247
Peres LA, da Cunha AD Jr (2013) Acute nephrotoxicity of cisplatin: molecular mechanisms. J Bras Nefrol 35:332–340
Pontillo A, Girardelli M, Kamada AJ, Pancotto JA, Donadi EA, Crovella S, Sandrin-Garcia P (2012) Polimorphisms in inflammasome genes are involved in the predisposition to systemic lupus erythematosus. Autoimmunity 45:271–278
Pourghasem M, Shafi H, Babazadeh Z (2015) Histological changes of kidney in diabetic nephropathy. Caspian J Intern Med 6:120–127
Prencipe G, Caiello I, Cherqui S, Whisenant T, Petrini S, Emma F, De Benedetti F (2014) Inflammasome activation by cystine crystals: implications for the pathogenesis of cystinosis. J Am Soc Nephrol 25:1163–1169
Puelles VG, Kanzaki G, Bertram JF (2016) Indirect estimation of nephron number: a new tool to predict outcomes in renal transplantation? Nephrol Dial Transplant 31:1378–1380
Pulskens WP, Butter LM, Teske GJ, Claessen N, Dessing MC, Flavell RA, Sutterwala FS, Florquin S, Leemans JC (2014) Nlrp3 prevents early renal interstitial edema and vascular permeability in unilateral ureteral obstruction. PLoS One 9:e85775

Rifai A (2007) IgA nephropathy: immune mechanisms beyond IgA mesangial deposition. Kidney Int 72:239–241

Rossaint J, Zarbock A (2016) Acute kidney injury: definition, diagnosis and epidemiology. Minerva Urol Nefrol 68:49–57

Rusai K, Huang H, Sayed N, Strobl M, Roos M, Schmaderer C, Heemann U, Lutz J (2008) Administration of interleukin-1 receptor antagonist ameliorates renal ischemia-reperfusion injury. Transpl Int 21:572–580

Sadat U, Usman A, Boyle JR, Hayes PD, Solomon RJ (2015) Contrast medium-induced acute kidney injury. Cardiorenal Med 5:219–228

Saemann MD, Weichhart T, Zeyda M, Staffler G, Schunn M, Stuhlmeier KM, Sobanov Y, Stulnig TM, Akira S, von Gabain A, von Ahsen U, Horl WH, Zlabinger GJ (2005) Tamm-Horsfall glycoprotein links innate immune cell activation with adaptive immunity via a Toll-like receptor-4-dependent mechanism. J Clin Invest 115:468–475

Salama AD, Chaudhry AN, Ryan JJ, Eren E, Levy JB, Pusey CD, Lightstone L, Lechler RI (2001) In Goodpasture's disease, CD4(+) T cells escape thymic deletion and are reactive with the autoantigen alpha3(IV)NC1. J Am Soc Nephrol 12:1908–1915

Schreiber A, Pham CT, Hu Y, Schneider W, Luft FC, Kettritz R (2012) Neutrophil serine proteases promote IL-1beta generation and injury in necrotizing crescentic glomerulonephritis. J Am Soc Nephrol 23:470–482

Schreiber A, Luft FC, Kettritz R (2015) Phagocyte NADPH oxidase restrains the inflammasome in ANCA-induced GN. J Am Soc Nephrol 26:411–424

Schrijvers BF, De Vriese AS, Flyvbjerg A (2004) From hyperglycemia to diabetic kidney disease: the role of metabolic, hemodynamic, intracellular factors and growth factors/cytokines. Endocr Rev 25:971–1010

Shahzad K, Bock F, Dong W, Wang H, Kopf S, Kohli S, Al-Dabet MM, Ranjan S, Wolter J, Wacker C, Biemann R, Stoyanov S, Reymann K, Soderkvist P, Gross O, Schwenger V, Pahernik S, Nawroth PP, Grone HJ, Madhusudhan T, Isermann B (2015) Nlrp3-inflammasome activation in non-myeloid-derived cells aggravates diabetic nephropathy. Kidney Int 87:74–84

Shen J, Wang L, Jiang N, Mou S, Zhang M, Gu L, Shao X, Wang Q, Qi C, Li S, Wang W, Che X, Ni Z (2016) NLRP3 inflammasome mediates contrast media-induced acute kidney injury by regulating cell apoptosis. Sci Rep 6:34682

Shigeoka AA, Mueller JL, Kambo A, Mathison JC, King AJ, Hall WF, Correia Jda S, Ulevitch RJ, Hoffman HM, McKay DB (2010) An inflammasome-independent role for epithelial-expressed Nlrp3 in renal ischemia-reperfusion injury. J Immunol 185:6277–6285

Shin MS, Kang Y, Lee N, Wahl ER, Kim SH, Kang KS, Lazova R, Kang I (2013) Self double-stranded (ds)DNA induces IL-1beta production from human monocytes by activating NLRP3 inflammasome in the presence of anti-dsDNA antibodies. J Immunol 190:1407–1415

Suarez-Fueyo A, Bradley SJ, Klatzmann D, Tsokos GC (2017) T cells and autoimmune kidney disease. Nat Rev Nephrol 13:329–343

Summers SA, Steinmetz OM, Li M, Kausman JY, Semple T, Edgtton KL, Borza DB, Braley H, Holdsworth SR, Kitching AR (2009) Th1 and Th17 cells induce proliferative glomerulonephritis. J Am Soc Nephrol 20:2518–2524

Summers SA, Steinmetz OM, Gan PY, Ooi JD, Odobasic D, Kitching AR, Holdsworth SR (2011) Toll-like receptor 2 induces Th17 myeloperoxidase autoimmunity while Toll-like receptor 9 drives Th1 autoimmunity in murine vasculitis. Arthritis Rheum 63:1124–1135

Takano Y, Yamauchi K, Hayakawa K, Hiramatsu N, Kasai A, Okamura M, Yokouchi M, Shitamura A, Yao J, Kitamura M (2007) Transcriptional suppression of nephrin in podocytes by macrophages: roles of inflammatory cytokines and involvement of the PI3K/Akt pathway. FEBS Lett 581:421–426

Tang WW, Feng L, Vannice JL, Wilson CB (1994) Interleukin-1 receptor antagonist ameliorates experimental anti-glomerular basement membrane antibody-associated glomerulonephritis. J Clin Invest 93:273–279

Tashiro M, Sasatomi Y, Watanabe R, Watanabe M, Miyake K, Abe Y, Yasuno T, Ito K, Ueki N, Hamauchi A, Noda R, Hisano S, Nakashima H (2016) IL-1beta promotes tubulointerstitial injury in MPO-ANCA-associated glomerulonephritis. Clin Nephrol 86:190–199

Taylor SR, Turner CM, Elliott JI, McDaid J, Hewitt R, Smith J, Pickering MC, Whitehouse DL, Cook HT, Burnstock G, Pusey CD, Unwin RJ, Tam FW (2009) P2X7 deficiency attenuates renal injury in experimental glomerulonephritis. J Am Soc Nephrol 20:1275–1281

Timoshanko JR, Kitching AR, Iwakura Y, Holdsworth SR, Tipping PG (2004a) Contributions of IL-1beta and IL-1alpha to crescentic glomerulonephritis in mice. J Am Soc Nephrol 15:910–918

Timoshanko JR, Kitching AR, Iwakura Y, Holdsworth SR, Tipping PG (2004b) Leukocyte-derived interleukin-1beta interacts with renal interleukin-1 receptor I to promote renal tumor necrosis factor and glomerular injury in murine crescentic glomerulonephritis. Am J Pathol 164: 1967–1977

Tsai YL, Hua KF, Chen A, Wei CW, Chen WS, Wu CY, Chu CL, Yu YL, Lo CW, Ka SM (2017) NLRP3 inflammasome: pathogenic role and potential therapeutic target for IgA nephropathy. Sci Rep 7:41123

Turner CM, Tam FW, Lai PC, Tarzi RM, Burnstock G, Pusey CD, Cook HT, Unwin RJ (2007) Increased expression of the pro-apoptotic ATP-sensitive P2X7 receptor in experimental and human glomerulonephritis. Nephrol Dial Transplant 22:386–395

Vilaysane A, Chun J, Seamone ME, Wang W, Chin R, Hirota S, Li Y, Clark SA, Tschopp J, Trpkov K, Hemmelgarn BR, Beck PL, Muruve DA (2010) The NLRP3 inflammasome promotes renal inflammation and contributes to CKD. J Am Soc Nephrol 21:1732–1744

Wang W, Wang X, Chun J, Vilaysane A, Clark S, French G, Bracey NA, Trpkov K, Bonni S, Duff HJ, Beck PL, Muruve DA (2013) Inflammasome-independent NLRP3 augments TGF-beta signaling in kidney epithelium. J Immunol 190:1239–1249

Wang J, Wen Y, Lv LL, Liu H, Tang RN, Ma KL, Liu BC (2015a) Involvement of endoplasmic reticulum stress in angiotensin II-induced NLRP3 inflammasome activation in human renal proximal tubular cells in vitro. Acta Pharmacol Sin 36:821–830

Wang P, Huang J, Li Y, Chang R, Wu H, Lin J, Huang Z (2015b) Exogenous carbon monoxide decreases sepsis-induced acute kidney injury and inhibits NLRP3 inflammasome activation in rats. Int J Mol Sci 16:20595–20608

Wang W, Luo R, Lin Y, Wang F, Zheng P, Levi M, Yang T, Li C (2015c) Aliskiren restores renal AQP2 expression during unilateral ureteral obstruction by inhibiting the inflammasome. Am J Physiol Ren Physiol 308:F910–F922

Wang L, Ma J, Guo C, Chen C, Yin Z, Zhang X, Chen X (2016) Danggui Buxue Tang attenuates tubulointerstitial fibrosis via suppressing NLRP3 inflammasome in a rat model of unilateral ureteral obstruction. Biomed Res Int 2016:9368483

Wen H, Gris D, Lei Y, Jha S, Zhang L, Huang MT, Brickey WJ, Ting JP (2011) Fatty acid-induced NLRP3-ASC inflammasome activation interferes with insulin signaling. Nat Immunol 12: 408–415

Wu J, Hicks J, Borillo J, Glass WF 2nd, Lou YH (2002) CD4(+) T cells specific to a glomerular basement membrane antigen mediate glomerulonephritis. J Clin Invest 109:517–524

Wu H, Craft ML, Wang P, Wyburn KR, Chen G, Ma J, Hambly B, Chadban SJ (2008) IL-18 contributes to renal damage after ischemia-reperfusion. J Am Soc Nephrol 19:2331–2341

Xiao J, Fu C, Zhang X, Zhu D, Chen W, Lu Y, Ye Z (2015a) Soluble monosodium urate, but not its crystal, induces toll like receptor 4-dependent immune activation in renal mesangial cells. Mol Immunol 66:310–318

Xiao J, Zhang XL, Fu C, Han R, Chen W, Lu Y, Ye Z (2015b) Soluble uric acid increases NALP3 inflammasome and interleukin-1beta expression in human primary renal proximal tubule epithelial cells through the Toll-like receptor 4-mediated pathway. Int J Mol Med 35:1347–1354

Xiong J, Wang Y, Shao N, Gao P, Tang H, Su H, Zhang C, Meng XF (2015) The expression and significance of NLRP3 inflammasome in patients with primary glomerular diseases. Kidney Blood Press Res 40:344–354

Yamamura M, Kawashima M, Taniai M, Yamauchi H, Tanimoto T, Kurimoto M, Morita Y, Ohmoto Y, Makino H (2001) Interferon-gamma-inducing activity of interleukin-18 in the joint with rheumatoid arthritis. Arthritis Rheum 44:275–285

Yuan F, Kolb R, Pandey G, Li W, Sun L, Liu F, Sutterwala FS, Liu Y, Zhang W (2016) Involvement of the NLRC4-inflammasome in diabetic nephropathy. PLoS One 11:e0164135

Zarbock A, Gomez H, Kellum JA (2014) Sepsis-induced acute kidney injury revisited: pathophysiology, prevention and future therapies. Curr Opin Crit Care 20:588–595

Zhang H, Hile KL, Asanuma H, Vanderbrink B, Franke EI, Campbell MT, Meldrum KK (2011) IL-18 mediates proapoptotic signaling in renal tubular cells through a Fas ligand-dependent mechanism. Am J Physiol Ren Physiol 301:F171–F178

Zhang C, Boini KM, Xia M, Abais JM, Li X, Liu Q, Li PL (2012) Activation of Nod-like receptor protein 3 inflammasomes turns on podocyte injury and glomerular sclerosis in hyperhomocysteinemia. Hypertension 60:154–162

Zhang W, Cai Y, Xu W, Yin Z, Gao X, Xiong S (2013) AIM2 facilitates the apoptotic DNA-induced systemic lupus erythematosus via arbitrating macrophage functional maturation. J Clin Immunol 33:925–937

Zhang Y, Yuan F, Cao X, Zhai Z, GangHuang X, Du Y, Wang J, Zhang Y, Huang JZ, Hou W (2014) P2X7 receptor blockade protects against cisplatin-induced nephrotoxicity in mice by decreasing the activities of inflammasome components, oxidative stress and caspase-3. Toxicol Appl Pharmacol 281:1–10

Zhao WY, Zhang L, Sui MX, Zhu YH, Zeng L (2016) Protective effects of sirtuin 3 in a murine model of sepsis-induced acute kidney injury. Sci Rep 6:33201

Zhen J, Zhang L, Pan J, Ma S, Yu X, Li X, Chen S, Du W (2014) AIM2 mediates inflammation-associated renal damage in hepatitis B virus-associated glomerulonephritis by regulating caspase-1, IL-1beta, and IL-18. Mediat Inflamm 2014:190860

Zheng L, Zhang J, Yuan X, Tang J, Qiu S, Peng Z, Yuan Q, Xie Y, Mei W, Tang Y, Meng J, Hu G, Tao L (2017) Fluorofenidone attenuates IL-1beta production by interacting with NLRP3 inflammasome in unilateral ureteral obstruction. Nephrology [Epub ahead of print]

Zhong S, Zhao L, Li Q, Yang P, Varghese Z, Moorhead JF, Chen Y, Ruan XZ (2015) Inflammatory stress exacerbated mesangial foam cell formation and renal injury via disrupting cellular cholesterol homeostasis. Inflammation 38:959–971

Zhou R, Tardivel A, Thorens B, Choi I, Tschopp J (2010) Thioredoxin-interacting protein links oxidative stress to inflammasome activation. Nat Immunol 11:136–140

Chapter 9
Pro-Inflammatory Actions of Red Blood Cell-Derived DAMPs

Viktória Jeney

Contents

Abstract Damage-associated molecular patterns (DAMPs) or alarmins are endogenous danger signals that are derived from damaged cells and extracellular matrix degradation, capable of triggering innate immune response to promote tissue damage repair. Hemolytic or hemorrhagic episodes are often associated with inflammation, even when infectious agents are absent, suggesting that damaged red blood cells (RBCs) release DAMPs.

Hemoglobin (Hb) composes 96% of the dry weight of RBCs; therefore upon hemolysis, tremendous amounts of Hb are released into the extracellular milieu. Hb oxidation occurs outside the protective environment of RBCs, leading to the formation of different Hb oxidation products and heme. Heme acts as a prototypic DAMP participating in toll-like receptor as well as intracellular nucleotide-binding oligomerization domain-like receptor signaling. Oxidized Hb forms also possess some

V. Jeney (✉)
Faculty of Medicine, Department of Internal Medicine, University of Debrecen, Debrecen, Hungary
e-mail: jeneyv@belklinika.com

M. D. Cordero, E. Alcocer-Gómez (eds.), *Inflammasomes: Clinical and Therapeutic Implications*, Experientia Supplementum 108,
https://doi.org/10.1007/978-3-319-89390-7_9

inflammatory actions independently of their heme releasing capability. Non-Hb-derived DAMPs such as ATP, interleukin-33, heat shock protein 70, as well as RBC membrane-derived microparticles might also contribute to the innate immune response triggered by hemolysis/hemorrhage.

In this chapter we will discuss the inflammatory properties of RBC-derived DAMPs with a particular focus on Hb derivatives, as well as therapeutic potential of the endogenous Hb and heme-binding proteins haptoglobin and hemopexin in the prevention of hemolysis/hemorrhage-associated inflammation.

Keywords Hemoglobin · Red blood cells · Inflammasome · DAMPs · Hemolysis · Hemorrhage

Abbreviations

ASC	Apoptosis-associated speck-like protein containing a caspase recruitment domain
ATP	Adenosine triphosphate
CO	Carbon monoxide
Cys	Cysteine
DAMPs	Damage-associated molecular patterns
FerrylHb	Ferrylhemoglobin
Hb	Hemoglobin
HO-1	Heme oxygenase-1
H_2O_2	Hydrogen-peroxide
Hp	Haptoglobin
Hsp	Heat shock protein
Hx	Hemopexin
ICAM-1	Intracellular adhesion molecule-1
ICH	Intracerebral hemorrhage
IL	Interleukin
LPS	Lipopolysaccharide
MetHb	Met(ferric) hemoglobin
Mhem macrophage	Hemorrhage-associated macrophage
MPs	Microparticles
MyD88	Myeloid differentiation primary response gene 88
NADPH	Nicotinamide adenine dinucleotide phosphate
NF-κB	Nuclear factor kappa B
NLR	NOD-like receptor
NLRP3	NLR family pyrin domain containing 3
NOD	Nucleotide-binding oligomerization domain
NRF2	Nuclear factor erythroid 2-related factor 2
PAMPs	Pathogen-associated molecular patterns
PPIX	Protoporphyrin IX

RBC	red blood cell
P2X7	P2X purinoceptor 7
TLR	Toll-like receptor
ROS	Reactive oxygen species
TNF-α	Tumor necrosis factor-alpha
TRIF	TIR-domain-containing adapter-inducing interferon-β
Tyr	Tyrosine
VCAM-1	Vascular cell adhesion molecule-1

9.1 Introduction

Damage-associated molecular patterns (DAMPs) or alarmins are endogenous danger signals that are derived from damaged cells and extracellular matrix degradation capable of triggering and/or exacerbating innate immune responses to promote tissue damage repair (Matzinger 1994). Hemolytic or hemorrhagic episodes are often associated with inflammation even when infectious agents are absent (Arruda et al. 2005), suggesting that damaged red blood cells (RBCs) release DAMPs (Mendonca et al. 2016).

The far most abundant protein in mature RBCs is hemoglobin (Hb) that composes 96% of the dry weight of RBCs; therefore upon hemolysis, tremendous amounts of Hb are released into the extracellular milieu. Once outside the protective environment of RBCs, Hb is prone to oxidation, in which process different Hb oxidation products form with diverse biological activities toward immune and nonimmune cells. Heme, the prosthetic group of Hb, is promptly released from oxidized Hb species and is the most studied RBC-derived alarmin (Soares and Bozza 2016). Heme is a strong prooxidant and is involved in toll-like receptor (TLR) as well as intracellular nucleotide-binding oligomerization domain (NOD)-like receptor (NLR) signaling [reviewed in Dutra and Bozza (2014), Soares and Bozza (2016)]. Besides heme, oxidized Hb forms also possess some inflammatory actions independently of their heme releasing capability [reviewed in Jeney et al. (2014)]. Non-Hb-derived DAMPs such as adenosine triphosphate (ATP), interleukin (IL)-33, heat shock protein (Hsp) 70, as well as RBC membrane-derived microparticles (MPs) might also contribute to the innate immune response triggered by hemolysis/hemorrhage.

Deleterious effects of extracellular Hb and heme are controlled by haptoglobin (Hp) and hemopexin (Hx), respectively. These acute phase proteins bind extracellular Hb and heme avidly and facilitate their removal from circulation through receptor-mediated endocytotic routes. Upon massive intravascular hemolysis, the scavenging capacities of Hp and Hx are overwhelmed. Along with this notion, Hp- and Hx-based therapeutic interventions could be beneficial in pathologies associated with hemolysis/hemorrhage.

9.1.1 Physiology of RBCs

RBCs are the most prevalent cells in the human body, structurally and functionally dedicated to transport oxygen and carbon dioxide throughout the organism. RBCs are formed in the bone marrow from pluripotent hematopoietic stem cells in the process of erythropoiesis. Differentiation takes place mainly in the bone marrow, until reticulocytes released into the bloodstream where they mature further 1–2 days into terminally differentiated RBCs. During differentiation RBCs loose nuclei and cytoplasmic organelles including mitochondria and ribosomes. The advantage of not having nuclei in mature RBCs is twofold: first, anucleated cells are more flexible assuring that they can squeeze through small blood capillaries; second, there is more space for Hb resulting in increased oxygen-binding capacity. On the other side of this trade-off, mature anucleated RBCs are unable to divide, and their rescuing mechanisms are limited. This explains the relatively short life-span (100–120 days) of RBCs in the circulation, and the enormous turnover of making and breaking RBCs (200 billion RBCs/day).

Circulating RBCs are continuously exposed to high levels of reactive oxygen species (ROS) of both endogenous and exogenous origin [reviewed in Mohanty et al. (2014)]. Each ml of blood contains 0.3 g of Hb, and auto-oxidation (Table 9.1, equation #1) of Hb is the major source of endogenous ROS in RBCs. Besides Hb auto-oxidation, nicotinamide adenine dinucleotide phosphate (NADPH) oxidases also contribute to endogenous ROS production in RBCs (George et al. 2013). To cope with this challenge, RBCs are equipped with a highly effective antioxidant defense system which includes enzymes such as Cu/Zn superoxide dismutase that convert superoxide anion to hydrogen peroxide (H_2O_2), catalase, glutathione peroxidase, and peroxiredoxins which decompose H_2O_2 to H_2O [reviewed in Siems et al. (2000), Jeney et al. (2013), Mohanty et al. (2014)]. Nonenzymatic low-molecular-weight scavengers such as glutathione and ascorbic acid also contribute to this

Table 9.1 Oxidative modifications of hemoglobin

	Formed species
(1) $Hb(Fe^{2+})O_2 \rightarrow Hb(Fe^{3+}) + O_2^{\bullet-}$	Methemoglobin
(2) $Hb(Fe^{2+})O_2 + H_2O_2 \rightarrow Hb(Fe^{4+} = O^{2-}) + H_2O + O_2$	Ferrylhemoglobin
(3) $Hb(Fe^{3+}) + H_2O_2 \rightarrow Hb^{\bullet+}(Fe^{4+} = O^{2-}) + H_2O$	Ferrylhemoglobin globin radical
(4) $Hb(Fe^{4+} = O^{2-}) + 2H^{+} \rightarrow Hb^{\bullet+}(Fe^{3+}) + H_2O$	Methemoglobin globin radical
(5) $Hb^{\bullet+}(Fe^{3+}) + Hb^{\bullet+}(Fe^{3+}) \rightarrow (Fe^{3+})\ ^{+}Hb\text{—}Hb^{+}(Fe^{3+})$	Covalently cross-linked methemoglobin multimer

Routes of hemoglobin oxidation. Auto-oxidation of Hb generates metHb and superoxide anions (equation 1). H_2O_2 triggers a two-electron oxidation of Hb leading to the formation of ferryl ($Fe^{4+} = O^{2-}$) Hb (equation 2). The reaction of metHb with H_2O_2 yields ferrylHb radical ($Hb^{\bullet+}(Fe^{4+} = O^{2-})$) in which the unpaired electron is associated with the globin or the porphyrin ring (equation 3). FerrylHb can trigger further production of globin radicals via an intramolecular electron transfer between the ferryl iron and specific amino acid residues of the globin chains resulting in the formation of metHb globin radical (equation 4). Termination reactions of globin- and porphyrin-centered radicals lead to the formation of globin-globin (equation 5) cross-links protection. Incomplete neutralization of ROS triggers RBC membrane damage and subsequent impairment of oxygen delivery to the tissues which eventually leads to tissue damage and inflammation.

Circulating RBCs lose 20% of their Hb content during their life-span via vesiculation (Willekens et al. 2003). Vesiculation is considered to be a self-protective mechanism of RBCs via which RBCs release membrane patches containing removal molecules including phosphatidylserine, immunoglobulin G, and senescent cell antigens (Willekens et al. 2008). Additionally, RBCs are able to get rid of intracellular inclusions, e.g., Heinz bodies via this mechanism (de Back et al. 2014), thereby postponing the premature loss of otherwise healthy RBCs from the circulation. RBC-derived vesicles are rapidly removed from the circulation by the mononuclear phagocyte system (Willekens et al. 2003).

At the end of their life-span, senescent RBCs are removed from the circulation by hemophagocytic macrophages, mainly in the spleen (Bratosin et al. 1998; de Back et al. 2014). Aged RBCs are smaller and denser because of the permanent loss of Hb and cell membrane via vesiculation and also characterized by decreased metabolic activity (Piomelli and Seaman 1993). At the terminal stage of RBC aging, "eat me" signals appear, and "don't eat me" signals disappear on the surface of senescent RBCs, and shortly after they are internalized by macrophages [reviewed in de Back et al. (2014)].

Different theories exist about the entity of the removal surface markers of senescent RBCs [reviewed in de Back et al. (2014)]. Phosphatidylserine, a phospholipid normally found in the inner membrane of RBCs, is a very likely candidate of being a removal signal, when it appears in the outer membrane of the RBCs (Boas et al. 1998). Phosphatidylserine is a general marker for apoptotic cells (Fernandez-Boyanapalli et al. 2009), and although RBCs cannot undergo a classical apoptosis because of the lack of nucleus and other cellular organelles, evidence suggest that aged or damaged RBCs can undergo a regulated process called eryptosis that is in many terms resembles to that of programmed cell death (Lang et al. 2005). Eryptosis is characterized by cell shrinkage, membrane blebbing, activation of proteases, and exposition of phosphatidylserine at the outer membrane leaflet of RBCs. Importantly, the removal of these phosphatidylserine-positive senescent, or terminally damaged RBCs by macrophages, is a non-inflammatory process and allows efficient and safe recycling of the RBC components, particularly the heme iron (Muckenthaler et al. 2017).

9.1.2 Hemolysis and Hemorrhage

Numerous pathologies are associated with hemolysis or hemorrhage characterized by uncontrolled destruction of RBCs. Hemolysis can occur in the vasculature but also in the extravascular space. Inherited or acquired conditions can cause hemolysis as listed in Table 9.2. Inherited hemolytic diseases are caused by mutations in genes encoding Hb, RBC membrane components, or certain enzymes in RBCs. The repertoire of acquired conditions associated with hemolysis is quite wide.

Table 9.2 Causes of hemolysis

Type	Cause of hemolysis	Example	Inflammasome activation
Inherited	RBC membrane abnormalities	Spherocytosis	Not reported
		Elliptocytosis	Not reported
	RBC metabolism abnormalities	G6PD deficiency	Not reported
		PK deficiency	Not reported
	Hemoglobinopathies	Thalassemias	Not reported
		Sickle cell disease	Yes (Cerqueira et al. 2011)
Acquired	Immune mediated (Autoimmune)	Warm antibody	Not reported
		Cold antibody	Not reported
	Immune mediated (Alloimmune)	Transfusion reaction	Controversial (Gibb et al. 2016, Land 2013)
		Hemolytic disease of the newborn	Not reported
	Mechanical, physical or chemical trauma	Microangiopathies	Not reported
		Prosthetic heart valves	Not reported
		Burns	Yes (Stanojcic et al. 2014)
		Heavy metal toxicity	Not reported
		Drug induced	Not reported
	Infections	Malaria	Controversial (Dostert et al. 2009, Reimer et al. 2010)

Auto- and alloimmune reactions, mechanical, physical, or chemical stress, and diverse infections can trigger substantial RBC lysis. RBCs outside the vasculature tend to lyse quickly; therefore hemorrhages are also associated with RBC lysis.

9.1.3 The Fate of Extracellular Hemoglobin

Hb is released in large amounts from lysing RBCs. Extracellular Hb exerts diverse unfavorable vasoactive effects. For example, extracellular Hb scavenges nitric oxide, an important vasodilator and signaling molecule in the vasculature [reviewed in Rother et al. (2005)]. Furthermore, once outside the protective environment of RBCs, Hb tends to oxidize. Auto-oxidation of Hb occurs resulting in metHb generation meanwhile superoxide anions are formed (Table 9.1, equation 1). Peroxides, such as H_2O_2 or lipid hydroperoxides, induce a two-electron oxidation of Hb leading to the formation of ferryl ($Fe^{4+} = O^{2-}$) Hb (Table 9.1, equation 2), whereas the reaction of metHb with H_2O_2 results in ferrylHb radical ($Hb^{\bullet+}(Fe^{4+} = O^{2-})$) in which the unpaired electron is located at either the globin chain or at the porphyrin ring (Table 9.1, equation 3) (Harel and Kanner 1988; Patel et al. 1996; Jia et al. 2007; Alayash et al. 2001). These high-valence iron compounds, i.e., ferrylHb and ferrylHb radical, are highly reactive intermediates that can decay by several ways

(Reeder et al. 2008). FerrylHb induces additional production of globin radicals via an intramolecular electron transfer between the ferryl iron and specific amino acid residues of the globin chains such as αTyr-24, αTyr-42, αHis-20, βTyr-35, βTyr-130, and βCys-93 leading to the formation of metHb globin radical (Table 9.1, equation 4) (Deterding et al. 2004; Ramirez et al. 2003; Jeney et al. 2013). Termination reactions of globin- and porphyrin-centered radicals lead to the formation of globin-globin (Table 9.1, equation 5) or porphyrin-globin crosslinks.

To prevent the deleterious effects of extracellular Hb, efficient mechanisms have evolved for its removal from the circulation. Hp, an acute-phase protein, is present in plasma in high amounts (0.41–1.65 mg/ml) with the special recognized function of capturing cell-free Hb [reviewed in Alayash (2011)]. The formation of the Hp-Hb complex is virtually irreversible, and Hp binding has multiple beneficial effects. First of all, Hp binding facilitates the removal of Hb from circulation through the CD163 macrophage scavenger receptor-mediated endocytosis (Kristiansen et al. 2001). Besides this effect, studies showed that Hb bound to Hp is less prone to H_2O_2-mediated oxidation than free Hb (Buehler et al. 2009; Banerjee et al. 2012; Miller et al. 1997). In fact, the Hb-Hp complex acts as a fairly efficient peroxidase (Kapralov et al. 2009). Further studies proved that Hp prevents H_2O_2-induced oxidation of amino acids in critical regions of Hb chains—i.e., α-Tyr42, β-Tyr145, and β-Cys93—and polymerization of Hb (Pimenova et al. 2010). The recent determination of the crystal structure of the porcine Hp-Hb complex revealed that Hb residues known to be prone to oxidative modifications are buried in the Hp-Hb interface, thereby explaining this direct protective role of Hp against H_2O_2-induced oxidation (Andersen et al. 2012).

Although the Hb/Hp/CD163 system is highly efficient in removing intravascular free Hb, it has some limitations. Plasma Hp can bind and clear approximately 3 g of Hb from the circulation which is less than 1% of the total amount of circulating Hb. In case of pronounced hemolysis, when more than 1% of RBCs disrupt, Hp is depleted from the circulation in which case free Hb is cleared (rather inefficiently) via a low-affinity pathway through CD163 (Schaer et al. 2006) and/or by renal excretion (Schaer et al. 2013; Murray et al. 1961). This latter is accompanied by generation of free iron and organ damage.

Another limitation of the Hp/CD163 system is that Hp and CD163 have decreased affinity for structurally altered (e.g., covalently cross-linked) Hb species that might form upon Hb oxidation. Recent studies have revealed that elimination of oxidized Hb species via both high-affinity and low-affinity pathways can be severely compromised (Schaer et al. 2006; Vallelian et al. 2008).

Upon massive hemolysis Hp is consumed, causing accumulation and oxidation of cell-free Hb that eventually lead to the release of the prosthetic heme group. Hx is an acute-phase plasma protein that binds heme with the highest affinity of any known heme-binding proteins (Hrkal et al. 1974). Hx-heme complexes are internalized via the scavenger receptor LDL receptor-related protein 1/CD91 (Hvidberg et al. 2005) mainly by hepatocytes and macrophages (Herz and Strickland 2001).

Following internalization of Hb or heme, cells and tissues upregulate heme oxygenase-1 (HO-1) and ferritin. HO-1 catabolizes free heme into equimolar amounts of Fe^{2+}, carbon monoxide (CO), and biliverdin (Tenhunen et al. 1968).

Liberated iron drives the upregulation of ferritin that is the main intracellular iron storage protein (Eisenstein et al. 1991).

9.1.4 Pro-inflammatory Actions of Hb-Derived Species

Massive intravascular hemolysis or hemorrhage result in the exhaustion of the endogenous defense system leading to the accumulation of oxidized Hb forms and free heme in the plasma or in the extravascular space (Pamplona et al. 2007; Larsen et al. 2010; Nagy et al. 2010). These Hb derivatives, particularly free heme, exert prooxidant activities [reviewed in Immenschuh et al. (2017), Jeney et al. (2013)]. Moreover, hemolytic or hemorrhagic episodes are often associated with inflammation even when infectious agents are absent (Arruda et al. 2005). Considerable effort has been made to define the mediators and the target cells involved in the hemolysis-/hemorrhage-induced inflammatory response. Accumulating evidence suggest that Hb-derived oxidized species possess diverse pro-inflammatory actions targeting different immune and non-immune cells (Table 9.3).

9.1.4.1 Macrophage Activation

Macrophages, the frontline cells of innate immunity, respond to a variety of pathogen-associated molecular patterns (PAMPs) and DAMPs. Lysis of RBCs leads to the release of different RBC components that can potentially behave as DAMPs and induce a sterile inflammatory response dependently of receptors such as TLRs or NOD-like receptors (Table 9.3).

Accumulating evidence suggests that heme that is released from oxidized Hb forms modulate macrophage phenotype. Bozza et al. showed that heme triggers tumor necrosis factor-alpha (TNF-α) secretion by macrophages in a TLR4-dependent manner (Figueiredo et al. 2007). The activation of TLR4 by heme is strictly dependent on its coordinated iron and the vinyl groups of the porphyrin ring (Figueiredo et al. 2007). Sustained exposure of macrophages to free heme triggers programmed necrosis that is dependent on autocrine production of TNF-α and ROS (Fortes et al. 2012). The pathogenic role of heme-mediated TLR4 activation was investigated in a murine model of intracerebral hemorrhage (ICH)-induced neuro-inflammation. In comparison to wild-type mice, TLR4$^{-/-}$ mice exhibited less inflammation, reduced cerebral edema, and lower neurological deficit scores, suggesting that heme-mediated TLR4 activation plays a critical role in ICH-associated neuro-inflammation (Lin et al. 2012). Gram et al. showed that after intraventricular hemorrhage, metHb forms and its level correlates to the expression of TNF-α (Gram et al. 2013). In agreement with this finding, a recent study of Kwon et al. revealed that metHb is an important endogenous activator of TLR4 that promotes widespread TLR4-mediated neuro-inflammation upon subarachnoid hemorrhage (Kwon et al. 2015).

Table 9.3 Pro-inflammatory actions of RBC-derived DAMPs

DAMP	Major finding	References
Heme	Heme triggers TLR4-dependent TNF-α secretion in macrophages	Figueiredo et al. (2007)
Heme	Heme-mediated activation of TLR4/MyD88/TRIF pathway plays a role in intracerebral hemorrhage	Lin et al. (2012)
MetHb	MetHb and TNF-α levels correlate in cerebrospinal fluid after intraventricular hemorrhage	Gram et al. (2013)
MetHb	MetHb promotes TLR4-dependent neuroinflammation upon subarachnoid hemorrhage	Kwon et al. (2015)
Heme	Heme induces NLRP3 activation and IL-1β secretion in LPS-primed macrophages	Dutra et al. (2014)
Heme	Heme triggers neutrophil recruitment, ROS production and IL-8 expression	Graca-Souza et al. (2002)
FerrylHb	FerrylHb triggers neutrophil recruitment in vivo independently of TLR4 activation	Silva et al. (2009)
Heme	Heme induces neutrophil extracellular trap formation	Chen et al. (2014)
Heme	Heme induces TLR4-dependent endothelial activation	Belcher et al. (2014)
FerrylHb	FerrylHb activates NF-κB, upregulates pro-inflammatory adhesion molecule expressions, and disrupts monolayer integrity in endothelial cells	Silva et al. (2009)
ATP	ATP activates P2X7 receptors leading to IL-1β secretion in LPS-primed macrophages	Perregaux et al. (2000)
ATP	ATP activates NF-κB, upregulates E-selectin expression, and induces deterioration of endothelial barrier function via acting on P2X7 receptors	McClenahan et al. (2009)
ATP	ATP induces NLRP3 activation and IL-1β secretion in LPS- or TNF-primed endothelial cells	Huck et al. (2015), Champaiboon et al. (2014)
ATP	ATP induces microparticle release, ROS formation, and apoptotic cell death in erythroid progenitor cells through activation of P2X7 receptors	Constantinescu et al. (2010), Wang and Sluyter (2013)
ATP	ATP triggers eicosanoid release, phosphatidylserine exposure, and lysis of mature RBCs through activation of P2X7 receptors	Jiang et al. (2006), Sluyter et al. (2007a, b)
HSP70	HSP70 activates macrophage IL-12 and E-selectin production in a TLR2/TLR4-dependent manner	Vabulas et al. (2002) Tsan and Gao (2004)
RBC MPs	RBC-derived MPs amplify thrombin-dependent activation of the complement system	Zecher et al. (2014)
RBC MPs	RBC-derived MPs enhance coagulation activation	van Beers et al. (2008)
RBC MPs	RBC-derived MPs activate endothelial cells via heme transfer	Camus et al. (2015)
RBC MPs	RBC-derived MPs are internalized by myeloid cells and induce pro-inflammatory cytokine production	Awojoodu et al. (2014)
RBC MPs	RBC-derived MPs contribute sickle cell disease-associated vascular dysfunction and cardiovascular complications	Tantawy et al. (2013b)
RBC MPs	RBC-derived MPs contribute to transfusion-induced inflammatory response	Cognasse et al. (2015)

Activation of the cytosolic NOD-like receptors results in the assembly of a caspase-1-activating scaffold. Active caspase-1 subsequently cleaves the pro-inflammatory IL-1 family of cytokines into their bioactive forms, IL-1β and IL-18, those can trigger pyroptosis, a type of inflammatory cell death [reviewed in Guo et al. (2015)]. The NLR family pyrin domain containing 3 (NLRP3) inflammasome, which belongs to the NOD-like receptor family, is the most extensively studied inflammasome, that is formed after the oligomerization of NLRP3, apoptosis-associated speck-like protein containing a caspase recruitment domain (ASC), and pro-caspase-1 (Schroder and Tschopp 2010).

Besides PAMPs, the NLRP3 inflammasome is activated in response to a wide variety of DAMPs including extracellular ATP, crystals of monosodium urate or cholesterol, β-amyloid fibers, the degradation of extracellular matrix components, and environmental or industrial particles and nanoparticles (Martinon et al. 2006; Mariathasan et al. 2006; Duewell et al. 2010; Halle et al. 2008; Babelova et al. 2009; Yazdi et al. 2010; Hornung et al. 2008).

Recently heme was added to the long list of NLRP3 activating danger signals. Dutra et al. showed that heme triggers active IL-1β production in lipopolysaccharide (LPS)-primed macrophages in an NLRP3- and caspase-1-dependent manner (Dutra et al. 2014). They also investigated the structural requirements of heme-mediated NLRP3 inflammasome activation. Heme analogs such as protoporphyrin IX (PPIX) that lacks the central iron atom or metal substitution derivatives such as CoPPIX and SnPPIX were unable to induce IL-1β secretion in LPS-primed macrophages (Dutra et al. 2014). Based on these observations, they came to the conclusion that NLRP3 activation by heme is strictly dependent on its coordinated iron, which is in conflict with the findings of Li et al. who reported that PPIX is as efficient in inducing IL-1β maturation and secretion as heme (Li et al. 2014).

9.1.4.2 Neutrophil Activation

Polymorphonuclear neutrophils are the first leukocytes migrating from the blood into injured or infected tissues. Neutrophils kill pathogens via various cytotoxic mechanisms and clear cellular debris; therefore they play a fundamental role in innate and adaptive immunity (Rosales et al. 2016). In the recent years, it has become evident that neutrophils not only sense PAMPs but can recognize and respond to endogenous DAMPs as well. In line of this notion, heme triggers neutrophil chemotaxis and activation, characterized by elevated ROS production and increased expression of the pro-inflammatory cytokine IL-8 (Graca-Souza et al. 2002). Heme-induced neutrophil recruitment is regulated through signaling pathways that are characteristic of chemoattractant molecules (Porto et al. 2007) but independent of TLR4-mediated signaling (Figueiredo et al. 2007). Besides heme, oxidized Hb (ferrylHb) is a very potent trigger of neutrophil infiltration in mice independently of TLR4 signaling (Silva et al. 2009). Additionally, Kono et al. showed that PPIX was as efficient as heme in inducing neutrophil ROS production, pointing out that this effect is independent of the coordinated iron present in heme (Kono et al. 2013).

Protoporphyrin ring-induced neutrophil activation was suggested to play a role in transfusion-related acute lung injury (Kono et al. 2013).

Additionally of ROS generation and the release of microbicidal molecules, neutrophils can release extracellular traps—a meshwork of chromatin fibers decorated by granular proteins—that represent an important strategy to immobilize and kill invading microorganisms (Brinkmann et al. 2004). Recently Chen et al. reported that heme is able to induce the formation of neutrophil extracellular traps and suggested that this mechanism contributes to vaso-occlusion crises in sickle cell disease (Chen et al. 2014).

9.1.4.3 Endothelial Cell Activation

Endothelium, the interface between blood and tissue, has a pivotal role in the inflammatory response mainly through the induction of the leukocyte adhesion cascade to facilitate transmigration of inflammatory cells to the inflamed tissue. Accordingly, inflammatory stimuli, such as IL-1, TNF-α, or LPS, upregulate cellular adhesion molecules including intracellular adhesion molecule-1 (ICAM-1), vascular cell adhesion molecule-1 (VCAM-1), and E selectin, in endothelial cells (Bevilacqua et al. 1985; Pohlman et al. 1986). Wagener et al. found that exposure of endothelial cells to heme upregulated the expressions of ICAM-1, VCAM-1, and E selectin, in a similar manner to that of IL-1, TNF-α, or LPS (Wagener et al. 1997). Recently Belcher et al. showed that heme activates endothelial cells in a TLR4-dependent manner and that this heme-mediated TLR4-dependent endothelial activation plays a pathogenic role in vaso-occlusion in a murine model of sickle cell disease (Belcher et al. 2014).

While searching for other mediators of hemolysis-associated inflammation, Silva et al. reported that ferrylHb but not native Hb or metHb triggers upregulation of the pro-inflammatory adhesion molecules ICAM-1, VCAM-1, and E-selectin (Silva et al. 2009). FerrylHb induced rearrangement of actin cytoskeleton in endothelial cells leading to the disruption of the endothelial monolayer integrity (Silva et al. 2009). FerrylHb-induced inflammatory response was dependent on actin polymerization and the activation of the c-Jun N-terminal kinase and the p38 mitogen-activated protein kinase signal transduction pathways (Silva et al. 2009). Silva et al. showed that induction of endothelial inflammatory response is a unique property of ferrylHb because neither Hb nor metHb triggered these effects (Silva et al. 2009). FerrylHb can release its prosthetic heme group (Potor et al. 2013), and one can ask whether ferrylHb-mediated inflammatory response is mediated by the released heme. Many lines of evidence suggest that in fact this is not the case. First of all, metHb, that can also release heme in a similar manner as ferrylHb, does not induce inflammatory response in endothelial cells (Silva et al. 2009). Second, ferrylHb-induced inflammatory response is not dependent on TLR4 signaling (Silva et al. 2009). These results suggest that heme and ferrylHb are two Hb-derived pro-inflammatory agonists that trigger endothelial activation via different signaling mechanisms.

9.1.5 Cytoprotective and Anti-inflammatory Actions of Hb-Derived Species

Interestingly enough, besides its prooxidant and pro-inflammatory actions, under special circumstances heme can induce cytoprotective and anti-inflammatory responses. These protective mechanisms largely rely on the heme-mediated upregulation of the HO-1/ferritin system [reviewed in Gozzelino et al. (2010)], and it mostly relies on the ability of HO-1 to degrade heme into CO, iron, and biliverdin, in which the latter is promptly converted to bilirubin. The subsequent upregulation of ferritin is essential to obtain the protective effect, as it can store the released iron in a catalytically inactive form (Balla et al. 1992). Additionally, the side products of heme degradation, i.e., bilirubin and CO, exert diverse antioxidant and anti-inflammatory actions (Gozzelino et al. 2010).

Along with these notions, a subset of macrophages, called hemorrhage-associated or Mhem macrophages with anti-inflammatory properties, were identified in atherosclerotic plaques with intraplaque hemorrhage (Boyle et al. 2009). Mhem macrophages are characterized by facilitated iron sequestration assured by elevated expressions of HO-1 and CD163 and at the same time protection from foam cell formation secured by induction of genes central to cholesterol efflux (Boyle et al. 2009, 2012). Boyle et al. also showed that Mhem macrophage polarization is driven by heme and identified two key transcription factors nuclear factor erythroid 2-related factor 2 (NRF2) and activating transcription factor 1 involved in this process (Boyle et al. 2011, 2012).

Endothelial cells can also benefit from the cytoprotective mechanism provided by the HO-1/ferritin system. In the early 1990s, Balla et al. showed that a brief exposure of sublethal concentration of heme made endothelial cells highly resistant to subsequent oxidant-mediated killing in which cytoprotection was relied on the upregulation of the HO-1/ferritin system (Balla et al. 1992). Since that initial work, many investigations targeted the multifunctional role and therapeutic potential of HO-1 in the vascular endothelium [reviewed in Calay and Mason (2014)].

9.1.6 Non-Hb-Derived RBC DAMPs

Although Hb is the far more abundant molecule in RBCs, there are other components in RBCs that can potentially become DAMPs following RBC lysis. For example ATP, a universal energy source, is present in RBCs in high concentration (~1.6 mmol/L). When present in the extracellular milieu, ATP becomes a signaling molecule that activates P2 receptors in diverse cells (Dubyak 1991). It has been shown that hypoxia, elevated shear stress, and reduced pH lead to ATP release from RBCs, although it is still a matter of debate whether it occurs via an active or passive process. Bergfeld et al. showed that under hypoxic conditions, RBCs release ATP in a regulated way through the plasma membrane protein band 4.5 (Bergfeld and

Forrester 1992). Recently Sridharan et al. proposed that pannexin 1, a channel-forming glycoprotein, is involved in hypoxia-mediated ATP release from RBCs (Sridharan et al. 2010). Regarding shear stress-induced ATP release, Wan et al. suggested that mechanosensitive ATP release is triggered by retraction of the spectrin-actin cytoskeleton network and influenced by membrane viscosity (Wan et al. 2008). Recently, Piezo1, a mechanically activated cation channel involved in physiological responses to touch, pressure, and stretch, was shown to regulate mechanosensitive release of ATP from RBCs via controlling the shear-induced calcium influx (Cinar et al. 2015). Contrary to the active process, Sikora et al. reported that hemolysis is the primary mechanism via which RBCs release ATP in response to hypoxia or mechanical stress (Sikora et al. 2014). Nevertheless, RBC-derived ATP can activate P2 purinergic receptors on vascular endothelial cells, resulting in the synthesis of powerful vasodilators such as nitric oxide and prostaglandins (Burnstock 2017). Via this mechanism RBCs actively participate in the regulation of microvascular blood flow and contribute to match oxygen delivery and local needs (Ellsworth et al. 1995).

Besides its vasoactive effects, activation of P2 purinergic receptors by ATP can trigger inflammatory responses in various immune and nonimmune cells (Idzko et al. 2014). For example, ATP activates P2X purinoceptor 7 (P2X7) and promotes IL-1β and IL-18 secretion in LPS-primed macrophages (Perregaux et al. 2000). Activation of P2X7 receptors by ATP on endothelial cells leads to nuclear factor kappa B (NF-κB) activation and subsequent upregulation of its target genes such as E-selectin (von Albertini et al. 1998). Extracellular ATP induces deterioration of endothelial barrier function and may trigger apoptotic cell death (McClenahan et al. 2009). ATP can induce activation of the NLRP3 inflammasome and subsequent release of low levels of IL-1β in endothelial cells primed with LPS or TNF-α (Huck et al. 2015; Champaiboon et al. 2014). Furthermore, both progenitor and mature RBCs express P2 purinergic receptors, and accumulating evidence suggest that extracellular ATP exerts various biological effects on these cells (Burnstock 2015; Sluyter 2015). ATP induces the release of MPs, ROS formation and apoptotic cell death in erythroid progenitor cells (Chahwala and Cantley 1984; Constantinescu et al. 2010; Wang and Sluyter 2013). Activation of P2 purinergic receptors in mature RBCs triggers eicosanoid release and phosphatidylserine exposure and eventually leads to hemolysis (Jiang et al. 2006; Sluyter et al. 2007a, b).

IL-33, the member of the IL-1 cytokine superfamily, is a well-known alarmin that is released upon stress and contributes to the pathogenesis of diverse inflammatory diseases through the activation of innate immune cells (Rider et al. 2017). Recently Wei et al. showed that RBCs contain IL-33 and that IL-33 is released in large amounts upon RBC lysis (Wei et al. 2015). They found association between plasma IL-33 levels and the degree of hemolysis in sickle cell disease patients with intravascular hemolysis (Wei et al. 2015). Similar association between plasma IL-33 concentration and hemolysis was reported in patients with autoimmune hemolytic anemia (Bu et al. 2015). Released IL-33 signals through ST2 receptors and enhances the functions of diverse lymphoid and myeloid immune cells [reviewed in Griesenauer and Paczesny (2017)].

Hsps are ubiquitously expressed proteins exerting diverse protective mechanisms during cellular stress. For example, both constitutive and inducible forms of the 70 kDa Hsp, Hsc70, and Hsp70, respectively, function as cytosolic chaperons during erythrocyte maturation. Although the expressions of Hsc70 and Hsp70 decrease significantly at the terminal stage of erythroid progenitor cell differentiation (Patterson et al. 2009), they are still present in mature RBCs (Gromov and Celis 1991). Vabulas et al. showed that extracellular Hsp70 activates macrophage IL-12 and E-selectin production via CD14/TLR2 and CD14/TLR4 receptor complex-mediated signal transduction pathways (Vabulas et al. 2002). However, recent evidence suggests that the reported cytokine effects of Hsp70 and other Hsps may be due to the contaminating LPS (Tsan and Gao 2004).

Microparticles (MPs) are small membrane-encapsulated vesicles present in body fluids. Blood MPs can originate from platelets, RBCs, leukocytes, or endothelial cells. They are shed from cells in response to cell activation, cell stress, or apoptosis, and besides the phospholipid bilayer, they contain cytosolic components of their parental cells. RBCs release MPs during their normal lifetime in which process they lose a substantial amount of Hb content and surface area (Willekens et al. 2003). Hemoglobinopathies, characterized by shortened life-span of RBCs, such as sickle cell disease and thalassemia major, are associated with accelerated formation of RBC-derived MPs (Tantawy et al. 2013a, b). Interestingly increased levels of RBC-derived MPs are present in patients with metabolic syndrome (Helal et al. 2011). Recently RBC-derived MPs attracted attention in transfusion medicine as well. For therapeutic interventions, packed RBCs are stored in the blood bank for up to 42 days. Storage is associated with diverse morphological and biochemical alterations of RBCs including reduced integrity of the RBC membrane and the formation of RBC-derived MPs (Kim-Shapiro et al. 2011; D'Alessandro et al. 2015). RBC-derived MPs exert diverse biological actions. For example, RBC-derived MPs scavenge nitric oxide (Donadee et al. 2011; Liu et al. 2013) and amplify systemic inflammation via thrombin-dependent activation of complement system (Zecher et al. 2014). Moreover, RBC-derived MPs enhance coagulation activation (van Beers et al. 2008) and are involved in endothelial activation via heme transfer (Camus et al. 2015). RBC-derived MPs are internalized by myeloid cells and induce pro-inflammatory cytokine secretion (Awojoodu et al. 2014). These mechanisms contribute significantly to sickle cell disease-associated vascular dysfunction and cardiovascular complications (Tantawy et al. 2013b) and involved in transfusion-induced inflammatory responses (Cognasse et al. 2015).

9.1.7 Therapeutic Interventions

Different therapeutic approaches were designed and investigated to limit the pathological consequences of massive hemolysis or hemorrhages. Some strategies are focusing on limiting the formation or fostering the elimination of RBC-derived prooxidant and pro-inflammatory molecules. For example, Pamplona et al. showed that CO—the product of heme catabolism—suppress the pathogenesis of experimental cerebral malaria. The effect is mediated by the binding of CO to Hb, preventing Hb oxidation and the generation of free heme, a molecule that plays a critical role in the pathogenesis of cerebral malaria (Pamplona et al. 2007). Recently the therapeutic potential of the natural plasma Hb and heme scavenger proteins, Hp and Hx, have been tested in preclinical animal studies and in small-scale human studies [reviewed in Schaer et al. (2013), Smith and McCulloh (2015)]. In humans Hp supplementation prevented hemoglobinuria or the development of acute kidney injury in a variety of hemolytic conditions [reviewed in Schaer et al. (2013)]. Vinchi et al. showed that Hx therapy improves cardiovascular function in mouse models of sickle cell anemia and β-thalassemia by preventing endothelial dysfunction (Vinchi et al. 2013) and inhibits heme-induced pro-inflammatory phenotypic change of macrophages in a mouse model of sickle cell disease (Vinchi et al. 2016).

Other therapeutic approaches against hemolysis-/hemorrhage-associated adverse effects rely on the induction of the natural antioxidant response. For example upregulation of the NRF2/HO-1 system suppresses the pathogenesis of severe malaria in mice, a pathology driven by RBC-derived heme (Pamplona et al. 2007; Ferreira et al. 2008; Seixas et al. 2009; Jeney et al. 2014). The protective mechanism provided by the NRF2/HO-1 system is very complex and relies on the effective removal of heme, the cytoprotective and anti-inflammatory actions of heme degradation products (bilirubin and CO), and the upregulation of the iron-sequestering protein, ferritin (Gozzelino et al. 2010).

9.2 Conclusions

The RBC is usually a blessing but sometimes a curse. It is a blessing, when it functions properly: circulates throughout the body about 170,000 times during its lifetime to deliver oxygen and remove carbon dioxide from cells and phagocytosed unperceivably at the end of its life-span by macrophages, and curse, when it is involved in pathophysiologic mischief upon hemorrhage or intravascular hemolysis.

Since the dogma breaking "danger model" introduced by Polly Matzinger in 1994 our understanding of how the immune system discriminates between dangerous and safe by recognition of pathogens or alarmins released by injured or stressed cells, underwent a fundamental revision. Diverse endogenous DAMPs were identified and their critical contributions were unquestionably verified in different pathologies. In the last decade, it became evident that upon hemolysis or hemorrhage RBCs release DAMPs that can activate immune and nonimmune cells via diverse signaling mechanisms. A lot of work needs to be done in the future to complete the colorful picture of RBC-derived DAMPs, their targeted cells, and the mechanisms of their actions. Fuller understanding of hemolysis/hemorrhage-associated inflammation could contribute to the development of novel therapeutics intended to interrupt these pathological events.

Acknowledgments This work was supported by grant from the National Research, Development and Innovation Office (NKFIH grant number: K116024).

Conflict of Interest The author has no conflict of interest.

References

Alayash AI (2011) Haptoglobin: old protein with new functions. Clin Chim Acta 412 (7–8):493–498. https://doi.org/10.1016/j.cca.2010.12.011

Alayash AI, Patel RP, Cashon RE (2001) Redox reactions of hemoglobin and myoglobin: biological and toxicological implications. Antioxid Redox Signal 3(2):313–327. https://doi.org/10.1089/152308601300185250

Andersen CB, Torvund-Jensen M, Nielsen MJ, de Oliveira CL, Hersleth HP, Andersen NH, Pedersen JS, Andersen GR, Moestrup SK (2012) Structure of the haptoglobin-haemoglobin complex. Nature 489(7416):456–459. https://doi.org/10.1038/nature11369

Arruda MA, Graca-Souza AV, Barja-Fidalgo C (2005) Heme and innate immunity: new insights for an old molecule. Mem Inst Oswaldo Cruz 100(7):799–803. doi:/S0074-02762005000700022

Awojoodu AO, Keegan PM, Lane AR, Zhang YY, Lynch KR, Platt MO, Botchwey EA (2014) Acid sphingomyelinase is activated in sickle cell erythrocytes and contributes to inflammatory microparticle generation in SCD. Blood 124(12):1941–1950. https://doi.org/10.1182/blood-2014-01-543652

Babelova A, Moreth K, Tsalastra-Greul W, Zeng-Brouwers J, Eickelberg O, Young MF, Bruckner P, Pfeilschifter J, Schaefer RM, Grone HJ, Schaefer L (2009) Biglycan, a danger signal that activates the NLRP3 inflammasome via toll-like and P2X receptors. J Biol Chem 284 (36):24035–24048. https://doi.org/10.1074/jbc.M109.014266

Balla G, Jacob HS, Balla J, Rosenberg M, Nath K, Apple F, Eaton JW, Vercellotti GM (1992) Ferritin: a cytoprotective antioxidant strategem of endothelium. J Biol Chem 267 (25):18148–18153

Banerjee S, Jia Y, Siburt CJ, Abraham B, Wood F, Bonaventura C, Henkens R, Crumbliss AL, Alayash AI (2012) Haptoglobin alters oxygenation and oxidation of hemoglobin and decreases propagation of peroxide-induced oxidative reactions. Free Radic Biol Med 53(6):1317–1326. https://doi.org/10.1016/j.freeradbiomed.2012.07.023

Belcher JD, Chen C, Nguyen J, Milbauer L, Abdulla F, Alayash AI, Smith A, Nath KA, Hebbel RP, Vercellotti GM (2014) Heme triggers TLR4 signaling leading to endothelial cell activation and vaso-occlusion in murine sickle cell disease. Blood 123(3):377–390. https://doi.org/10.1182/blood-2013-04-495887

Bergfeld GR, Forrester T (1992) Release of Atp from human erythrocytes in response to a brief period of hypoxia and hypercapnia. Cardiovasc Res 26(1):40–47. https://doi.org/10.1093/Cvr/26.1.40

Bevilacqua MP, Pober JS, Wheeler ME, Cotran RS, Gimbrone MA Jr (1985) Interleukin 1 acts on cultured human vascular endothelium to increase the adhesion of polymorphonuclear leukocytes, monocytes, and related leukocyte cell lines. J Clin Invest 76(5):2003–2011. https://doi.org/10.1172/JCI112200

Boas FE, Forman L, Beutler E (1998) Phosphatidylserine exposure and red cell viability in red cell aging and in hemolytic anemia. Proc Natl Acad Sci U S A 95(6):3077–3081

Boyle JJ, Harrington HA, Piper E, Elderfield K, Stark J, Landis RC, Haskard DO (2009) Coronary intraplaque hemorrhage evokes a novel atheroprotective macrophage phenotype. Am J Pathol 174(3):1097–1108. https://doi.org/10.2353/ajpath.2009.080431

Boyle JJ, Johns M, Lo J, Chiodini A, Ambrose N, Evans PC, Mason JC, Haskard DO (2011) Heme induces heme oxygenase 1 via Nrf2: role in the homeostatic macrophage response to intraplaque hemorrhage. Arterioscler Thromb Vasc Biol 31(11):2685–2691. https://doi.org/10.1161/ATVBAHA.111.225813

Boyle JJ, Johns M, Kampfer T, Nguyen AT, Game L, Schaer DJ, Mason JC, Haskard DO (2012) Activating transcription factor 1 directs Mhem atheroprotective macrophages through coordinated iron handling and foam cell protection. Circ Res 110(1):20–33. https://doi.org/10.1161/CIRCRESAHA.111.247577

Bratosin D, Mazurier J, Tissier JP, Estaquier J, Huart JJ, Ameisen JC, Aminoff D, Montreuil J (1998) Cellular and molecular mechanisms of senescent erythrocyte phagocytosis by macrophages. A review. Biochimie 80(2):173–195

Brinkmann V, Reichard U, Goosmann C, Fauler B, Uhlemann Y, Weiss DS, Weinrauch Y, Zychlinsky A (2004) Neutrophil extracellular traps kill bacteria. Science 303 (5663):1532–1535. https://doi.org/10.1126/science.1092385

Bu XM, Zhang TL, Wang CH, Ren T, Wen ZK (2015) IL-33 reflects dynamics of disease activity in patients with autoimmune hemolytic anemia by regulating autoantibody production. J Transl Med 13:381. https://doi.org/10.1186/S12967-015-0745-0

Buehler PW, Abraham B, Vallelian F, Linnemayr C, Pereira CP, Cipollo JF, Jia Y, Mikolajczyk M, Boretti FS, Schoedon G, Alayash AI, Schaer DJ (2009) Haptoglobin preserves the CD163 hemoglobin scavenger pathway by shielding hemoglobin from peroxidative modification. Blood 113(11):2578–2586. https://doi.org/10.1182/blood-2008-08-174466

Burnstock G (2015) Blood cells: an historical account of the roles of purinergic signalling. Purinergic Signal 11(4):411–434. https://doi.org/10.1007/s11302-015-9462-7

Burnstock G (2017) Purinergic signaling in the cardiovascular system. Circ Res 120(1):207–228. https://doi.org/10.1161/Circresaha.116.309726

Calay D, Mason JC (2014) The multifunctional role and therapeutic potential of HO-1 in the vascular endothelium. Antioxid Redox Signal 20(11):1789–1809. https://doi.org/10.1089/ars.2013.5659

Camus SM, De Moraes JA, Bonnin P, Abbyad P, Le Jeune S, Lionnet F, Loufrani L, Grimaud L, Lambry JC, Charue D, Kiger L, Renard JM, Larroque C, Le Clesiau H, Tedgui A, Bruneval P, Barja-Fidalgo C, Alexandrou A, Tharaux PL, Boulanger CM, Blanc-Brude OP (2015) Circulating cell membrane microparticles transfer heme to endothelial cells and trigger vasoocclusions in sickle cell disease. Blood 125(24):3805–3814. https://doi.org/10.1182/blood-2014-07-589283

Cerqueira BA, Boas WV, Zanette AD, Reis MG, Goncalves MS (2011) Increased concentrations of IL-18 and uric acid in sickle cell anemia: contribution of hemolysis, endothelial activation and the inflammasome. Cytokine 56(2):471–476. https://doi.org/10.1016/j.cyto.2011.08.013

Chahwala SB, Cantley LC (1984) Extracellular ATP induces ion fluxes and inhibits growth of Friend erythroleukemia cells. J Biol Chem 259(22):13717–13722

Champaiboon C, Poolgesorn M, Wisitrasameewong W, Sa-Ard-Iam N, Rerkyen P, Mahanonda R (2014) Differential inflammasome activation by Porphyromonas gingivalis and cholesterol crystals in human macrophages and coronary artery endothelial cells. Atherosclerosis 235 (1):38–44. https://doi.org/10.1016/j.atherosclerosis.2014.04.007

Chen G, Zhang D, Fuchs TA, Manwani D, Wagner DD, Frenette PS (2014) Heme-induced neutrophil extracellular traps contribute to the pathogenesis of sickle cell disease. Blood 123 (24):3818–3827. https://doi.org/10.1182/blood-2013-10-529982

Cinar E, Zhou S, DeCourcey J, Wang YX, Waugh RE, Wan J (2015) Piezo1 regulates mechanotransductive release of ATP from human RBCs. Proc Natl Acad Sci U S A 112 (38):11783–11788. https://doi.org/10.1073/pnas.1507309112

Cognasse F, Hamzeh-Cognasse H, Laradi S, Chou ML, Seghatchian J, Burnouf T, Boulanger C, Garraud O, Amabile N (2015) The role of microparticles in inflammation and transfusion: A concise review. Transfus Apher Sci 53(2):159–167. https://doi.org/10.1016/j.transci.2015.10.013

Constantinescu P, Wang B, Kovacevic K, Jalilian I, Bosman GJ, Wiley JS, Sluyter R (2010) P2X7 receptor activation induces cell death and microparticle release in murine erythroleukemia cells. Biochim Biophys Acta 1798(9):1797–1804. https://doi.org/10.1016/j.bbamem.2010.06.002

D'Alessandro A, Kriebardis AG, Rinalducci S, Antonelou MH, Hansen KC, Papassideri IS, Zolla L (2015) An update on red blood cell storage lesions, as gleaned through biochemistry and omics technologies. Transfusion 55(1):205–219. https://doi.org/10.1111/trf.12804
de Back DZ, Kostova EB, van Kraaij M, van den Berg TK, van Bruggen R (2014) Of macrophages and red blood cells; a complex love story. Front Physiol 5:9. https://doi.org/10.3389/fphys.2014.00009
Deterding LJ, Ramirez DC, Dubin JR, Mason RP, Tomer KB (2004) Identification of free radicals on hemoglobin from its self-peroxidation using mass spectrometry and immuno-spin trapping: observation of a histidinyl radical. J Biol Chem 279(12):11600–11607. https://doi.org/10.1074/jbc.M310704200
Donadee C, Raat NJH, Kanias T, Tejero J, Lee JS, Kelley EE, Zhao XJ, Liu C, Reynolds H, Azarov I, Frizzell S, Meyer EM, Donnenberg AD, Qu LR, Triulzi D, Kim-Shapiro DB, Gladwin MT (2011) Nitric oxide scavenging by red blood cell microparticles and cell-free hemoglobin as a mechanism for the red cell storage lesion. Circulation 124(4):465–U294. https://doi.org/10.1161/Circulationaha.110.008698
Dostert C, Guarda G, Romero JF, Menu P, Gross O, Tardivel A, Suva ML, Stehle JC, Kopf M, Stamenkovic I, Corradin G, Tschopp J (2009) Malarial hemozoin is a Nalp3 inflammasome activating danger signal. PLoS One 4(8):e6510. https://doi.org/10.1371/journal.pone.0006510
Dubyak GR (1991) Signal transduction by P2-purinergic receptors for extracellular ATP. Am J Respir Cell Mol Biol 4(4):295–300. https://doi.org/10.1165/ajrcmb/4.4.295
Duewell P, Kono H, Rayner KJ, Sirois CM, Vladimer G, Bauernfeind FG, Abela GS, Franchi L, Nunez G, Schnurr M, Espevik T, Lien E, Fitzgerald KA, Rock KL, Moore KJ, Wright SD, Hornung V, Latz E (2010) NLRP3 inflammasomes are required for atherogenesis and activated by cholesterol crystals. Nature 464(7293):1357–1361. https://doi.org/10.1038/nature08938
Dutra FF, Bozza MT (2014) Heme on innate immunity and inflammation. Front Pharmacol 5:115. https://doi.org/10.3389/fphar.2014.00115
Dutra FF, Alves LS, Rodrigues D, Fernandez PL, de Oliveira RB, Golenbock DT, Zamboni DS, Bozza MT (2014) Hemolysis-induced lethality involves inflammasome activation by heme. Proc Natl Acad Sci U S A 111(39):E4110–E4118. https://doi.org/10.1073/pnas.1405023111
Eisenstein RS, Garcia-Mayol D, Pettingell W, Munro HN (1991) Regulation of ferritin and heme oxygenase synthesis in rat fibroblasts by different forms of iron. Proc Natl Acad Sci U S A 88 (3):688–692
Ellsworth ML, Forrester T, Ellis CG, Dietrich HH (1995) The erythrocyte as a regulator of vascular tone. Am J Physiol Heart C 269(6):H2155–H2161
Fernandez-Boyanapalli RF, Frasch SC, McPhillips K, Vandivier RW, Harry BL, Riches DW, Henson PM, Bratton DL (2009) Impaired apoptotic cell clearance in CGD due to altered macrophage programming is reversed by phosphatidylserine-dependent production of IL-4. Blood 113(9):2047–2055. https://doi.org/10.1182/blood-2008-05-160564
Ferreira A, Balla J, Jeney V, Balla G, Soares MP (2008) A central role for free heme in the pathogenesis of severe malaria: the missing link? J Mol Med 86(10):1097–1111. https://doi.org/10.1007/s00109-008-0368-5
Figueiredo RT, Fernandez PL, Mourao-Sa DS, Porto BN, Dutra FF, Alves LS, Oliveira MF, Oliveira PL, Graca-Souza AV, Bozza MT (2007) Characterization of heme as activator of Toll-like receptor 4. J Biol Chem 282(28):20221–20229. https://doi.org/10.1074/jbc.M610737200
Fortes GB, Alves LS, de Oliveira R, Dutra FF, Rodrigues D, Fernandez PL, Souto-Padron T, De Rosa MJ, Kelliher M, Golenbock D, Chan FK, Bozza MT (2012) Heme induces programmed necrosis on macrophages through autocrine TNF and ROS production. Blood 119 (10):2368–2375. https://doi.org/10.1182/blood-2011-08-375303
George A, Pushkaran S, Konstantinidis DG, Koochaki S, Malik P, Mohandas N, Zheng Y, Joiner CH, Kalfa TA (2013) Erythrocyte NADPH oxidase activity modulated by Rac GTPases, PKC, and plasma cytokines contributes to oxidative stress in sickle cell disease. Blood 121 (11):2099–2107. https://doi.org/10.1182/blood-2012-07-441188

Gibb DR, Calabro S, Liu D, Tormey CA, Spitalnik SL, Zimring JC, Hendrickson JE, Hod EA, Eisenbarth SC (2016) The Nlrp3 inflammasome does not regulate alloimmunization to transfused red blood cells in mice. EBioMedicine 9:77–86. https://doi.org/10.1016/j.ebiom.2016.06.008

Gozzelino R, Jeney V, Soares MP (2010) Mechanisms of cell protection by heme oxygenase-1. Annu Rev Pharmacol Toxicol 50:323–354. https://doi.org/10.1146/annurev.pharmtox.010909.105600

Graca-Souza AV, Arruda MAB, de Freitas MS, Barja-Fidalgo C, Oliveira PL (2002) Neutrophil activation by heme: implications for inflammatory processes. Blood 99(11):4160–4165. https://doi.org/10.1182/blood.V99.11.4160

Gram M, Sveinsdottir S, Ruscher K, Hansson SR, Cinthio M, Akerstrom B, Ley D (2013) Hemoglobin induces inflammation after preterm intraventricular hemorrhage by methemoglobin formation. J Neuroinflammation 10:100. https://doi.org/10.1186/1742-2094-10-100

Griesenauer B, Paczesny S (2017) The ST2/IL-33 axis in immune cells during inflammatory diseases. Front Immunol 8:475. https://doi.org/10.3389/Fimmu.2017.00475

Gromov PS, Celis JE (1991) Identification of two molecular chaperons (HSX70, HSC70) in mature human erythrocytes. Exp Cell Res 195(2):556–559

Guo H, Callaway JB, Ting JP (2015) Inflammasomes: mechanism of action, role in disease, and therapeutics. Nat Med 21(7):677–687. https://doi.org/10.1038/nm.3893

Halle A, Hornung V, Petzold GC, Stewart CR, Monks BG, Reinheckel T, Fitzgerald KA, Latz E, Moore KJ, Golenbock DT (2008) The NALP3 inflammasome is involved in the innate immune response to amyloid-beta. Nat Immunol 9(8):857–865. https://doi.org/10.1038/ni.1636

Harel S, Kanner J (1988) The generation of ferryl or hydroxyl radicals during interaction of haemproteins with hydrogen peroxide. Free Radic Res Commun 5(1):21–33

Helal O, Defoort C, Robert S, Marin C, Lesavre N, Lopez-Miranda J, Riserus U, Basu S, Lovegrove J, McMonagle J, Roche HM, Dignat-George F, Lairon D (2011) Increased levels of microparticles originating from endothelial cells, platelets and erythrocytes in subjects with metabolic syndrome: relationship with oxidative stress. Nutr Metab Cardiovasc Dis 21 (9):665–671. https://doi.org/10.1016/j.numecd.2010.01.004

Herz J, Strickland DK (2001) LRP: a multifunctional scavenger and signaling receptor. J Clin Invest 108(6):779–784. https://doi.org/10.1172/JCI13992

Hornung V, Bauernfeind F, Halle A, Samstad EO, Kono H, Rock KL, Fitzgerald KA, Latz E (2008) Silica crystals and aluminum salts activate the NALP3 inflammasome through phagosomal destabilization. Nat Immunol 9(8):847–856. https://doi.org/10.1038/ni.1631

Hrkal Z, Vodrazka Z, Kalousek I (1974) Transfer of heme from ferrihemoglobin and ferrihemoglobin isolated chains to hemopexin. Eur J Biochem/FEBS 43(1):73–78

Huck O, Elkaim R, Davideau JL, Tenenbaum H (2015) Porphyromonas gingivalis-impaired innate immune response via NLRP3 proteolysis in endothelial cells. Innate Immun 21(1):65–72. https://doi.org/10.1177/1753425914523459

Hvidberg V, Maniecki MB, Jacobsen C, Hojrup P, Moller HJ, Moestrup SK (2005) Identification of the receptor scavenging hemopexin-heme complexes. Blood 106(7):2572–2579. https://doi.org/10.1182/blood-2005-03-1185

Idzko M, Ferrari D, Eltzschig HK (2014) Nucleotide signalling during inflammation. Nature 509 (7500):310–317. https://doi.org/10.1038/nature13085

Immenschuh S, Vijayan V, Janciauskiene S, Gueler F (2017) Heme as a target for therapeutic interventions. Front Pharmacol 8:146. https://doi.org/10.3389/Fphar.2017.00146

Jeney V, Eaton JW, Balla G, Balla J (2013) Natural history of the bruise: formation, elimination, and biological effects of oxidized hemoglobin. Oxidative Med Cell Longev 2013:703571. https://doi.org/10.1155/2013/703571

Jeney V, Balla G, Balla J (2014) Red blood cell, hemoglobin and heme in the progression of atherosclerosis. Front Physiol 5:379. https://doi.org/10.3389/fphys.2014.00379

Jia Y, Buehler PW, Boykins RA, Venable RM, Alayash AI (2007) Structural basis of peroxide-mediated changes in human hemoglobin: a novel oxidative pathway. J Biol Chem 282 (7):4894–4907. https://doi.org/10.1074/jbc.M609955200

Jiang HL, Zhu AG, Wong PYK, McGiff JC (2006) Stimulation of rat erythrocyte P2X7 receptor induces the release of epoxyeicosatrienoic acids. FASEB J 20(4):A483–A483
Kapralov A, Vlasova II, Feng W, Maeda A, Walson K, Tyurin VA, Huang Z, Aneja RK, Carcillo J, Bayir H, Kagan VE (2009) Peroxidase activity of hemoglobin-haptoglobin complexes: covalent aggregation and oxidative stress in plasma and macrophages. J Biol Chem 284 (44):30395–30407. https://doi.org/10.1074/jbc.M109.045567
Kim-Shapiro DB, Lee J, Gladwin MT (2011) Storage lesion: role of red blood cell breakdown. Transfusion 51(4):844–851. https://doi.org/10.1111/j.1537-2995.2011.03100.x
Kono M, Saigo K, Takagi Y, Kawauchi S, Wada A, Hashimoto M, Sugimoto T, Takenokuchi M, Morikawa T, Funakoshi K (2013) Morphological and flow-cytometric analysis of haemin-induced human neutrophil activation: implications for transfusion-related acute lung injury. Blood transfus 11(1):53–60. https://doi.org/10.2450/2012.0141-11
Kristiansen M, Graversen JH, Jacobsen C, Sonne O, Hoffman HJ, Law SK, Moestrup SK (2001) Identification of the haemoglobin scavenger receptor. Nature 409(6817):198–201. https://doi.org/10.1038/35051594
Kwon MS, Woo SK, Kurland DB, Yoon SH, Palmer AF, Banerjee U, Iqbal S, Ivanova S, Gerzanich V, Simard JM (2015) Methemoglobin is an endogenous toll-like receptor 4 ligand-relevance to subarachnoid hemorrhage. Int J Mol Sci 16(3):5028–5046. https://doi.org/10.3390/ijms16035028
Land WG (2013) Transfusion-related acute lung injury: the work of DAMPs. Transfus Med Hemother 40(1):3–13. https://doi.org/10.1159/000345688
Lang KS, Lang PA, Bauer C, Duranton C, Wieder T, Huber SM, Lang F (2005) Mechanisms of suicidal erythrocyte death. Cell Physiol Biochem 15(5):195–202. https://doi.org/10.1159/000086406
Larsen R, Gozzelino R, Jeney V, Tokaji L, Bozza FA, Japiassu AM, Bonaparte D, Cavalcante MM, Chora A, Ferreira A, Marguti I, Cardoso S, Sepulveda N, Smith A, Soares MP (2010) A central role for free heme in the pathogenesis of severe sepsis. Sci Transl Med 2(51):51ra71. https://doi.org/10.1126/scitranslmed.3001118
Li Q, Fu W, Yao J, Ji Z, Wang Y, Zhou Z, Yan J, Li W (2014) Heme induces IL-1beta secretion through activating NLRP3 in kidney inflammation. Cell Biochem Biophys 69(3):495–502. https://doi.org/10.1007/s12013-014-9823-9
Lin S, Yin Q, Zhong Q, Lv FL, Zhou Y, Li JQ, Wang JZ, Su BY, Yang QW (2012) Heme activates TLR4-mediated inflammatory injury via MyD88/TRIF signaling pathway in intracerebral hemorrhage. J Neuroinflammation 9:46. https://doi.org/10.1186/1742-2094-9-46
Liu C, Zhao WX, Christ GJ, Gladwin MT, Kim-Shapiro DB (2013) Nitric oxide scavenging by red cell microparticles. Free Radic Biol Med 65:1164–1173. https://doi.org/10.1016/j.freeradbiomed.2013.09.002
Mariathasan S, Weiss DS, Newton K, McBride J, O'Rourke K, Roose-Girma M, Lee WP, Weinrauch Y, Monack DM, Dixit VM (2006) Cryopyrin activates the inflammasome in response to toxins and ATP. Nature 440(7081):228–232. https://doi.org/10.1038/nature04515
Martinon F, Petrilli V, Mayor A, Tardivel A, Tschopp J (2006) Gout-associated uric acid crystals activate the NALP3 inflammasome. Nature 440(7081):237–241. https://doi.org/10.1038/nature04516
Matzinger P (1994) Tolerance, danger, and the extended family. Annu Rev Immunol 12:991–1045. https://doi.org/10.1146/annurev.iy.12.040194.005015
McClenahan D, Hillenbrand K, Kapur A, Carlton D, Czuprynski C (2009) Effects of extracellular ATP on bovine lung endothelial and epithelial cell monolayer morphologies, apoptoses, and permeabilities. Clin Vaccine Immunol 16(1):43–48. https://doi.org/10.1128/Cvi.00282-08
Mendonca R, Silveira AA, Conran N (2016) Red cell DAMPs and inflammation. Inflamm Res 65 (9):665–678. https://doi.org/10.1007/s00011-016-0955-9
Miller YI, Altamentova SM, Shaklai N (1997) Oxidation of low-density lipoprotein by hemoglobin stems from a heme-initiated globin radical: antioxidant role of haptoglobin. Biochemistry 36 (40):12189–12198. https://doi.org/10.1021/bi970258a

Mohanty JG, Nagababu E, Rifkind JM (2014) Red blood cell oxidative stress impairs oxygen delivery and induces red blood cell aging. Front Physiol 5:84. https://doi.org/10.3389/fphys.2014.00084

Muckenthaler MU, Rivella S, Hentze MW, Galy B (2017) A red carpet for iron metabolism. Cell 168(3):344–361. https://doi.org/10.1016/j.cell.2016.12.034

Murray RK, Connell GE, Pert JH (1961) The role of haptoglobin in the clearance and distribution of extracorpuscular hemoglobin. Blood 17:45–53

Nagy E, Eaton JW, Jeney V, Soares MP, Varga Z, Galajda Z, Szentmiklosi J, Mehes G, Csonka T, Smith A, Vercellotti GM, Balla G, Balla J (2010) Red cells, hemoglobin, heme, iron, and atherogenesis. Arterioscler Thromb Vasc Biol 30(7):1347–1353. https://doi.org/10.1161/ATVBAHA.110.206433

Pamplona A, Ferreira A, Balla J, Jeney V, Balla G, Epiphanio S, Chora A, Rodrigues CD, Gregoire IP, Cunha-Rodrigues M, Portugal S, Soares MP, Mota MM (2007) Heme oxygenase-1 and carbon monoxide suppress the pathogenesis of experimental cerebral malaria. Nat Med 13 (6):703–710. https://doi.org/10.1038/nm1586

Patel RP, Svistunenko DA, Darley-Usmar VM, Symons MC, Wilson MT (1996) Redox cycling of human methaemoglobin by H2O2 yields persistent ferryl iron and protein based radicals. Free Radic Res 25(2):117–123

Patterson ST, Li J, Kang JA, Wickrema A, Williams DB, Reithmeier RAF (2009) Loss of specific chaperones involved in membrane glycoprotein biosynthesis during the maturation of human erythroid progenitor cells. J Biol Chem 284(21):14547–14557. https://doi.org/10.1074/jbc.M809076200

Perregaux DG, McNiff P, Laliberte R, Conklyn M, Gabel CA (2000) ATP acts as an agonist to promote stimulus-induced secretion of IL-1 beta and IL-18 in human blood. J Immunol 165 (8):4615–4623

Pimenova T, Pereira CP, Gehrig P, Buehler PW, Schaer DJ, Zenobi R (2010) Quantitative mass spectrometry defines an oxidative hotspot in hemoglobin that is specifically protected by haptoglobin. J Proteome Res 9(8):4061–4070. https://doi.org/10.1021/pr100252e

Piomelli S, Seaman C (1993) Mechanism of red blood cell aging: relationship of cell density and cell age. Am J Hematol 42(1):46–52

Pohlman TH, Stanness KA, Beatty PG, Ochs HD, Harlan JM (1986) An endothelial cell surface factor(s) induced in vitro by lipopolysaccharide, interleukin 1, and tumor necrosis factor-alpha increases neutrophil adherence by a CDw18-dependent mechanism. J Immunol 136 (12):4548–4553

Porto BN, Alves LS, Fernandez PL, Dutra TP, Figueiredo RT, Graca-Souza AV, Bozza MT (2007) Heme induces neutrophil migration and reactive oxygen species generation through signaling pathways characteristic of chemotactic receptors. J Biol Chem 282(33):24430–24436. https://doi.org/10.1074/jbc.M703570200

Potor L, Banyai E, Becs G, Soares MP, Balla G, Balla J, Jeney V (2013) Atherogenesis may involve the prooxidant and proinflammatory effects of ferryl hemoglobin. Oxidative Med Cell Longev 2013:676425. https://doi.org/10.1155/2013/676425

Ramirez DC, Chen YR, Mason RP (2003) Immunochemical detection of hemoglobin-derived radicals formed by reaction with hydrogen peroxide: involvement of a protein-tyrosyl radical. Free Radic Biol Med 34(7):830–839

Reeder BJ, Cutruzzola F, Bigotti MG, Hider RC, Wilson MT (2008) Tyrosine as a redox-active center in electron transfer to ferryl heme in globins. Free Radic Biol Med 44(3):274–283. https://doi.org/10.1016/j.freeradbiomed.2007.06.030

Reimer T, Shaw MH, Franchi L, Coban C, Ishii KJ, Akira S, Horii T, Rodriguez A, Nunez G (2010) Experimental cerebral malaria progresses independently of the Nlrp3 inflammasome. Eur J Immunol 40(3):764–769. https://doi.org/10.1002/eji.200939996

Rider P, Voronov E, Dinarello CA, Apte RN, Cohen I (2017) Alarmins: feel the stress. J Immunol 198(4):1395–1402. https://doi.org/10.4049/jimmunol.1601342

Rosales C, Demaurex N, Lowell CA, Uribe-Querol E (2016) Neutrophils: their role in innate and adaptive immunity. J Immunol Res 2016:1469780. https://doi.org/10.1155/2016/1469780

Rother RP, Bell L, Hillmen P, Gladwin MT (2005) The clinical sequelae of intravascular hemolysis and extracellular plasma hemoglobin: a novel mechanism of human disease. JAMA 293 (13):1653–1662. https://doi.org/10.1001/jama.293.13.1653

Schaer DJ, Schaer CA, Buehler PW, Boykins RA, Schoedon G, Alayash AI, Schaffner A (2006) CD163 is the macrophage scavenger receptor for native and chemically modified hemoglobins in the absence of haptoglobin. Blood 107(1):373–380. https://doi.org/10.1182/blood-2005-03-1014

Schaer DJ, Buehler PW, Alayash AI, Belcher JD, Vercellotti GM (2013) Hemolysis and free hemoglobin revisited: exploring hemoglobin and hemin scavengers as a novel class of therapeutic proteins. Blood 121(8):1276–1284. https://doi.org/10.1182/blood-2012-11-451229

Schroder K, Tschopp J (2010) The inflammasomes. Cell 140(6):821–832. https://doi.org/10.1016/j.cell.2010.01.040

Seixas E, Gozzelino R, Chora A, Ferreira A, Silva G, Larsen R, Rebelo S, Penido C, Smith NR, Coutinho A, Soares MP (2009) Heme oxygenase-1 affords protection against noncerebral forms of severe malaria. Proc Natl Acad Sci U S A 106(37):15837–15842. https://doi.org/10.1073/pnas.0903419106

Siems WG, Sommerburg O, Grune T (2000) Erythrocyte free radical and energy metabolism. Clin Nephrol 53(1 Suppl):S9–S17

Sikora J, Orlov SN, Furuya K, Grygorczyk R (2014) Hemolysis is a primary ATP-release mechanism in human erythrocytes. Blood 124(13):2150–2157. https://doi.org/10.1182/blood-2014-05-572024

Silva G, Jeney V, Chora A, Larsen R, Balla J, Soares MP (2009) Oxidized hemoglobin is an endogenous proinflammatory agonist that targets vascular endothelial cells. J Biol Chem 284 (43):29582–29595. https://doi.org/10.1074/jbc.M109.045344

Sluyter R (2015) P2X and P2Y receptor signaling in red blood cells. Front Mol Biosci 2:60. https://doi.org/10.3389/fmolb.2015.00060

Sluyter R, Shemon AN, Hughes WE, Stevenson RO, Georgiou JG, Eslick GD, Taylor RM, Wiley JS (2007a) Canine erythrocytes express the P2X(7) receptor: greatly increased function compared with human erythrocytes. Am J Physiol Reg I 293(5):R2090–R2098. https://doi.org/10.1152/ajpregu.00166.2007

Sluyter R, Shemon AN, Wiley JS (2007b) P2X(7) receptor activation causes phosphatidylserine exposure in human erythrocytes. Biochem Biophys Res Commun 355(1):169–173. https://doi.org/10.1016/j.bbrc.2007.01.124

Smith A, McCulloh RJ (2015) Hemopexin and haptoglobin: allies against heme toxicity from hemoglobin not contenders. Front Physiol 6:187. https://doi.org/10.3389/fphys.2015.00187

Soares MP, Bozza MT (2016) Red alert: labile heme is an alarmin. Curr Opin Immunol 38:94–100. https://doi.org/10.1016/j.coi.2015.11.006

Sridharan M, Adderley SP, Bowles EA, Egan TM, Stephenson AH, Ellsworth ML, Sprague RS (2010) Pannexin 1 is the conduit for low oxygen tension-induced ATP release from human erythrocytes. Am J Physiol Heart C 299(4):H1146–H1152. https://doi.org/10.1152/ajpheart.00301.2010

Stanojcic M, Chen P, Harrison RA, Wang V, Antonyshyn J, Zuniga-Pflucker JC, Jeschke MG (2014) Leukocyte infiltration and activation of the NLRP3 inflammasome in white adipose tissue following thermal injury. Crit Care Med 42(6):1357–1364. https://doi.org/10.1097/CCM.0000000000000209

Tantawy AA, Adly AA, Ismail EA, Habeeb NM (2013a) Flow cytometric assessment of circulating platelet and erythrocytes microparticles in young thalassemia major patients: relation to pulmonary hypertension and aortic wall stiffness. Eur J Haematol 90(6):508–518. https://doi.org/10.1111/ejh.12108

Tantawy AA, Adly AA, Ismail EA, Habeeb NM, Farouk A (2013b) Circulating platelet and erythrocyte microparticles in young children and adolescents with sickle cell disease: relation to cardiovascular complications. Platelets 24(8):605–614. https://doi.org/10.3109/09537104.2012.749397

Tenhunen R, Marver HS, Schmid R (1968) The enzymatic conversion of heme to bilirubin by microsomal heme oxygenase. Proc Natl Acad Sci U S A 61(2):748–755

Tsan MF, Gao BC (2004) Cytokine function of heat shock proteins. Am J Physiol Cell Physiol 286 (4):C739–C744. https://doi.org/10.1152/ajpcell.00364.2003

Vabulas RM, Ahmad-Nejad P, Ghose S, Kirschning CJ, Issels RD, Wagner H (2002) HSP70 as endogenous stimulus of the toll/interleukin-1 receptor signal pathway. J Biol Chem 277 (17):15107–15112. https://doi.org/10.1074/jbc.M111204200

Vallelian F, Pimenova T, Pereira CP, Abraham B, Mikolajczyk MG, Schoedon G, Zenobi R, Alayash AI, Buehler PW, Schaer DJ (2008) The reaction of hydrogen peroxide with hemoglobin induces extensive alpha-globin crosslinking and impairs the interaction of hemoglobin with endogenous scavenger pathways. Free Radic Biol Med 45(8):1150–1158. https://doi.org/10.1016/j.freeradbiomed.2008.07.013

van Beers EJ, Schaap MC, Berckmans RJ, Nieuwland R, Sturk A, Meijers JCM, Biemond BJ (2008) Circulating erythrocyte-derived microparticles are associated with coagulation activation in sickle cell disease. Blood 112(11):52–53

Vinchi F, De Franceschi L, Ghigo A, Townes T, Cimino J, Silengo L, Hirsch E, Altruda F, Tolosano E (2013) Hemopexin therapy improves cardiovascular function by preventing heme-induced endothelial toxicity in mouse models of hemolytic diseases. Circulation 127 (12):1317–1329. https://doi.org/10.1161/CIRCULATIONAHA.112.130179

Vinchi F, Costa da Silva M, Ingoglia G, Petrillo S, Brinkman N, Zuercher A, Cerwenka A, Tolosano E, Muckenthaler MU (2016) Hemopexin therapy reverts heme-induced proinflammatory phenotypic switching of macrophages in a mouse model of sickle cell disease. Blood 127(4):473–486. https://doi.org/10.1182/blood-2015-08-663245

von Albertini M, Palmetshofer A, Kaczmarek E, Koziak K, Stroka D, Grey ST, Stuhlmeier KM, Robson SC (1998) Extracellular ATP and ADP activate transcription factor NF-kappa B and induce endothelial cell apoptosis. Biochem Biophys Res Commun 248(3):822–829. https://doi.org/10.1006/bbrc.1998.9055

Wagener FA, Feldman E, de Witte T, Abraham NG (1997) Heme induces the expression of adhesion molecules ICAM-1, VCAM-1, and E selectin in vascular endothelial cells. Proc Soc Exp Biol Med 216(3):456–463

Wan JD, Ristenpart WD, Stone HA (2008) Dynamics of shear-induced ATP release from red blood cells. Proc Natl Acad Sci U S A 105(43):16432–16437. https://doi.org/10.1073/pnas.0805779105

Wang B, Sluyter R (2013) P2X7 receptor activation induces reactive oxygen species formation in erythroid cells. Purinergic Signal 9(1):101–112. https://doi.org/10.1007/s11302-012-9335-2

Wei JX, Zhao J, Schrott V, Zhang YZ, Gladwin M, Bullock G, Zhao YT (2015) Red blood cells store and release interleukin-33. J Investig Med 63(6):806–810. https://doi.org/10.1097/Jim.0000000000000213

Willekens FL, Roerdinkholder-Stoelwinder B, Groenen-Dopp YA, Bos HJ, Bosman GJ, van den Bos AG, Verkleij AJ, Werre JM (2003) Hemoglobin loss from erythrocytes in vivo results from spleen-facilitated vesiculation. Blood 101(2):747–751. https://doi.org/10.1182/blood-2002-02-0500

Willekens FL, Werre JM, Groenen-Dopp YA, Roerdinkholder-Stoelwinder B, de Pauw B, Bosman GJ (2008) Erythrocyte vesiculation: a self-protective mechanism? Br J Haematol 141 (4):549–556. https://doi.org/10.1111/j.1365-2141.2008.07055.x

Yazdi AS, Guarda G, Riteau N, Drexler SK, Tardivel A, Couillin I, Tschopp J (2010) Nanoparticles activate the NLR pyrin domain containing 3 (Nlrp3) inflammasome and cause pulmonary inflammation through release of IL-1 alpha and IL-1 beta. Proc Natl Acad Sci U S A 107 (45):19449–19454. https://doi.org/10.1073/pnas.1008155107

Zecher D, Cumpelik A, Schifferli JA (2014) Erythrocyte-derived microvesicles amplify systemic inflammation by thrombin-dependent activation of complement. Arterioscler Thromb Vasc 34 (2):313–320. https://doi.org/10.1161/Atvbaha.113.302378

Chapter 10
Role of Inflammasomes in the Development of Gastrointestinal Diseases

Mazhar A. Kanak, Yoshitaro Shindo, Pavithra SaiKumar, and Bashoo Naziruddin

Contents

Abstract Many diseases of the gastrointestinal tract have been attributed to chronic inflammation, and a few have identified the role of inflammasomes in their pathogenesis. Inflammasomes are a group of protein complexes comprising of several intracellular proteins that link the sensing of microbial products and metabolic stress to the proteolytic activation of the proinflammatory cytokines. Recent studies have implicated activation of several families of NOD-like receptors (NLRs) which are major components of inflammasomes in the development and exacerbation of many diseases of human systems. In this chapter, we discuss the role of inflammasomes in

M. A. Kanak · Y. Shindo
Division of Transplantation, Department of Surgery, Virginia Commonwealth University Medical Center, Richmond, VA, USA

P. SaiKumar
Acic, San Francisco, CA, USA

B. Naziruddin (✉)
Sammons Cancer Center, Baylor Simmons Transplant Institute, Dallas, TX, USA
e-mail: Bashoo.Naziruddin@bswhealth.org

M. D. Cordero, E. Alcocer-Gómez (eds.), *Inflammasomes: Clinical and Therapeutic Implications*, Experientia Supplementum 108,
https://doi.org/10.1007/978-3-319-89390-7_10

some of the most prevalent diseases of the gastrointestinal tract and highlight potential targets for treatment.

Keywords Inflammasome · Gastritis · Gastric cancer · Inflammatory bowel disease · Colorectal cancer · Fatty liver disease · Hepatitis · Liver cirrhosis · Hepatocellular carcinoma · Pancreatitis · Pancreatic cancer

Acute infection or injury to an organ system of the human body often leads to activation of several inflammatory pathways and signals, followed by a response to subside and recuperate from the injury. In some cases, the cycle of damage and repair is recurrent, resulting in a chronic inflammatory state that may affect normal physiology. Recently, attention has focused on the role of a new group of protein complexes called inflammasomes in chronic inflammatory diseases. Inflammasomes are composed of pattern recognition receptors such as NOD-like receptors (NLRs) or absent in melanoma 2 (AIM2)-like receptors (ALRs), the adaptor protein ASC, and procaspase-1 (Liu et al. 2015; Guo et al. 2015). These subunits activate caspase-1, which mediates IL-1β and IL-18 release and inflammatory cell death called pyroptosis (Guo et al. 2015) (Fig. 10.1). Inflammasome in the intestine plays an important role in mediating host defense against infection and maintaining tissue homeostasis beyond immune and inflammatory responses (Chen and Nunez 2011; Nunes and de Souza 2013; Lei and Maloy 2016). Inflammasome activation is regulated by the innate immune system in

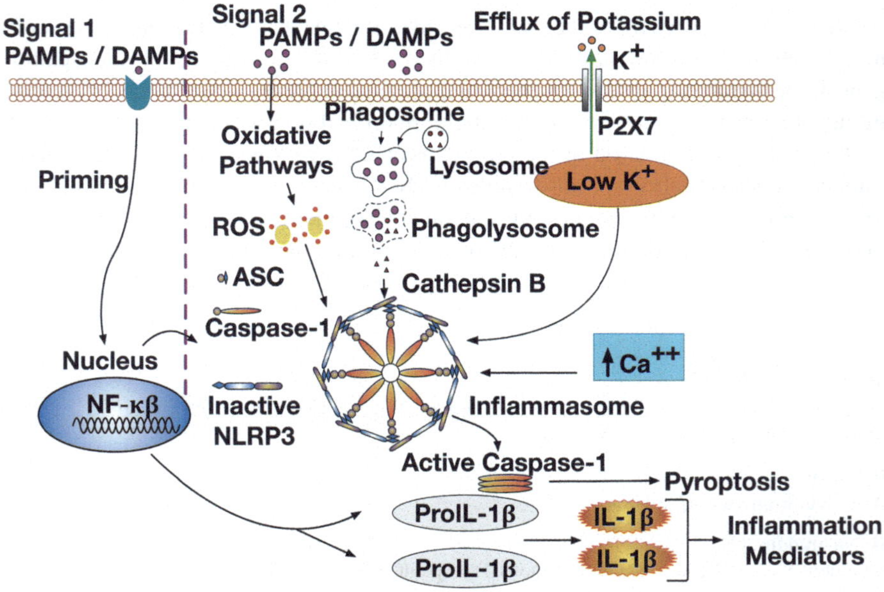

Fig. 10.1 Schematic representation of the activation and function of inflammasome

response to a pathogenic signal called pathogen-associated molecular patterns (PAMPs) or a tissue damage signal called damage-associated molecular patterns (DAMPs) binding to the surface pattern recognition receptors.

Many diseases have been attributed to chronic inflammation, and a few have identified the role of inflammasomes in their pathogenesis. In this chapter, we discuss the role of inflammasomes in some of the most prevalent diseases of the gastrointestinal tract: gastritis, gastric cancer, inflammatory bowel disease (IBD), colorectal cancer (CRC), fatty liver disease, hepatitis, liver cirrhosis, hepatocellular carcinoma (HCC), pancreatitis, and pancreatic cancer.

10.1 Gastritis

Gastritis is one of the most common diseases of the gastrointestinal tract that has long-term morbidity and mortality effects. Although many etiologies exist, *Helicobacter pylori* infection is the most common cause for the chronic inflammation in the lining of the stomach. With prevalence being as high as 50% in developing countries, about one-third of the adult population is infected in developed countries (Eusebi et al. 2014; Suerbaum and Michetti 2002). It is well documented that *H. pylori* infection can lead to chronic gastric and duodenal ulcers and is an important risk factor for gastric adenocarcinoma and mucosa-associated lymphoid tissue lymphoma (Peek and Blaser 2002).

H. pylori, a gram-negative spiral bacterium that colonizes the gastrointestinal tract, is the only bacterium to be classified as a class 1 carcinogen by the World Health Organization (Kusters et al. 2006; Vogiatzi et al. 2007). The organism possesses virulence factors such as cytotoxin-associated gene A (CagA), vacuolating cytotoxin A (VacA), outer inflammatory protein A (OipA), duodenal ulcer-promoting gene A (DupA), and induced by contact with epithelium (IceA), all of which play an important role in disrupting the host's innate and adaptive immunity (Shiota et al. 2013).

H. pylori is recognized by the toll-like receptors (TLRs) as a PAMP, and stimulation of TLRs leads to a cascade of events resulting in the activation of inflammasome molecules. NLRP3 has been shown to be involved in the pathogenesis of gastritis (Semper et al. 2014). The events of inflammasome formation are similar to those in other diseases in which caspase, IL-1β, and IL-18 play an important role in the inflammatory process (El-Omar et al. 2000). Gain of function of IL-1β can lead to increased susceptibility for chronic gastric inflammation, which in the long term will be a precursor for gastric carcinoma (El-Omar et al. 2000; Fox et al. 2003). Various pathways have been suggested for inflammasome stimulation. TLR2 stimulation has been found to be indispensable for the formation of NLRP3, as deficiency of TLR2 failed to produce caspase and other cytokines (Koch and Muller 2015). Although other noncanonical pathways such as TRIF, IRF3/7, and type 1 interferon (IFN) are present, they play a minor role (Rathinam et al. 2012).

Various host factors such as mucus epithelial barrier, acidic pH, β defensins, and lactoferrin provide the first line of innate immunity. MUC1, a cell surface-associated mucin lining the stomach, is a component of the physical barrier that limits the colonization of organisms such as *H. pylori* (Bafna et al. 2010). Research studies have shown that MUC1 has an anti-inflammatory role in *H. pylori* infection in addition to being a physical barrier. Ng and Sutton (2016) demonstrated that MUC1 inhibits NLRP3 formation by suppressing the nuclear factor kappa-light-chain-enhancer of activated B-cells (NFκB) pathway. Mouse models have shown that MUC1-deficient mice had severe, rapid progression of *H. pylori* gastritis in comparison to the wild-type mice (McGuckin et al. 2007).

Studies have focused on Cag pathogenicity island (Cag PAI) and its influence on the pathogenesis of the disease. Cag PAI is a 40-kb region with 30 prominent coding regions. It encodes for type 4 secretion system (T4SS), which can inject CagA and other virulence factors into the host. CagL forms an essential part of the pilus T4SS (Backert and Tegtmeyer 2017). Kim et al. (2013) indicated that Cag PAI and CagL (not VacA and CagA) have the most important role in production of IL-1β. Koch and Muller (2015) showed that there are two other factors specific to *H. pylori*—lipopolysaccharide (LPS) and urease—that contribute to production of inflammatory cytokines: LPS/TLR4/MyD88 through pro IL-1β and urease/TLR4/MyD88 by formation of NLRP3. However, there are controversies about whether LPS is a strong factor; interestingly, urease, which is an enzyme that protects the organism against the acidic pH of the stomach, has proven to be a critical stimulator of NLRP3 (Koch and Muller 2015).

Withaferin A (WA), one of the steroidal lactones (withanolides), has been shown to have anti-inflammatory and anticancer properties. These withanolides were used in ancient Ayurvedic medications for treatment of chronic inflammatory diseases (Maitra et al. 2009; Hahm and Singh 2013). Kim et al. (2015) demonstrated that WA alleviates procaspase, caspase, and IL-1β induced by *H. pylori* and inhibits the NLRP3 activators. This leads to new modalities in the treatment of *H. pylori* that is targeting the NLRP3 pathway.

10.2 Gastric Cancer

Gastric cancer is the fifth most common malignancy and the third leading cause of cancer-related deaths worldwide, accounting for 8.8% of cancer deaths every year (Liang 2016). Gastric cancer is relatively common in East Asia, Eastern Europe, and South America, whereas it is rare in North America and most parts of Africa (Karimi et al. 2014).

H. pylori infection is the strongest known risk factor for gastric cancer (Suh and Yang 2015). *H. pylori*-infected individuals develop gastric diseases such as chronic active gastritis, peptic ulcers, B-cell lymphoma of mucosa-associated lymphoid tissue, and adenocarcinoma (Kim et al. 2015; Gobert and Wilson 2017). Although the molecular mechanisms involved in the oncogenesis and progression of gastric cancer are not yet fully understood, it has been reported that IL-1β induced through inflammasome activation is associated with the development of gastric cancer after *H. pylori* infection (Shigematsu et al. 2013; Kameoka et al. 2016; Yin et al. 2016).

IL-1β is a proinflammatory cytokine that has various roles in inflammatory injury, epigenetic changes, bone marrow cell recruitment, and the promotion of angiogenesis (Yin et al. 2016). Stomach-specific expression of human IL-1β in transgenic mice was shown to cause spontaneous gastric inflammation and cancer, which resulted in recruitment of myeloid-derived suppressor cells to the stomach (Tu et al. 2008). Another study reported that *H. pylori* infection induced IL-1β expression and resulted in severe inflammation in wild-type mice, while recruitment of neutrophils and macrophages by *H. pylori* infection was suppressed in IL-1β$^{-/-}$ mice. Moreover, the number of gastric tumors induced by *H. pylori* infection was significantly diminished in IL-1β$^{-/-}$ mice (Shigematsu et al. 2013). Recently, IL-1β has been shown to increase the expression of cAMP response element-binding protein and CCAAT/enhancer-binding protein beta through ERK1/2 kinase signaling, causing proliferation of gastric cancer cells both in vitro and in vivo (Resende et al. 2016). IL-1β also inhibits gastric acid secretion, which leads to gastric atrophy for gastric cancer development (Sun et al. 2016). It has been reported that IL-1β polymorphism is associated with increased gastric cancer risk in the presence of *H. pylori* infection (Ying et al. 2016).

H. pylori LPS induced pro-IL-1β by TLR4-mediated NFκB activation, whereas *H. pylori* urease B subunit induced NLRP3 transcription by TLR2-mediated NFκB activation (Koch and Muller 2015). In mouse bone marrow-derived dendritic cells infected with *H. pylori*, *cag*PAI was demonstrated to induce production of pro-IL-1β and NLRP3 via TLR2 and Nod2 (Kim et al. 2013). It was also demonstrated that infected *NOD2*$^{-/-}$, *Tlr2*$^{-/-}$, and double-deficient dendritic cells significantly reduced NLRP3 expression (Kim et al. 2013). Furthermore, it has been reported that IL-1β production in *H. pylori*-infected phorbol-12-myristate-13-acetate (PMA)-differentiated THP-1 cells was dependent on caspase-1 and NLRP3, which required reactive oxygen species (ROS), extracellular adenosine triphosphate (ATP), K^+ efflux, and Ca^{2+} signaling (Kameoka et al. 2016).

It has been reported that WA inhibited *H. pylori*-induced IL-1β production by regulating NFκB and NLRP3 inflammasome activation (Kim et al. 2015). WA suppressed *H. pylori*-induced gene expression of pro-IL-1β and NLRP3 in murine bone marrow-derived dendritic cells and THP-1 cells by inhibiting NFκB activation, suggesting an inhibition of the priming signal. Furthermore, WA suppressed caspase-1 activation and IL-1β maturation by blocking signal 2 such as ATP, nigericin, and monosodium urate crystals in LPS-primed murine macrophages, indicating that WA can inhibit NLRP3 inflammasome directly (Kim et al. 2015). The therapeutic strategy of targeting inflammasome might be effective for gastric cancer in the future.

10.3 Inflammatory Bowel Disease

IBD occurs due to chronic intestinal inflammation that leads to tissue damage by uncharacteristic production of proinflammatory molecules by the immune system. Recent analysis of the pathogenesis of IBD has demonstrated the dysregulation or involvement of cytokines, chemokines, inflammasomes, miRNAs, DAMPs, antimicrobial peptides, and neuropeptides (Park et al. 2017). An increase in the level of cytokines produced by innate immune cells and nonimmune cells such as epithelial and stromal cells is a hallmark of IBD. A significant rise in the level of IL-1β produced by the lamina propria was seen in ulcerative colitis and Crohn's disease (Neurath 2014). The severity of Crohn's disease was also correlated with an increase in the inflammasome protein complex (Opipari and Franchi 2015).

In a mouse model of ulcerative colitis, it was demonstrated that Galectin-3 expression increased the production of IL-1β by macrophages and induced activation of NLRP3 (Simovic Markovic et al. 2016). Analysis of mucosal macrophages and colonic epithelial cells of IBD patients revealed an increase in the activity of NFκB (Neurath et al. 1996). A polymorphism in the NLRP3 gene has also been recently linked to the pathogenesis of IBD. Mice lacking the NLRP3 gene are more prone to develop experimental colitis (Hirota et al. 2011), and this may be due to the increase in the intestinal barrier permeability that causes overgrowth of commensal bacteria that may lead to bacteremia (presence of bacteria in the blood) (Zaki et al. 2010a). This study supports the protective role of NLRP3 inflammasome in maintaining homeostasis of the commensal species. On the contrary, another study reported that NLRP3 knockout mice also developed a less severe colitis (Bauer et al. 2010), and colonic tissue produced lower levels of proinflammatory cytokines when induced with dextran sodium sulfate (DSS) or 2,4,6-trinitrobenzenesulfonic acid (Wang et al. 2016). It has been suggested that NLRP3-deficient mice have altered intestinal microflora that may have caused a discrepancy in the results obtained in different lab conditions (Bauer et al. 2012). The NLRP3 deletion studies have shown a very critical role of NLRP3 in maintaining gut equilibrium and homeostasis during inflammation.

In addition to the canonical NLRP3 inflammasome activation, a noncanonical caspase-11-induced NLRP3 inflammasome has been shown to be activated in response to gram-negative bacteria (Pellegrini et al. 2017). Mice lacking caspase-11 showed increased susceptibility to DSS-induced colitis, morbidity, tissue damage, and a decreased expression of cytokines IL-18, IL-22, and IL-1α compared to wild-type counterparts (Williams et al. 2015). Although the data suggest a protective role of caspase-11-induced NLRP3 inflammasome in experimental colitis, further studies are needed to establish the molecular connection between caspase-11 and NLRP3 regulation and bolster this claim.

Patients with mutations in the IL-10R gene spontaneously develop Crohn's disease at a very early age, suggesting a significance of IL-10 signaling in the pathogenesis of Crohn's disease (Glocker et al. 2009). Similarly, mice lacking the IL-10 gene have been shown to develop colitis spontaneously under specific pathogen-free conditions. It was shown that $IL10^{-/-}$ mice had increased expression and activity of NLRP3 inflammasome even before the onset of colitis. The activity of NLRP3 progressively increased with the severity of the disease, and significant elevation of proinflammatory cytokines/chemokines was observed. Treatment with the NLRP3 inhibitor glyburide significantly reversed this pathologic condition. RNA and protein levels of both NLRP3 and ASC were significantly increased in the colonic mucosa of IL-10-deficient mice during the subclinical stage. NOD2 expression is elevated in a later stage of the disease, escalating the damage. In the conclusion of the study, it was discussed that baseline NLRP3 activity is needed for homeostasis, and IL-10 maintains this; however, in the lack of IL-10, persistent activation of NLRP3 causes inflammation in the early stages of colitis, which is exacerbated by the increased expression of NOD2, leading to a complete disease pathology (Liu et al. 2016).

Although animal models have shown the relevance and involvement of NLRP3 inflammasome in the pathogenesis of colitis, patient data demonstrating the involvement of inflammasome in ulcerative colitis is still lacking. But in the case of Crohn's disease, evidence of increased caspase-1 activity and elevation of IL-1β and IL-18 correlating with disease severity may suggest a role of inflammasome in its pathogenesis (de Souza and Fiocchi 2016). Further investigation to clearly delineate the role of NLRP3 inflammasome in the pathophysiology of ulcerative colitis and Crohn's disease is currently ongoing.

10.4 Colorectal Cancer

CRC is the third most common cancer and the fourth leading cause of cancer deaths; with one to two million new cases diagnosed every year (Marmol et al. 2017). The morbidity of CRC is relatively high in Europe, Oceania, and North America, whereas it is low in some parts of Asia and Africa (Fan et al. 2016). Many genetic and environmental factors play an important role in the pathogenesis of CRC (Peters et al. 2015). Chronic inflammation is also a risk factor for CRC and links to tumor growth, proliferation, and metastasis (Ryan et al. 2014; Janakiram and Rao 2014). In addition, IBD, ulcerative colitis, and Crohn's disease are associated with an increased risk of CRC. It is estimated that 1–6 deaths in patients with ulcerative colitis and 1 of 12 deaths in patients with Crohn's disease are caused by CRC (Andersen and Jess 2013).

Several studies of patients with CRC revealed significantly decreased expression of NLRs and AIM2 in CRC samples compared with normal controls (Liu et al. 2015; Dihlmann et al. 2014; Choubey 2016). Therefore, inflammasome in CRC is considered a tumor suppressor that maintains tissue homeostasis against tumorigenesis, although the molecular mechanisms remain unclear (Liu et al. 2015; Choubey 2016).

It has been reported that *Nlrp3*$^{-/-}$, *ASC*$^{-/-}$, and *caspase-1*$^{-/-}$ mice were more susceptible to DSS-induced colitis as well as azoxymethane/DSS-induced colon tumorigenesis than wild-type mice (Zaki et al. 2010a; Allen et al. 2010; Dupaul-Chicoine et al. 2010). Bone marrow reconstitution experiments demonstrated that NLRP3 activated in the hematopoietic cells, rather than intestinal epithelial cells or stromal cells, is more important for host protection against tumorigenesis (Allen et al. 2010).

NLRP6, which is highly expressed in intestinal epithelial cells, also plays a critical role in protection against colon tumorigenesis (Chen et al. 2011; Wang et al. 2015). *Nlrp6*$^{-/-}$ mice treated with azoxymethane/DSS were more susceptible to colorectal carcinogenesis than wild-type mice, which resulted in increased inflammatory responses and decreased IL-18 production within the colon (Chen et al.

2011). Similar to NLRP3, bone marrow chimera studies showed that NLRP6 activated in hematopoietic cells is critical for mediating protection against colon tumorigenesis (Chen et al. 2011). Furthermore, NLRP6 is considered to regulate intestinal microbiota and maintain colonic homeostasis (Wlodarska et al. 2014). In azoxymethane/DSS treatment, wild-type mice cohoused with *Nlrp6*$^{-/-}$ mice experienced significantly greater tumorigenesis than singly housed wild-type mice. The alteration in the microbiota of *Nlrp6*$^{-/-}$ mice transferred into cohoused wild-type mice—an effect dependent on local induction of IL-6 secretion, which resulted in promoting epithelial cell proliferation (Hu et al. 2013). The protective effect of NLRP3 and NLRP6 is associated with inflammasome-mediated release of IL-18, which contributes to epithelial barrier repair against intestinal inflammation and colitis (Oficjalska et al. 2015; Nowarski et al. 2015). *Il18*$^{-/-}$ and *Il18r*$^{-/-}$ mice were more susceptible to DSS-induced colitis and colorectal carcinogenesis than wild-type mice (Takagi et al. 2003; Salcedo et al. 2010). Further, mice with a deficiency of *Myd88*, which resulted in inactivation of IL-18, were also more susceptible to colitis and tumorigenesis (Salcedo et al. 2010). In addition, administration of exogenous IL-18 rescued the colitis susceptibility and suppressed colitis-induced tumorigenesis from inflammasome-deficient mice, indicating that this inflammasome-mediated cytokine might be an effective treatment strategy against certain cases of CRC (Karki et al. 2017; Zaki et al. 2010b; Dupaul-Chicoine et al. 2010).

Casp1$^{-/-}$ and *Nlrc4*$^{-/-}$ mice had significantly increased and more aggressive tumors than wild-type mice (Hu et al. 2010). Interestingly, there were no differences in inflammation severity between inflammasome-deficient mice and wild-type mice. These effects in tumorigenesis were independent of inflammation and were associated with increased epithelial proliferation and reduced apoptosis of tumor cells.

AIM2 is a sensor of double-stranded DNA and another component of inflammasome (Choubey 2016; Lasry et al. 2016; Liu et al. 2015). Similar to NLR inflammasome, AIM2 is believed to mediate a host defense against infection and an inflammatory response and to regulate tissue homeostasis within the intestine (Liu et al. 2015; Choubey 2016; Lasry et al. 2016). It has been reported that CRC tissues reduced AIM2 expression compared with adjacent normal tissues. Furthermore, lack of AIM2 expression was closely associated with a poor outcome for CRC patients (Dihlmann et al. 2014). *Aim2*$^{-/-}$ mice treated with azoxymethane/DSS developed significantly more colorectal tumors than wild-type mice. However, there was no difference in the production of inflammasome-associated proinflammatory cytokines between *Aim2*$^{-/-}$ mice and wild-type mice (Wilson et al. 2015; Man et al. 2015). Therefore, AIM2 is believed to protect against colon tumorigenesis by regulating stem cells proliferation in an inflammasome-independent manner. It has been reported that AIM2 reduced colon tumorigenesis by inhibiting the activation of Akt, a master regulator of cellular survival, through interaction with DNA-dependent protein kinases (Wilson et al. 2015). Another study indicated that AIM2 inhibited expansion of tumor-initiating stem cells through the Wnt pathway (Man et al. 2015). Further, *Aim2*$^{-/-}$ mice cohoused with wild-type mice experienced less tumorigenesis than singly housed *Aim2*$^{-/-}$ mice, indicating that regulation of genetic and environmental factors might be essential for CRC treatment (Man et al. 2015).

10.5 Fatty Liver Disease

Fatty liver disease has been classified into two types: alcoholic liver disease (ALD) and nonalcoholic fatty liver disease (NAFLD). Overconsumption of alcohol is the cause of ALD, and the disease progresses from fatty liver to alcoholic steatohepatitis and eventually cirrhosis. Alcohol-related cirrhosis is the cause of death in 50% of people with cirrhosis and accounts for 1% of all deaths worldwide (Masarone et al. 2016).

A growing number of reports have suggested that increased gut permeability and alteration of the gut microbiota are involved in the pathogenesis of both ALD and NAFLD (Szabo et al. 2010). An increased level of proinflammatory cytokines IL-1β, tumor necrosis factor (TNF)-α, and IL-8 in serum and liver, along with elevated expression of NLRP3 and CASP-1 with infiltration of neutrophils, monocytes, and macrophages, is seen in the liver of patients with ALD (McClain et al. 1986; O'Shea et al. 2010; Peng et al. 2014). In animal models of ALD, there was upregulation of inflammasome molecules NLRP3, ASC, procaspase-1, and IL-1β in the liver. Furthermore, blockade of IL-1R1 or global deficiency of ASC or caspase-1 resulted in amelioration of ALD (Petrasek et al. 2012; Szabo and Petrasek 2015). It was also reported that the alcohol-induced hepatocyte death was mediated by the NLRP3 inflammasome (Xiao et al. 2014). It is presumed that alcohol consumption causes hepatocyte death, alters the gut microbiota, and increases gut permeability, causing the portal blood to carry microbes into the liver. The PAMPs from the microbes activate TLR4 (Inokuchi et al. 2011) and through NFκB signaling regulate the production of pro-IL-1β and pro-IL-18. The second signal for the inflammasome activation may be uric acid or ATP that may have accumulated due to the change in metabolism caused by alcohol-induced mitochondrial dysfunctions (Hoek et al. 2002; Petrasek et al. 2015). Alternatively, hepatocyte death also releases DAMPs such as high-mobility group box protein-1 (HMGB-1) that may also provide the secondary signal for activation of inflammasomes (Ge et al. 2014). The maturation of IL-1β by inflammasomes in turn activates the inflammatory signaling pathways in the Kupffer cells and causes production of TNFα and MCP-1. MCP-1 potentiates the pathology of the disease, and activation of hepatic stellate cells (HSC) by IL-1β eventually leads to cirrhosis (Szabo and Petrasek 2015).

NAFLD is the most common chronic liver disease in developed countries (Bellentani 2017) and in the world, estimated to affect up to 30% of the adult population and 70% to 80% of obese and diabetic individuals (Chalasani et al. 2012). Nonalcoholic steatohepatitis (NASH) is the second leading etiology of end-stage liver disease among adults awaiting liver transplantation in the United States and is projected to become the most common indication for liver transplantation in the next decade (Wong et al. 2015). Furthermore, NASH is an emerging risk factor for type 2 diabetes, cardiovascular disease, and end-stage kidney disease (Chalasani et al. 2012; Musso et al. 2014). Despite its high prevalence and high morbidity, the exact pathogenesis of NASH remains debated.

The mechanism for the development of NAFLD is complex and multifactorial. But, just like ALD, it progresses to NASH and then to cirrhosis. The early stage of NAFLD is characterized by the accumulation of triglyceride in the liver cells, which is termed hepatic steatosis. During this stage, most patients remain asymptomatic. However, steatosis can progress to NASH, which has marked inflammation in the liver cells. NASH can further lead to cirrhosis (inflammation along with marked fibrosis), end-stage liver disease, and HCC. Though many theories have been postulated, the factors responsible for the step-by-step progression of the disease remain undiscovered.

A "two-hit hypothesis" was proposed to explain the pathogenesis of NAFLD/ NASH. The first hit is the accumulation of triglycerides in the liver followed by lipid peroxidation and oxidative stress (second hit), which turns on the inflammatory cascade (Namikawa et al. 2006). Some other models propose a "third hit": oxidative stress-induced cell death decreases the replication of mature hepatocytes, leading to the accumulation of an immature, abnormal progenitor cell population, leading to liver cirrhosis and HCC. Buzzetti et al. explained that the two- or three-hit hypothesis is a simple theory, while the actual disease process is complex, wherein multiple factors act synergistically for disease progression. A few of the known factors of this "multiple-hit hypothesis" are intestinal dysbiosis, increased intestinal permeability, endoplasmic reticulum stress, mitochondrial dysfunction, gut-liver axis, and defective innate immunity (Buzzetti et al. 2016). While various factors are involved in pathogenesis, the inflammasome-induced inflammatory cascade is vital in the disease process.

The activation of inflammasome in fatty liver disease is similar to its activation in any other diseases, with many factors playing a role. In addition to external antigens like microbes (PAMPs), immune cells are exposed to endogenous sterile stimuli known as DAMPs, which are released when tissue is injured. There are many DAMPs, including ATP, uric acid, cholesterol crystals, amyloid, and calcium pyrophosphate crystals (Chen and Nunez 2010; Warren et al. 2010). The cholesterol crystals undergo phagocytosis by macrophages, which stimulates the release of lysosomal protease cathepsin B leading to activation of inflammasome (Latz 2010). While NLRP3 involvement in the pathogenesis of this disease is clear, the role of AIM2 and NLRP6 involvement is still being examined.

As NLRP3, ASC, and caspase-1 play an important role in the inflammatory cascade of NASH, various animal models were studied to see the result of gain or loss of function of these mediators. A high-fat diet was fed to NLRP3 knocked-out mice, and they were protected from hepatomegaly, steatosis, and cirrhosis, whereas mice with NLRP3 showed an inflammatory reaction in the liver with fibrosis (Wree et al. 2014). Similar results were obtained when experiments were performed on ASC-deficient mice (Dixon et al. 2012, 2013). It is to be noted that specific inhibition of NLRP3 in Kupffer cells alleviates the inflammatory process in NASH, whereas generalized NLRP3 inhibition (inclusive of the gut) has proven detrimental by causing dysbiosis and further stimulating inflammation by portal transmission of the TLRs from the gut (Henao-Mejia et al. 2012).

All these experiments have paved the way for new treatment options, which are still in the trial stage. Various strategies have been implemented to prevent inflammasome formation, including use of cholesterol-lowering drugs, use of the xanthine oxidase inhibitor to prevent DAMPs (cholesterol, uric acid), inhibition of saturated fatty acid-induced TLR activation with ethyl pyruvate, and use of phenylmethimazole (Ioannou et al. 2015; Xu et al. 2015; Miura et al. 2013). A few compounds, such as isoliquiritigenin, arglabin, and auranofin, that have been discovered to inhibit NLRP3 inhibitors and a few caspase inhibitors are undergoing Phase 2 trials (Honda et al. 2014; Abderrazak et al. 2015; Isakov et al. 2014). However, since inhibition of inflammasomes occurs in the liver and not in the intestines, these therapies need to be tissue specific.

Interestingly, the human gut harbors approximately 10–100 trillion microorganisms, mainly bacteria—an amount that greatly exceeds the number of human cells (Ursell et al. 2012). Gut dysbiosis (altered gut microbiota) has been involved in the pathogenesis of obesity-related diseases such as metabolic syndrome and NAFLD (Cox et al. 2014). Early life is a critical period for the growth of normal commensal bacteria, and various factors such as mode of delivery, breastfeeding, weaning, and use of antibiotics in newborns can alter the gut flora and predispose an individual to diseases such as metabolic syndrome and NASH (Reinhardt et al. 2009). The gut-derived microbial products (LPS) stimulate the TLRs in the intestine, which are transported to the liver via the portal system. The transported TLRs initiate inflammasome formation, resulting in the production of proinflammatory and profibrotic cytokines IL-1, IL-6, and TNF (Than and Newsome 2015; Mencin et al. 2009; Federico et al. 2016; Friedman 2007; Takaki et al. 2014; Tyrer et al. 2011). Through human descriptive studies, Wigg et al. (2001) demonstrated a connection between NASH and small intestinal bacterial overgrowth through C14 xylose and lactulose breath test.

Both the multiple-hit and the earlier two-hit hypothesis cite accumulation of fat in the liver cells as an important initiation step in the disease process. In addition to insulin resistance, dietary intake is a major contributor in hepatic steatosis. Researchers have revealed interesting facts about the types of fatty acids and their role in the disease process. The notorious saturated fatty acids such as palmitic acid are known to cause hypercholesterolemia and be a major risk factor for atherosclerosis, coronary artery disease, and stroke. These saturated fatty acids can directly stimulate NLRP3 inflammasome formation and can initiate the process of NASH (Wen et al. 2011; Csak et al. 2011). Polyunsaturated fatty acids (PUFA) were considered healthy, but interestingly studies have proven that ω-3 PUFAs are anti-inflammatory, whereas ω-6 PUFAs are proinflammatory, with the ability to stimulate inflammasome formation as well as cytokine production (Wree et al. 2013). Animal models have proven that ω-3 PUFA suppressed the LPS-mediated priming and inhibited the formation of NLRP3 inflammasome (Sui et al. 2016). Further, studies have shown that a high ω-6/ω-3 ratio diet increases the severity of steatohepatitis. In addition, a diet rich in fructose and cholesterol can contribute to disease progression. A high-fructose diet increases de novo lipogenesis and insulin resistance and can stimulate inflammation, thereby increasing the severity of disease (Abdelmalek et al. 2010).

It is interesting to note that genes play a critical role in the development of NASH and the progression of its severity. Human genome-wide association studies have shown that a variant of patatin-like phospholipase domain containing three genes, PNPLAr3, rs738409, and I148M, is a strong predictor of NASH and steatosis. In the future, the detection of these genes can help us identify individuals who are at high risk for NASH and provide prophylactic care to prevent the onset of the severe irreversible stage of the disease (Romeo et al. 2008; Sookoian and Pirola 2011; Abdelmalek et al. 2010).

The role of inflammasome in the pathogenesis of NASH has also been extensively investigated. Methionine-choline-deficient-induced animal models for NASH were found to have increased activation of AIM2 and NLRP3 via the TLR9-MyD88 pathway (Csak et al. 2014). Upregulation of nlrp3, Asc, and Casp-1 mRNA is seen, but activation of inflammasome is not detected in early stages of the NAFLD animal model. Further investigation of the role of inflammasome molecules in the pathogenesis of ALD and NAFLD is ongoing.

10.6 Hepatitis

Hepatitis refers to inflammation of the liver. This inflammation can be short-lived (acute hepatitis) or persistent and progressive (chronic hepatitis) (Negash and Gale 2015). Although the most common cause of hepatitis is a viral infection, other infections, metabolic disorders, or exposure to toxic substances such as alcohol and drugs can also cause hepatitis (Negash and Gale 2015; Neuman et al. 2017). There are several types of hepatitis viruses, including hepatitis A (HAV), B (HBV), C (HCV), D (HDV), and E (HEV) viruses (Dey and Banerjee 2016). HAV, HBV, and HCV are responsible for most cases of viral hepatitis. HAV is a common cause of acute hepatitis, while HCV infections tend to establish chronic persistent infection. Acute HBV infection is effectively controlled in more than 90% of adults, although it can be chronic persistent after neonatal infection (Shin et al. 2016). Inflammation caused by these chronic infections leads to an increased risk of liver cirrhosis and HCC (Alavi et al. 2016).

Reports on the involvement of inflammasome in HAV infection are lacking, and in-depth study will be required. Inflammatory pathways by hepatitis viruses such as HBV and HCV involve activation of inflammasomes, leading to caspase-1 activation and production of IL-1β and IL-18 (Negash and Gale 2015). It has been reported that HBV and HCV trigger AIM2 and NLRP3 inflammasome activation, respectively (Pan et al. 2016; Farag et al. 2017). The presence of IL-1β during HBV infection has been proposed to be beneficial for antiviral activity by activation-induced cytidine deaminase (AID) (Watashi et al. 2013). Moreover, hepatitis B e antigen (HBeAg) has been shown to inhibit IL-18 activity, and hepatitis B c antigen (HBcAg) has been shown to induce IL-18 expression (Jegaskanda et al. 2014; Manigold et al. 2003). Higher levels of AIM2 inflammasome were reported in patients with acute HBV infection than in those with chronic infection. Moreover, this inflammasome was also linked with an increase in IL-1β and IL-18, which are required for viral clearance (Wu et al. 2013). Evaluation of the activity of NLRC4,

NLRP1, and NLRP3 in chronic HBV infection revealed no significant activity by these inflammasome molecules (Askari et al. 2016). Chronic HBV infection may be a result of reduced inflammasome activity resulting in poor antiviral activity. This may be due to HBeAg preventing activation of NFκB signaling and ROS production, thereby preventing the NLRP3 inflammasome activation and IL-1β production required for viral clearance (Yu et al. 2017). A recent finding revealed the presence of AIM2 in the cytoplasm of hepatocytes, and upon HBV infection, the expression of IL-18 was also increased in vitro. Further, it was also shown that patients with chronic HBV infection had increased expression of AIM2 in hepatocytes compared with controls. Both AIM2 and IL-18 expression has been correlated with infection severity and liver injury (Pan et al. 2016).

HCV infection may be sensed by several pattern recognition receptors including retinoic acid-inducible gene I and TLRs 2, 3, 4, 7, 8, and 9. Stimulation of TLRs by HCV leads to activation of an IFN response (Saha and Szabo 2014; Heim 2013). Chronic hepatitis from HCV infection has been characterized by dysregulated inflammatory cytokine production and impaired T-cell activity (Negash and Gale 2015).

The increase in intrahepatic IL-1β expression has been associated with cirrhosis in patients with chronic HCV infection (Chattergoon et al. 2014). It has been identified that macrophages are the major contributors of IL-1β in response to HCV infection (Negash et al. 2013). An in vitro study showed that human HCV-infected hepatoma cells produced IL-1β by caspase-1-mediated processing, and ROS was partially involved in the induction of inflammasome complex (Burdette et al. 2012). Furthermore, it was also reported that hepatic macrophages phagocytose HCV and induce inflammasome activity, which results in the production of IL-1β. This study also elaborated that compared with Kupffer cells, hepatocytes produce much less IL-1β in response to HCV infection (Negash et al. 2013).

Just like IL-1β, IL-18 is also involved in the pathogenesis of HCV. HCV-infected patients have increased serum levels of IL-18, which has been correlated with liver damage. It was shown that ROS-mediated NLRP3 inflammasome activation by HCV RNA in myeloid cells is the cause for increased plasma levels of IL-1β and IL-18 levels in HCV-infected patients (Chen et al. 2014). The production of IL-18 by monocytes is required for the activation of NK cells and production of an antiviral IFNγ response upon HCV infection. However, the monocyte-derived TNFα and NK cell-derived IFNγ-mediated antiviral response is dampened in patients with chronic HCV infection (Serti et al. 2014).

HCV infection causes hepatosteatosis by the accumulation of lipid droplets in infected cells. It has been shown that inflammasome components such as NLRP3, CARD, and caspase-1 are required for lipid droplet formation. Moreover, inflammasome activation and caspase-1 activity degrade insulin-induced proteins and activated sterol regulatory element-binding protein for the upregulation of lipogenic genes. This study demonstrates a unique role of inflammasome activation in the pathogenesis of HCV infection (McRae et al. 2016).

HDV infection is the most serious form of viral hepatitis (Grabowski and Wedemeyer 2010; Alvarado-Mora et al. 2013). HDV infection, which is usually concurrent with HBV infection, may result in a rapid progression of hepatitis, which can usually be attenuated by IFNα therapy (Negash and Gale 2015). The mechanism of infection and the role of inflammasome in the pathogenesis of this disease need further investigation. Similarly, the pathogenesis of hepatitis caused by HEV needs to be further studied to identify the role of inflammasomes in HEV infection.

10.7 Liver Cirrhosis

Cirrhosis is histologically defined as the replacement of the normal architecture of the liver with regenerating nodules surrounded by fibrous bands. Cirrhosis has various causes, of which viral hepatitis and alcohol have the highest prevalence. The Centers for Disease Control and Prevention (2017) reported that HCV is more common in the United States; 60% to 70% of HCV patients develop chronic liver disease, and 5% to 20% progress to cirrhosis over 20–30 years. Other conditions that can lead to cirrhosis are NAFLD, primary biliary cirrhosis, primary sclerosing cholangitis, hemochromatosis, Wilson's disease, and alpha 1 antitrypsin deficiency. Cirrhosis will progress to decompensated liver failure, with complications such as portal hypertension and HCC. Cirrhosis is also the leading indication for liver transplantation in the United States (Schuppan and Afdhal 2008). Although the causes of the disease are well documented, the exact molecular pathway for the progression from inflammation (hepatitis) to fibrosis (cirrhosis) remains unknown.

Deposition of excess extracellular matrix, type 1 and type 3 collagen is the hallmark of fibrosis (van Dijk et al. 2015; Tacke and Trautwein 2015). HSC, present in the perisinusoidal spaces, plays the key role of initiation, activation, and progression of fibrosis (Josan et al. 2015; Li et al. 2015). Though the primary source of myofibroblasts and fibroblasts is stellate cells, it is believed that bone marrow-derived fibroblasts and circulating mesenchymal cells also contribute to this process (Zhang et al. 2016). Studies have proven the direct role of NLRP3 inflammasome in HSC activation, enhanced by production of α-smooth muscle actin, fibrogenic cytokines like transforming growth factor-β, connective tissue growth factor, as well as extracellular matrix proteins (procollagen 1, TIMP-1) (Watanabe et al. 2009). In addition to NLRP3, ROS and various cytokines produced as a result of inflammasome formation (most importantly IL-1β and IL-18) play a role in the conversion of the quiescent HSC to activated HSC (Basaranoglu et al. 2013). In one study, tissue inhibitors of matrix metalloproteinase (TIMP), which is a marker of HSC activation, were measured after feeding mice with a choline-deficient amino acid-defined diet, which induces steatosis followed by fibrosis. Levels of TIMP increased in wild-type mice, in comparison to a significant decrease in NLRP3 knocked-out mice, thereby proving that a deficiency of NLRP3 protects mice from fibrosis (Wree et al. 2014).

Cell-to-cell interaction helps create a strong inflammatory microenvironment, which increases fibrinogenesis (Ford et al. 2015; Coulouarn et al. 2012). Studies have shown that LPS-stimulated Kupffer cells play a role in fibrogenesis by activating the adjacent HSC via inflammasome-mediated cytokine production (Fallowfield 2011). Jiang et al. (2017) showed that ATP-mediated P2XR stimulation of inflammasome is another way to stimulate the HSC and further proved that blockage of P2XR with A438079 inhibits both the stellate cells and deposition of extracellular matrix. This finding opens the door for new treatment options. It has been postulated that farnesoid receptor stimulation plays an important role in halting the disease process. It has also been clinically proven that the farnesoid receptor agonist obeticholic acid has antifibrotic effects and improves fat-induced cirrhosis (Neuschwander-Tetri et al. 2015; Khalid et al. 2015).

Through animal models, it has been shown that tetramethylpyrazine, a natural product, has inhibitory effects over the inflammatory pathway to fibrosis and can disrupt the NLRP3/caspase-1-mediated production of cytokines IL-1β and IL-18, thereby ameliorating fibrosis. This study was conducted with platelet-derived growth factor, which was a potent stimulator of inflammation and fibrogenesis in HSC (Wu et al. 2015). Currently transplant is the only definitive treatment for cirrhosis, and approved medical interventions are yet to be discovered for this irreversible disease.

10.8 Hepatocellular Carcinoma

HCC is the most common primary malignant tumor of the liver and is the second leading cause of cancer-related death worldwide (Park et al. 2015). More than 80% of HCC cases occur in East Asia and sub-Saharan Africa (>20 per 100,000 population), whereas South and Central America and Northern Europe have a low incidence of HCC (<5 per 100,000 population) (Zhu et al. 2016). Most HCC cases occur in patients with advanced fibrosis and liver cirrhosis, with major risk factors including chronic infection with HBV, HCV, ALD, and NASH (Sanyal et al. 2010; Sasaki et al. 2017). Although the exact mechanism of the oncogenesis and progression of HCC is complicated and remains unknown, inflammation is considered one of the most important factors in HCC development (Bishayee 2014). It has been reported that inflammation stimulates angiogenesis, DNA damage, and malignant tumor cell growth (Dondeti et al. 2016).

Inflammasomes such as NLRP3, NLRC4, and AIM2 have been shown to play an important role via inflammatory pathways in HCC (Fan et al. 2014; Sonohara et al. 2017; Ma et al. 2016). One clinical investigation identified that the expression level of NLRP3 was significantly reduced in the liver tissue of patients with HCC, and the expression inversely correlated with disease severity (Wei et al. 2014). Expression of inflammasome was low in normal hepatocytes, but inflammatory events resulted in upregulation of inflammasome molecules, and significant downregulation was seen in malignant liver cancer tissues. An initial insult to the hepatocytes causes activation of inflammatory signals and also induces formation of the inflammasome complex, which may play a role in the establishment of homeostasis. However, a chronic inflammatory condition and extensive damage result in fibrosis, and compensatory hepatocyte proliferation and liver regeneration are prone to mutations causing dysplasia and tumor development originating from precursor cells lacking inflammasome. HCC patients with downregulated inflammasome molecules are prone to advanced clinical stages and may favor HCC progression (Wei et al. 2014).

Similar to NLRP3 (Wei et al. 2014), estrogen receptor (ER)-β has been shown to be significantly downregulated in liver cancer cells; therefore, the link between ERβ and NLRP3 inflammasome in liver cancer cells was investigated. This study identified a strong correlation between ERβ and NLRP3 expression in the liver. The treatment of HCC cells with 17β-estradiol resulted in inhibition of proliferation, migration, and the colonizing ability of HCC cells, and this effect was reversed when NLRP3 inhibitor was added. The study also demonstrated that only pyroptosis and not apoptosis by NLRP3 inflammasome induced death of HCC cells (Wei et al. 2015). In a contrasting study, it was reported that luteoloside, a naturally occurring flavonoid, suppresses the proliferation of HCC by inhibiting the expression of NLRP3 inflammasome and intracellular ROS accumulation (Fan et al. 2014). Luteoloside was able to significantly inhibit proliferation, migration, invasion, and metastasis of HCC cells both in vitro and in vivo. Expression of NLRP3 inflammasome, caspase-1, and IL-1β was significantly downregulated by luteoloside in HCC cell lines (Fan et al. 2014). The contrasting roles of NLRP3 in the pathogenesis of HCC may be due to the different models used in these studies.

In a research study investigating the anticancer property of poly(amidoamine) (PAMAM) dendrimers, it was demonstrated that autophagy, oxidative stress, and inflammasomes were involved in the cytotoxicity of HCC cell lines. The PAMAM dendrimers induced activation of inflammasomes in HCC cells and may have played a role in cytotoxicity (Li et al. 2014). In a recent study examining the gene expression pattern of noncancerous tissue to determine recurrence of HCC after surgical resection, NLRP3, NLRC4, and AIM2 were overexpressed in noncancerous adjacent tissue compared to controls and HCC tissue. High expression of the inflammasome molecules in the surrounding noncancerous tissue was associated with poor overall survival after tumor resection (Sonohara et al. 2017). All these studies suggest that inflammasome components play a critical role in the pathogenesis of HCC. However, the contrasting reports demand a thorough analysis of the role of inflammasome in the development of the disease. Further understanding will result in proper guidance toward development of therapeutic approaches for HCC.

10.9 Pancreatitis

Acute pancreatitis (AP) is an inflammatory condition of the pancreas, accompanied by abdominal and back pain and elevations of pancreatic enzymes. It is one of the most common gastrointestinal diseases, resulting in 275,000 hospital admissions every year in the United States (Forsmark et al. 2016). The most common causes of AP are gallstones and excessive alcohol consumption (Greenberg et al. 2016). Based on severity, AP ranges from a mild self-limited condition to a life-threatening situation with a high incidence of systemic complications (He et al. 2016). Although the molecular mechanisms of AP are still poorly understood, inflammatory cell infiltration and inflammatory mediators may play a key role (Dong et al. 2016).

Chronic pancreatitis (CP) is a progressive fibroinflammatory disease of the pancreas that alters its normal structure and functions (Wang et al. 2013). Patients with CP usually present with severe abdominal pain and exocrine and endocrine dysfunction (Stram et al. 2016). In the advanced stage of CP, the quality of life of patients is impaired by intractable pain, malabsorption, and diabetes mellitus (Duggan et al. 2014). Alcoholic pancreatitis is the leading cause of CP, followed by idiopathic pancreatitis (Hobbs et al. 2016). It has been reported that NFκB plays a significant role in the pathogenesis of CP, although the downstream mechanisms associated with disease progression are not clearly understood (Kanak et al. 2017).

The pathogenesis of AP and CP is still unclear. However, several molecular mechanisms have been linked to the development of the disease. These include anomalous calcium signaling, oxidative stress, an increase in intracellular ROS, NFκB signaling, endoplasmic reticulum stress, autophagy, and altered intracellular and extracellular pH (Sah et al. 2013). We recently identified that inflammasomes may play a critical role in the development of CP (Kanak et al. 2017). Earlier reports documented the involvement of inflammasomes on AP (Hoque et al. 2011; Ren et al. 2014).

AP is initiated by premature activation of digestive enzymes within pancreatic acinar cells, resulting in self-digestion, cellular inflammation, and damage (Kang et al. 2016; Dong et al. 2016). Intracellular contents released from damaged cells serve as DAMPs, including nuclear DNA, mitochondrial DNA, and ATP (Hoque et al. 2011, 2012). These DAMPs can stimulate TLR4 and TLR9 expressed in pancreatic macrophages to trigger transcriptional pathways, leading to the activation of NFκB and pro-IL-1β transcription. DAMPS can also stimulate plasma membrane purinergic receptor P2X7, cytosolic receptors of NLRs, which mediate IL-1β maturation through inflammasome components (Hoque et al. 2011).

In addition to the NLR family NLRP3 inflammasome, several DAMPs activate AIM2 inflammasome in peripheral blood mononuclear cells (PBMCs) with AP (Algaba-Chueca et al. 2017). Production of IL-1β leads to further cytokine production, recruitment of immune cells, and pyroptosis (Hoque et al. 2011; Tait et al. 2014). It has been reported that antagonists of TLR9 and P2X7 markedly reduced the inflammatory response in AP (Hoque et al. 2011; Perez et al. 2015). Using animal models, it was shown that knocking down key components of the inflammasome pathway, such as NLRP3, ASC, caspase-1, TLR-9, and purinergic receptor P2X7, reduced inflammation and IL-1β expression upon administration of cerulein to induce AP (Hoque et al. 2011).

Previous studies showed that mature IL-18, which is processed by inflammasomes, is required for the development of pancreatitis (Yuan et al. 2007). In addition, an increased level of IL-18 was also noted in the serum of patients suffering from both AP and CP (Hirota et al. 2006; Ueda et al. 2006). The increased level of IL-18 along with TNFα correlated with the severity of pancreatitis (Endo et al. 2001). Treatment with inhibitor of caspase-1 maturation drugs resulted in reduced expression of IL-18 and amelioration of pancreatitis in animal models of pancreatitis (Zhang et al. 2007).

HMGB-1 and genomic DNA are DAMPs involved in the activation of inflammasomes, since these are elevated in patients with AP (Kocsis et al. 2009). Blocking HMGB-1 using pharmacological inhibitors has been shown to reduce necrosis and inflammation in animal models of pancreatitis (Sawa et al. 2006).

Acinar cells are the major source of inflammatory mediators during the onset of pancreatitis. When mice fed with alcohol were treated with LPS, there was an increase in the inflammatory mediators. It is also suggested that IL-18 and caspase-1 may contribute to aggravated acinar cell injury in alcohol-induced pancreatitis (Gu et al. 2013).

Acute injury to the pancreas caused by cerulein administration was reduced when lactate was injected in mice. Lactate has been shown to negatively affect TLR4 signaling and also reduce NLRP3 inflammasome activation and IL-1β production (Hoque et al. 2014). Rat models of cerulein and alcohol-induced pancreatitis had an increased level of myeloperoxidase that may be responsible for the formation of ROS. The study also showed upregulation of ASC and increased expression of caspase-1, which were dampened by treatment with rutin, an anti-inflammatory flavonoid molecule (Aruna et al. 2014).

Another study showed that NLRP3 is required for developing severe AP upon cerulein injection to obese mice. Lean mice developed a mild AP compared to the obese mice. According to the authors, inhibition of NLRP3 using glyburide had no effect on AP development in lean mice but significantly reduced the severity in obese mice. It was reasoned that since the mice are obese, there is already some activity of NLRP3 inflammasome, which is exacerbated by cerulein administration to develop AP (York et al. 2014).

ROS plays a vital role in the activation of NLRP3 inflammasome, and elimination of ROS by antioxidants has alleviated inflammasome activity and prevented IL-1β maturation (Rubartelli 2012). Similarly, the involvement of ROS in the pathogenesis of AP has been reported, and treatment of mice with hydrogen-rich saline to abolish ROS activity was investigated. NLRP3 was overexpressed upon cerulein administration, and treatment with hydrogen-rich saline reduced ROS and NLRP3 activity significantly (Ren et al. 2014).

A more recent study demonstrated that receptor for advanced glycation end products (RAGE) is involved in the activation of AIM2 inflammasome in experimental AP. It was observed that knockout of the AIM2 gene protected mice from experimental AP. Additionally, mice lacking RAGE did not have a strong inflammatory response during induction of experimental AP (Kang et al. 2016). This study was supported by clinical evidence that showed that patients with new-onset AP demonstrated increased expression of inflammasome molecules, including AIM2, in the PBMC. Production of IL-1β and IL-18 by PBMCs was increased in AP patients compared to healthy controls. AIM2 may be responsible for causing systemic inflammation, as patients suffering from transient or persistent organ failure due to AP demonstrated further elevation of AIM2 expression, showing a strong correlation of AIM2 expression with severity of disease (Algaba-Chueca et al. 2017).

In a recent study from our group, we showed for the first time that inflammasome may be involved in the pathogenesis of CP (Kanak et al. 2017). We also demonstrated that inhibition of NFκB using the small molecule inhibitor WA significantly downregulated inflammasome genes and reduced the severity of pancreatitis. It is suggested that due to chronic inflammation, immune cells infiltrating the pancreas are activated via TLR signaling by binding of DAMPs such as HMGB-1, genomic DNA, or ATP released by damaged acinar cells (Fig. 10.2). This may result in assembly and activation of inflammasomes that cause further damage by production of IL-1β and IL-18. An analytical study showed serum HMGB-1 to be increased in patients with pancreatitis, with a strong correlation with disease progression (Lin et al. 2015). Upon cerulein administration, we observed significant upregulation of HMGB-1, which was inhibited by WA treatment. Messenger RNA levels of key inflammasome components NLRP3, ASC, IL-1β, and IL-18 were also significantly upregulated, suggesting increased inflammasome signaling in cerulein-induced CP. WA was able to strongly inhibit the expression levels of all inflammasome molecules. Given the evidence, it is believed that the NLRP3 inflammasome may be partly involved in the pathogenesis of CP, but the extent of involvement needs to be investigated (Kanak et al. 2017). Further investigation is needed to determine the extent of NLRP3 inflammasome activity involved in the pathophysiology of CP.

A comparative study of pancreatic cancer, periodontitis, and CP investigated the polymorphisms in the NLRP3 and NLRP2 inflammasomes. This study revealed that F359L polymorphism in the NLRP2 inflammasome was higher in CP patients (Miskiewicz et al. 2015). These studies show evidence of the involvement of inflammasomes in the pathogenesis of CP, but further investigation may be needed to evaluate the role of inflammasomes in the development of the disease. These studies may pave the way for novel therapeutic strategies to treat CP.

10.10 Pancreatic Cancer

Pancreatic cancer represents the fourth leading cause of cancer-related death in the United States (Ko 2016). The prognosis for patients with pancreatic cancer is extremely poor, with an overall 5-year survival of only 6% (Yamamoto et al. 2015). The main reason is that pancreatic cancer is an aggressive malignancy and difficult to diagnose at an early stage. Smoking, positive family history and genetics, diabetes mellitus, obesity, dietary factors, alcohol use, and physical inactivity are considered to be risk factors (Ilic and Ilic 2016). Pancreatic cancers are usually associated with somatic mutation of the *KRAS* oncogene and inactivation of tumor suppressor genes such as *TP53*, *CDKN2A*, and *SMAD4* (Huang et al. 2015). Although a combination of genetic background and environmental factors is needed for the development of pancreatic cancer, chronic inflammation such as pancreatitis is also considered to be a major risk factor (Chang et al. 2016; Zambirinis et al. 2014).

IL-1β, which is often detected in pancreatic cancer, is a proinflammatory cytokine and has various roles such as cell growth, differentiation, tissue repair, and regulation of immune response (Okamoto et al. 2010). IL-1β is associated with the

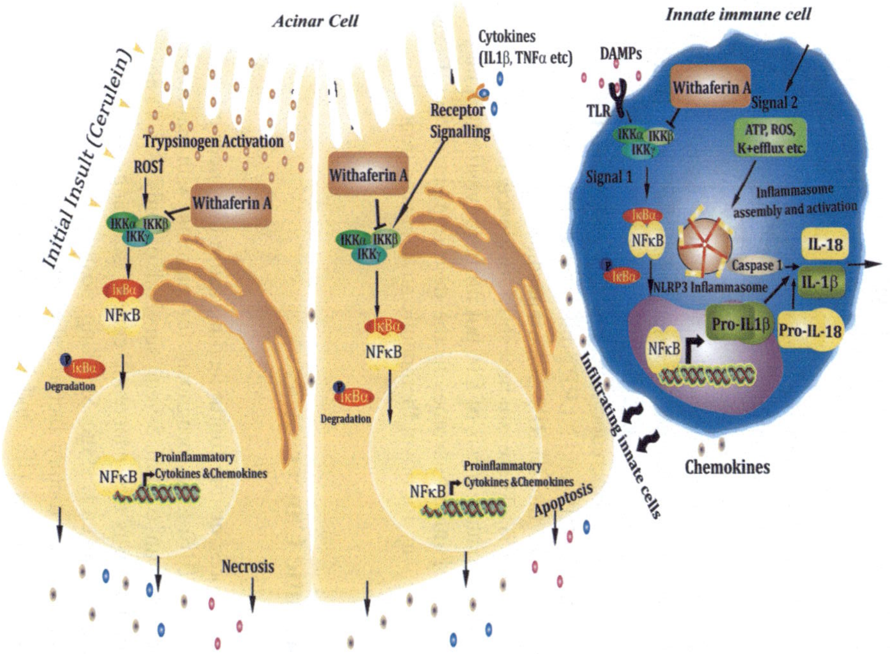

Fig. 10.2 Schematic representation of the signaling pathways operative in the incidence of chronic pancreatitis and the blockage by Withaferin A

invasiveness, metastasis, and chemoresistance of pancreatic cancer (Okamoto et al. 2010; Muerkoster et al. 2006). Intracellular contents released by dead and dying tumor cells or present in the tumor microenvironment are considered to activate NFκB and inflammasome, which induce activation and release of proinflammatory cytokines (De Monte et al. 2011). A recent study reported that polymorphism in the NLRP3 gene *Q705K* is more prevalent in patients with pancreatic cancer (Miskiewicz et al. 2015). Carbone et al. (2009) reported that increased serum levels of cytokine IL-18 processed by the inflammasome complexes were associated with poor survival in patients with pancreatic carcinoma. The authors reported an increased level of precursor IL-18 with no functional properties, which was produced and secreted by pancreatic tumors. However, after treatment of pancreatic cancer cells with 5-fluorouracil, which is a polyadjuvant chemotherapy used to reduce malignancy, it was shown that mature forms of IL-18 were increased in the serum of patients as a result of caspase-dependent processing of pro-IL-18 (Carbone et al. 2005). This study described the involvement of IL-18 but did not address the mechanism of caspase activity and IL-18 processing. A thorough analysis of the role of inflammasome in the progression and malignancy of pancreatic cancer is needed.

10.11 Conclusion

Studies over the past decade have highlighted the role of various inflammasomes in gastrointestinal diseases. Inflammasome-mediated cellular processes are important during microbial infections and also in regulation of metabolic processes and immune responses. In most of the gastrointestinal disease including cancer, the evidence linking the involvement of inflammasomes in disease development in vivo remains preliminary and awaits further confirmation. It is important to clearly understand the balance between beneficial and detrimental inflammasome activation. This knowledge is critical to the development of novel therapeutic strategies including the development of small molecules that directly target inflammasome components rather than inhibitors of proinflammatory cytokines.

References

Abdelmalek MF, Suzuki A, Guy C, Unalp-Arida A, Colvin R, Johnson RJ, Diehl AM, Network NSCR (2010) Increased fructose consumption is associated with fibrosis severity in patients with nonalcoholic fatty liver disease. Hepatology 51(6):1961–1971. https://doi.org/10.1002/hep.23535

Abderrazak A, Couchie D, Mahmood DF, Elhage R, Vindis C, Laffargue M, Mateo V, Buchele B, Ayala MR, El Gaafary M, Syrovets T, Slimane MN, Friguet B, Fulop T, Simmet T, El Hadri K, Rouis M (2015) Anti-inflammatory and antiatherogenic effects of the NLRP3 inflammasome inhibitor arglabin in ApoE2.Ki mice fed a high-fat diet. Circulation 131(12):1061–1070. https://doi.org/10.1161/CIRCULATIONAHA.114.013730

Alavi M, Law MG, Grebely J, Amin J, Hajarizadeh B, George J, Dore GJ (2016) Time to decompensated cirrhosis and hepatocellular carcinoma after an HBV or HCV notification: a population-based study. J Hepatol 65(5):879–887. https://doi.org/10.1016/j.jhep.2016.06.025

Algaba-Chueca F, de-Madaria E, Lozano-Ruiz B, Martinez-Cardona C, Quesada-Vazquez N, Bachiller V, Tarin F, Such J, Frances R, Zapater P, Gonzalez-Navajas JM (2017) The expression and activation of the AIM2 inflammasome correlates with inflammation and disease severity in patients with acute pancreatitis. Pancreatology. https://doi.org/10.1016/j.pan.2017.03.006

Allen IC, TeKippe EM, Woodford RM, Uronis JM, Holl EK, Rogers AB, Herfarth HH, Jobin C, Ting JP (2010) The NLRP3 inflammasome functions as a negative regulator of tumorigenesis during colitis-associated cancer. J Exp Med 207(5):1045–1056. https://doi.org/10.1084/jem.20100050

Alvarado-Mora MV, Locarnini S, Rizzetto M, Pinho JR (2013) An update on HDV: virology, pathogenesis and treatment. Antivir Ther 18(3 Pt B):541–548. https://doi.org/10.3851/IMP2598

Andersen NN, Jess T (2013) Has the risk of colorectal cancer in inflammatory bowel disease decreased? World J Gastroenterol 19(43):7561–7568. https://doi.org/10.3748/wjg.v19.i43.7561

Aruna R, Geetha A, Suguna P (2014) Rutin modulates ASC expression in NLRP3 inflammasome: a study in alcohol and cerulein-induced rat model of pancreatitis. Mol Cell Biochem 396 (1–2):269–280. https://doi.org/10.1007/s11010-014-2162-8

Askari A, Nosratabadi R, Khaleghinia M, Zainodini N, Kennedy D, Shabani Z, Kazemi Arababadi M (2016) Evaluation of NLRC4, NLRP1, and NLRP3, as components of inflammasomes, in chronic hepatitis B virus–infected patients. Viral Immunol 29(9):496–501. https://doi.org/10.1089/vim.2016.0045

Backert S, Tegtmeyer N (2017) Type IV secretion and signal transduction of *Helicobacter pylori* CagA through interactions with host cell receptors. *Toxins* (Basel) 9(4). https://doi.org/10.3390/toxins9040115

Bafna S, Kaur S, Batra SK (2010) Membrane-bound mucins: the mechanistic basis for alterations in the growth and survival of cancer cells. *Oncogene* 29(20):2893–2904. https://doi.org/10.1038/onc.2010.87

Basaranoglu M, Basaranoglu G, Senturk H (2013) From fatty liver to fibrosis: a tale of "second hit". World J Gastroenterol 19(8):1158–1165. https://doi.org/10.3748/wjg.v19.i8.1158

Bauer C, Duewell P, Mayer C, Lehr HA, Fitzgerald KA, Dauer M, Tschopp J, Endres S, Latz E, Schnurr M (2010) Colitis induced in mice with dextran sulfate sodium (DSS) is mediated by the NLRP3 inflammasome. Gut 59(9):1192–1199. https://doi.org/10.1136/gut.2009.197822

Bauer C, Duewell P, Lehr HA, Endres S, Schnurr M (2012) Protective and aggravating effects of Nlrp3 inflammasome activation in IBD models: influence of genetic and environmental factors. Dig Dis 30(Suppl 1):82–90. https://doi.org/10.1159/000341681

Bellentani S (2017) The epidemiology of non-alcoholic fatty liver disease. Liver Int 37(Suppl 1:81–84. https://doi.org/10.1111/liv.13299

Bishayee A (2014) The role of inflammation and liver cancer. Adv Exp Med Biol 816:401–435. https://doi.org/10.1007/978-3-0348-0837-8_16

Burdette D, Haskett A, Presser L, McRae S, Iqbal J, Waris G (2012) Hepatitis C virus activates interleukin-1beta via caspase-1-inflammasome complex. J Gen Virol 93(Pt 2):235–246. https://doi.org/10.1099/vir.0.034033-0

Buzzetti E, Pinzani M, Tsochatzis EA (2016) The multiple-hit pathogenesis of non-alcoholic fatty liver disease (NAFLD). Metabolism 65(8):1038–1048. https://doi.org/10.1016/j.metabol.2015.12.012

Carbone A, Rodeck U, Mauri FA, Sozzi M, Gaspari F, Smirne C, Prati A, Addeo A, Novarino A, Robecchi A, Bertetto O, Emanuelli G, Bellone G (2005) Human pancreatic carcinoma cells secrete bioactive interleukin-18 after treatment with 5-fluorouracil: implications for anti-tumor immune response. Cancer Biol Ther 4(2):231–241

Carbone A, Vizio B, Novarino A, Mauri FA, Geuna M, Robino C, Brondino G, Prati A, Giacobino A, Campra D, Chiarle R, Fronda GR, Ciuffreda L, Bellone G (2009) IL-18 paradox in pancreatic carcinoma: elevated serum levels of free IL-18 are correlated with poor survival. J Immunother 32(9):920–931. https://doi.org/10.1097/CJI.0b013e3181b29168

Centers for Disease Control and Prevention (2017) Hepatitis C FAQs for health professionals. https://www.cdc.gov/hepatitis/hcv/hcvfaq.htm. Accessed 23 May 2017

Chalasani N, Younossi Z, Lavine JE, Diehl AM, Brunt EM, Cusi K, Charlton M, Sanyal AJ (2012) The diagnosis and management of non-alcoholic fatty liver disease: practice guideline by the American Association for the Study of Liver Diseases, American College of Gastroenterology, and the American Gastroenterological Association. Hepatology 55(6):2005–2023. https://doi.org/10.1002/hep.25762

Chang JH, Jiang Y, Pillarisetty VG (2016) Role of immune cells in pancreatic cancer from bench to clinical application: an updated review. Medicine 95(49):e5541. https://doi.org/10.1097/md.0000000000005541

Chattergoon MA, Latanich R, Quinn J, Winter ME, Buckheit RW 3rd, Blankson JN, Pardoll D, Cox AL (2014) HIV and HCV activate the inflammasome in monocytes and macrophages via endosomal toll-like receptors without induction of type 1 interferon. PLoS Pathog 10(5): e1004082. https://doi.org/10.1371/journal.ppat.1004082

Chen GY, Nunez G (2010) Sterile inflammation: sensing and reacting to damage. Nat Rev Immunol 10(12):826–837. https://doi.org/10.1038/nri2873

Chen GY, Nunez G (2011) Inflammasomes in intestinal inflammation and cancer. Gastroenterology 141(6):1986–1999. https://doi.org/10.1053/j.gastro.2011.10.002

Chen GY, Liu M, Wang F, Bertin J, Nunez G (2011) A functional role for Nlrp6 in intestinal inflammation and tumorigenesis. J Immunol 186(12):7187–7194. https://doi.org/10.4049/jimmunol.1100412

Chen W, Xu Y, Li H, Tao W, Xiang Y, Huang B, Niu J, Zhong J, Meng G (2014) HCV genomic RNA activates the NLRP3 inflammasome in human myeloid cells. PLoS One 9(1):e84953. https://doi.org/10.1371/journal.pone.0084953

Choubey D (2016) Absent in melanoma 2 proteins in the development of cancer. Cell Mol Life Sci 73(23):4383–4395. https://doi.org/10.1007/s00018-016-2296-9

Coulouarn C, Corlu A, Glaise D, Guenon I, Thorgeirsson SS, Clement B (2012) Hepatocyte-stellate cell cross-talk in the liver engenders a permissive inflammatory microenvironment that drives progression in hepatocellular carcinoma. Cancer Res 72(10):2533–2542. https://doi.org/10.1158/0008-5472.CAN-11-3317

Cox LM, Yamanishi S, Sohn J, Alekseyenko AV, Leung JM, Cho I, Kim SG, Li H, Gao Z, Mahana D, Zarate Rodriguez JG, Rogers AB, Robine N, Loke P, Blaser MJ (2014) Altering the intestinal microbiota during a critical developmental window has lasting metabolic consequences. Cell 158(4):705–721. https://doi.org/10.1016/j.cell.2014.05.052

Csak T, Ganz M, Pespisa J, Kodys K, Dolganiuc A, Szabo G (2011) Fatty acid and endotoxin activate inflammasomes in mouse hepatocytes that release danger signals to stimulate immune cells. Hepatology 54(1):133–144. https://doi.org/10.1002/hep.24341

Csak T, Pillai A, Ganz M, Lippai D, Petrasek J, Park JK, Kodys K, Dolganiuc A, Kurt-Jones EA, Szabo G (2014) Both bone marrow–derived and non–bone marrow-derived cells contribute to AIM2 and NLRP3 inflammasome activation in a MyD88-dependent manner in dietary steatohepatitis. Liver Int 34(9):1402–1413. https://doi.org/10.1111/liv.12537

De Monte L, Reni M, Tassi E, Clavenna D, Papa I, Recalde H, Braga M, Di Carlo V, Doglioni C, Protti MP (2011) Intratumor T helper type 2 cell infiltrate correlates with cancer-associated fibroblast thymic stromal lymphopoietin production and reduced survival in pancreatic cancer. J Exp Med 208(3):469–478. https://doi.org/10.1084/jem.20101876

de Souza HS, Fiocchi C (2016) Immunopathogenesis of IBD: current state of the art. Nat Rev Gastroenterol Hepatol 13(1):13–27. https://doi.org/10.1038/nrgastro.2015.186

Dey D, Banerjee M (2016) Inhibitor-based therapeutics for treatment of viral hepatitis. J Clin Transl Hepatol 4(3):248–257. https://doi.org/10.14218/jcth.2016.00025

Dihlmann S, Tao S, Echterdiek F, Herpel E, Jansen L, Chang-Claude J, Brenner H, Hoffmeister M, Kloor M (2014) Lack of absent in melanoma 2 (AIM2) expression in tumor cells is closely associated with poor survival in colorectal cancer patients. Int J Cancer 135(10):2387–2396. https://doi.org/10.1002/ijc.28891

Dixon LJ, Berk M, Thapaliya S, Papouchado BG, Feldstein AE (2012) Caspase-1-mediated regulation of fibrogenesis in diet-induced steatohepatitis. Lab Investig 92(5):713–723. https://doi.org/10.1038/labinvest.2012.45

Dixon LJ, Flask CA, Papouchado BG, Feldstein AE, Nagy LE (2013) Caspase-1 as a central regulator of high fat diet–induced non-alcoholic steatohepatitis. PLoS One 8(2):e56100. https://doi.org/10.1371/journal.pone.0056100

Dondeti MF, El-Maadawy EA, Talaat RM (2016) Hepatitis-related hepatocellular carcinoma: insights into cytokine gene polymorphisms. World J Gastroenterol 22(30):6800–6816. https://doi.org/10.3748/wjg.v22.i30.6800

Dong Z, Shang H, Chen YQ, Pan LL, Bhatia M, Sun J (2016) Sulforaphane protects pancreatic acinar cell injury by modulating Nrf2-mediated oxidative stress and NLRP3 inflammatory pathway. Oxidative Med Cell Longev 2016:7864150. https://doi.org/10.1155/2016/7864150

Duggan SN, Smyth ND, Murphy A, Macnaughton D, O'Keefe SJ, Conlon KC (2014) High prevalence of osteoporosis in patients with chronic pancreatitis: a systematic review and meta-analysis. Clin Gastroenterol Hepatol 12(2):219–228. https://doi.org/10.1016/j.cgh.2013.06.016

Dupaul-Chicoine J, Yeretssian G, Doiron K, Bergstrom KS, McIntire CR, LeBlanc PM, Meunier C, Turbide C, Gros P, Beauchemin N, Vallance BA, Saleh M (2010) Control of intestinal homeostasis, colitis, and colitis-associated colorectal cancer by the inflammatory caspases. Immunity 32(3):367–378. https://doi.org/10.1016/j.immuni.2010.02.012

El-Omar EM, Carrington M, Chow WH, McColl KE, Bream JH, Young HA, Herrera J, Lissowska J, Yuan CC, Rothman N, Lanyon G, Martin M, Fraumeni JF Jr, Rabkin CS (2000) Interleukin-1 polymorphisms associated with increased risk of gastric cancer. Nature 404 (6776):398–402. https://doi.org/10.1038/35006081

Endo S, Inoue Y, Fujino Y, Wakabayashi G, Inada K, Sato S (2001) Interleukin 18 levels reflect the severity of acute pancreatitis. Res Commun Mol Pathol Pharmacol 110(5–6):285–291

Eusebi LH, Zagari RM, Bazzoli F (2014) Epidemiology of *Helicobacter pylori* infection. Helicobacter 19(Suppl 1):1–5. https://doi.org/10.1111/hel.12165

Fallowfield JA (2011) Therapeutic targets in liver fibrosis. Am J Physiol Gastrointest Liver Physiol 300(5):G709–G715. https://doi.org/10.1152/ajpgi.00451.2010

Fan SH, Wang YY, Lu J, Zheng YL, Wu DM, Li MQ, Hu B, Zhang ZF, Cheng W, Shan Q (2014) Luteoloside suppresses proliferation and metastasis of hepatocellular carcinoma cells by inhibition of NLRP3 inflammasome. PLoS One 9(2):e89961. https://doi.org/10.1371/journal.pone.0089961

Fan H, Zhu JH, Yao XQ (2016) Prognostic significance of B7-H3 expression in patients with colorectal cancer: a meta-analysis. Pak J Med Sci 32(6):1568–1573. https://doi.org/10.12669/pjms.326.11511

Farag NS, Breitinger U, El-Azizi M, Breitinger HG (2017) The p7 viroporin of the hepatitis C virus contributes to liver inflammation by stimulating production of interleukin-1beta. Biochim Biophys Acta 1863(3):712–720. https://doi.org/10.1016/j.bbadis.2016.12.006

Federico A, Dallio M, Godos J, Loguercio C, Salomone F (2016) Targeting gut-liver axis for the treatment of nonalcoholic steatohepatitis: translational and clinical evidence. Transl Res 167 (1):116–124. https://doi.org/10.1016/j.trsl.2015.08.002

Ford AJ, Jain G, Rajagopalan P (2015) Designing a fibrotic microenvironment to investigate changes in human liver sinusoidal endothelial cell function. Acta Biomater 24:220–227. https://doi.org/10.1016/j.actbio.2015.06.028

Forsmark CE, Vege SS, Wilcox CM (2016) Acute pancreatitis. N Engl J Med 375(20):1972–1981. https://doi.org/10.1056/NEJMra1505202

Fox JG, Wang TC, Rogers AB, Poutahidis T, Ge Z, Taylor N, Dangler CA, Israel DA, Krishna U, Gaus K, Peek RM Jr (2003) Host and microbial constituents influence *Helicobacter pylori*–induced cancer in a murine model of hypergastrinemia. Gastroenterology 124(7):1879–1890

Friedman SL (2007) A deer in the headlights: BAMBI meets liver fibrosis. Nat Med 1 (11):1281–1282. https://doi.org/10.1038/nm1107-1281

Ge X, Antoine DJ, Lu Y, Arriazu E, Leung TM, Klepper AL, Branch AD, Fiel MI, Nieto N (2014) High mobility group box-1 (HMGB1) participates in the pathogenesis of alcoholic liver disease (ALD). J Biol Chem 289(33):22672–22691. https://doi.org/10.1074/jbc.M114.552141

Glocker EO, Kotlarz D, Boztug K, Gertz EM, Schaffer AA, Noyan F, Perro M, Diestelhorst J, Allroth A, Murugan D, Hatscher N, Pfeifer D, Sykora KW, Sauer M, Kreipe H, Lacher M, Nustede R, Woellner C, Baumann U, Salzer U, Koletzko S, Shah N, Segal AW, Sauerbrey A, Buderus S, Snapper SB, Grimbacher B, Klein C (2009) Inflammatory bowel disease and mutations affecting the interleukin-10 receptor. N Engl J Med 361(21):2033–2045. https://doi.org/10.1056/NEJMoa0907206

Gobert AP, Wilson KT (2017) Human and *Helicobacter pylori* interactions determine the outcome of gastric diseases. Curr Top Microbiol Immunol 400:27–52. https://doi.org/10.1007/978-3-319-50520-6_2

Grabowski J, Wedemeyer H (2010) Hepatitis delta: immunopathogenesis and clinical challenges. Dig Dis 28(1):133–138. https://doi.org/10.1159/000282076

Greenberg JA, Hsu J, Bawazeer M, Marshall J, Friedrich JO, Nathens A, Coburn N, May GR, Pearsall E, McLeod RS (2016) Clinical practice guideline: management of acute pancreatitis. Can J Surg 59(2):128–140

Gu H, Werner J, Bergmann F, Whitcomb DC, Buchler MW, Fortunato F (2013) Necro-inflammatory response of pancreatic acinar cells in the pathogenesis of acute alcoholic pancreatitis. Cell Death Dis 4:e816. https://doi.org/10.1038/cddis.2013.354

Guo H, Callaway JB, Ting JP (2015) Inflammasomes: mechanism of action, role in disease, and therapeutics. Nat Med 21(7):677–687. https://doi.org/10.1038/nm.3893

Hahm ER, Singh SV (2013) Withaferin A-induced apoptosis in human breast cancer cells is associated with suppression of inhibitor of apoptosis family protein expression. Cancer Lett 334(1):101–108. https://doi.org/10.1016/j.canlet.2012.08.026

He Z, Hua J, Qian D, Gong J, Lin S, Xu C, Wei G, Meng H, Yang T, Zhou B, Song Z (2016) Intravenous hMSCs ameliorate acute pancreatitis in mice via secretion of tumor necrosis factor-alpha stimulated gene/protein 6. Sci Rep 6:38438. https://doi.org/10.1038/srep38438

Heim MH (2013) Innate immunity and HCV. J Hepatol 58(3):564–574. https://doi.org/10.1016/j.jhep.2012.10.005

Henao-Mejia J, Elinav E, Jin C, Hao L, Mehal WZ, Strowig T, Thaiss CA, Kau AL, Eisenbarth SC, Jurczak MJ, Camporez JP, Shulman GI, Gordon JI, Hoffman HM, Flavell RA (2012) Inflammasome-mediated dysbiosis regulates progression of NAFLD and obesity. Nature 482 (7384):179–185. https://doi.org/10.1038/nature10809

Hirota M, Shibata M, Baba H (2006) Significance of elevated serum IL-18 levels in patients with acute pancreatitis. J Gastroenterol 41(2):182–183. https://doi.org/10.1007/s00535-006-1769-2

Hirota SA, Ng J, Lueng A, Khajah M, Parhar K, Li Y, Lam V, Potentier MS, Ng K, Bawa M, McCafferty DM, Rioux KP, Ghosh S, Xavier RJ, Colgan SP, Tschopp J, Muruve D, MacDonald JA, Beck PL (2011) NLRP3 inflammasome plays a key role in the regulation of intestinal homeostasis. Inflamm Bowel Dis 17(6):1359–1372. https://doi.org/10.1002/ibd.21478

Hobbs PM, Johnson WG, Graham DY (2016) Management of pain in chronic pancreatitis with emphasis on exogenous pancreatic enzymes. World J Gastrointest Pharmacol Ther 7 (3):370–386. https://doi.org/10.4292/wjgpt.v7.i3.370

Hoek JB, Cahill A, Pastorino JG (2002) Alcohol and mitochondria: a dysfunctional relationship. Gastroenterology 122(7):2049–2063

Honda H, Nagai Y, Matsunaga T, Okamoto N, Watanabe Y, Tsuneyama K, Hayashi H, Fujii I, Ikutani M, Hirai Y, Muraguchi A, Takatsu K (2014) Isoliquiritigenin is a potent inhibitor of NLRP3 inflammasome activation and diet-induced adipose tissue inflammation. J Leukoc Biol 96(6):1087–1100. https://doi.org/10.1189/jlb.3A0114-005RR

Hoque R, Sohail M, Malik A, Sarwar S, Luo Y, Shah A, Barrat F, Flavell R, Gorelick F, Husain S, Mehal W (2011) TLR9 and the NLRP3 inflammasome link acinar cell death with inflammation in acute pancreatitis. Gastroenterology 141(1):358–369. https://doi.org/10.1053/j.gastro.2011.03.041

Hoque R, Malik AF, Gorelick F, Mehal WZ (2012) Sterile inflammatory response in acute pancreatitis. Pancreas 41(3):353–357. https://doi.org/10.1097/MPA.0b013e3182321500

Hoque R, Farooq A, Ghani A, Gorelick F, Mehal WZ (2014) Lactate reduces liver and pancreatic injury in Toll-like receptor- and inflammasome-mediated inflammation via GPR81-mediated suppression of innate immunity. Gastroenterology 146(7):1763–1774. https://doi.org/10.1053/j.gastro.2014.03.014

Hu B, Elinav E, Huber S, Booth CJ, Strowig T, Jin C, Eisenbarth SC, Flavell RA (2010) Inflammation-induced tumorigenesis in the colon is regulated by caspase-1 and NLRC4. Proc Natl Acad Sci U S A 107(50):21635–21640. https://doi.org/10.1073/pnas.1016814108

Hu B, Elinav E, Huber S, Strowig T, Hao L, Hafemann A, Jin C, Wunderlich C, Wunderlich T, Eisenbarth SC, Flavell RA (2013) Microbiota-induced activation of epithelial IL-6 signaling links inflammasome-driven inflammation with transmissible cancer. Proc Natl Acad Sci U S A 110(24):9862–9867. https://doi.org/10.1073/pnas.1307575110

Huang J, Lohr JM, Nilsson M, Segersvard R, Matsson H, Verbeke C, Heuchel R, Kere J, Iafrate AJ, Zheng Z, Ye W (2015) Variant profiling of candidate genes in pancreatic ductal adenocarcinoma. Clin Chem 61(11):1408–1416. https://doi.org/10.1373/clinchem.2015.238543

Ilic M, Ilic I (2016) Epidemiology of pancreatic cancer. World J Gastroenterol 22(44):9694–9705. https://doi.org/10.3748/wjg.v22.i44.9694

Inokuchi S, Tsukamoto H, Park E, Liu ZX, Brenner DA, Seki E (2011) Toll-like receptor 4 mediates alcohol-induced steatohepatitis through bone marrow-derived and endogenous liver cells in mice. Alcohol Clin Exp Res 35(8):1509–1518. https://doi.org/10.1111/j.1530-0277.2011.01487.x

Ioannou GN, Van Rooyen DM, Savard C, Haigh WG, Yeh MM, Teoh NC, Farrell GC (2015) Cholesterol-lowering drugs cause dissolution of cholesterol crystals and disperse Kupffer cell crown-like structures during resolution of NASH. J Lipid Res 56(2):277–285. https://doi.org/10.1194/jlr.M053785

Isakov E, Weisman-Shomer P, Benhar M (2014) Suppression of the pro-inflammatory NLRP3/interleukin-1beta pathway in macrophages by the thioredoxin reductase inhibitor auranofin. Biochim Biophys Acta 1840(10):3153–3161. https://doi.org/10.1016/j.bbagen.2014.07.012

Janakiram NB, Rao CV (2014) The role of inflammation in colon cancer. Adv Exp Med Biol 816:25–52. https://doi.org/10.1007/978-3-0348-0837-8_2

Jegaskanda S, Ahn SH, Skinner N, Thompson AJ, Ngyuen T, Holmes J, De Rose R, Navis M, Winnall WR, Kramski M, Bernardi G, Bayliss J, Colledge D, Sozzi V, Visvanathan K, Locarnini SA, Kent SJ, Revill PA (2014) Downregulation of interleukin-18–mediated cell signaling and interferon gamma expression by the hepatitis B virus E antigen. J Virol 88(18):10412–10420. https://doi.org/10.1128/JVI.00111-14

Jiang S, Zhang Y, Zheng JH, Li X, Yao YL, Wu YL, Song SZ, Sun P, Nan JX, Lian LH (2017) Potentiation of hepatic stellate cell activation by extracellular ATP is dependent on P2X7R-mediated NLRP3 inflammasome activation. Pharmacol Res 117:82–93. https://doi.org/10.1016/j.phrs.2016.11.040

Josan S, Billingsley K, Orduna J, Park JM, Luong R, Yu L, Hurd R, Pfefferbaum A, Spielman D, Mayer D (2015) Assessing inflammatory liver injury in an acute CCl4 model using dynamic 3D metabolic imaging of hyperpolarized [1-(13)C]pyruvate. NMR Biomed 28(12):1671–1677. https://doi.org/10.1002/nbm.3431

Kameoka S, Kameyama T, Hayashi T, Sato S, Ohnishi N, Hayashi T, Murata-Kamiya N, Higashi H, Hatakeyama M, Takaoka A (2016) *Helicobacter pylori* induces IL-1beta protein through the inflammasome activation in differentiated macrophagic cells. Biomed Res 37(1):21–27. https://doi.org/10.2220/biomedres.37.21

Kanak MA, Shahbazov R, Yoshimatsu G, Levy MF, Lawrence MC, Naziruddin B (2017) A small molecule inhibitor of NFkappaB blocks ER stress and the NLRP3 inflammasome and prevents progression of pancreatitis. J Gastroenterol 52(3):352–365. https://doi.org/10.1007/s00535-016-1238-5

Kang R, Chen R, Xie M, Cao L, Lotze MT, Tang D, Zeh HJ 3rd (2016) The receptor for advanced glycation end products activates the AIM2 inflammasome in acute pancreatitis. J Immunol 196(10):4331–4337. https://doi.org/10.4049/jimmunol.1502340

Karimi P, Islami F, Anandasabapathy S, Freedman ND, Kamangar F (2014) Gastric cancer: descriptive epidemiology, risk factors, screening, and prevention. Cancer Epidemiol Biomark Prev 23(5):700–713. https://doi.org/10.1158/1055-9965.epi-13-1057

Karki R, Man SM, Kanneganti TD (2017) Inflammasomes and cancer. Cancer Immunol Res 5 (2):94–99. https://doi.org/10.1158/2326-6066.cir-16-0269

Khalid Q, Bailey I, Patel VB (2015) Non-alcoholic fatty liver disease: the effect of bile acids and farnesoid X receptor agonists on pathophysiology and treatment. Liver Res 1(2):32–40. https://doi.org/10.17140/LROJ-1-106

Kim DJ, Park JH, Franchi L, Backert S, Nunez G (2013) The Cag pathogenicity island and interaction between TLR2/NOD2 and NLRP3 regulate IL-1beta production in *Helicobacter pylori* infected dendritic cells. Eur J Immunol 43(10):2650–2658. https://doi.org/10.1002/eji.201243281

Kim JE, Lee JY, Kang MJ, Jeong YJ, Choi JA, Oh SM, Lee KB, Park JH (2015) Withaferin A inhibits *Helicobacter pylori*–induced production of IL-1beta in dendritic cells by regulating NF-kappaB and NLRP3 inflammasome activation. Immune Netw 15(6):269–277. https://doi.org/10.4110/in.2015.15.6.269

Ko AH (2016) Nanomedicine developments in the treatment of metastatic pancreatic cancer: focus on nanoliposomal irinotecan. Int J Nanomedicine 11:1225–1235. https://doi.org/10.2147/ijn.s88084

Koch KN, Muller A (2015) *Helicobacter pylori* activates the TLR2/NLRP3/caspase-1/IL-18 axis to induce regulatory T-cells, establish persistent infection and promote tolerance to allergens. Gut Microbes 6(6):382–387. https://doi.org/10.1080/19490976.2015.1105427

Kocsis AK, Szabolcs A, Hofner P, Takacs T, Farkas G, Boda K, Mandi Y (2009) Plasma concentrations of high-mobility group box protein 1, soluble receptor for advanced glycation end-products and circulating DNA in patients with acute pancreatitis. Pancreatology 9 (4):383–391. https://doi.org/10.1159/000181172

Kusters JG, van Vliet AH, Kuipers EJ (2006) Pathogenesis of *Helicobacter pylori* infection. Clin Microbiol Rev 19(3):449–490. https://doi.org/10.1128/CMR.00054-05

Lasry A, Zinger A, Ben-Neriah Y (2016) Inflammatory networks underlying colorectal cancer. Nat Immunol 17(3):230–240. https://doi.org/10.1038/ni.3384

Latz E (2010) The inflammasomes: mechanisms of activation and function. Curr Opin Immunol 22 (1):28–33. https://doi.org/10.1016/j.coi.2009.12.004

Lei A, Maloy KJ (2016) Colon cancer in the land of NOD: NLRX1 as an intrinsic tumor suppressor. Trends Immunol 37(9):569–570. https://doi.org/10.1016/j.it.2016.07.004

Li Y, Wang S, Wang Z, Qian X, Fan J, Zeng X, Sun Y, Song P, Feng M, Ju D (2014) Cationic poly (amidoamine) dendrimers induced cyto-protective autophagy in hepatocellular carcinoma cells. Nanotechnology 25(36):365101. https://doi.org/10.1088/0957-4484/25/36/365101

Li D, He L, Guo H, Chen H, Shan H (2015) Targeting activated hepatic stellate cells (aHSCs) for liver fibrosis imaging. EJNMMI Res 5(1):71. https://doi.org/10.1186/s13550-015-0151-x

Liang H (2016) Intraperitoneal chemotherapy for locally advanced gastric cancer to prevent and treat peritoneal carcinomatosis. Transl Gastroenterol Hepatol 1:62. https://doi.org/10.21037/tgh.2016.07.04

Lin Y, Lin LJ, Jin Y, Cao Y, Zhang Y, Zheng CQ, Liu JL, Yang SL (2015) Correlation between serum levels of high mobility group box-1 protein and pancreatitis: a meta-analysis. Biomed Res Int 2015:430185. https://doi.org/10.1155/2015/430185

Liu R, Truax AD, Chen L, Hu P, Li Z, Chen J, Song C, Chen L, Ting JP (2015) Expression profile of innate immune receptors, NLRs and AIM2, in human colorectal cancer: correlation with cancer stages and inflammasome components. Oncotarget 6(32):33456–33469. https://doi.org/10.18632/oncotarget.5587

Liu L, Dong Y, Ye M, Jin S, Yang J, Joosse ME, Sun Y, Zhang J, Lazarev M, Brant SR, Safar B, Marohn M, Mezey E, Li X (2016) The pathogenic role of NLRP3 inflammasome activation in inflammatory bowel diseases of both mice and humans. J Crohns Colitis. https://doi.org/10.1093/ecco-jcc/jjw219

Ma X, Guo P, Qiu Y, Mu K, Zhu L, Zhao W, Li T, Han L (2016) Loss of AIM2 expression promotes hepatocarcinoma progression through activation of mTOR-S6K1 pathway. Oncotarget 7(24):36185–36197. https://doi.org/10.18632/oncotarget.9154

Maitra R, Porter MA, Huang S, Gilmour BP (2009) Inhibition of NFkappaB by the natural product Withaferin A in cellular models of cystic fibrosis inflammation. J Inflamm (Lond) 6:15. https://doi.org/10.1186/1476-9255-6-15

Man SM, Zhu Q, Zhu L, Liu Z, Karki R, Malik A, Sharma D, Li L, Malireddi RK, Gurung P, Neale G, Olsen SR, Carter RA, McGoldrick DJ, Wu G, Finkelstein D, Vogel P, Gilbertson RJ, Kanneganti TD (2015) Critical role for the DNA sensor AIM2 in stem cell proliferation and cancer. Cell 162(1):45–58. https://doi.org/10.1016/j.cell.2015.06.001

Manigold T, Bocker U, Chen J, Gundt J, Traber P, Singer MV, Rossol S (2003) Hepatitis B core antigen is a potent inductor of interleukin-18 in peripheral blood mononuclear cells of healthy controls and patients with hepatitis B infection. J Med Virol 71(1):31–40. https://doi.org/10.1002/jmv.10445

Marmol I, Sanchez-de-Diego C, Pradilla Dieste A, Cerrada E, Rodriguez Yoldi MJ (2017) Colorectal carcinoma: a general overview and future perspectives in colorectal cancer. Int J Mol Sci 18(1). https://doi.org/10.3390/ijms18010197

Masarone M, Rosato V, Dallio M, Abenavoli L, Federico A, Loguercio C, Persico M (2016) Epidemiology and natural history of alcoholic liver disease. Rev Recent Clin Trials 11 (3):167–174

McClain CJ, Cohen DA, Dinarello CA, Cannon JG, Shedlofsky SI, Kaplan AM (1986) Serum interleukin-1 (IL-1) activity in alcoholic hepatitis. Life Sci 39(16):1479–1485

McGuckin MA, Every AL, Skene CD, Linden SK, Chionh YT, Swierczak A, McAuley J, Harbour S, Kaparakis M, Ferrero R, Sutton P (2007) Muc1 mucin limits both *Helicobacter pylori* colonization of the murine gastric mucosa and associated gastritis. Gastroenterology 133 (4):1210–1218. https://doi.org/10.1053/j.gastro.2007.07.003

McRae S, Iqbal J, Sarkar-Dutta M, Lane S, Nagaraj A, Ali N, Waris G (2016) The hepatitis C virus-induced NLRP3 inflammasome activates the sterol regulatory element-binding protein (SREBP) and regulates lipid metabolism. J Biol Chem 291(7):3254–3267. https://doi.org/10.1074/jbc.M115.694059

Mencin A, Kluwe J, Schwabe RF (2009) Toll-like receptors as targets in chronic liver diseases. Gut 58(5):704–720. https://doi.org/10.1136/gut.2008.156307

Miskiewicz A, Szparecki G, Durlik M, Rydzewska G, Ziobrowski I, Gorska R (2015) The Q705K and F359L single-nucleotide polymorphisms of NOD-like receptor signaling pathway: association with chronic pancreatitis, pancreatic cancer, and periodontitis. Arch Immunol Ther Exp (Warsz) 63(6):485–494. https://doi.org/10.1007/s00005-015-0355-9

Miura K, Yang L, van Rooijen N, Brenner DA, Ohnishi H, Seki E (2013) Toll-like receptor 2 and palmitic acid cooperatively contribute to the development of nonalcoholic steatohepatitis through inflammasome activation in mice. Hepatology 57(2):577–589. https://doi.org/10.1002/hep.26081

Muerkoster SS, Lust J, Arlt A, Hasler R, Witt M, Sebens T, Schreiber S, Folsch UR, Schafer H (2006) Acquired chemoresistance in pancreatic carcinoma cells: induced secretion of IL-1beta and NO lead to inactivation of caspases. Oncogene 25(28):3973–3981. https://doi.org/10.1038/sj.onc.1209423

Musso G, Gambino R, Tabibian JH, Ekstedt M, Kechagias S, Hamaguchi M, Hultcrantz R, Hagstrom H, Yoon SK, Charatcharoenwitthaya P, George J, Barrera F, Hafliethadottir S, Bjornsson ES, Armstrong MJ, Hopkins LJ, Gao X, Francque S, Verrijken A, Yilmaz Y, Lindor KD, Charlton M, Haring R, Lerch MM, Rettig R, Volzke H, Ryu S, Li G, Wong LL, Machado M, Cortez-Pinto H, Yasui K, Cassader M (2014) Association of non-alcoholic fatty liver disease with chronic kidney disease: a systematic review and meta-analysis. PLoS Med 11 (7):e1001680. https://doi.org/10.1371/journal.pmed.1001680

Namikawa C, Zhang H, Toi M (2006) Non-alcoholic steatohepatitis: pathology and pathogenesis. Nihon Rinsho 64(6):1107–1113

Negash AA, Gale M Jr (2015) Hepatitis regulation by the inflammasome signaling pathway. Immunol Rev 265(1):143–155. https://doi.org/10.1111/imr.12279

Negash AA, Ramos HJ, Crochet N, Lau DT, Doehle B, Papic N, Delker DA, Jo J, Bertoletti A, Hagedorn CH, Jr GM (2013) IL-1beta production through the NLRP3 inflammasome by hepatic macrophages links hepatitis C virus infection with liver inflammation and disease. PLoS Pathog 9(4):e1003330. https://doi.org/10.1371/journal.ppat.1003330

Neuman MG, French SW, Zakhari S, Malnick S, Seitz HK, Cohen LB, Salaspuro M, Voinea-Griffin A, Barasch A, Kirpich IA, Thomes PG, Schrum LW, Donohue TM Jr, Kharbanda KK, Cruz M, Opris M (2017) Alcohol, microbiome, life style influence alcohol and non-alcoholic organ damage. Exp Mol Pathol 102(1):162–180. https://doi.org/10.1016/j.yexmp.2017.01.003

Neurath MF (2014) Cytokines in inflammatory bowel disease. Nat Rev Immunol 14(5):329–342. https://doi.org/10.1038/nri3661

Neurath MF, Pettersson S, Meyer zum Buschenfelde KH, Strober W (1996) Local administration of antisense phosphorothioate oligonucleotides to the p65 subunit of NF-kappa B abrogates established experimental colitis in mice. Nat Med 2(9):998–1004

Neuschwander-Tetri BA, Loomba R, Sanyal AJ, Lavine JE, Van Natta ML, Abdelmalek MF, Chalasani N, Dasarathy S, Diehl AM, Hameed B, Kowdley KV, McCullough A, Terrault N, Clark JM, Tonascia J, Brunt EM, Kleiner DE, Doo E, Network NCR (2015) Farnesoid X nuclear receptor ligand obeticholic acid for non-cirrhotic, non-alcoholic steatohepatitis (FLINT): a multicentre, randomised, placebo-controlled trial. Lancet 385(9972):956–965. https://doi.org/10.1016/S0140-6736(14)61933-4

Ng GZ, Sutton P (2016) The MUC1 mucin specifically inhibits activation of the NLRP3 inflammasome. Genes Immun 17(3):203–206. https://doi.org/10.1038/gene.2016.10

Nowarski R, Jackson R, Gagliani N, de Zoete MR, Palm NW, Bailis W, Low JS, Harman CC, Graham M, Elinav E, Flavell RA (2015) Epithelial IL-18 equilibrium controls barrier function in colitis. Cell 163(6):1444–1456. https://doi.org/10.1016/j.cell.2015.10.072

Nunes T, de Souza HS (2013) Inflammasome in intestinal inflammation and cancer. Mediat Inflamm 2013:654963. https://doi.org/10.1155/2013/654963

O'Shea RS, Dasarathy S, McCullough AJ; Practice Guideline Committee of the American Association for the Study of Liver Disease; Practice Parameters Committee of the American College of Gastroenterology (2010) Alcoholic liver disease. Hepatology 51(1):307–328. https://doi.org/10.1002/hep.23258

Oficjalska K, Raverdeau M, Aviello G, Wade SC, Hickey A, Sheehan KM, Corr SC, Kay EW, O'Neill LA, Mills KH, Creagh EM (2015) Protective role for caspase-11 during acute experimental murine colitis. J Immunol 194(3):1252–1260. https://doi.org/10.4049/jimmunol.1400501

Okamoto M, Liu W, Luo Y, Tanaka A, Cai X, Norris DA, Dinarello CA, Fujita M (2010) Constitutively active inflammasome in human melanoma cells mediating autoinflammation via caspase-1 processing and secretion of interleukin-1beta. J Biol Chem 285(9):6477–6488. https://doi.org/10.1074/jbc.M109.064907

Opipari A, Franchi L (2015) Role of inflammasomes in intestinal inflammation and Crohn's disease. Inflamm Bowel Dis 21(1):173–181. https://doi.org/10.1097/MIB.0000000000000230

Pan X, Xu H, Zheng C, Li M, Zou X, Cao H, Xu Q (2016) Human hepatocytes express absent in melanoma 2 and respond to hepatitis B virus with interleukin-18 expression. Virus Genes 52 (4):445–452. https://doi.org/10.1007/s11262-016-1327-9

Park JW, Chen M, Colombo M, Roberts LR, Schwartz M, Chen PJ, Kudo M, Johnson P, Wagner S, Orsini LS, Sherman M (2015) Global patterns of hepatocellular carcinoma management from diagnosis to death: the BRIDGE Study. Liver Int 35(9):2155–2166. https://doi.org/10.1111/liv.12818

Park JH, Peyrin-Biroulet L, Eisenhut M, Shin JI (2017) IBD immunopathogenesis: a comprehensive review of inflammatory molecules. Autoimmun Rev 16(4):416–426. https://doi.org/10.1016/j.autrev.2017.02.013

Peek RM Jr, Blaser MJ (2002) *Helicobacter pylori* and gastrointestinal tract adenocarcinomas. Nat Rev Cancer 2(1):28–37. https://doi.org/10.1038/nrc703

Pellegrini C, Antonioli L, Lopez-Castejon G, Blandizzi C, Fornai M (2017) Canonical and non-canonical activation of NLRP3 inflammasome at the crossroad between immune tolerance and intestinal inflammation. Front Immunol 8:36. https://doi.org/10.3389/fimmu.2017.00036

Peng Y, French BA, Tillman B, Morgan TR, French SW (2014) The inflammasome in alcoholic hepatitis: its relationship with Mallory-Denk body formation. Exp Mol Pathol 97(2):305–313. https://doi.org/10.1016/j.yexmp.2014.08.006

Perez S, Pereda J, Sabater L, Sastre J (2015) Redox signaling in acute pancreatitis. Redox Biol 5:1–14. https://doi.org/10.1016/j.redox.2015.01.014

Peters U, Bien S, Zubair N (2015) Genetic architecture of colorectal cancer. Gut 64(10):1623–1636. https://doi.org/10.1136/gutjnl-2013-306705

Petrasek J, Bala S, Csak T, Lippai D, Kodys K, Menashy V, Barrieau M, Min SY, Kurt-Jones EA, Szabo G (2012) IL-1 receptor antagonist ameliorates inflammasome-dependent alcoholic steatohepatitis in mice. J Clin Invest 122(10):3476–3489. https://doi.org/10.1172/JCI60777

Petrasek J, Iracheta-Vellve A, Saha B, Satishchandran A, Kodys K, Fitzgerald KA, Kurt-Jones EA, Szabo G (2015) Metabolic danger signals, uric acid and ATP, mediate inflammatory cross-talk between hepatocytes and immune cells in alcoholic liver disease. J Leukoc Biol 98(2):249–256. https://doi.org/10.1189/jlb.3AB1214-590R

Rathinam VA, Vanaja SK, Waggoner L, Sokolovska A, Becker C, Stuart LM, Leong JM, Fitzgerald KA (2012) TRIF licenses caspase-11–dependent NLRP3 inflammasome activation by gram-negative bacteria. Cell 150(3):606–619. https://doi.org/10.1016/j.cell.2012.07.007

Reinhardt C, Reigstad CS, Backhed F (2009) Intestinal microbiota during infancy and its implications for obesity. J Pediatr Gastroenterol Nutr 48(3):249–256

Ren JD, Ma J, Hou J, Xiao WJ, Jin WH, Wu J, Fan KH (2014) Hydrogen-rich saline inhibits NLRP3 inflammasome activation and attenuates experimental acute pancreatitis in mice. Mediat Inflamm 2014:930894. https://doi.org/10.1155/2014/930894

Resende C, Regalo G, Duraes C, Pinto MT, Wen X, Figueiredo C, Carneiro F, Machado JC (2016) Interleukin-1B signalling leads to increased survival of gastric carcinoma cells through a CREB-C/EBPbeta-associated mechanism. Gastric Cancer 19(1):74–84. https://doi.org/10.1007/s10120-014-0448-x

Romeo S, Kozlitina J, Xing C, Pertsemlidis A, Cox D, Pennacchio LA, Boerwinkle E, Cohen JC, Hobbs HH (2008) Genetic variation in PNPLA3 confers susceptibility to nonalcoholic fatty liver disease. Nat Genet 40(12):1461–1465. https://doi.org/10.1038/ng.257

Rubartelli A (2012) Redox control of NLRP3 inflammasome activation in health and disease. J Leukoc Biol 92(5):951–958. https://doi.org/10.1189/jlb.0512265

Ryan BM, Zanetti KA, Robles AI, Schetter AJ, Goodman J, Hayes RB, Huang WY, Gunter MJ, Yeager M, Burdette L, Berndt SI, Harris CC (2014) Germline variation in NCF4, an innate immunity gene, is associated with an increased risk of colorectal cancer. Int J Cancer 134 (6):1399–1407. https://doi.org/10.1002/ijc.28457

Sah RP, Dawra RK, Saluja AK (2013) New insights into the pathogenesis of pancreatitis. Curr Opin Gastroenterol 29(5):523–530. https://doi.org/10.1097/MOG.0b013e328363e399

Saha B, Szabo G (2014) Innate immune cell networking in hepatitis C virus infection. J Leukoc Biol 96(5):757–766. https://doi.org/10.1189/jlb.4MR0314-141R

Salcedo R, Worschech A, Cardone M, Jones Y, Gyulai Z, Dai RM, Wang E, Ma W, Haines D, O'HUigin C, Marincola FM, Trinchieri G (2010) MyD88-mediated signaling prevents development of adenocarcinomas of the colon: role of interleukin 18. J Exp Med 207(8):1625–1636. https://doi.org/10.1084/jem.20100199

Sanyal AJ, Yoon SK, Lencioni R (2010) The etiology of hepatocellular carcinoma and consequences for treatment. Oncologist 15(Suppl 4):14–22. https://doi.org/10.1634/theoncologist.2010-S4-14

Sasaki K, Shindoh J, Nishioka Y, Margonis GA, Sugawara T, Andreatos N, Hashimoto M, Pawlik TM (2017) Impact of viral etiology on postoperative de novo recurrence after hepatectomy for hepatocellular carcinoma in cirrhotic patients. J Gastrointest Surg 21(3):487–495. https://doi.org/10.1007/s11605-016-3344-3

Sawa H, Ueda T, Takeyama Y, Yasuda T, Shinzeki M, Nakajima T, Kuroda Y (2006) Blockade of high mobility group box-1 protein attenuates experimental severe acute pancreatitis. World J Gastroenterol 12(47):7666–7670

Schuppan D, Afdhal NH (2008) Liver cirrhosis. Lancet 371(9615):838–851. https://doi.org/10.1016/S0140-6736(08)60383-9

Semper RP, Mejias-Luque R, Gross C, Anderl F, Muller A, Vieth M, Busch DH, Prazeres da Costa C, Ruland J, Gross O, Gerhard M (2014) *Helicobacter pylori*–induced IL-1beta secretion in innate immune cells is regulated by the NLRP3 inflammasome and requires the cag pathogenicity island. J Immunol 193(7):3566–3576. https://doi.org/10.4049/jimmunol.1400362

Serti E, Werner JM, Chattergoon M, Cox AL, Lohmann V, Rehermann B (2014) Monocytes activate natural killer cells via inflammasome-induced interleukin 18 in response to hepatitis C virus replication. Gastroenterology 147(1):209–U343. https://doi.org/10.1053/j.gastro.2014.03.046

Shigematsu Y, Niwa T, Rehnberg E, Toyoda T, Yoshida S, Mori A, Wakabayashi M, Iwakura Y, Ichinose M, Kim YJ, Ushijima T (2013) Interleukin-1beta induced by *Helicobacter pylori* infection enhances mouse gastric carcinogenesis. Cancer Lett 340(1):141–147. https://doi.org/10.1016/j.canlet.2013.07.034

Shin EC, Sung PS, Park SH (2016) Immune responses and immunopathology in acute and chronic viral hepatitis. Nat Rev Immunol 16(8):509–523. https://doi.org/10.1038/nri.2016.69

Shiota S, Suzuki R, Yamaoka Y (2013) The significance of virulence factors in *Helicobacter pylori*. J Dig Dis 14(7):341–349. https://doi.org/10.1111/1751-2980.12054

Simovic Markovic B, Nikolic A, Gazdic M, Bojic S, Vucicevic L, Kosic M, Mitrovic S, Milosavljevic M, Besra G, Trajkovic V, Arsenijevic N, Lukic ML, Volarevic V (2016) Galectin-3 plays an important pro-inflammatory role in the induction phase of acute colitis by promoting activation of NLRP3 inflammasome and production of IL-1beta in macrophages. J Crohns Colitis 10(5):593–606. https://doi.org/10.1093/ecco-jcc/jjw013

Sonohara F, Inokawa Y, Kanda M, Nishikawa Y, Yamada S, Fujii T, Sugimoto H, Kodera Y, Nomoto S (2017) Association of inflammasome components in background liver with poor prognosis after curatively-resected hepatocellular carcinoma. Anticancer Res 37(1):293–300. https://doi.org/10.21873/anticanres.11320

Sookoian S, Pirola CJ (2011) Meta-analysis of the influence of I148M variant of patatin-like phospholipase domain containing 3 gene (PNPLA3) on the susceptibility and histological severity of nonalcoholic fatty liver disease. Hepatology 53(6):1883–1894. https://doi.org/10.1002/hep.24283

Stram M, Liu S, Singhi AD (2016) Chronic pancreatitis. Surg Pathol Clin 9(4):643–659. https://doi.org/10.1016/j.path.2016.05.008

Suerbaum S, Michetti P (2002) *Helicobacter pylori* infection. N Engl J Med 347(15):1175–1186. https://doi.org/10.1056/NEJMra020542

Suh YS, Yang HK (2015) Screening and early detection of gastric cancer: east versus west. Surg Clin North Am 95(5):1053–1066. https://doi.org/10.1016/j.suc.2015.05.012

Sui YH, Luo WJ, Xu QY, Hua J (2016) Dietary saturated fatty acid and polyunsaturated fatty acid oppositely affect hepatic NOD-like receptor protein 3 inflammasome through regulating nuclear factor-kappa B activation. World J Gastroenterol 22(8):2533–2544. https://doi.org/10.3748/wjg.v22.i8.2533

Sun Z, Meng Y, Liu G, Jiang Y, Meng Q, Hu S (2016) Effect of interleukin-1beta and tumor necrosis factor alpha gene silencing on mouse gastric cancer cell proliferation and migration. Oncol Lett 11(4):2559–2565. https://doi.org/10.3892/ol.2016.4253

Szabo G, Petrasek J (2015) Inflammasome activation and function in liver disease. Nat Rev Gastroenterol Hepatol 12(7):387–400. https://doi.org/10.1038/nrgastro.2015.94

Szabo G, Bala S, Petrasek J, Gattu A (2010) Gut-liver axis and sensing microbes. Dig Dis 28 (6):737–744. https://doi.org/10.1159/000324281

Tacke F, Trautwein C (2015) Mechanisms of liver fibrosis resolution. J Hepatol 63(4):1038–1039. https://doi.org/10.1016/j.jhep.2015.03.039

Tait SW, Ichim G, Green DR (2014) Die another way – non-apoptotic mechanisms of cell death. J Cell Sci 127(Pt 10):2135–2144. https://doi.org/10.1242/jcs.093575

Takagi H, Kanai T, Okazawa A, Kishi Y, Sato T, Takaishi H, Inoue N, Ogata H, Iwao Y, Hoshino K, Takeda K, Akira S, Watanabe M, Ishii H, Hibi T (2003) Contrasting action of IL-12 and IL-18 in the development of dextran sodium sulphate colitis in mice. Scand J Gastroenterol 38(8):837–844

Takaki A, Kawai D, Yamamoto K (2014) Molecular mechanisms and new treatment strategies for non-alcoholic steatohepatitis (NASH). Int J Mol Sci 15(5):7352–7379. https://doi.org/10.3390/ijms15057352

Than NN, Newsome PN (2015) A concise review of non-alcoholic fatty liver disease. Atherosclerosis 239(1):192–202. https://doi.org/10.1016/j.atherosclerosis.2015.01.001

Tu S, Bhagat G, Cui G, Takaishi S, Kurt-Jones EA, Rickman B, Betz KS, Penz-Oesterreicher M, Bjorkdahl O, Fox JG, Wang TC (2008) Overexpression of interleukin-1beta induces gastric inflammation and cancer and mobilizes myeloid-derived suppressor cells in mice. Cancer Cell 14(5):408–419. https://doi.org/10.1016/j.ccr.2008.10.011

Tyrer PC, Bean EG, Ruth Foxwell A, Pavli P (2011) Effects of bacterial products on enterocyte-macrophage interactions in vitro. Biochem Biophys Res Commun 413(2):336–341. https://doi.org/10.1016/j.bbrc.2011.08.100

Ueda T, Takeyama Y, Yasuda T, Matsumura N, Sawa H, Nakajima T, Ajiki T, Fujino Y, Suzuki Y, Kuroda Y (2006) Significant elevation of serum interleukin-18 levels in patients with acute pancreatitis. J Gastroenterol 41(2):158–165. https://doi.org/10.1007/s00535-005-1735-4

Ursell LK, Clemente JC, Rideout JR, Gevers D, Caporaso JG, Knight R (2012) The interpersonal and intrapersonal diversity of human-associated microbiota in key body sites. J Allergy Clin Immunol 129(5):1204–1208. https://doi.org/10.1016/j.jaci.2012.03.010

van Dijk F, Olinga P, Poelstra K, Beljaars L (2015) Targeted therapies in liver fibrosis: combining the best parts of platelet-derived growth factor BB and interferon gamma. Front Med 2:72. https://doi.org/10.3389/fmed.2015.00072

Vogiatzi P, Cassone M, Luzzi I, Lucchetti C, Jr OL, Giordano A (2007) *Helicobacter pylori* as a class I carcinogen: physiopathology and management strategies. J Cell Biochem 102 (2):264–273. https://doi.org/10.1002/jcb.21375

Wang H, Desai KD, Dong H, Owzarski S, Romagnuolo J, Morgan KA, Adams DB (2013) Prior surgery determines islet yield and insulin requirement in patients with chronic pancreatitis. Transplantation 95(8):1051–1057. https://doi.org/10.1097/TP.0b013e3182845fbb

Wang P, Zhu S, Yang L, Cui S, Pan W, Jackson R, Zheng Y, Rongvaux A, Sun Q, Yang G, Gao S, Lin R, You F, Flavell R, Fikrig E (2015) Nlrp6 regulates intestinal antiviral innate immunity. Science 350(6262):826–830. https://doi.org/10.1126/science.aab3145

Wang Y, Wang H, Qian C, Tang J, Zhou W, Liu X, You Q, Hu R (2016) 3-(2-Oxo-2-phenylethylidene)-2,3,6,7-tetrahydro-1H-pyrazino[2,1-a]isoquinolin-4(1 1bH)-one (compound 1), a novel potent Nrf2/ARE inducer, protects against DSS-induced colitis via inhibiting NLRP3 inflammasome. Biochem Pharmacol 101:71–86. https://doi.org/10.1016/j.bcp.2015.11.015

Warren SE, Armstrong A, Hamilton MK, Mao DP, Leaf IA, Miao EA, Aderem A (2010) Cutting edge: cytosolic bacterial DNA activates the inflammasome via Aim2. J Immunol 185 (2):818–821. https://doi.org/10.4049/jimmunol.1000724

Watanabe A, Sohail MA, Gomes DA, Hashmi A, Nagata J, Sutterwala FS, Mahmood S, Jhandier MN, Shi Y, Flavell RA, Mehal WZ (2009) Inflammasome-mediated regulation of hepatic stellate cells. Am J Physiol Gastrointest Liver Physiol 296(6):G1248–G1257. https://doi.org/10.1152/ajpgi.90223.2008

Watashi K, Liang G, Iwamoto M, Marusawa H, Uchida N, Daito T, Kitamura K, Muramatsu M, Ohashi H, Kiyohara T, Suzuki R, Li J, Tong S, Tanaka Y, Murata K, Aizaki H, Wakita T (2013) Interleukin-1 and tumor necrosis factor-alpha trigger restriction of hepatitis B virus infection via a cytidine deaminase activation-induced cytidine deaminase (AID). J Biol Chem 288 (44):31715–31727. https://doi.org/10.1074/jbc.M113.501122

Wei Q, Mu K, Li T, Zhang Y, Yang Z, Jia X, Zhao W, Huai W, Guo P, Han L (2014) Deregulation of the NLRP3 inflammasome in hepatic parenchymal cells during liver cancer progression. Lab Investig 94(1):52–62. https://doi.org/10.1038/labinvest.2013.126

Wei Q, Guo P, Mu K, Zhang Y, Zhao W, Huai W, Qiu Y, Li T, Ma X, Liu Y, Chen X, Han L (2015) Estrogen suppresses hepatocellular carcinoma cells through ERbeta-mediated upregulation of the NLRP3 inflammasome. Lab Investig 95(7):804–816. https://doi.org/10.1038/labinvest.2015.63

Wen H, Gris D, Lei Y, Jha S, Zhang L, Huang MT, Brickey WJ, Ting JP (2011) Fatty acid-induced NLRP3-ASC inflammasome activation interferes with insulin signaling. Nat Immunol 12 (5):408–415. https://doi.org/10.1038/ni.2022

Wigg AJ, Roberts-Thomson IC, Dymock RB, McCarthy PJ, Grose RH, Cummins AG (2001) The role of small intestinal bacterial overgrowth, intestinal permeability, endotoxaemia, and tumour necrosis factor alpha in the pathogenesis of non-alcoholic steatohepatitis. Gut 48(2):206–211

Williams TM, Leeth RA, Rothschild DE, McDaniel DK, Coutermarsh-Ott SL, Simmons AE, Kable KH, Heid B, Allen IC (2015) Caspase-11 attenuates gastrointestinal inflammation and experimental colitis pathogenesis. Am J Physiol Gastrointest Liver Physiol 308(2):G139–G150. https://doi.org/10.1152/ajpgi.00234.2014

Wilson JE, Petrucelli AS, Chen L, Koblansky AA, Truax AD, Oyama Y, Rogers AB, Brickey WJ, Wang Y, Schneider M, Muhlbauer M, Chou WC, Barker BR, Jobin C, Allbritton NL, Ramsden DA, Davis BK, Ting JP (2015) Inflammasome-independent role of AIM2 in suppressing colon tumorigenesis via DNA-PK and Akt. Nat Med 21(8):906–913. https://doi.org/10.1038/nm.3908

Wlodarska M, Thaiss CA, Nowarski R, Henao-Mejia J, Zhang JP, Brown EM, Frankel G, Levy M, Katz MN, Philbrick WM, Elinav E, Finlay BB, Flavell RA (2014) NLRP6 inflammasome orchestrates the colonic host-microbial interface by regulating goblet cell mucus secretion. Cell 156(5):1045–1059. https://doi.org/10.1016/j.cell.2014.01.026

Wong RJ, Aguilar M, Cheung R, Perumpail RB, Harrison SA, Younossi ZM, Ahmed A (2015) Nonalcoholic steatohepatitis is the second leading etiology of liver disease among adults awaiting liver transplantation in the United States. Gastroenterology 148(3):547–555. https://doi.org/10.1053/j.gastro.2014.11.039

Wree A, Broderick L, Canbay A, Hoffman HM, Feldstein AE (2013) From NAFLD to NASH to cirrhosis—new insights into disease mechanisms. Nat Rev Gastroenterol Hepatol 10 (11):627–636. https://doi.org/10.1038/nrgastro.2013.149

Wree A, McGeough MD, Pena CA, Schlattjan M, Li H, Inzaugarat ME, Messer K, Canbay A, Hoffman HM, Feldstein AE (2014) NLRP3 inflammasome activation is required for fibrosis development in NAFLD. J Mol Med 92(10):1069–1082. https://doi.org/10.1007/s00109-014-1170-1

Wu DL, Xu GH, Lu SM, Ma BL, Miao NZ, Liu XB, Cheng YP, Feng JH, Liu ZG, Feng D, Na L, Li WQ, Zhao YR (2013) Correlation of AIM2 expression in peripheral blood mononuclear cells from humans with acute and chronic hepatitis B. Hum Immunol 74(5):514–521. https://doi.org/10.1016/j.humimm.2013.01.022

Wu X, Zhang F, Xiong X, Lu C, Lian N, Lu Y, Zheng S (2015) Tetramethylpyrazine reduces inflammation in liver fibrosis and inhibits inflammatory cytokine expression in hepatic stellate cells by modulating NLRP3 inflammasome pathway. IUBMB Life 67(4):312–321. https://doi.org/10.1002/iub.1348

Xiao J, Zhu Y, Liu Y, Tipoe GL, Xing F, So KF (2014) Lycium barbarum polysaccharide attenuates alcoholic cellular injury through TXNIP-NLRP3 inflammasome pathway. Int J Biol Macromol 69:73–78. https://doi.org/10.1016/j.ijbiomac.2014.05.034

Xu C, Wan X, Xu L, Weng H, Yan M, Miao M, Sun Y, Xu G, Dooley S, Li Y, Yu C (2015) Xanthine oxidase in non-alcoholic fatty liver disease and hyperuricemia: one stone hits two birds. J Hepatol 62(6):1412–1419. https://doi.org/10.1016/j.jhep.2015.01.019

Yamamoto T, Yagi S, Kinoshita H, Sakamoto Y, Okada K, Uryuhara K, Morimoto T, Kaihara S, Hosotani R (2015) Long-term survival after resection of pancreatic cancer: a single-center retrospective analysis. World J Gastroenterol 21(1):262–268. https://doi.org/10.3748/wjg.v21.i1.262

Yin S, Lan C, Pei H, Zhu Z (2016) Expression of interleukin 1beta in gastric cancer tissue and its effects on gastric cancer. Onco Targets Ther 9:31–35. https://doi.org/10.2147/ott.s94277

Ying HY, Yu BW, Yang Z, Yang SS, Bo LH, Shan XY, Wang HJ, Zhu YJ, Wu XS (2016) Interleukin-1B 31 C>T polymorphism combined with *Helicobacter pylori*–modified gastric cancer susceptibility: evidence from 37 studies. J Cell Mol Med 20(3):526–536. https://doi.org/10.1111/jcmm.12737

York JM, Castellanos KJ, Cabay RJ, Fantuzzi G (2014) Inhibition of the nucleotide-binding domain, leucine-rich containing family, pyrin-domain containing 3 inflammasome reduces the severity of experimentally induced acute pancreatitis in obese mice. Transl Res 164(4):259–269. https://doi.org/10.1016/j.trsl.2014.06.012

Yu X, Lan P, Hou X, Han Q, Lu N, Li T, Jiao C, Zhang J, Zhang C, Tian Z (2017) HBV inhibits LPS-induced NLRP3 inflammasome activation and IL-1beta production via suppressing the NF-kappaB pathway and ROS production. J Hepatol 66(4):693–702. https://doi.org/10.1016/j.jhep.2016.12.018

Yuan BS, Zhu RM, Braddock M, Zhang XH, Shi W, Zheng MH (2007) Interleukin-18: a pro-inflammatory cytokine that plays an important role in acute pancreatitis. Expert Opin Ther Targets 11(10):1261–1271. https://doi.org/10.1517/14728222.11.10.1261

Zaki MH, Boyd KL, Vogel P, Kastan MB, Lamkanfi M, Kanneganti TD (2010a) The NLRP3 inflammasome protects against loss of epithelial integrity and mortality during experimental colitis. Immunity 32(3):379–391. https://doi.org/10.1016/j.immuni.2010.03.003

Zaki MH, Vogel P, Body-Malapel M, Lamkanfi M, Kanneganti TD (2010b) IL-18 production downstream of the Nlrp3 inflammasome confers protection against colorectal tumor formation. J Immunol 185(8):4912–4920. https://doi.org/10.4049/jimmunol.1002046

Zambirinis CP, Pushalkar S, Saxena D, Miller G (2014) Pancreatic cancer, inflammation, and microbiome. Cancer J 20(3):195–202. https://doi.org/10.1097/ppo.0000000000000045

Zhang XH, Zhu RM, Xu WA, Wan HJ, Lu H (2007) Therapeutic effects of caspase-1 inhibitors on acute lung injury in experimental severe acute pancreatitis. World J Gastroenterol 13 (4):623–627

Zhang CY, Yuan WG, He P, Lei JH, Wang CX (2016) Liver fibrosis and hepatic stellate cells: etiology, pathological hallmarks and therapeutic targets. World J Gastroenterol 22 (48):10512–10522. https://doi.org/10.3748/wjg.v22.i48.10512

Zhu RX, Seto WK, Lai CL, Yuen MF (2016) Epidemiology of hepatocellular carcinoma in the Asia-Pacific region. Gut Liver 10(3):332–339. https://doi.org/10.5009/gnl15257

Chapter 11
Inflammasomes in Bone Diseases

Gabriel Mbalaviele and Deborah J. Veis

Contents

Abstract Unresolved inflammation is harmful to any tissues in the organism. Bone in particular is vulnerable to inflammatory assaults because its integrity depends on the activity of osteoclasts, which arise from myeloid precursors. Osteoclasts are responsible for bone resorption in normal and disease conditions. Increased osteolysis is a common feature of inflammatory disorders and a risk factor for bone fractures. Thus, bone is impacted negatively not only by local and systemic inflammatory mediators, but also directly, by alterations affecting myelopoiesis and lineage allocations. Such perturbations are characteristics of dysregulated inflammasomes, which are key regulators of innate immunity. In this review, we discuss the role of inflammasomes in bone diseases caused by sterile or non-sterile inflammation.

G. Mbalaviele (✉)
Division of Bone and Mineral Diseases, Department of Medicine, Washington University School of Medicine, St. Louis, MO, USA
e-mail: gmbalaviele@wustl.edu

D. J. Veis
Division of Bone and Mineral Diseases, Department of Medicine, Washington University School of Medicine, St. Louis, MO, USA

Division of Anatomic and Molecular Pathology, Department of Pathology and Immunology, Washington University School of Medicine, St. Louis, MO, USA

M. D. Cordero, E. Alcocer-Gómez (eds.), *Inflammasomes: Clinical and Therapeutic Implications*, Experientia Supplementum 108,
https://doi.org/10.1007/978-3-319-89390-7_11

Keywords Inflammation · Inflammasome · NLRP3 · PARP1 · Osteoclast · Bone · DAMPs · PAMPs · IL-1 · Infection

11.1 Introduction

Inflammasomes are multifunctional intracellular protein complexes that protect against infections upon recognition of microbial structures known as pathogen-associated molecular patterns (PAMPs). They also restore tissue integrity after injury upon sensing debris from damaged tissues, signals termed danger-associated molecular patterns (DAMPs) (Guo et al. 2015; Lukens et al. 2012; Martin 2016; Schroder and Tschopp 2010). Some receptors such as nucleotide-binding oligomerization domain-like receptors (NLRs, e.g., NLRP3) and absent in melanoma 2-like receptors (ALRs, e.g., AIM 2) associate with apoptosis-associated speck-like protein containing a CARD (ASC) upon recognition or sensing of specific PAMPs or DAMPs. ASC then forms polymers that facilitate the recruitment of pro-caspase-1, which is converted into active caspase-1 through an ill-defined proximity-enabled reaction. Prominent caspase-1-mediated responses include maturation of interleukin (IL)-1β and IL-18 and pyroptosis (Guo et al. 2015). Caspase-1 also processes several other substrates including glycolytic enzymes (Shao et al. 2007; Wang et al. 2016), and caspase-7, which subsequently cleaves poly(ADP-ribose) polymerase 1 (PARP1) (Erener et al. 2012). Inflammasomes are functional in various cell types, including myeloid cells (Guarda et al. 2011), osteoclasts (Bonar et al. 2012; Qu et al. 2015; Scianaro et al. 2014; Kim et al. 2017), osteoblasts (Bonar et al. 2012; McCall et al. 2008), and chondrocytes (Feldmann et al. 2002). However, the skeletal impact of over-reactive inflammasomes is not well understood.

Bone is laid down and modeled during development by osteoblasts (Kronenberg 2003). After development, the amount of bone that is resorbed by osteoclasts is replenished fully by osteoblasts to maintain homeostasis (Raisz 2005). Perturbation of this balance in inflammatory states causes bone loss (Novack and Mbalaviele 2016; Walsh and Gravallese 2010; Mbalaviele 2017). In this book chapter, we review known and potential roles for the inflammasomes in skeletal diseases characterized by sterile or non-sterile inflammation (Fig. 11.1).

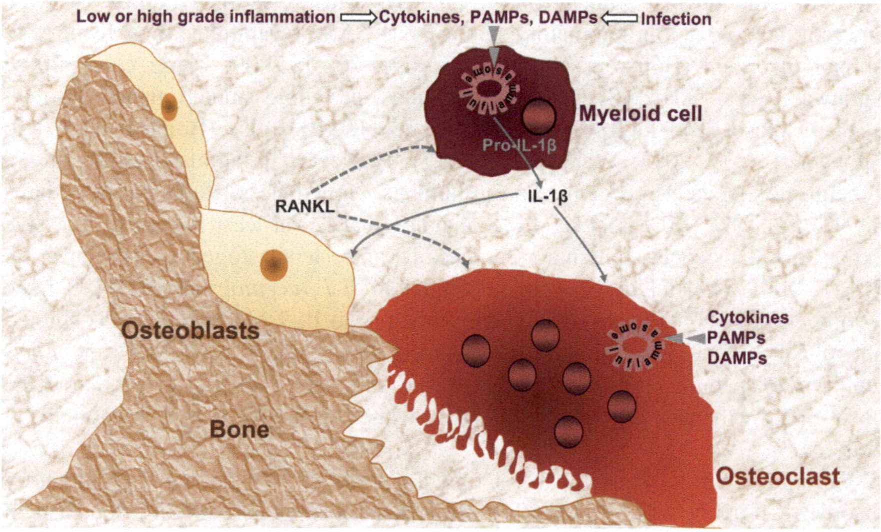

Fig. 11.1 Inflammasomes and bone resorption. Combined actions of cytokines, PAMPs and DAMPs produced by inflammation (open arrows) lead to inflammasome activation (arrowheads), and subsequent IL-1β secretion by myeloid cells (solid arrows), some of which form osteoclasts. Potential endogenous DAMPs include bone degradation products (not represented). Inflammasomes can also be activated by gain-of-function mutations independently of PAMPs and DAMPs. IL-1β in turn stimulates osteoclast differentiation and bone resorption directly and indirectly through induction of osteoclastogenic factors (e.g., RANKL) by osteoblasts (dashed arrows). Inflammasome activation also leads to PARP1 degradation (not depicted), a process that promotes osteoclastogenesis

11.2 Role of the Inflammasomes in Sterile Osteolysis

11.2.1 The NLRP3 Inflammasome

NLRP3-activating mutations cause cryopyrin-associated periodic syndromes (CAPS), which are autoinflammatory disorders associated with excessive IL-1β and IL-18 production, recurrent fever, and urticaria-like rash (de Jesus et al. 2015; Hoffman and Broderick 2016). Skeletal anomalies including osteopenia, bone deformities, bulky epiphyses, leg length discrepancy, and short stature are prominent features of neonatal-onset multisystem inflammatory disease (NOMID), the most severe manifestation of the CAPS spectrum (de Jesus et al. 2015; Hoffman and Broderick 2016). Epiphyseal outgrowths in these patients are abnormally calcified and associated with disorganized and hypocellular growth plates (Feldmann et al. 2002; Aksentijevich et al. 2002; Zaki et al. 2012; Hill et al. 2007). IL-1β blocking agents (e.g., anakinra, rilonacept, and canakinumab) relieve inflammatory symptoms, but anakinra, for example, appears to have limited or no efficacy against skeletal lesions (Anton et al. 2015; Sibley et al. 2012; Neven et al. 2010; Rigante et al. 2011); the causes of drug resistance remain unknown.

Murine models of CAPS, including NOMID, Muckle-Wells syndrome (MWS), and familial cold autoinflammatory syndrome (FCAS), recapitulate most features of human disorders (Bonar et al. 2012; Brydges et al. 2009; Meng et al. 2009; Snouwaert et al. 2016). Skeletal complications, including stunted growth, growth plate dysplasia, and low bone mass occur in NOMID mice (Bonar et al. 2012; Snouwaert et al. 2016; Wang et al. 2017). However, inflammasopathies are in general more severe in rodents than in humans. Nonetheless, studies in mice have revealed that NOMID-associated osteopenia is caused at least in part by massive expansion of osteoclast precursors and exuberant osteoclastogenesis (Bonar et al. 2012). While skeletal defects are prevented in *Il-1 receptor* null NOMID mice (Wang et al. 2017), human patients are refractory to IL-1 blocking agents as noted above. Since IL-1 signaling is blocked during development in NOMID mice, it is tempting to speculate that early diagnosis and treatment with IL-1 drugs before the onset of growth plate dysplasia may prevent the development of this anomaly in children.

Chronic recurrent multifocal osteomyelitis (CRMO) is an autoinflammatory disorder. Components of the NLRP3 inflammasome are expressed in osteoclasts in bone specimens from CRMO patients (Scianaro et al. 2014), implying that this inflammasome may exert osteoclast autonomous functions as discussed below. Levels of anti-inflammatory cytokines (e.g., IL-10) are decreased whereas those of pro-inflammatory cytokines (e.g., IL-1β, IL-6 and TNF-α) are increased in this disease, a cytokine imbalance that presumably leads to NLRP3 inflammasome hyper-activation and exaggerated osteoclastogenesis (Hofmann et al. 2016). Therapeutic options include NSAIDs, bisphosphonates, and blockers of TNF-α or IL-1 (Hofmann et al. 2016). Neutrophils in CRMO mice overproduce IL-1β, which enhances osteoclastogenesis and bone resorption (Cassel et al. 2014; Lukens et al. 2014; Chitu et al. 2012). Caspase-8 and NLRP3 inflammasome-dependent caspase-1 play a redundant role in IL-1β maturation in neutrophils (Cassel et al. 2014; Lukens et al. 2014; Gurung et al. 2016), implying that selective inhibitors of inflammasomes may have limited efficacy in the treatment of CRMO.

Postmenopausal osteoporosis is another condition in which levels of IL-1β and TNF-α are elevated. We find that NLRP3 deficiency attenuates bone loss in a mouse model of this condition. Our data indicate that bone degradation products released in this condition of accelerated bone turnover activate the inflammasome (Alippe et al. 2017). We also find that genetic activation of this inflammasome in osteoclasts using *Cre* driven by the *cathepsin K* promoter causes osteopenia in mice, without altering IL-1β production and osteoclast number (Qu et al. 2015). Increased bone resorption in mutant cells correlates with cytoskeletal changes (Qu et al. 2015) (Fig. 11.1). Thus, this inflammasome can cause bone resorption in the absence of high grade inflammation. Indeed, NLRP3 inflammasome-dependent low grade age-related sterile inflammation is also linked to chronic diseases in mice, including bone loss (Youm et al. 2013). Inflammasomes are also involved in hyper-multinucleation of murine osteoclasts caused by purinergic receptor P2X5 signaling (Kim et al. 2017). On the other hand, a recent report indicates that NLRP12, which also regulates caspase-1 and nucleates a functional inflammasome (Silveira et al. 2017; Janowski and Sutterwala 2016), inhibits osteoclastogenesis through suppression of NF-κB signaling (Krauss et al. 2015). Thus, various inflammasomes regulate bone metabolism. It will be therefore important to determine whether selective pharmacological inhibition of some of these proteins (Youm et al. 2015; Coll et al. 2015) provides superior efficacy than IL-1 blocking agents, which are ineffective in reducing the risk of fractures in osteoporotic patients.

Pro-inflammatory cytokines, including IL-1β, IL-6, IL-17, and TNF-α, are also upregulated in rheumatoid arthritis (RA) and likely account for the high incidence of osteoporosis in RA patients as they promote bone resorption while inhibiting bone formation (Maruotti et al. 2014). The NLRP3 inflammasome pathway is activated in RA patients (Rosengren et al. 2005), but its role in arthritis development in preclinical models is conflicting. Indeed, NLRP3 and caspase-1 are dispensable for collagen-induced arthritis and antigen-induced arthritis (Ippagunta et al. 2010; Kolly et al. 2010), yet NLRP3 deficiency protects joints from destruction in arthritis induced by A20 ablation (Vande Walle et al. 2014). These inconsistent observations may reflect redundancy in inflammasome functions. In fact, not only NLRP3 but also AIM 2 is upregulated in the synovium of IL-10-deficient mice exposed to antigen-induced arthritis (Greenhill et al. 2014), and osteoclast differentiation from bone marrow cells isolated from these mutant mice is inhibited by tool compound inhibitors of NLRP3 and AIM 2 inflammasomes (Greenhill et al. 2014). Moreover, arthritis induced by DNase II deficiency, which is associated with accrual of self DNA, is attenuated by AIM 2 ablation (Baum et al. 2015). Thus, several inflammasomes are involved in joint destruction in inflammatory arthritis, but the action of the NLRP3 inflammasome in particular appears to be mouse-model dependent.

11.2.2 The NLRC4 Inflammasome

Sensing bacterial type III and IV secretion systems and flagellin via NLR family apoptosis inhibitory proteins (NAIPs) has been the only known function of the NLRC4 inflammasome (Miao et al. 2010; Zhao et al. 2011) until recently when gain-of-function mutations in *NLRC4* were identified in patients. These patients exhibit elevated serum levels of IL-1β and IL-18 accompanied by recurrent fever flares, cytopenia, high ferritin levels, hemophagocytosis, and splenomegaly (Canna et al. 2014; Romberg et al. 2014). This phenotype is reminiscent of the macrophage activation syndrome (MAS) (Canna and Nigrovic 2016). MAS is a frequent complication of systemic juvenile idiopathic arthritis (sJIA), a disease that interferes with normal skeletal development and bone mass acquisition (Maruotti et al. 2014; Bechtold and Simon 2014). Besides glucocorticoids, nonsteroidal anti-inflammatory drugs (NSAIDs), IL-6, or TNF-α biologics with or without methotrexate, IL-1 blockers are also therapeutic options for sJIA, thus indicating a critical role of the cytokine in this disease (Tarp et al. 2016). Consistent with the human phenotype, transgenic mice expressing active NLRC4 mutant produce high levels of IL-1β and develop arthritis (Kitamura et al. 2014). Whether NLRC4 can modulate osteoclastogenesis in a cell autonomous manner remains to be determined.

11.2.3 The Pyrin Inflammasome

Activating mutations in *MEFV*, which encodes pyrin, cause familial Mediterranean fever (FMF), an autoinflammatory disease that is associated with excessive production of IL-1β, recurrent fever episodes, arthritis, and decreased bone mineral density (BMD) (Ozen and Bilginer 2014; Ben-Zvi and Livneh 2011). IL-6, IL-8, and IL-12 are also elevated in these patients. Current treatments include oral colchicine and IL-1 biologics (de Jesus et al. 2015). The efficacy of IL-1 blockers in colchicine-resistant FMF patients underscores the pathogenic action of this cytokine (Ben-Zvi et al. 2017). MEFV mice develop severe systemic inflammation, indicating that pyrin assembles a pro-inflammatory inflammasome (Chae et al. 2011). These mice are also runted and exhibit massive cartilage and bone erosion. While there are several mechanisms of decreased BMD in FMF patients, including chronic inflammation and steroid-based therapies, data from MEFV mice suggest that inflammation drives skeletal manifestations in this disorder.

11.3 Role of the Inflammasomes in Infection-Associated Osteolysis

Osteolysis is common in infectious diseases such as periodontitis and osteomyelitis. Indeed, severe bone loss is associated with infection by *Porphyromonas gingivalis*, the major oral bacterial species implicated in the pathogenesis of periodontitis and a potent activator of the NLRP3 and AIM 2 inflammasomes (Bostanci et al. 2009, 2011; Park et al. 2014). Although most studies have examined expression of NLRP3 and cytokines in either whole tissue or macrophage lineage cells, osteoblasts may also play a role in bone loss. These bone forming cells can be infected by *P. gingivalis* and induced to express NLRP3 (Yoshida et al. 2017), presumably contributing to the elevated IL-1β levels. IL-1β released into the local environment can synergize with RANKL and other inflammatory cytokines to increase osteoclastogenesis and thereby destroy alveolar bone (Kassem et al. 2015; Han et al. 2013; Akiyama et al. 2014). Emphasizing the importance of the inflammasome, *Nlrp3*-deficient mice are protected from *P. gingivalis*-induced alveolar bone loss (Yamaguchi et al. 2016).

Skeletal infection outside of the oral cavity, known as osteomyelitis, is most commonly caused by *Staphylococcus aureus*, a pathogen sensed by the NLRP3 and NLRC5 inflammasomes (Miller et al. 2007; Munoz-Planillo et al. 2009; Davis et al. 2011). Although much work has been done to understand the activation and downstream consequences of inflammasome activation by *S. aureus* in other contexts, this interaction has not been specifically examined in osteomyelitis. Nevertheless, it is likely that inflammasome activation in bone plays a role in *S. aureus*-associated bone loss, as IL-1β is one of the many cytokines upregulated in this condition (Yoshii et al. 2002). *S. aureus* bacterial products including hemolysins, bacterial lipoproteins, and Panton-Valentine leukocidin all activate NLRP3 (Holzinger et al. 2012). Further studies will be needed to determine the details of inflammasome activation during bone infection and the implications for disease progression and bacterial clearance.

11.4 Conclusion

Bone loss is a frequent outcome of various inflammatory conditions, which are associated with elevated IL-1β levels. While IL-1 blocking agents spare bone from destruction in various preclinical models, the efficacy of these drugs in human conditions is not uniform. Since the inflammasomes regulate several inflammatory pathways, including IL-1β, IL-18, and pyroptosis, inhibition of inflammasomes should in theory supersede the efficacy of IL-1 blockers, a proposition that will be tested certainly in the near future. In the meantime, there are still many unresolved questions regarding inflammasome biology in bone, including: (1) Are inflammasome actions in bone limited to maturation of IL-1β? (2) What is the repertoire of functional NLRs and other inflammasome components in bone? (3) What are the DAMPs that activate the inflammasomes in bone environment in the absence of infection? (4) Do inflammasomes have autonomous actions in the osteoblast lineage? The answers to the questions will inform a rationale therapeutic targeting of inflammasomes for the treatment of inflammatory osteolysis.

Acknowledgments and Conflict of Interest G.M. is supported by NIH/NIAMS AR064755 and AR068972 grants and is co-founder of Confluence Life Sciences, Inc. D.J.V is supported by NIH/NIAMS AR070030 and the Shriners Hospitals for Children.

References

Akiyama T, Miyamoto Y, Yoshimura K, Yamada A, Takami M, Suzawa T, Hoshino M, Imamura T, Akiyama C, Yasuhara R et al (2014) Porphyromonas gingivalis-derived lysine gingipain enhances osteoclast differentiation induced by tumor necrosis factor-alpha and interleukin-1beta but suppresses that by interleukin-17A: importance of proteolytic degradation of osteoprotegerin by lysine gingipain. J Biol Chem 289:15621–15630

Aksentijevich I, Nowak M, Mallah M, Chae JJ, Watford WT, Hofmann SR, Stein L, Russo R, Goldsmith D, Dent P et al (2002) De novo CIAS1 mutations, cytokine activation, and evidence for genetic heterogeneity in patients with neonatal-onset multisystem inflammatory disease (NOMID): a new member of the expanding family of pyrin-associated autoinflammatory diseases. Arthritis Rheum 46:3340–3348

Alippe Y, Wang C, Ricci B, Xiao J, Chao Q, Zou W, Novack DV, Abu-Amer Y, Civitelli R, Mbalaviele G (2017) Bone matrix components activate the NLRP3 inflammasome and promote osteoclast differentiation. Sci Rep 7:6630

Anton J, Calvo I, Fernandez-Martin J, Gamir ML, Merino R, Jimenez-Trevino S, Sevilla B, Cabades F, Bou R, Arostegui JI (2015) Efficacy and safety of canakinumab in cryopyrin-associated periodic syndromes: results from a Spanish cohort. Clin Exp Rheumatol 33:S67–S71

Baum R, Sharma S, Carpenter S, Li QZ, Busto P, Fitzgerald KA, Marshak-Rothstein A, Gravallese EM (2015) Cutting edge: AIM2 and endosomal TLRs differentially regulate arthritis and autoantibody production in DNase II-deficient mice. J Immunol 194:873–877

Bechtold S, Simon D (2014) Growth abnormalities in children and adolescents with juvenile idiopathic arthritis. Rheumatol Int 34:1483–1488

Ben-Zvi I, Livneh A (2011) Chronic inflammation in FMF: markers, risk factors, outcomes and therapy. Nat Rev Rheumatol 7:105–112

Ben-Zvi I, Kukuy O, Giat E, Pras E, Feld O, Kivity S, Lecturer S, Perski O, Bornstein G, Grossman C et al (2017) Anakinra for colchicine resistant familial Mediterranean fever – a randomized, double blind, placebo-controlled trial. Arthritis Rheumatol 69:854–862

Bonar SL, Brydges SD, Mueller JL, McGeough MD, Pena C, Chen D, Grimston SK, Hickman-Brecks CL, Ravindran S, McAlinden A et al (2012) Constitutively activated NLRP3 inflammasome causes inflammation and abnormal skeletal development in mice. PLoS One 7: e35979

Bostanci N, Emingil G, Saygan B, Turkoglu O, Atilla G, Curtis MA, Belibasakis GN (2009) Expression and regulation of the NALP3 inflammasome complex in periodontal diseases. Clin Exp Immunol 157:415–422

Bostanci N, Meier A, Guggenheim B, Belibasakis GN (2011) Regulation of NLRP3 and AIM2 inflammasome gene expression levels in gingival fibroblasts by oral biofilms. Cell Immunol 270:88–93

Brydges SD, Mueller JL, McGeough MD, Pena CA, Misaghi A, Gandhi C, Putnam CD, Boyle DL, Firestein GS, Horner AA et al (2009) Inflammasome-mediated disease animal models reveal roles for innate but not adaptive immunity. Immunity 30:875–887

Canna SW, Nigrovic PA (2016) Editorial: 21st century storm chasers: defining macrophage activation syndrome. Arthritis Rheumatol 68:557–560

Canna SW, de Jesus AA, Gouni S, Brooks SR, Marrero B, Liu Y, DiMattia MA, Zaal KJ, Sanchez GA, Kim H et al (2014) An activating NLRC4 inflammasome mutation causes autoinflammation with recurrent macrophage activation syndrome. Nat Genet 46:1140–1146

Cassel SL, Janczy JR, Bing X, Wilson SP, Olivier AK, Otero JE, Iwakura Y, Shayakhmetov DM, Bassuk AG, Abu-Amer Y et al (2014) Inflammasome-independent IL-1beta mediates autoinflammatory disease in Pstpip2-deficient mice. Proc Natl Acad Sci U S A 111:1072–1077

Chae JJ, Cho YH, Lee GS, Cheng J, Liu PP, Feigenbaum L, Katz SI, Kastner DL (2011) Gain-of-function Pyrin mutations induce NLRP3 protein-independent interleukin-1beta activation and severe autoinflammation in mice. Immunity 34:755–768

Chitu V, Nacu V, Charles JF, Henne WM, McMahon HT, Nandi S, Ketchum H, Harris R, Nakamura MC, Stanley ER (2012) PSTPIP2 deficiency in mice causes osteopenia and increased differentiation of multipotent myeloid precursors into osteoclasts. Blood 120:3126–3135

Coll RC, Robertson AA, Chae JJ, Higgins SC, Munoz-Planillo R, Inserra MC, Vetter I, Dungan LS, Monks BG, Stutz A et al (2015) A small-molecule inhibitor of the NLRP3 inflammasome for the treatment of inflammatory diseases. Nat Med 21:248–255

Davis BK, Roberts RA, Huang MT, Willingham SB, Conti BJ, Brickey WJ, Barker BR, Kwan M, Taxman DJ, Accavitti-Loper MA et al (2011) Cutting edge: NLRC5-dependent activation of the inflammasome. J Immunol 186:1333–1337

de Jesus AA, Canna SW, Liu Y, Goldbach-Mansky R (2015) Molecular mechanisms in genetically defined autoinflammatory diseases: disorders of amplified danger signaling. Annu Rev Immunol 33:823–874

Erener S, Petrilli V, Kassner I, Minotti R, Castillo R, Santoro R, Hassa PO, Tschopp J, Hottiger MO (2012) Inflammasome-activated caspase 7 cleaves PARP1 to enhance the expression of a subset of NF-kappaB target genes. Mol Cell 46:200–211

Feldmann J, Prieur AM, Quartier P, Berquin P, Certain S, Cortis E, Teillac-Hamel D, Fischer A, de Saint Basile G (2002) Chronic infantile neurological cutaneous and articular syndrome is caused by mutations in CIAS1, a gene highly expressed in polymorphonuclear cells and chondrocytes. Am J Hum Genet 71:198–203

Greenhill CJ, Jones GW, Nowell MA, Newton Z, Harvey AK, Moideen AN, Collins FL, Bloom AC, Coll RC, Robertson AA et al (2014) Interleukin-10 regulates the inflammasome-driven augmentation of inflammatory arthritis and joint destruction. Arthritis Res Ther 16:419

Guarda G, Zenger M, Yazdi AS, Schroder K, Ferrero I, Menu P, Tardivel A, Mattmann C, Tschopp J (2011) Differential expression of NLRP3 among hematopoietic cells. J Immunol 186:2529–2534

Guo H, Callaway JB, Ting JP (2015) Inflammasomes: mechanism of action, role in disease, and therapeutics. Nat Med 21:677–687

Gurung P, Burton A, Kanneganti TD (2016) NLRP3 inflammasome plays a redundant role with caspase 8 to promote IL-1beta-mediated osteomyelitis. Proc Natl Acad Sci U S A 113:4452–4457

Han X, Lin X, Yu X, Lin J, Kawai T, LaRosa KB, Taubman MA (2013) Porphyromonas gingivalis infection-associated periodontal bone resorption is dependent on receptor activator of NF-kappaB ligand. Infect Immun 81:1502–1509

Hill SC, Namde M, Dwyer A, Poznanski A, Canna S, Goldbach-Mansky R (2007) Arthropathy of neonatal onset multisystem inflammatory disease (NOMID/CINCA). Pediatr Radiol 37:145–152

Hoffman HM, Broderick L (2016) The role of the inflammasome in patients with autoinflammatory diseases. J Allergy Clin Immunol 138:3–14

Hofmann SR, Schnabel A, Rosen-Wolff A, Morbach H, Girschick HJ, Hedrich CM (2016) Chronic nonbacterial osteomyelitis: pathophysiological concepts and current treatment strategies. J Rheumatol 43:1956–1964

Holzinger D, Gieldon L, Mysore V, Nippe N, Taxman DJ, Duncan JA, Broglie PM, Marketon K, Austermann J, Vogl T et al (2012) Staphylococcus aureus Panton-Valentine leukocidin induces an inflammatory response in human phagocytes via the NLRP3 inflammasome. J Leukoc Biol 92:1069–1081

Ippagunta SK, Brand DD, Luo J, Boyd KL, Calabrese C, Stienstra R, Van de Veerdonk FL, Netea MG, Joosten LA, Lamkanfi M et al (2010) Inflammasome-independent role of apoptosis-associated speck-like protein containing a CARD (ASC) in T cell priming is critical for collagen-induced arthritis. J Biol Chem 285:12454–12462

Janowski AM, Sutterwala FS (2016) Atypical Inflammasomes. Methods Mol Biol 1417:45–62

Kassem A, Henning P, Lundberg P, Souza PP, Lindholm C, Lerner UH (2015) Porphyromonas gingivalis stimulates bone resorption by enhancing RANKL (receptor activator of NF-kappaB Ligand) through activation of toll-like receptor 2 in osteoblasts. J Biol Chem 290:20147–20158

Kim H, Walsh MC, Takegahara N, Middleton SA, Shin HI, Kim J, Choi Y (2017) The purinergic receptor P2X5 regulates inflammasome activity and hyper-multinucleation of murine osteoclasts. Sci Rep 7:196

Kitamura A, Sasaki Y, Abe T, Kano H, Yasutomo K (2014) An inherited mutation in NLRC4 causes autoinflammation in human and mice. J Exp Med 211:2385–2396

Kolly L, Busso N, Palmer G, Talabot-Ayer D, Chobaz V, So A (2010) Expression and function of the NALP3 inflammasome in rheumatoid synovium. Immunology 129:178–185

Krauss JL, Zeng R, Hickman-Brecks CL, Wilson JE, Ting JP, Novack DV (2015) NLRP12 provides a critical checkpoint for osteoclast differentiation. Proc Natl Acad Sci U S A 112:10455–10460

Kronenberg HM (2003) Developmental regulation of the growth plate. Nature 423:332–336

Lukens JR, Gross JM, Kanneganti TD (2012) IL-1 family cytokines trigger sterile inflammatory disease. Front Immunol 3:315

Lukens JR, Gross JM, Calabrese C, Iwakura Y, Lamkanfi M, Vogel P, Kanneganti T-D (2014) Critical role for inflammasome-independent IL-1β production in osteomyelitis. Proc Natl Acad Sci U S A 111:1066–1071

Martin SJ (2016) Cell death and inflammation: the case for IL-1 family cytokines as the canonical DAMPs of the immune system. FEBS J 283:2599–2615

Maruotti N, Corrado A, Cantatore FP (2014) Osteoporosis and rheumatic diseases. Reumatismo 66:125–135

Mbalaviele G, Novack DV, Schett G, Teitelbaum SL (2017) Inflammatory osteolysis: a conspiracy against bone. J Clin Investig 127(6):2030–2039

McCall SH, Sahraei M, Young AB, Worley CS, Duncan JA, Ting JP, Marriott I (2008) Osteoblasts express NLRP3, a nucleotide-binding domain and leucine-rich repeat region containing receptor implicated in bacterially induced cell death. J Bone Miner Res 23:30–40

Meng G, Zhang F, Fuss I, Kitani A, Strober W (2009) A mutation in the Nlrp3 gene causing inflammasome hyperactivation potentiates Th17 cell-dominant immune responses. Immunity 30:860–874

Miao EA, Mao DP, Yudkovsky N, Bonneau R, Lorang CG, Warren SE, Leaf IA, Aderem A (2010) Innate immune detection of the type III secretion apparatus through the NLRC4 inflammasome. Proc Natl Acad Sci U S A 107:3076–3080

Miller LS, Pietras EM, Uricchio LH, Hirano K, Rao S, Lin H, O'Connell RM, Iwakura Y, Cheung AL, Cheng G et al (2007) Inflammasome-mediated production of IL-1beta is required for neutrophil recruitment against Staphylococcus aureus in vivo. J Immunol 179:6933–6942

Munoz-Planillo R, Franchi L, Miller LS, Nunez G (2009) A critical role for hemolysins and bacterial lipoproteins in Staphylococcus aureus-induced activation of the Nlrp3 inflammasome. J Immunol 183:3942–3948

Neven B, Marvillet I, Terrada C, Ferster A, Boddaert N, Couloignier V, Pinto G, Pagnier A, Bodemer C, Bodaghi B et al (2010) Long-term efficacy of the interleukin-1 receptor antagonist anakinra in ten patients with neonatal-onset multisystem inflammatory disease/chronic infantile neurologic, cutaneous, articular syndrome. Arthritis Rheum 62:258–267

Novack DV, Mbalaviele G (2016) Osteoclasts-key players in skeletal health and disease. Microbiol Spectr 4

Ozen S, Bilginer Y (2014) A clinical guide to autoinflammatory diseases: familial Mediterranean fever and next-of-kin. Nat Rev Rheumatol 10:135–147

Park E, Na HS, Song YR, Shin SY, Kim YM, Chung J (2014) Activation of NLRP3 and AIM2 inflammasomes by Porphyromonas gingivalis infection. Infect Immun 82:112–123

Qu C, Bonar SL, Hickman-Brecks CL, Abu-Amer S, McGeough MD, Pena CA, Broderick L, Yang C, Grimston SK, Kading J et al (2015) NLRP3 mediates osteolysis through inflammation-dependent and -independent mechanisms. FASEB J 29:1269–1279

Raisz LG (2005) Pathogenesis of osteoporosis: concepts, conflicts, and prospects. J Clin Invest 115:3318–3325

Rigante D, Leone A, Marrocco R, Laino ME, Stabile A (2011) Long-term response after 6-year treatment with anakinra and onset of focal bone erosion in neonatal-onset multisystem inflammatory disease (NOMID/CINCA). Rheumatol Int 31:1661–1664

Romberg N, Al Moussawi K, Nelson-Williams C, Stiegler AL, Loring E, Choi M, Overton J, Meffre E, Khokha MK, Huttner AJ et al (2014) Mutation of NLRC4 causes a syndrome of enterocolitis and autoinflammation. Nat Genet 46:1135–1139

Rosengren S, Hoffman HM, Bugbee W, Boyle DL (2005) Expression and regulation of cryopyrin and related proteins in rheumatoid arthritis synovium. Ann Rheum Dis 64:708–714

Schroder K, Tschopp J (2010) The inflammasomes. Cell 140:821–832

Scianaro R, Insalaco A, Bracci Laudiero L, De Vito R, Pezzullo M, Teti A, De Benedetti F, Prencipe G (2014) Deregulation of the IL-1beta axis in chronic recurrent multifocal osteomyelitis. Pediatr Rheumatol Online J 12:30

Shao W, Yeretssian G, Doiron K, Hussain SN, Saleh M (2007) The caspase-1 digestome identifies the glycolysis pathway as a target during infection and septic shock. J Biol Chem 282:36321–36329

Sibley CH, Plass N, Snow J, Wiggs EA, Brewer CC, King KA, Zalewski C, Kim HJ, Bishop R, Hill S et al (2012) Sustained response and prevention of damage progression in patients with neonatal-onset multisystem inflammatory disease treated with anakinra: a cohort study to determine three- and five-year outcomes. Arthritis Rheum 64:2375–2386

Silveira TN, Gomes MT, Oliveira LS, Campos PC, Machado GG, Oliveira SC (2017) NLRP12 negatively regulates proinflammatory cytokine production and host defense against Brucella abortus. Eur J Immunol, vol 47, pp 51–59

Snouwaert JN, Nguyen M, Repenning PW, Dye R, Livingston EW, Kovarova M, Moy SS, Brigman BE, Bateman TA, Ting JP et al (2016) An NLRP3 mutation causes arthropathy and osteoporosis in humanized mice. Cell Rep 17:3077–3088

Tarp S, Amarilyo G, Foeldvari I, Christensen R, Woo JM, Cohen N, Pope TD, Furst DE (2016) Efficacy and safety of biological agents for systemic juvenile idiopathic arthritis: a systematic review and meta-analysis of randomized trials. Rheumatology (Oxford) 55:669–679

Vande Walle L, Van Opdenbosch N, Jacques P, Fossoul A, Verheugen E, Vogel P, Beyaert R, Elewaut D, Kanneganti TD, van Loo G et al (2014) Negative regulation of the NLRP3 inflammasome by A20 protects against arthritis. Nature 512:69–73

Walsh NC, Gravallese EM (2010) Bone remodeling in rheumatic disease: a question of balance. Immunol Rev 233:301–312

Wang L, Fu H, Nanayakkara G, Li Y, Shao Y, Johnson C, Cheng J, Yang WY, Yang F, Lavallee M et al (2016) Novel extracellular and nuclear caspase-1 and inflammasomes propagate inflammation and regulate gene expression: a comprehensive database mining study. J Hematol Oncol 9:122

Wang C, Xu CX, Alippe Y, Qu C, Xiao J, Schipani E, Civitelli R, Abu-Amer Y, Mbalaviele G (2017) Chronic inflammation triggered by the NLRP3 inflammasome in myeloid cells promotes growth plate dysplasia by mesenchymal cells. Sci Rep 7(1):4880

Yamaguchi Y, Kurita-Ochiai T, Kobayashi R, Suzuki T, Ando T (2016) Regulation of the NLRP3 inflammasome in Porphyromonas gingivalis-accelerated periodontal disease. Inflamm Res 66:1–7

Yoshida K, Okamura H, Hiroshima Y, Abe K, Kido JI, Shinohara Y, Ozaki K (2017) PKR induces the expression of NLRP3 by regulating the NF-kappaB pathway in Porphyromonas gingivalis-infected osteoblasts. Exp Cell Res 354:57–64

Yoshii T, Magara S, Miyai D, Nishimura H, Kuroki E, Furudoi S, Komori T, Ohbayashi C (2002) Local levels of interleukin-1beta, -4, -6 and tumor necrosis factor alpha in an experimental model of murine osteomyelitis due to staphylococcus aureus. Cytokine 19:59–65

Youm YH, Grant RW, McCabe LR, Albarado DC, Nguyen KY, Ravussin A, Pistell P, Newman S, Carter R, Laque A et al (2013) Canonical Nlrp3 inflammasome links systemic low-grade inflammation to functional decline in aging. Cell Metab 18:519–532

Youm YH, Nguyen KY, Grant RW, Goldberg EL, Bodogai M, Kim D, D'Agostino D, Planavsky N, Lupfer C, Kanneganti TD et al (2015) The ketone metabolite beta-hydroxybutyrate blocks NLRP3 inflammasome-mediated inflammatory disease. Nat Med 21:263–269

Zaki FM, Sridharan R, Pei TS, Ibrahim S, Ping TS (2012) NOMID: the radiographic and MRI features and review of literature. J Radiol Case Rep 6:1–8

Zhao Y, Yang J, Shi J, Gong YN, Lu Q, Xu H, Liu L, Shao F (2011) The NLRC4 inflammasome receptors for bacterial flagellin and type III secretion apparatus. Nature 477:596–600

Chapter 12
Inflammasome and Cancer

Zhiyu Wang, Neng Wang, Yifeng Zheng, and Shengqi Wang

Contents

Abstract The current chapter focuses on the role of inflammasome in cancer prevention and development. Emerging evidence suggested that inflammasome is closely correlated with elevated levels of IL-1β and IL-18, activation of NF-κB signaling, enhanced mitochondrial oxidative stress, and activation of autophagic process in cancer. Meanwhile, inflammasome component NOD-like receptors (NLRs) are also involved in carcinogenesis and closely correlated to chemoresponse and prognosis. Although several lines indicated the duplex role of inflammasome in cancer development, the phenomenon might be attributed to NLR difference, cell and tissue type, cancer stage, and specific experimental conditions. Designation of inflammasome targeting strategy has become a novel tool for cancer prevention or treatment.

Z. Wang (✉) · N. Wang
Integrative Breast Cancer Research Laboratory, Guangdong Provincial Hospital of Chinese Medicine, Guangzhou University of Chinese Medicine, Guangzhou, China

The Research Centre for Integrative Medicine, Discipline of Integrated Chinese and Western Medicine, Guangzhou University of Chinese Medicine, Guangdong, China

Y. Zheng · S. Wang
Integrative Breast Cancer Research Laboratory, Guangdong Provincial Hospital of Chinese Medicine, Guangzhou University of Chinese Medicine, Guangzhou, China

M. D. Cordero, E. Alcocer-Gómez (eds.), *Inflammasomes: Clinical and Therapeutic Implications*, Experientia Supplementum 108,
https://doi.org/10.1007/978-3-319-89390-7_12

Keywords Inflammasome · NOD-like receptors · Inflammation · Cancer prevention · IL-1β/IL-18

Abbreviations

ALRs	AIM-like receptors
AOM-DSS	Azoxymethane-dextran sodium sulfate
BIR	Baculovirus inhibitor of apoptosis repeat domain
CARD	Caspase recruitment and activation domain
CASR	Calcium-sensing receptor
DAMPs	Danger-associated molecular patterns
EMT	Epithelial-mesenchymal transition
ER	Endoplasmic reticulum
IL	Interleukin
NLRs	NOD-like receptors
PAMPs	Pathogen-associated molecular patterns
PANX1	Prompts pannexin-1
PRRs	Pattern recognition receptors
PYD	Pyrin domain
ROS	Reactive oxygen species
TLRs	Toll-like receptors
TRPM2	Transient receptor potential melastatin 2

12.1 Introduction

Inflammation is recognized as a major hallmark of cancer. As early as 1863, Rudolph Virchow speculated on a link between cancer and inflammation based on the observation of leukocyte infiltration in human breast cancer (Balkwill and Mantovani 2001; R V 1863). It is generally accepted that up to 25% of malignancies are related to chronic inflammation, chronic infection, or both (Karin 2006; Hussain and Harris 2007; Mantovani et al. 2008). Numerous studies provide evidence that chronic inflammation facilitates resistance to growth inhibition, independent neoangiogenesis, apoptotic evasion, malignant transformation, and metastatic potential obtainment (Shalapour and Karin 2015). During tumor initiation, oxidative molecules including reactive oxygen species and reactive nitrogen species induced by tumor-infiltrating immune cells induce epigenetic alterations in oncogenes or tumor suppressive genes, thereby promoting carcinogenesis (Reuter et al. 2010; Khansari et al. 2009; Bartsch and Nair 2006). On the other hand, during tumor progression and metastasis, cytokines or chemokines secreted by immune cells lead to an increase in cell survival, motility, and invasiveness, such as epithelial-mesenchymal transition (EMT) (Coffelt and Cancer 2014; Yang 2010; Cohen et al. 2015). Elucidating the molecular network between inflammation and cancer risk is of great significance for cancer prevention and treatment.

Once invaded by harmful microbes or foreign particles, germline-encoded pattern recognition receptors (PRRs) constitute the first line of defense. The PRR superfamily includes members of the Toll-like receptors (TLRs), nucleotide-binding oligomerization domain-containing receptors (NOD-like receptors, NLRs), retinoic - acid-inducible gene (RIG) I-like RNA helicases, C-type lectins, and AIM-like receptors (ALRs) (Takeda and Akira 2005; Huysamen and Brown 2009; Yoneyama and Fujita 2007). The molecular targets of PRRs usually include pathogen-associated molecular patterns (PAMPs) and danger-associated molecular patterns (DAMPs). Binding of PAMPs or DAMPs to these receptors leads to an initiation of the host's immune response by activation of inflammatory cells and a number of transcription factors such as NF-κB, STAT, and FOXO (Kono and Rock 2008; Gross et al. 2011). The multimeric inflammasome complex senses all these processes.

Jurg Tschopp was the first to identify the inflammasome in 2002 (Martinon et al. 2002). Its structure consists of an assembly, either of the NLR proteins, NLRP1, NLRP3, NLRC4, NLRP6, and NAIP5 or the DNA-sensing complex of AIM2, a member of the interferon-inducible HIN-200 protein family (Lechtenberg et al. 2014). Activation of inflammasome leads to NLR oligomerization and subsequent interaction with the adaptor protein ASC and the CARD domain of caspase-1. Caspase-1, in turn, regulates the maturation of pro-inflammatory cytokines interleukin-1β (IL-1β) and IL-18 or the rapid inflammatory form of cell death called pyroptosis (Kanneganti 2010; Shin and Brodsky 2015; Rathinam et al. 2012). Notably, the level of IL-1β and IL-18 was found to be significantly elevated in various types of malignancies. These cytokines can facilitate pro-carcinogenic activity by triggering the secretion of VEGF, FGF2, and STAT3 and subsequently support cancer survival and distant metastasis (Tas et al. 2015; Fabbi et al. 2015; Kim et al. 2013). Therefore, elucidating the molecular network of inflammasomes has become a novel strategy for cancer prevention research.

12.2 Inflammasome Cascade Signaling

Compared to TLRs that are usually located on the membrane, NLRs are intracellular molecules and classified into 22 and 34 isoforms in human and mouse genome, respectively. The NLRs are characterized by a tripartite structure, consisting of a carboxy-terminal leucine-rich repeat domain, a central nucleotide-binding oligomerization domain, and a variable N-terminal protein-protein interaction domain, which can be either a Pyrin domain (PYD), a caspase recruitment and activation domain (CARD), or a baculovirus inhibitor of apoptosis repeat domain (BIR) (Fig. 12.1) (Franchi et al. 2009; Siegel 2006). The common NLRs and their functions and ligands are summarized in Table 12.1. The leucine-rich repeat domain appears to act as a ligand-sensing component of NLRs; however, the molecular basis of ligand-binding mechanisms of NLRs is poorly understood (Suarez and Buelvas 2015). The nucleotide-binding oligomerization domain facilitates recruitment of pro-caspase-1 via interactions between pro-caspase-1 and adaptor protein ASC, which take place in the CARD domain (de Alba 2009). The PYD domain of ASC interacts with NLRs and its CARD domain binds directly with pro-caspase-1. Once pro-caspase-1 is recruited to the inflammasome, it will be cleaved into a p35 and p10 fragments in a

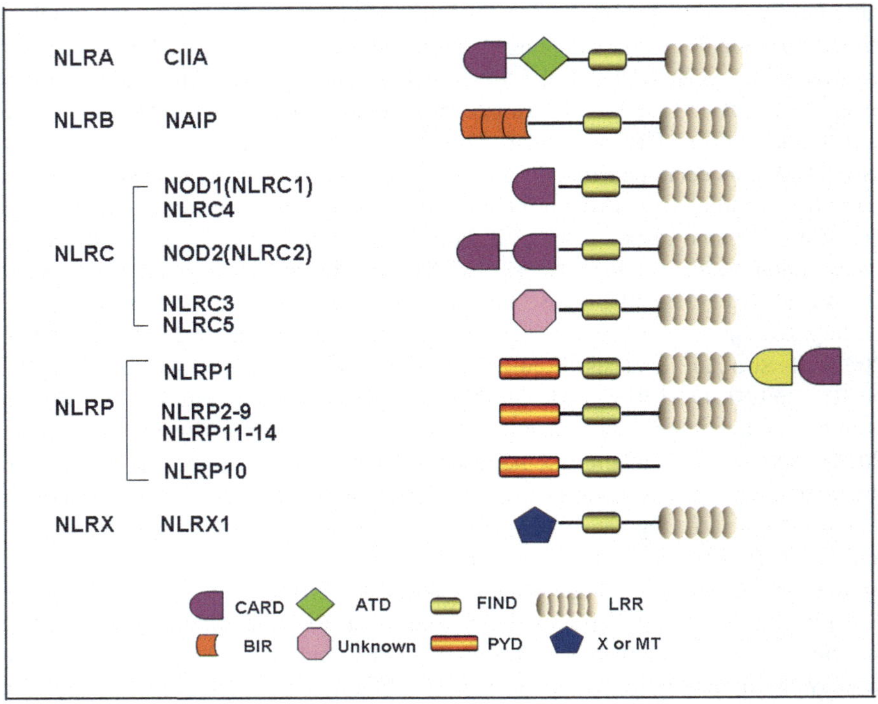

Fig. 12.1 Schematic representation of the basic structure of individual NLR domain. Human NLRs were classified into five categories including NLRA, NLRB, NLRC, NLRP, and NLRX. All 22 human NLRs contain a central NACHT domain and a C-terminal ligand-sensing domain LRR, with the exception of NLRP10. The N-terminal domain of each NLR is specific and responsible for ascribing different biofunctions. *CARD* caspase association and recruitment domain, *ATD* acidic transactivation domain, *FIND* function to find domain, *PYD* pyrin domain, *BIR* baculovirus inhibitor of apoptosis repeat domain, *LRR* leucine-rich repeats, *MT* targets NLRX1 to the mitochondria but no sequence homology with traditional mitochondrial targeting sequence has been reported

proximity-induced multimerization manner. The p35 fragment will subsequently be processed into the CARD and a p20 subunit. The p10 fragment together with two molecules of p20 will finally form an active caspase-1 enzyme, which converts pro-IL-1β and pro-IL-18 into their active forms (Fig. 12.2) (Bryan et al. 2009; Ippagunta et al. 2011). Furthermore, pyroptosis could also be induced following caspase-1 activation, and its activation is considered to be a critical mechanism fighting against Gram-negative and Gram-positive bacteria (Miao et al. 2010).

Alternatively, inflammasome could also be activated through a noncanonical pathway, which involves caspase-11 or caspase-8. Caspase-11 was found to be necessary for the maturation of IL-1β and IL-18 in enteric bacteria such as *Escherichia coli*, *Citrobacter rodentium*, and *Vibrio cholera*. After recruitment to the inflammasome, pro-caspase-11 is cleaved into the p26 subunit and subsequently interacts with caspase-1 (Kayagaki et al. 2011). Studies show that caspase-8 was necessary for the inflammasome activation in LPS-primed macrophages and

Table 12.1 Inflammasome- and non-inflammasome-forming NLRs, functions, and their ligands

NLRs	NLR family	Functions	Ligands
NLRP3	NLRP	Interacts with caspase-1 and ASC; activates NF-κB signaling and IL-1β/IL-18 release	Muramyl dipeptide, LPS, bacterial and viral DNA/RNA, silica, amyloid-β fibrils, extracellular ATP
NLRC4	LRC	Interacts with caspase-1, ASC, and NAIP; elevates NF-κB signaling and IL-1β/IL-18 release	Flagellin from *Salmonella*, *Legionella*, *Listeria*, *Pseudomonas*
NAIP	NLRB	Formation of NAIP/NLRC4 inflammasome complex	Flagellin from *Legionella*
NLRP6	NLRP	Inflammasome complex formation with ASC and caspase-1; activates NF-κB signaling and IL-1β/IL-18 release	Ligands unknown
NLRP1	NLRP	Inflammasome complex formation with ASC and caspase-1	Muramyl dipeptide, *Toxoplasma gondii* and *Bacillus anthracis* lethal toxin
NLRP12	NLRP	Inhibits IRAK1, TRAF3, and NIK; attenuates both canonical and noncanonical NF-κB signaling	Ligands unknown
NLRX1	NLRX	Inhibits TRAF6 and attenuates canonical NF-κB signaling	Viral RNA
NLRC3	NLRC	Inhibits TRAF6 and attenuates canonical NF-κB signaling	Ligands unknown
NOD1	NLRC	Interacts with RIP2 and recruits RICK and CARD9	GM-tripeptide γ-d-Glu-DAP(iEDAP) d-lactyl-l-Ala-γ-Glu-meso-DAP-Gly (FK156) heptanolyl-γ-Glu-meso-DAP-Ala
NOD2	NLRC	Recruits RIP2 and activates NF-κB and MAPK pathways; negatively regulated by CARD8	Muramyl dipeptide MurNAc-l-Ala-g-d-Glu-l-Lys (M-TRlys)

dendritic cells (Gurung and Kanneganti 2015); however, the detailed interaction mode and mechanisms are poorly understood.

12.3 Regulation of Inflammasome Activation

Given that IL-1β, IL-18, and pyroptotic death response have the potential to damage the host, tight control of inflammasome activation is of great significance for the prevention of disease progression. According to the 2-signal model of inflammasome induction, NF-κB is thought to serve as the first signal that primes NLR and pro-IL-1β expression (Gross et al. 2011; Rathinam et al. 2012; Hoesel and Schmid 2013). Constitutive activation of NF-κB is shown in a wide variety of tumor types, such as

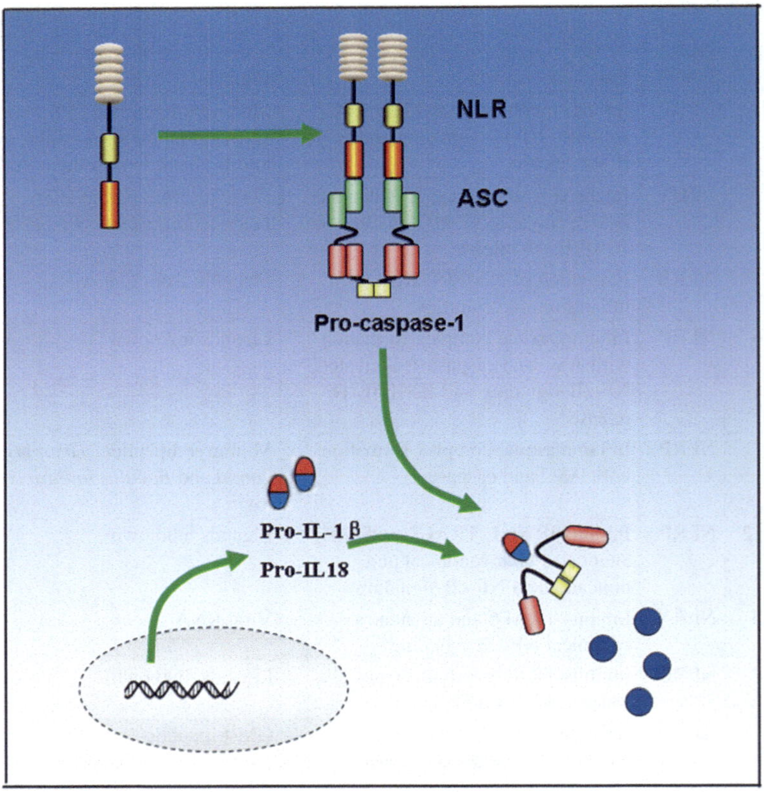

Fig. 12.2 Basic mechanisms of activation of the main NLR inflammasome. The recognition of PAMPs and/or DAMPs leads to NOD domain oligomerization, which in turn facilitates recruitment of pro-caspase-1 via the CARD domain interactions between pro-caspase-1 and adaptor protein ASC. Pro-caspase-1 will be then cleaved and converts pro-IL-1β and pro-IL-18 into their active forms to amplify the inflammatory response. On the other hand, caspase-1 can lead to cell pyroptosis with the consequence of membrane rupture and release of alarmins such as IL-1α and HMGB1

lymphoma, liver cancer, lung cancer, breast cancer, etc. (Fan et al. 2013; DiDonato et al. 2012). Besides, NF-κB is activated in response to carcinogenic processes such as tobacco, stress, obesity, alcohol, infectious agents, irradiation, etc (Fan et al. 2013; DiDonato et al. 2012). Furthermore, NF-κB controls the expression of the genes linked with proliferation, invasion, angiogenesis, and metastasis of cancer (Prasad et al. 2010). Besides, NF-κB activation further upregulates a series of inflammatory factors, such as TNFα, IL-6, IL-1, and IL-8, which constitute a positive feedback loop to induce cellular and DNA damage and to promote cell proliferation and transformation (Fan et al. 2013). A previous study also demonstrated that NLRP3 promoter contains putative NF-κB binding site and NF-κB inhibition resulted in a significant reduction of NLRP3 expression (Qiao et al. 2012). Meanwhile, mounting evidence suggested that NLRP3 inflammasome formation is positively associated with NF-κB activity following drug treatment such as LPS, CPT-11, FGF-21, etc. (Li et al. 2015a; Liu 2015; Xiang et al. 2015). All these findings implied that the pro-tumorigenic ability of NF-κB might be attributed to inflammasome activation.

Similar to NF-κB, type I interferon is also important for inflammasome activation. AIM2 inflammasome activation following *F. tularensis* infection requires type I interferon stimuli, whereas macrophages deficient in the type I interferon secretion result in reduced response of AIM2 inflammasome (Fernandes-Alnemri et al. 2010; Jones et al. 2010; Henry et al. 2007). Although the precise mechanism of interferon signaling remains unclear, it has been proposed that type I interferon activates AIM2 inflammasome by generating cytosolic DNA from *F. tularensis* (Jones et al. 2010; Henry et al. 2007). However, type I interferon is also reported to inhibit inflammasome activation by two distinct mechanisms including the alteration of intracellular pro-IL-1β concentration and inhibition of caspase-1 activation (Guarda et al. 2011). The reduction of pro-IL-1β is determined by the capacity of type I interferon to induce the production of the anti-inflammatory cytokine, IL-10. IL-10 activation by STAT3 signaling pathway can inhibit the synthesis of pro-IL-1β and pro-IL-1α (Dickensheets and Donnelly 1997). In addition, type I interferon is capable of suppressing caspase-1 activity *by* activation of the transcription factor, STAT1, subsequently inhibiting NLRP3 and NLRP1 inflammasome (Fig. 12.3) (Detjen et al. 2001). Both in vitro and in vivo experiments further confirmed that IFN-β could suppress NLRP3 inflammasome, but the exact molecular mechanism that guides the preferential targeting of NLRP3 and NLRP1 inflammasome by type I interferon remains to be identified (Malhotra et al. 2015). These data provide a duplex role of type I interferon in inflammasome modulation, which might be dependent on infected organisms or cell type and inflammation status. Besides the cross talk between cytokines and inflammasomes, recent studies also suggest that the effector and memory T cells can block the activation of caspase-1 and IL-1β in macrophages and dendritic cells, mediated by CD40L, OX40L, and RANKL, which are all members of the TNF superfamily of ligands expressed on activated cells (Masters et al. 2010; Guarda et al. 2009). Interestingly, although it is recognized that the T cells only target NLRP1 and NLRP3 inflammasomes, the underlying molecular mechanism of how TNF ligands mediate the inhibition of caspase-1-dependent production of IL-1β is unknown and needs further investigation.

Evolutionarily, autophagy is a cell-protective mechanism against harmful stress that facilitates catabolic processes and inhibits anabolic metabolism. Growing evidence indicates that autophagy is a critical process participating in cancer initiation and metastasis, growth, and drug resistance (Rebecca and Amaravadi 2016; Jiang et al. 2015). Intriguingly, recent reports have also indicated that autophagy regulates various aspects of the immune response, such as antigen presentation, cell death, and cytokine secretion in immune cells (Pan et al. 2016). In autophagy-deficient *Atg16*$^{-/-}$ mice, it was observed that the levels of IL-1β and IL-18 were significantly elevated following LPS treatment. However, the elevated IL-1β and IL-18 expression was not due to enhanced transcriptional activity, but instead was attributed to over-activation of caspase-1 (Saitoh et al. 2008). Subsequent mechanistic studies demonstrated that the augmented caspase-1 activity might be due to the failure of autophagy-deficient cells to clear damaged mitochondrion (Saitoh et al. 2008). When autophagy is

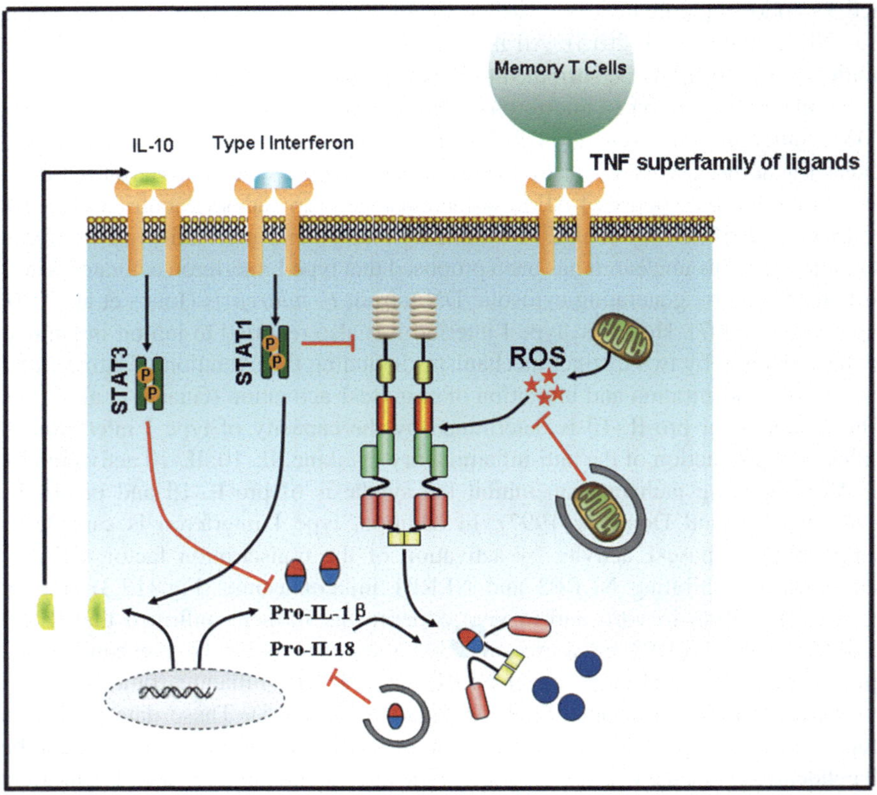

Fig. 12.3 Main signaling involved in the regulation of inflammasome activation. Type I interferon signaling triggers the production of IL-10, which in turn acts on cells in an autocrine or paracrine manner to suppress the intracellular concentration of pro-IL-1β via the stat3 pathway. ROS burst from damaged mitochondrion could drive activation of inflammasome, but autophagy could block the accelerated IL-1β/IL-18 production via degrading the damaged mitochondrion and sequestering intracellular stores of pro-IL-1β and IL-18. Meanwhile, effector and memory T cells could also suppress inflammasome activation via a cognate mechanism mediated by TNF superfamily and their receptors

inhibited, excessive reactive oxygen species (ROS) will accumulate in the damaged mitochondrion, resulting in the release of mitochondrion DNA into the cytoplasm that finally triggers activation of NLRP3 inflammasome (Fig. 12.3) (Zhou et al. 2011; Nakahira et al. 2011). However, detailed mechanisms accounting for how mitochondrial ROS or mitochondrial DNA activates inflammasome are still unclear. Meanwhile, an inflammasome-independent mechanism of autophagy-mediated regulation of IL-1β expression was recently identified. Autophagosome could degrade pro-IL-1β, thereby restraining the substrate for caspase-1 processing (Fig. 12.3) (Harris et al. 2011; Crisan et al. 2011). Alternatively, autophagy inhibition could also activate the transcription of pro-IL-1β in human peripheral blood mononuclear cells (Crisan et al. 2011). Recently, autophagy was reported to affect inflammasome activity by influencing IL-1β translocation from the endoplasmic reticulum and Golgi apparatus (Dupont et al. 2011). Lastly, autophagy machinery is believed to participate in clearing large inflammasome complexes from cells in order to prevent excessive cell damage by IL-1β and IL-18 (Martins et al. 2015). Therefore, the autophagic process regulates inflammasome activation at several levels.

12.4 Role of NOD-Like Receptors in Cancer Development

The abnormal activation of inflammasome is linked to various types of human disease, such as cryopyrinopathies, gout, asbestosis, silicosis, Alzheimer's disease, and autoimmune diseases (Lamkanfi et al. 2011; Hoffman and Brydges 2011). To date, more than 70 inherited mutations have been identified associating with cryopyrinopathy occurrence, a large majority of which are situated within and around NLRP3 NACHT domain (Masters et al. 2009; Dowds et al. 2004). These mutations are therefore believed to induce conformational changes that render NLRP3 constitutively active, resulting in continuous caspase-1 activation and release of IL-1β and IL-18 (Dowds et al. 2004). Besides, decreased NLRP3 expression and reduced IL-1β production have recently been linked with increased susceptibility to Crohn's disease in humans (Li et al. 2004; Villani et al. 2009). Moreover, inflammasome deregulation was also recorded to contribute to the pathogenesis of experimental autoimmune encephalomyelitis (Shaw et al. 2010). Significantly, accumulating evidence also suggested that NLRs are closely correlated to cancer occurrence, but conflicting evidence also exists, which might be due to the dual functions of inflammasomes in promoting carcinogenic inflammation or eliminating malignant cells via the pyroptosis death pathway.

12.4.1 NLRP3 Signaling and Its Duplex Role

NLRP3 is the most well-studied member of NLR family. It can be activated by a wide range of signals including infected pathogens, endogenous or environmental origins (Sutterwala et al. 2014). Based on current findings, three distinct mechanisms have been proposed to account for NLRP3 activation, including potassium efflux, phagolysosomal destabilization, and mitochondrial ROS burst. Various bacterial pathogens can secrete pore-forming toxins (e.g., nigericin from *Streptomyces hygroscopicus*, listeriolysin O from *Listeria monocytogenes*, pneumolysin from *Streptococcus pneumoniae*, alpha-hemolysin from *Escherichia coli*) and subsequently activate the NLRP3 inflammasome by increasing potassium efflux (Munoz-Planillo et al. 2009; Meixenberger et al. 2010; Kim et al. 2010; Willingham et al. 2009; Duncan et al. 2009; Allen et al. 2009; Hise et al. 2009). In addition, bacterial and viral RNA is also reported to be an initial factor contributing to NLRP3 inflammasome assembly (Li et al. 2015b). Moreover, extracellular ATP released from phagocytosed dying cells acts on purinergic receptor P2X7 and induces pannexin-1 (PANX1) channels to promote potassium efflux and results in NLRP3 activation (Mariathasan et al. 2006; Kanneganti et al. 2007; Piccini et al. 2008). On the other hand, intracellular uptake of crystalline and particulate matters is also capable of causing lysosomal destabilization and release of cathepsin B, a sensor of NLRP3 (Cassel et al. 2008; Dostert et al. 2008; Eisenbarth et al. 2008; Halle et al. 2008; Hornung et al. 2008). Lysosome rupture-induced NLRP3 activation was also observed in cathepsin B-deficient cells, a phenomenon that may be attributed to potassium efflux (Fig. 12.4) (Dostert et al. 2009).

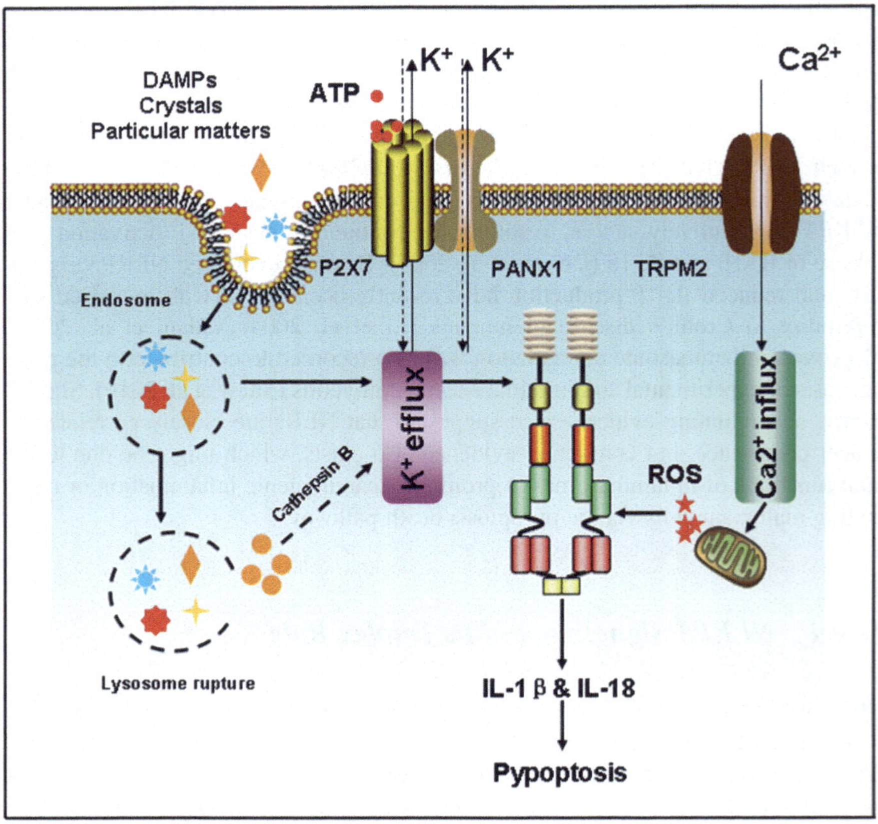

Fig. 12.4 Simplified mechanisms for NLRP3 inflammasome activation. Three distinct machineries have been proposed to account for NLRP3 activation, including K^+ efflux, phagolysosomal destabilization, and mitochondrial ROS burst. Extracellular ATP released from dying cells acts on purinergic receptor P2X7 and prompts pannexin-1 (PANX1) channels to enhance K^+ efflux and result in NLRP3 activation. Meanwhile, PAMPs such as pore-forming toxins are also capable to facilitate K^+ efflux and activate NLRP3 inflammasome. Besides, K^+ efflux could be activated by crystals or particular maters, which enter the cells via endocytosis and trigger NLRP3 inflammasome via cathepsin B following lysosome rupture. Finally, intracellular Ca^{2+} accumulation could result in mitochondrion damage and lead to ROS burst, which may activate the NLRP3 inflammasome either directly or by inducing K^+ efflux. Following NLRP3 activation, IL-1β and IL-18 will be greatly produced and result in inflammation or pyroptotic cell death

Notably, mitochondrial ROS generation is considered to be one of the most important mechanisms of NLRP3 activation. Pharmacological inhibition of mitochondrial ROS burst has been shown to prevent NLRP3-inflammasome formation (Cassel et al. 2008; Dostert et al. 2008; Zhong et al. 2013). Although the detailed molecular mechanisms of ROS-mediated NLRP3 activation remain largely unclear, calcium influx mediated by the transient receptor potential melastatin 2 (TRPM2) has been suggested to be a possible reason (Zhong et al. 2013). Extracellular calcium has been shown to activate the calcium-sensing receptor (CASR) and thus lead to the release of calcium stores from the endoplasmic reticulum (ER), eventually triggering the formation of NLRP3 inflammasome (Fig. 12.4) (Lee et al. 2012; Murakami et al. 2012; Rossol et al. 2012). On the other hand, mitochondrial ROS burst is an upstream event leading to the loss of mitochondrial membrane potential, a pivotal event in inducing intrinsic apoptosis. Interestingly, overexpression of the antiapoptotic protein, BCL2, was shown to limit the activation of NLRP3 inflammasome, indicating that apoptosis-regulated proteins might be closely correlated with NLRP3 activation (Bruey et al. 2007; Faustin et al. 2009). More recently, cIAP1, cIAP2, and XIAP have also been linked with inflammasome activation. cIAP1 and cIAP2 were found to enhance inflammasome activation by ubiquitinating and stabilizing caspase-1 and consequently promoting Il-1β release, whereas concurrent inhibition of cIAP1, cIAP2, and XIAP was shown to limit caspase-1 activation (Hawkins et al. 1996; Labbe et al. 2011; Vince et al. 2012). Overall, these studies place the mitochondria as a potential player for inflammasome activation. However, the precise role of mitochondria in mediating NLRP3 inflammasome formation and subsequent promotion of carcinogenesis awaits clarification.

With regard to pro-tumorigenic ability, NLRP3 polymorphism is shown to be associated with melanoma susceptibility, colorectal cancer prognosis, and overall survival of myeloma (Cook et al. 2010). In a Swedish case-control study, NLRP3 variant (rs35829419) was significantly more common in male patients than in controls (OR, 2.22; CI, 1.27–3.86) and showed strong association with nodular melanoma (OR, 2.89; CI: 1.33–6.30) (Verma et al. 2012). It has been suggested that NLRP3 activation could suppress NK and T cell-mediated antitumor actions in a sarcoma mouse model and metastatic melanoma, whereas the population of myeloid-derived suppressor cells and Tregs were increased (Chow et al. 2012). Consistently, NLRP3 silencing resulted in a fivefold reduction in the number of tumor-associated myeloid-derived suppressor cells found in host mice, and NLRP3 $^{-/-}$ MDSCs were less efficient to reach the tumor site, demonstrating the critical role of NLRP3 in preventing cancer occurrence by modulating host immunity (van Deventer et al. 2010). Furthermore, it was found that NLRP3-deficient mice generated less pulmonary metastasis in an orthotopic transplant mouse model of mammary adenocarcinoma (Ghiringhelli et al. 2009). In addition, chemotherapeutic agents, gemcitabine and 5-fluorouracil, were shown to activate NLRP3-mediated inflammasome formation in myeloid-derived suppressor cells, leading to IL-1β production that is capable of inducing IL-17 secretion from $CD4^+$ T cells and blunting the anticancer efficacy of chemotherapeutic drugs (Bruchard et al. 2013). Accordingly, gemcitabine and 5-fluorouracil exert increased antitumor effects when tumors were established in NLRP3$^{-/-}$ or caspase-1$^{-/-}$ mice, and NLRP3 activation by chemotherapeutic drugs is considered to be a positive regulator to promote cancer growth (Bruchard et al. 2013). All these findings suggest the pro-tumorigenic role of NLRP3 in cancer development.

Although several lines of evidence have indicated the pro-carcinogenic activities of NLRP3, its role in cancer development remains controversial. NLRP3$^{-/-}$ mice were shown to be more susceptible to cancer, and the number of colon polyps in the AOM-DSS mouse model and the accelerated tumor growth in the carcinogenesis model was accompanied with drastically low levels of colonic IL-18, suggesting that NLRP3 may play a protective role against neoplasia formation and IL-18 might be closely associated with colon cancer initiation (Zaki et al. 2010a; Dupaul-Chicoine et al. 2010). Of note, IL-18 knockout mice generated more tumors than controls after administering AOM-DSS (azoxymethane-dextran sodium sulfate), whereas injection of recombinant IL-18 successfully restrained disease progression, which might be associated with MyD88-related pathway (Zaki et al. 2010b). Similar anticarcinogenic role of NLRP3 was also observed in hepatocellular carcinoma. Both mRNA and protein levels of NLRP3 were significantly downregulated in the hepatic parenchymal cells derived from liver cancer biopsies compared to noncancerous samples (Wei et al. 2014). In this context, it is logical to deduce that NLRP3 may play a duplex role in controlling cancer growth. Thus, on one hand, NLRP3 could promote tumor cell survival through activation of NF-κB-stat1/3 pathway or by limiting cytotoxic immune cell infiltration, but on the other, it could suppress malignant progression by triggering mitochondrial apoptotic pathway or by enhancing immune cytokine levels in the tumor microenvironment. In addition, the role of NLRP3 in cancer development might be tissue or cell dependent. For example, NLRP3 exhibits a protective role for colon cancer, but pro-carcinogenic effects for gastric and prostate malignancies (Xu et al. 2013). Therefore, a much more comprehensive analysis of NLRP3 using conditional knockout models and pharmacological activators or inhibitors is needed to decode the precise effects of NLRP3 on cancer development in the future.

12.4.2 Other NLRs in Carcinogenesis

Besides NLRP3, a number of NLRs have also been shown to be associated with cancer progression. NLRC4 was identified as a downstream transcriptional target of p53, indicating the tumor suppressive role of NLRC4 (Sadasivam et al. 2005). Mice lacking NLRC4 had significantly increased tumor numbers and burden compared to the wild-type controls in the AOM-DSS colon cancer model, but no differences in inflammation severity were noted, implying that tumor regulation by NLRC4 might be mostly cell intrinsic and not through downregulation of inflammation (Hu et al. 2011; Hu et al. 2010). Similar to NLRC4, both NLRP6 and NLRP12 were also found to play a critical role in AOM-DSS tumorigenesis. A significant increase in tumor number and burden was observed in NLRP6-deficient mice compared to wild-type controls after chemical induction. But unlike NLRC4, NLRP6-mediated protection against tumor formation is attributed to hematopoietic cells rather than intestinal epithelial cells, because similar numbers of tumor were observed between NLRP6-deficient mice and irradiated wild-type mice that were transplanted with NLRP6-deficient bone marrow (Chen et al. 2011; Elinav et al. 2011; Normand et al. 2011). By contrast, NLRP6-deficient mice that received wild-type bone marrow transplant were shown to have reduced tumorigenic ability, similar to that of wild-type animals (Elinav et al. 2011). In addition, genetic profiling of tumors from wild-type and NLRP6-deficient mice exhibited a significant increase in the number of genes in the Wnt and Notch signaling cascade from a set of 1,884 genes, supporting a novel role of NLRP6 in controlling intestinal proliferation (Normand et al. 2011). Of note, pro-inflammatory cytokines such as TNFα and IL-6 were elevated in the tumor microenvironment, whereas the level of IL-18 was significantly reduced. Meanwhile, IL-18 silencing in NLRP6-deficient mice has been associated with increased colon cancer development, indicating the pivotal role of cytokines in mediating the anticarcinogenic activities of NLRP6 (Chen et al. 2011). Similar to NLRP6, NLRP12 was also considered to be a tumor suppressive molecule as shown in ex vivo and in vivo carcinogenic animal models. Mice lacking NLRP12 were found to be more susceptible to DSS injury, accompanied by increased body weight loss, enhanced pathology scores coupled with severe inflammatory cell infiltration, and high levels of cytokine production (Jeru et al. 2008; Arthur et al. 2010; Borghini et al. 2011). The AOM-DSS mouse model also revealed that NLRP12-deficient mice had accelerated colon tumor development and progression, which was demonstrated with over-activation of NF-κB signaling pathway and enhanced gene expression such as CXCL12 and CXCl13 (Zaki et al. 2011; Allen et al. 2012). Taken together, the NLRP6/12-mediated protective mechanisms against tumorigenesis provide a complex network involving interactions between hematopoietic cells, cytokines, and epithelial cells and further show that experimental validation is needed to pinpoint the precise signaling transduction mode underlying their anticarcinogenic effects.

12.4.3 Double-Edged Swords of Pyroptosis

Pyroptosis is a critical self-protection mechanism responding to pathogen invasion by inducing pro-inflammatory cell death. Unlike apoptosis, pyroptosis is characterized by cytoplasmic swelling and early cellular membrane rupture, which happens following caspase activation, nuclear condensation, and DNA fragmentation (Bergsbaken et al. 2009). Although the precise mechanisms underlying pyroptosis induction still remain elusive, the products released from dead cells may limit malignant cell survival and proliferation by activating the innate immune response. Increasing evidence validates that dying tumor cells following chemotherapy might activate the NLRP3 inflammasome of dendritic cells via P2X7 purinergic receptors, thus priming tumor-specific interferon-γ-producing T lymphocytes to limit cancer growth (Zitvogel et al. 2012). Moreover, mice lacking P2X7 or NLRP3 failed to prime interferon-γ-producing $CD8^+$ T cells after chemotherapy, and anthracycline-treated breast cancer patients with P2X7 mutation developed metastatic lesions more rapidly than normal individuals (Ghiringhelli et al. 2009). Notably, a novel therapeutic strategy is in development to foster dendritic cell-mediated antitumor immunity via acceleration of pyroptosis of cancer cells by oncolytic viruses (Li et al. 2008). However, several studies have also indicated that pyroptosis might contribute to tumorigenic ability after inflammasome activation (Masters et al. 2012; Wree et al. 2014). These conflicting findings may be attributed to differences in the redox status of model cells and specific molecules involved in the process. For example, the reduced form of HMGB1 released from dying cells could trigger dendritic cells to induce antitumor immune response, while the oxidized form of HMGB1 would be unable to activate the immune response (Apetoh et al. 2007; Kazama et al. 2008). In addition, the role of pyroptosis in cancer development might critically depend on the cell type. Pyroptosis of immune cells might bring harmful consequences to tumor immunoediting, while cancer cell pyroptosis would improve anticancer immunity. Overall, impaired pyroptosis has been considered to be a potential mechanism linking chronic inflammation to cancer initiation, and pyroptosis targeting is becoming a novel strategy to prevent cancer and improve cancer therapeutic efficacy.

12.5 Conclusions

Remarkable advancements in recent years have greatly increased our understanding of NLR function and the associated inflammasome in host defense and disease pathogenesis. NLR-containing inflammasomes are not only important for fighting against bacteria, fungi, and viruses but also appear to be a critical step in mediating cancer initiation and progression. Inflammasome activation would create a pro-inflammatory microenvironment for inducing malignant transformation and suppress local immunity caused by NK or T cells. In addition, chemotherapeutic agents were found to activate inflammasome defense, which had positive feedback to support cancer growth. Notably, inflammasome-related autophagy is also believed to significantly contribute to cancer drug resistance and metastasis. All these findings greatly highlight the role of inflammasome as a novel target to prevent and treat cancer.

Despite mounting evidence listed above suggesting the potential of the inflammasome as a promising marker for cancer prevention, contrary data also exists to imply that inflammasome signaling could behave as a kind of anticancer mechanism. Mice lacking NLRP3 or NLRC4 show higher susceptibility to colon cancer following AOM-DSS treatment, and aberrant inflammasome formation leads to inhibition of tumor suppressor genes such as p53 and over-activation of oncogenes such as Wnt. What's more, inflammasome-mediated pyroptosis is also considered to play a critical role in recruiting dendritic cells to limit cancer growth. Based on the conflicting evidence, a number of questions remain unanswered. Whether or not inflammasome-related carcinogenesis is cell dependent is an important question. A second question is whether a specific NLR would exhibit different bioactivity correlating with a cancer stage. Meanwhile, there are 22 NLR members in humans, and it is unknown how these NLR molecules are activated or how they interact with each other. The complex network awaits elucidation. There is also a lot of interest to identify novel ligand-receptor binding molecules, novel signaling pathway, and novel targets for cancer prevention or therapy. The inflammasome is becoming a significant research topic in tumor microenvironment field, and there is every likelihood that it could be developed as important biomarker for cancer diagnosis or prognosis prediction. Meanwhile, drug discovery targeting inflammasome modulation is also expected to improve cancer therapeutic efficacy to successfully reduce cancer risk. Taken together and given the emerging role of inflammasome in cancer development, understanding its signaling network and pathological significance might bring novel strategies for malignancy therapy and prevention.

Acknowledgments This work was supported by the National Natural science Foundation of China (81573651, 81703749 and 81703764), Guangdong Science and Technology Department (2016A030306025), Pearl River S&T Nova Program of Guangzhou (201506010098), Combined Scientific Project Funded by Guangdong Provincial Science and Technology Agency and Guangdong Provincial Academy of Traditional Chinese Medicine (2014A020221047), Guangdong Highlevel university construction project (A1-AFD018161Z1510, A1-AFD01817Z11102, A1-AFD01817Z11101), Guangdong High-level Personnel of Special Support Program (A1-3002-16-111-003) and The PhD Start-up Fund of Natural Science Foundation of Guangdong Province, China (2017A030310213).

Conflict of Interests No potential conflicts of interest were disclosed.

References

Allen IC, Scull MA, Moore CB, Holl EK, McElvania-TeKippe E, Taxman DJ, Guthrie EH, Pickles RJ, Ting JP (2009) The NLRP3 inflammasome mediates in vivo innate immunity to influenza A virus through recognition of viral RNA. Immunity 30(4):556–565

Allen IC, Wilson JE, Schneider M, Lich JD, Roberts RA, Arthur JC, Woodford RM, Davis BK, Uronis JM, Herfarth HH, Jobin C, Rogers AB, Ting JP (2012) NLRP12 suppresses colon inflammation and tumorigenesis through the negative regulation of noncanonical NF-kappaB signaling. Immunity 36(5):742–754

Apetoh L, Ghiringhelli F, Tesniere A, Obeid M, Ortiz C, Criollo A, Mignot G, Maiuri MC, Ullrich E, Saulnier P, Yang H, Amigorena S, Ryffel B, Barrat FJ, Saftig P, Levi F et al (2007) Toll-like receptor 4-dependent contribution of the immune system to anticancer chemotherapy and radiotherapy. Nat Med 13(9):1050–1059

Arthur JC, Lich JD, Ye Z, Allen IC, Gris D, Wilson JE, Schneider M, Roney KE, O'Connor BP, Moore CB, Morrison A, Sutterwala FS, Bertin J, Koller BH, Liu Z, Ting JP (2010) Cutting edge: NLRP12 controls dendritic and myeloid cell migration to affect contact hypersensitivity. J Immunol 185(8):4515–4519

Balkwill F, Mantovani A (2001) Inflammation and cancer: back to Virchow? Lancet 357(9255): 539–545

Bartsch H, Nair J (2006) Chronic inflammation and oxidative stress in the genesis and perpetuation of cancer: role of lipid peroxidation, DNA damage, and repair. Langenbeck's Arch Surg/ Deutsche Gesellschaft fur Chirurgie 391(5):499–510

Bergsbaken T, Fink SL, Cookson BT (2009) Pyroptosis: host cell death and inflammation. Nat Rev Microbiol 7(2):99–109

Borghini S, Tassi S, Chiesa S, Caroli F, Carta S, Caorsi R, Fiore M, Delfino L, Lasiglie D, Ferraris C, Traggiai E, Di Duca M, Santamaria G, D'Osualdo A, Tosca M, Martini A et al (2011) Clinical presentation and pathogenesis of cold-induced autoinflammatory disease in a family with recurrence of an NLRP12 mutation. Arthritis Rheum 63(3):830–839

Bruchard M, Mignot G, Derangere V, Chalmin F, Chevriaux A, Vegran F, Boireau W, Simon B, Ryffel B, Connat JL, Kanellopoulos J, Martin F, Rebe C, Apetoh L, Ghiringhelli F (2013) Chemotherapy-triggered cathepsin B release in myeloid-derived suppressor cells activates the Nlrp3 inflammasome and promotes tumor growth. Nat Med 19(1):57–64

Bruey JM, Bruey-Sedano N, Luciano F, Zhai D, Balpai R, Xu C, Kress CL, Bailly-Maitre B, Li X, Osterman A, Matsuzawa S, Terskikh AV, Faustin B, Reed JC (2007) Bcl-2 and Bcl-XL regulate proinflammatory caspase-1 activation by interaction with NALP1. Cell 129(1):45–56

Bryan NB, Dorfleutner A, Rojanasakul Y, Stehlik C (2009) Activation of inflammasomes requires intracellular redistribution of the apoptotic speck-like protein containing a caspase recruitment domain. J Immunol 182(5):3173–3182

Cassel SL, Eisenbarth SC, Iyer SS, Sadler JJ, Colegio OR, Tephly LA, Carter AB, Rothman PB, Flavell RA, Sutterwala FS (2008) The Nalp3 inflammasome is essential for the development of silicosis. Proc Natl Acad Sci U S A 105(26):9035–9040

Chen GY, Liu M, Wang F, Bertin J, Nunez G (2011) A functional role for Nlrp6 in intestinal inflammation and tumorigenesis. J Immunol 186(12):7187–7194

Chow MT, Sceneay J, Paget C, Wong CS, Duret H, Tschopp J, Moller A, Smyth MJ (2012) NLRP3 suppresses NK cell-mediated responses to carcinogen-induced tumors and metastases. Cancer Res 72(22):5721–5732

Coffelt SB, Cancer d VKE (2014) Inflammation lights the way to metastasis. Nature 507(7490): 48–49

Cohen EN, Gao H, Anfossi S, Mego M, Reddy NG, Debeb B, Giordano A, Tin S, Wu Q, Garza RJ, Cristofanilli M, Mani SA, Croix DA, Ueno NT, Woodward WA, Luthra R et al (2015) Inflammation mediated metastasis: immune induced epithelial-to-mesenchymal transition in inflammatory breast cancer Cells. PLoS One 10(7):e0132710

Cook GP, Savic S, Wittmann M, McDermott MF (2010) The NLRP3 inflammasome, a target for therapy in diverse disease states. Eur J Immunol 40(3):631–634

Crisan TO, Plantinga TS, van de Veerdonk FL, Farcas MF, Stoffels M, Kullberg BJ, van der Meer JW, Joosten LA, Netea MG (2011) Inflammasome-independent modulation of cytokine response by autophagy in human cells. PLoS One 6(4):e18666

de Alba E (2009) Structure and interdomain dynamics of apoptosis-associated speck-like protein containing a CARD (ASC). J Biol Chem 284(47):32932–32941

Detjen KM, Farwig K, Welzel M, Wiedenmann B, Rosewicz S (2001) Interferon gamma inhibits growth of human pancreatic carcinoma cells via caspase-1 dependent induction of apoptosis. Gut 49(2):251–262

Dickensheets HL, Donnelly RP (1997) IFN-gamma and IL-10 inhibit induction of IL-1 receptor type I and type II gene expression by IL-4 and IL-13 in human monocytes. J Immunol 159(12):6226–6233

DiDonato JA, Mercurio F, Karin M (2012) NF-kappaB and the link between inflammation and cancer. Immunol Rev 246(1):379–400

Dostert C, Petrilli V, Van Bruggen R, Steele C, Mossman BT, Tschopp J (2008) Innate immune activation through Nalp3 inflammasome sensing of asbestos and silica. Science 320(5876): 674–677

Dostert C, Guarda G, Romero JF, Menu P, Gross O, Tardivel A, Suva ML, Stehle JC, Kopf M, Stamenkovic I, Corradin G, Tschopp J (2009) Malarial hemozoin is a Nalp3 inflammasome activating danger signal. PLoS One 4(8):e6510

Dowds TA, Masumoto J, Zhu L, Inohara N, Nunez G (2004) Cryopyrin-induced interleukin 1beta secretion in monocytic cells: enhanced activity of disease-associated mutants and requirement for ASC. J Biol Chem 279(21):21924–21928

Duncan JA, Gao X, Huang MT, O'Connor BP, Thomas CE, Willingham SB, Bergstralh DT, Jarvis GA, Sparling PF, Ting JP (2009) Neisseria gonorrhoeae activates the proteinase cathepsin B to mediate the signaling activities of the NLRP3 and ASC-containing inflammasome. J Immunol 182(10):6460–6469

Dupaul-Chicoine J, Yeretssian G, Doiron K, Bergstrom KS, McIntire CR, LeBlanc PM, Meunier C, Turbide C, Gros P, Beauchemin N, Vallance BA, Saleh M (2010) Control of intestinal homeostasis, colitis, and colitis-associated colorectal cancer by the inflammatory caspases. Immunity 32(3):367–378

Dupont N, Jiang S, Pilli M, Ornatowski W, Bhattacharya D, Deretic V (2011) Autophagy-based unconventional secretory pathway for extracellular delivery of IL-1beta. EMBO J 30(23): 4701–4711

Eisenbarth SC, Colegio OR, O'Connor W, Sutterwala FS, Flavell RA (2008) Crucial role for the Nalp3 inflammasome in the immunostimulatory properties of aluminium adjuvants. Nature 453(7198):1122–1126

Elinav E, Strowig T, Kau AL, Henao-Mejia J, Thaiss CA, Booth CJ, Peaper DR, Bertin J, Eisenbarth SC, Gordon JI, Flavell RA (2011) NLRP6 inflammasome regulates colonic microbial ecology and risk for colitis. Cell 145(5):745–757

Fabbi M, Carbotti G, Ferrini S (2015) Context-dependent role of IL-18 in cancer biology and counter-regulation by IL-18BP. J Leukoc Biol 97(4):665–675

Fan Y, Mao R, Yang J (2013) NF-kappaB and STAT3 signaling pathways collaboratively link inflammation to cancer. Protein Cell 4(3):176–185

Faustin B, Chen Y, Zhai D, Le Negrate G, Lartigue L, Satterthwait A, Reed JC (2009) Mechanism of Bcl-2 and Bcl-X(L) inhibition of NLRP1 inflammasome: loop domain-dependent suppression of ATP binding and oligomerization. Proc Natl Acad Sci U S A 106(10):3935–3940

Fernandes-Alnemri T, Yu JW, Juliana C, Solorzano L, Kang S, Wu J, Datta P, McCormick M, Huang L, McDermott E, Eisenlohr L, Landel CP, Alnemri ES (2010) The AIM2 inflammasome is critical for innate immunity to Francisella tularensis. Nat Immunol 11(5):385–393

Franchi L, Warner N, Viani K, Nunez G (2009) Function of nod-like receptors in microbial recognition and host defense. Immunol Rev 227(1):106–128

Ghiringhelli F, Apetoh L, Tesniere A, Aymeric L, Ma Y, Ortiz C, Vermaelen K, Panaretakis T, Mignot G, Ullrich E, Perfettini JL, Schlemmer F, Tasdemir E, Uhl M, Genin P, Civas A et al (2009) Activation of the NLRP3 inflammasome in dendritic cells induces IL-1beta-dependent adaptive immunity against tumors. Nat Med 15(10):1170–1178

Gross O, Thomas CJ, Guarda G, Tschopp J (2011) The inflammasome: an integrated view. Immunol Rev 243(1):136–151

Guarda G, Dostert C, Staehli F, Cabalzar K, Castillo R, Tardivel A, Schneider P, Tschopp J (2009) T cells dampen innate immune responses through inhibition of NLRP1 and NLRP3 inflammasomes. Nature 460(7252):269–273

Guarda G, Braun M, Staehli F, Tardivel A, Mattmann C, Forster I, Farlik M, Decker T, Du Pasquier RA, Romero P, Tschopp J (2011) Type I interferon inhibits interleukin-1 production and inflammasome activation. Immunity 34(2):213–223

Gurung P, Kanneganti TD (2015) Novel roles for caspase-8 in IL-1beta and inflammasome regulation. Am J Pathol 185(1):17–25

Halle A, Hornung V, Petzold GC, Stewart CR, Monks BG, Reinheckel T, Fitzgerald KA, Latz E, Moore KJ, Golenbock DT (2008) The NALP3 inflammasome is involved in the innate immune response to amyloid-beta. Nat Immunol 9(8):857–865

Harris J, Hartman M, Roche C, Zeng SG, O'Shea A, Sharp FA, Lambe EM, Creagh EM, Golenbock DT, Tschopp J, Kornfeld H, Fitzgerald KA, Lavelle EC (2011) Autophagy controls IL-1beta secretion by targeting pro-IL-1beta for degradation. J Biol Chem 286(11):9587–9597

Hawkins CJ, Uren AG, Hacker G, Medcalf RL, Vaux DL (1996) Inhibition of interleukin 1 beta-converting enzyme-mediated apoptosis of mammalian cells by baculovirus IAP. Proc Natl Acad Sci U S A 93(24):13786–13790

Henry T, Brotcke A, Weiss DS, Thompson LJ, Monack DM (2007) Type I interferon signaling is required for activation of the inflammasome during Francisella infection. J Exp Med 204(5): 987–994

Hise AG, Tomalka J, Ganesan S, Patel K, Hall BA, Brown GD, Fitzgerald KA (2009) An essential role for the NLRP3 inflammasome in host defense against the human fungal pathogen Candida albicans. Cell Host Microbe 5(5):487–497

Hoesel B, Schmid JA (2013) The complexity of NF-kappaB signaling in inflammation and cancer. Mol Cancer 12:86

Hoffman HM, Brydges SD (2011) Genetic and molecular basis of inflammasome-mediated disease. J Biol Chem 286(13):10889–10896

Hornung V, Bauernfeind F, Halle A, Samstad EO, Kono H, Rock KL, Fitzgerald KA, Latz E (2008) Silica crystals and aluminum salts activate the NALP3 inflammasome through phagosomal destabilization. Nat Immunol 9(8):847–856

Hu B, Elinav E, Huber S, Booth CJ, Strowig T, Jin C, Eisenbarth SC, Flavell RA (2010) Inflammation-induced tumorigenesis in the colon is regulated by caspase-1 and NLRC4. Proc Natl Acad Sci U S A 107(50):21635–21640

Hu B, Elinav E, Flavell RA (2011) Inflammasome-mediated suppression of inflammation-induced colorectal cancer progression is mediated by direct regulation of epithelial cell proliferation. Cell Cycle 10(12):1936–1939

Hussain SP, Harris CC (2007) Inflammation and cancer: an ancient link with novel potentials. Int J Cancer 121(11):2373–2380

Huysamen C, Brown GD (2009) The fungal pattern recognition receptor, Dectin-1, and the associated cluster of C-type lectin-like receptors. FEMS Microbiol Lett 290(2):121–128

Ippagunta SK, Malireddi RK, Shaw PJ, Neale GA, Vande Walle L, Green DR, Fukui Y, Lamkanfi M, Kanneganti TD (2011) The inflammasome adaptor ASC regulates the function of adaptive immune cells by controlling Dock2-mediated Rac activation and actin polymerization. Nat Immunol 12(10):1010–1016

Jeru I, Duquesnoy P, Fernandes-Alnemri T, Cochet E, Yu JW, Lackmy-Port-Lis M, Grimprel E, Landman-Parker J, Hentgen V, Marlin S, McElreavey K, Sarkisian T, Grateau G, Alnemri ES, Amselem S (2008) Mutations in NALP12 cause hereditary periodic fever syndromes. Proc Natl Acad Sci U S A 105(5):1614–1619

Jiang X, Overholtzer M, Thompson CB (2015) Autophagy in cellular metabolism and cancer. J Clin Invest 125(1):47–54

Jones JW, Kayagaki N, Broz P, Henry T, Newton K, O'Rourke K, Chan S, Dong J (2010) Qu Y, Roose-Girma M, Dixit VM and Monack DM. Absent in melanoma 2 is required for innate immune recognition of Francisella tularensis. Proc Natl Acad Sci U S A 107(21):9771–9776

Kanneganti TD (2010) Central roles of NLRs and inflammasomes in viral infection. Nat Rev Immunol 10(10):688–698

Kanneganti TD, Lamkanfi M, Kim YG, Chen G, Park JH, Franchi L, Vandenabeele P, Nunez G (2007) Pannexin-1-mediated recognition of bacterial molecules activates the cryopyrin inflammasome independent of Toll-like receptor signaling. Immunity 26(4):433–443

Karin M (2006) Nuclear factor-kappaB in cancer development and progression. Nature 441(7092): 431–436

Kayagaki N, Warming S, Lamkanfi M, Vande Walle L, Louie S, Dong J, Newton K, Qu Y, Liu J, Heldens S, Zhang J, Lee WP, Roose-Girma M, Dixit VM (2011) Non-canonical inflammasome activation targets caspase-11. Nature 479(7371):117–121

Kazama H, Ricci JE, Herndon JM, Hoppe G, Green DR, Ferguson TA (2008) Induction of immunological tolerance by apoptotic cells requires caspase-dependent oxidation of high-mobility group box-1 protein. Immunity 29(1):21–32

Khansari N, Shakiba Y, Mahmoudi M (2009) Chronic inflammation and oxidative stress as a major cause of age-related diseases and cancer. Recent Patents Inflamm Allergy Drug Discov 3(1): 73–80

Kim S, Bauernfeind F, Ablasser A, Hartmann G, Fitzgerald KA, Latz E, Hornung V (2010) Listeria monocytogenes is sensed by the NLRP3 and AIM2 inflammasome. Eur J Immunol 40(6):1545–1551

Kim JW, Koh Y, Kim DW, Ahn YO, Kim TM, Han SW, Oh DY, Lee SH, Im SA, Kim TY, Heo DS, Bang YJ (2013) Clinical Implications of VEGF, TGF-beta1, and IL-1beta in Patients with Advanced Non-small Cell Lung Cancer. Cancer Res Treat 45(4):325–333

Kono H, Rock KL (2008) How dying cells alert the immune system to danger. Nat Rev Immunol 8(4):279–289

Labbe K, McIntire CR, Doiron K, Leblanc PM, Saleh M (2011) Cellular inhibitors of apoptosis proteins cIAP1 and cIAP2 are required for efficient caspase-1 activation by the inflammasome. Immunity 35(6):897–907

Lamkanfi M, Vande Walle L, Kanneganti TD (2011) Deregulated inflammasome signaling in disease. Immunol Rev 243(1):163–173

Lechtenberg BC, Mace PD, Riedl SJ (2014) Structural mechanisms in NLR inflammasome signaling. Curr Opin Struct Biol 29:17–25

Lee GS, Subramanian N, Kim AI, Aksentijevich I, Goldbach-Mansky R, Sacks DB, Germain RN, Kastner DL, Chae JJ (2012) The calcium-sensing receptor regulates the NLRP3 inflammasome through Ca2+ and cAMP. Nature 492(7427):123–127

Li J, Moran T, Swanson E, Julian C, Harris J, Bonen DK, Hedl M, Nicolae DL, Abraham C, Cho JH (2004) Regulation of IL-8 and IL-1beta expression in Crohn's disease associated NOD2/CARD15 mutations. Hum Mol Genet 13(16):1715–1725

Li QX, Liu G, Wong-Staal F (2008) Oncolytic virotherapy as a personalized cancer vaccine. Int J Cancer 123(3):493–499

Li Q, Zhang X, Wang W, Li L, Xu Q, Wu X, Gu Y (2015a) CPT-11 activates NLRP3 inflammasome through JNK and NF-kappaB signalings. Toxicol Appl Pharmacol 289(2): 133–141

Li J, Hu L, Liu Y, Huang L, Mu Y, Cai X, Weng C (2015b) DDX19A senses viral RNA and mediates NLRP3-dependent inflammasome activation. J Immunol 195(12):5732–5749

Liu MH (2015) FGF-21 alleviates diabetes-associated vascular complications: Inhibiting NF-kappaB/NLRP3 inflammasome-mediated inflammation? Int J Cardiol 185:320–321

Malhotra S, Rio J, Urcelay E, Nurtdinov R, Bustamante MF, Fernandez O, Oliver B, Zettl U, Brassat D, Killestein J, Lechner-Scott J, Drulovic J, Chan A, Martinelli-Boneschi F, Garcia-Merino A, Montalban X et al (2015) NLRP3 inflammasome is associated with the response to IFN-beta in patients with multiple sclerosis. Brain 138(Pt 3):644–652

Mantovani A, Allavena P, Sica A, Balkwill F (2008) Cancer-related inflammation. Nature 454(7203):436–444

Mariathasan S, Weiss DS, Newton K, McBride J, O'Rourke K, Roose-Girma M, Lee WP, Weinrauch Y, Monack DM, Dixit VM (2006) Cryopyrin activates the inflammasome in response to toxins and ATP. Nature 440(7081):228–232

Martinon F, Burns K, Tschopp J (2002) The inflammasome: a molecular platform triggering activation of inflammatory caspases and processing of proIL-beta. Mol Cell 10(2):417–426

Martins JD, Liberal J, Silva A, Ferreira I, Neves BM, Cruz MT (2015) Autophagy and inflammasome interplay. DNA Cell Biol 34(4):274–281

Masters SL, Simon A, Aksentijevich I, Kastner DL (2009) Horror autoinflammaticus: the molecular pathophysiology of autoinflammatory disease (*). Annu Rev Immunol 27:621–668

Masters SL, Mielke LA, Cornish AL, Sutton CE, O'Donnell J, Cengia LH, Roberts AW, Wicks IP, Mills KH, Croker BA (2010) Regulation of interleukin-1beta by interferon-gamma is species specific, limited by suppressor of cytokine signalling 1 and influences interleukin-17 production. EMBO Rep 11(8):640–646

Masters SL, Gerlic M, Metcalf D, Preston S, Pellegrini M, O'Donnell JA, McArthur K, Baldwin TM, Chevrier S, Nowell CJ, Cengia LH, Henley KJ, Collinge JE, Kastner DL, Feigenbaum L, Hilton DJ et al (2012) NLRP1 inflammasome activation induces pyroptosis of hematopoietic progenitor cells. Immunity 37(6):1009–1023

Meixenberger K, Pache F, Eitel J, Schmeck B, Hippenstiel S, Slevogt H, N'Guessan P, Witzenrath M, Netea MG, Chakraborty T, Suttorp N, Opitz B (2010) Listeria monocytogenes-infected human peripheral blood mononuclear cells produce IL-1beta, depending on listeriolysin O and NLRP3. J Immunol 184(2):922–930

Miao EA, Leaf IA, Treuting PM, Mao DP, Dors M, Sarkar A, Warren SE, Wewers MD, Aderem A (2010) Caspase-1-induced pyroptosis is an innate immune effector mechanism against intracellular bacteria. Nat Immunol 11(12):1136–1142

Munoz-Planillo R, Franchi L, Miller LS, Nunez G (2009) A critical role for hemolysins and bacterial lipoproteins in Staphylococcus aureus-induced activation of the Nlrp3 inflammasome. J Immunol 183(6):3942–3948

Murakami T, Ockinger J, Yu J, Byles V, McColl A, Hofer AM, Horng T (2012) Critical role for calcium mobilization in activation of the NLRP3 inflammasome. Proc Natl Acad Sci U S A 109(28):11282–11287

Nakahira K, Haspel JA, Rathinam VA, Lee SJ, Dolinay T, Lam HC, Englert JA, Rabinovitch M, Cernadas M, Kim HP, Fitzgerald KA, Ryter SW, Choi AM (2011) Autophagy proteins regulate innate immune responses by inhibiting the release of mitochondrial DNA mediated by the NALP3 inflammasome. Nat Immunol 12(3):222–230

Normand S, Delanoye-Crespin A, Bressenot A, Huot L, Grandjean T, Peyrin-Biroulet L, Lemoine Y, Hot D, Chamaillard M (2011) Nod-like receptor pyrin domain-containing protein 6 (NLRP6) controls epithelial self-renewal and colorectal carcinogenesis upon injury. Proc Natl Acad Sci U S A 108(23):9601–9606

Pan H, Chen L, Xu Y, Han W, Lou F, Fei W, Liu S, Jing Z, Sui X (2016) Autophagy-associated immune responses and cancer immunotherapy. Oncotarget 7(16):21235–21246

Piccini A, Carta S, Tassi S, Lasiglie D, Fossati G, Rubartelli A (2008) ATP is released by monocytes stimulated with pathogen-sensing receptor ligands and induces IL-1beta and IL-18 secretion in an autocrine way. Proc Natl Acad Sci U S A 105(23):8067–8072

Prasad S, Ravindran J, Aggarwal BB (2010) NF-kappaB and cancer: how intimate is this relationship. Mol Cell Biochem 336(1-2):25–37

Qiao Y, Wang P, Qi J, Zhang L, Gao C (2012) TLR-induced NF-kappaB activation regulates NLRP3 expression in murine macrophages. FEBS Lett 586(7):1022–1026

R V (1863) Diekrankhaften Geschwülste. August Hirschwald, Berlin

Rathinam VA, Vanaja SK, Fitzgerald KA (2012) Regulation of inflammasome signaling. Nat Immunol 13(4):333–342

Rebecca VW, Amaravadi RK (2016) Emerging strategies to effectively target autophagy in cancer. Oncogene 35(1):1–11

Reuter S, Gupta SC, Chaturvedi MM, Aggarwal BB (2010) Oxidative stress, inflammation, and cancer: how are they linked? Free Radic Biol Med 49(11):1603–1616

Rossol M, Pierer M, Raulien N, Quandt D, Meusch U, Rothe K, Schubert K, Schoneberg T, Schaefer M, Krugel U, Smajilovic S, Brauner-Osborne H, Baerwald C, Wagner U (2012) Extracellular Ca2+ is a danger signal activating the NLRP3 inflammasome through G protein-coupled calcium sensing receptors. Nat Commun 3:1329

Sadasivam S, Gupta S, Radha V, Batta K, Kundu TK, Swarup G (2005) Caspase-1 activator Ipaf is a p53-inducible gene involved in apoptosis. Oncogene 24(4):627–636

Saitoh T, Fujita N, Jang MH, Uematsu S, Yang BG, Satoh T, Omori H, Noda T, Yamamoto N, Komatsu M, Tanaka K, Kawai T, Tsujimura T, Takeuchi O, Yoshimori T, Akira S (2008) Loss of the autophagy protein Atg16L1 enhances endotoxin-induced IL-1beta production. Nature 456(7219):264–268

Shalapour S, Karin M (2015) Immunity, inflammation, and cancer: an eternal fight between good and evil. J Clin Invest 125(9):3347–3355

Shaw PJ, Lukens JR, Burns S, Chi H, McGargill MA, Kanneganti TD (2010) Cutting edge: critical role for PYCARD/ASC in the development of experimental autoimmune encephalomyelitis. J Immunol 184(9):4610–4614

Shin S, Brodsky IE (2015) The inflammasome: learning from bacterial evasion strategies. Semin Immunol 27(2):102–110

Siegel RM (2006) Caspases at the crossroads of immune-cell life and death. Nat Rev Immunol 6(4): 308–317

Suarez R, Buelvas N (2015) Inflammasome: activation mechanisms. Investig Clin 56(1):74–99

Sutterwala FS, Haasken S, Cassel SL (2014) Mechanism of NLRP3 inflammasome activation. Ann N Y Acad Sci 1319:82–95

Takeda K, Akira S (2005) Toll-like receptors in innate immunity. Int Immunol 17(1):1–14

Tas F, Tilgen Yasasever C, Karabulut S, Tastekin D, Duranyildiz D (2015) Clinical significance of serum interleukin-18 (IL-18) levels in patients with gastric cancer. Biomed Pharmacother 70: 19–23

van Deventer HW, Burgents JE, Wu QP, Woodford RM, Brickey WJ, Allen IC, McElvania-Tekippe E, Serody JS, Ting JP (2010) The inflammasome component NLRP3 impairs antitumor vaccine by enhancing the accumulation of tumor-associated myeloid-derived suppressor cells. Cancer Res 70(24):10161–10169

Verma D, Bivik C, Farahani E, Synnerstad I, Fredrikson M, Enerback C, Rosdahl I, Soderkvist P (2012) Inflammasome polymorphisms confer susceptibility to sporadic malignant melanoma. Pigment Cell Melanoma Res 25(4):506–513

Villani AC, Lemire M, Fortin G, Louis E, Silverberg MS, Collette C, Baba N, Libioulle C, Belaiche J, Bitton A, Gaudet D, Cohen A, Langelier D, Fortin PR, Wither JE, Sarfati M et al (2009) Common variants in the NLRP3 region contribute to Crohn's disease susceptibility. Nat Genet 41(1):71–76

Vince JE, Wong WW, Gentle I, Lawlor KE, Allam R, O'Reilly L, Mason K, Gross O, Ma S, Guarda G, Anderton H, Castillo R, Hacker G, Silke J, Tschopp J (2012) Inhibitor of apoptosis proteins limit RIP3 kinase-dependent interleukin-1 activation. Immunity 36(2):215–227

Wei Q, Mu K, Li T, Zhang Y, Yang Z, Jia X, Zhao W, Huai W, Guo P, Han L (2014) Deregulation of the NLRP3 inflammasome in hepatic parenchymal cells during liver cancer progression. Lab Invest 94(1):52–62

Willingham SB, Allen IC, Bergstralh DT, Brickey WJ, Huang MT, Taxman DJ, Duncan JA, Ting JP (2009) NLRP3 (NALP3, Cryopyrin) facilitates in vivo caspase-1 activation, necrosis, and HMGB1 release via inflammasome-dependent and -independent pathways. J Immunol 183(3):2008–2015

Wree A, Eguchi A, McGeough MD, Pena CA, Johnson CD, Canbay A, Hoffman HM, Feldstein AE (2014) NLRP3 inflammasome activation results in hepatocyte pyroptosis, liver inflammation, and fibrosis in mice. Hepatology 59(3):898–910

Xiang P, Chen T, Mou Y, Wu H, Xie P, Lu G, Gong X, Hu Q, Zhang Y, Ji H (2015) NZ suppresses TLR4/NF-kappaB signalings and NLRP3 inflammasome activation in LPS-induced RAW264.7 macrophages. Inflamm Res 64(10):799–808

Xu Y, Li H, Chen W, Yao X, Xing Y, Wang X, Zhong J, Meng G (2013) Mycoplasma hyorhinis activates the NLRP3 inflammasome and promotes migration and invasion of gastric cancer cells. PLoS One 8(11):e77955

Yang L (2010) TGFbeta and cancer metastasis: an inflammation link. Cancer Metastasis Rev 29(2): 263–271

Yoneyama M, Fujita T (2007) RIG-I family RNA helicases: cytoplasmic sensor for antiviral innate immunity. Cytokine Growth Factor Rev 18(5–6):545–551

Zaki MH, Boyd KL, Vogel P, Kastan MB, Lamkanfi M, Kanneganti TD (2010a) The NLRP3 inflammasome protects against loss of epithelial integrity and mortality during experimental colitis. Immunity 32(3):379–391

Zaki MH, Vogel P, Body-Malapel M, Lamkanfi M, Kanneganti TD (2010b) IL-18 production downstream of the Nlrp3 inflammasome confers protection against colorectal tumor formation. J Immunol 185(8):4912–4920

Zaki MH, Vogel P, Malireddi RK, Body-Malapel M, Anand PK, Bertin J, Green DR, Lamkanfi M, Kanneganti TD (2011) The NOD-like receptor NLRP12 attenuates colon inflammation and tumorigenesis. Cancer Cell 20(5):649–660

Zhong Z, Zhai Y, Liang S, Mori Y, Han R, Sutterwala FS, Qiao L (2013) TRPM2 links oxidative stress to NLRP3 inflammasome activation. Nat Commun 4:1611

Zhou R, Yazdi AS, Menu P, Tschopp J (2011) A role for mitochondria in NLRP3 inflammasome activation. Nature 469(7329):221–225

Zitvogel L, Kepp O, Galluzzi L, Kroemer G (2012) Inflammasomes in carcinogenesis and anticancer immune responses. Nat Immunol 13(4):343–351

Chapter 13
Aging and the Inflammasomes

Fabiola Marín-Aguilar, Jesús Ruiz-Cabello, and Mario D. Cordero

Contents

Abstract The inflammasomes are innate immune system sensors that control the activation of caspase-1 and induce inflammation in response to infectious microbes and molecules originating from host proteins, leading to the release of

F. Marín-Aguilar
Research Laboratory, Oral Medicine Department, University of Sevilla, Sevilla, Spain

J. Ruiz-Cabello
CIC biomaGUNE, San Sebastian-Donostia, Spain, Madrid, Spain

CIC biomaGUNE, Madrid, Spain

IKERBASQUE, Basque Foundation for Science, Biscay, Spain

CIBER de Enfermedades Respiratorias (CIBERES), Madrid, Spain

Universidad Complutense Madrid, Madrid, Spain

M. D. Cordero (✉)
Department of Physiology, Institute of Nutrition and Food Technology "José Mataix", Biomedical Research Center (CIBM), University of Granada, Armilla, Spain
e-mail: mdcormor@ugr.es

M. D. Cordero, E. Alcocer-Gómez (eds.), *Inflammasomes: Clinical and Therapeutic Implications*, Experientia Supplementum 108,
https://doi.org/10.1007/978-3-319-89390-7_13

pro-inflammatory cytokines, Il1b and IL18, and a particular inflammatory type of cell death termed pyroptosis. It is broadly considered that chronic inflammation may be a common link in age-related diseases, aging being the greatest risk factor for the development of chronic diseases. In this sense, we discuss the role of inflammasomes in non-infectious inflammation and their interest in aging and age-related diseases.

Keywords Inflammasomes · Aging · Age-related diseases · Chronic inflammatory diseases

Why does aging occur? Aging constitutes one of the biggest concerns for the human being and to ask why do we age is to enter the field of evolutionary biology, which is crucial to understand health and disease (Kirkwood 2005). Nowadays, the challenge for researchers lies in explaining the reasons for aging instead of the obvious drawbacks of the natural process. Aging is commonly characterized as a progressive and generalized impairment of function, resulting in an increased vulnerability to environmental and genetic factors (Ljubuncic and Reznick 2009; Jin 2010; Lipsky and King 2015). Getting older is in fact, a highly medically relevant and enigmatic biological process, because despite considering increased longevity a remarkable achievement for humankind, aging is the major risk factor for the development of chronic diseases and represents an extraordinary financial burden on the health care systems (Leon and Gustafsson 2016). It is estimated that in 15 years, a large percentage of the population will be aged 65 or older (North and Sinclair 2012). The aging process is linked to an accumulation of mutations and genomic instability resulting in a progressive functional and structural decline in multiple organs. Therefore, far from being considered an illness in itself, aging is the greatest risk factor for the onset of chronic age-related diseases such as cardiovascular disease, cancer, diabetes, Alzheimer's disease (AD), and a tendency to infection (Strowig et al. 2012).

However, and despite the current advances in medicine to date, some questions concerning the span of human life and health remain. Over the years, various principles to explain the reasons for aging have been proposed, but the mechanisms are still unclear, although many of the theories have arisen from the necessity of explaining how the aging process takes place (Kirkwood 2005).

13.1 The Science of Elderly

Biological aging is initiated from the time of the birth of organisms and refers to a progressive manifestation of accumulated cellular damage determined by both genetic and environmental factors. The process of aging is complicated and can be elucidated by many theories. One of the best-known and most conventional approaches to date is the mitochondrial free radical theory of aging, nowadays termed the "oxidative stress theory" (Chandrasekaran et al. 2017). The oxidative stress hypothesis interprets aging at a molecular level, explaining that aging occurs because of an imbalance between the production of reactive oxygen species (ROS) and the capacity of the biological system to repair the outcoming damage, resulting

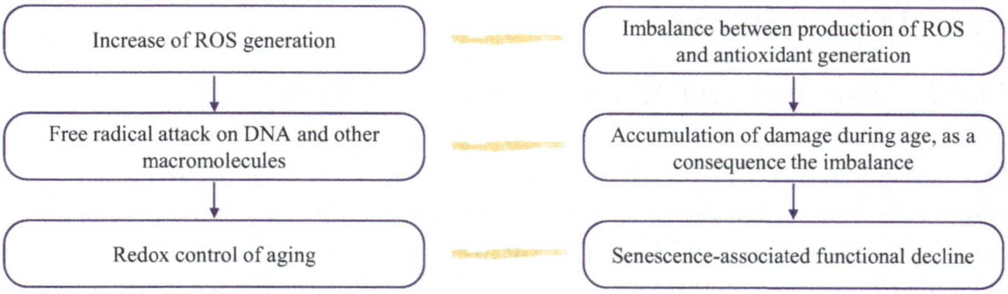

Fig. 13.1 Graphic image of the oxidative stress theory of aging. The primary element of this hypothesis is the increase in oxidant flux and the concomitant failure of antioxidant mechanisms, causing structural damage to macromolecules that accumulates with age, leading to the typical decline occurring in the elderly

in the failure to maintain the mitochondrial integrity and DNA repair (Fig. 13.1). ROS molecules include superoxide ($O_2^{\bullet -}$), hydroxyl radical ($OH^{\bullet}$), hydroperoxyl radical ($HO_2^{\bullet}$), nitric oxide ($NO^{\bullet}$), nitrogen dioxide ($NO_2^{\bullet}$), and peroxyl ($ROO^{\bullet}$), produced either as by-products during the mitochondrial electron transport of aerobic respiration or by oxidoreductase enzymes.

Nowadays, several aging mechanisms exist that are widely acknowledged. However, what is very restricting is that in practice, most research is focused on unique mechanisms of theories that indicate that molecular and cellular lesions hypothesized do occur as we age, but there are no data that demonstrate the theory itself to be sufficient to account for age-related disease. Therefore, recent initiatives state that there is a need to develop a net of aging theories considering the contribution of various mechanisms together, allowing the interaction of different processes (Kirkwood et al. 2003).

Although many theories could explain the aging process (Chandrasekaran et al. 2017), recent studies have shown that immunological inflammation may be closely linked to aging. As we age, the adaptive immunity response significantly declines (Goldberg and Dixit 2015). This concept is known as immunosenescence, where the innate immunity response is markedly activated, leading to the senescent low-level, chronic inflammatory phenotype known as "inflammaging." The concept "inflammaging" describes the systemic low-grade inflammatory process that contributes to the development of chronic diseases and degenerative changes during the aging process. Franceschi et al. (2000) coined the word "inflammaging" in 2000, referring to a progressive increase in proinflammatory status, a significant characteristic of aging. This fact can be reflected in diseases where chronic and abnormal inflammation exists (Franceschi et al. 2000). Age-related inflammation in various organs may lead to a considerable decline, even in the absence of any disease. For example, chronic inflammation is simultaneously associated with aging and with age-related diseases, such as diabetes, atherosclerosis, cancer or neurodegenerative diseases.

It is believed that a systemic increase in inflammation contributes to incremented disease prevalence and severity during aging (Franceschi and Campisi 2014), because aging is associated with an increment in IL-18, IL-1b, and IL-6. Notably, Il-1b and IL-18 are produced after inflammasome-dependent caspase-1 activation, and interestingly, many of the endogenous signals that have been described as inflammasome activators are known to accumulate as we age.

13.2 Inflammasomes

13.2.1 Structure and Mechanisms of Activation

The "inflammasome," a term coined by Schroder and Tschopp in 2010, is a group of multimeric proteins that assemble in the cytosol when pathogenic microorganisms or sterile stressors are present, and is also involved in the onset and development of the inflammatory response. Stimuli related to infection are known as pathogen-associated molecular patterns (PAMPs); however, those referring to endogenous cellular stress derived from host proteins, are danger-associated molecular patterns (DAMPs) (Strowig et al. 2012). The inflammasome assembly culminates in the activation of caspase-1. Subsequently, active caspase-1 cleavage triggers a signaling cascade that leads to release of type I interferon (IFN alpha and beta) and pro-inflammatory cytokines (IL-1b and IL-18), finally inducing an inflammatory type of cell death termed pyroptosis (de Zoete et al. 2014; Guo et al. 2015). Dysregulation of inflammasomes has been linked with many autoinflammatory and autoimmune diseases, including metabolic disorders (type 2 diabetes [T2D] mellitus, obesity, atherosclerosis) and neurodegenerative diseases (multiple sclerosis, AD, Parkinson's disease). Therefore, the current understanding of inflammasome activation and its involvement in inflammatory pathological conditions have drawn scientific community attention to developing potential therapies targeting inflammasomes (Youm et al. 2013).

Structurally, inflammasomes consist of an intracellular sensor protein, which usually is a NOD-like receptor (NLR), the adaptor protein apoptosis-associated speck-like protein containing a caspase recruitment domain (ASC), and the proinflammatory caspase-1 precursor (Schroder and Tschopp 2010). The first family of sensor proteins discovered to form inflammasomes was the nucleotide-binding domain and leucine-rich repeat-containing receptor, the NLR family consisting of 22 genes in humans and 33 in mice (Ting et al. 2008). NLRs are classified according to their domain structure. All NLRs, except NLRP10, contain a leucine-rich repeat (LRR) domain, which is thought to provide the critical structural framework for molecular interactions, and a signaling domain (Ting et al. 2008) that enables the recruitment of caspase-1, directly through caspase recruitment domain (CARD)–CARD interactions, such as NLRC4 inflammasome (Guo et al. 2015) or indirectly, through a PYRIN domain that is shown to bind to ASC. Apart from the NLR family, non-NLR proteins can also assemble to conform to inflammasomes and possess an HIN-200 DNA-binding domain instead of an LRR, such as the AIM2-like receptor (ALR) family. Then, as mentioned before, upon sensing certain stimuli, NLR or AIM2 can oligomerize to become a caspase-1-activated scaffold.

The most interesting and relevant question related to the inflammasome field is connected to the specific signals that lead to the assembly of the different NLRs or ALRs into active complexes, as the stimuli leading to activation of the different inflammasomes consist of a wide range of variable and selective activators.

The NLRs share a similar structure consisting of three domains:

1. A central nucleotide binding oligomerization domain NOD or NACHT.
2. A C-terminal domain LRR present in all members of the family (except for NLRP10) and believed to be used for PAMPs recognition.
3. An N-terminal domain of recruitment of effector molecules, which determines the classification of the different NLRs (Fig. 13.2).

a

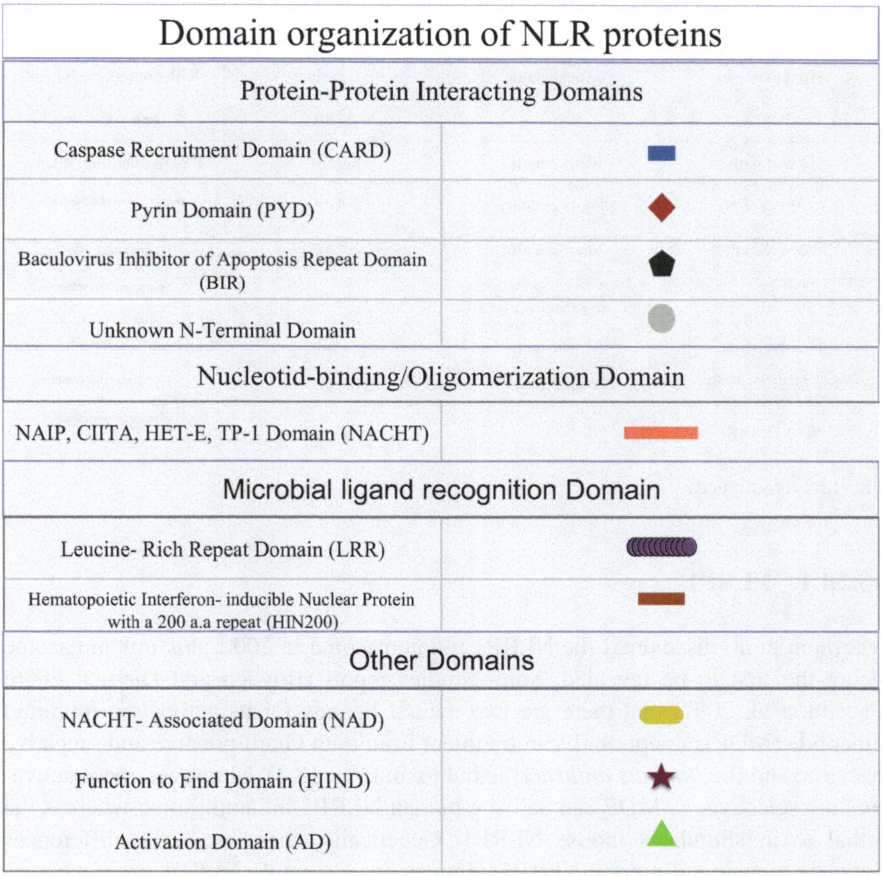

Fig. 13.2 (**a**) Domain key for inflammasome classification. Inflammasome names are based on the protein forming the scaffold. NLR proteins are categorized into four subfamilies, depending on the type of N-terminal domain; the NLRA subfamily refers to the acidic transactivating domain: CIITA. (**b**) Classification of different NLR proteins according to their domain structure. From the NLRB subfamily is for baculovirus inhibitor apoptosis repeats present in NLR family apoptosis-inhibiting proteins; NLRCs contain a CARD domain: NLRC1, NLRC2, NLRC3, NLRC5, and N-terminal domain in NLRP subfamily is PYD: NLRP1, NLRP2, NLRP3, NLRP4, NLRP5, NLRP6, NLRP7, NLRP8, NLRP9, NLRP10, NLRP11, NLRP12, NLRP13, and NLRP14. An additional subfamily has arisen, NLRX, but its N-terminal domain is still unknown

b

Domain organization of NLR proteins			
Human name	Mouse name	Family	CARD-containing NLRs
CIITA (NLRA)	Cllta (Nlra)	NLRA	
NOD1 (NLRC1)	Nod1 (Nlrc1)	NLRC	
NOD2 (NLRC2)	Nod2 (Nlrc2)	NLRC	
IPAF (NLRC4)	Ipaf (Nlrc4)	NLRC	
Human name	Mouse name	Family	BIR-containing NLRs
NAIP	Naip (1-7)	NLRB	
Human name	Mouse name	Family	PYD-containing NLRs
NLRP1 (NALP1)	Nlrp1a-c (Nalp1)	NLRP	
NLRP10 (NALP10)	Nlrp10 (Nalp10)	NLRP	
NLRP2-9 (NALP2-9), NLRP11-14 (NALP11-14)	Nlrp2, Nlrp3 (Nalp3), Nlrp4a-g (Nalp4a-g), Nlrp5, Nlrp6, Nlrp9a-c (Nalp9a-c), Nlrp12 (Nalp12), Nlrp14 (Nalp14)	NLRP	
Human name	Mouse name	Family	Unknown N-terminal domain
NLRC3 (NOD3), NLRC5 (NOD27)	Nlrc3, Nlrc5	NLRC	
NLRX1 (NOD9)	Nlrx1	NLRX	

Fig. 13.2 (continued)

13.2.1.1 NLRP1

Martinon et al. discovered the NLRP1 inflammasome in 2002, this inflammasome being the first to be revealed. Some studies report (Boyden and Dietrich 2006; Faustin et al. 2007) that there are two natural ligands for its activation: muramyl dipeptide (MDP), a peptidoglycan fragment from both Gram-positive and -negative bacteria, and the *Bacillus anthracis* lethal toxin (Fig. 13.3). Moreover, these activators are selective, as MDP can activate human NLRP1 inflammasome whereas the lethal toxin stimulates mouse NLRP1. Genetically, there are some differences between human and murine NLRP1. Humans have a single NLRP1 gene, whereas mice have a group of three homologous genes, Nlrp1a, Nlrp1b, Nlrp1c (Fig. 13.2). The activation of the Nlrp1 inflammasome is not directly through its LRR motif. Lethal toxin consists of a zinc metalloprotease lethal factor, which is responsible for Nlrp1b cleavage, and subsequently, this cleavage of Nlrp1 itself will make macrophages susceptible to pyroptosis (Levinsohn et al. 2012). However, human NLRP1 binds directly to MDP, inducing a conformational change that allows the binding of ATP as well. When ATP hydrolysis occurs, NLRP1 oligomerizes, promoting caspase-1 recruitment and activation (Faustin et al. 2007). It has been reported that in addition to caspase-1, caspase-5 is also involved in the assembly of the NLRP1 inflammasome complex (Martinon et al. 2002).

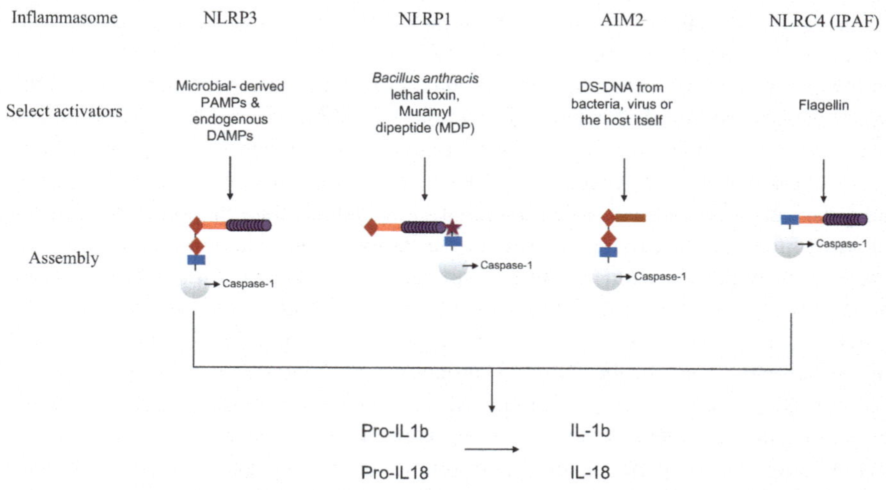

Fig. 13.3 Examples of oligomerized inflammasome complexes. Models for inflammasome-selective activation and assembly. NLRs are characterized by a NACHT domain with or without an N-terminal PYD domain and a variable number of LRRs. AIM2 contains a N-terminal PYD domain followed by a DNA-binding HIN-200 domain. The PYD domain of NLRP3 and AIM2 recruit the adaptor protein ASC via homotypic binding to its PYD domain, allowing indirect recruitment of caspase-1 through interaction with the CARD domain. NLRP1 and NLRC4 directly recruit caspase-1 through a CARD domain. Activation of the inflammasome leads to maturation and secretion of IL-1β and IL-18 and inflammatory cell death by pyroptosis. *AIM2* absent in melanoma 2, *ASC* adaptor protein apoptosis-associated speck-like protein containing a caspase recruitment domain *CARD* caspase recruitment domain, *DAMP* danger-associated molecular pattern, *FIND* domain with function to find, *LRR* leucine-rich repeat, *NACHT* nucleotide-binding and oligomerization domain, *NLR* NOD-like receptor, *PAMP* pathogen-associated molecular pattern, *PYD* pyrin domain

13.2.1.2 NLRP3

The most in depth studied inflammasome complex is the NLRP3, in part because of its gamut of well-known activators. The NLRP3 inflammasome requires two signals for activation (Fig. 13.3). The first signal depends on NF-κB activation of NLRP3 (Bauernfeind et al. 2009) and IL-1B (Barker et al. 2011). The second signal consists of sensing a broad range of PAMPs and stress-associated signals or host-derived DAMPs to trigger complex assembly. Most of the DAMPs that activate the inflammasome include ROS (Dostert et al. 2008), ATP (Mariathasan et al. 2006), uric acid crystals (Martinon et al. 2006), and endogenous host metabolic products that stimulate caspase-1 cleavage in an NLRP3-dependent mechanism, and, importantly, are shown to increase during aging (Goldberg and Dixit 2015). NLRP3 is expressed in myeloid cells (Guarda et al. 2011), including macrophages, which use PRRs to recognize PAMPs and initiate the inflammatory signal pathway, this process being crucial to controlling pathogenic propagation.

Several studies have focused on age-related changes in myeloid cells concerning infection. In general, lipopolysaccharide (LPS) stimulation from an aged host results in lower tumor necrosis factor (TNF)-alpha and IL-6 secretion compared with adult macrophages. Although this helps to explain why old hosts exhibit poor control of bacterial spread and increased susceptibility to bacterial infections, the production of these cytokines is independent of NLRP3 activation. Nowadays, it is becoming progressively evident that sterile inflammation, which is inflammation in the absence of overt infection, is a more consistent contributor to age-related inflammation and disease. A wide spectrum of sterile particles can stimulate inflammation; some examples of these particles include silica dioxide, asbestos, cholesterol crystals, or amyloid-β fibrils (Rock et al. 2010). Moreover, as we age, it has been reported that the basal elevation of the NLRP3 inflammasome interferes with the specific up-regulation of caspase-1 that is required for a successful immune response against infections in mice (Knrone et al. 2013). This fact emphasizes the importance of maintaining an adequate balance between tissue homeostasis and host defense during infection.

Although the Nlrp3 inflammasome is the most thoroughly studied NLR, its complex activation has been shown to be even more complicated with the discovery of noncanonical inflammasome activation (Kayagaki et al. 2011). Here, the authors show that caspase-11 (also known as caspase-4) is critical for caspase-1 activation and IL-1β production in C57BL/6 Casp11 gene-targeted mice macrophages infected with *Escherichia coli*, *Citrobacter rodentium* or *Vibrio cholerae*, and they also realized that the published Casp1(−/−) mice lacked both caspase-11 and caspase-1. Thus, they concluded that Casp11(−/−) macrophages secreted IL-1β, usually in response to ATP and monosodium urate, indicating that caspase-11 is engaged by a noncanonical inflammasome.

13.2.1.3 NLRP6 and NLRP12

NLRP6 inflammasome and NLRP12 present several characteristics in common. NLRP6 is mainly expressed in nonhematopoietic cells. Specifically, NLRP6 is highly expressed in intestinal epithelial and goblet cells (Wlodarska et al. 2014) where it is involved in the vital role of maintaining intestinal homeostasis (Hu et al. 2013). Recent studies suggest that NLRP6 might form inflammasome complexes, as its combined expression with ASC results in caspase-1 activation (Grenier et al. 2002). Both inflammasomes, NLRP6 and NLRP12, seem to maintain intestinal homeostasis through negative MAPK and NF-κB inflammatory pathway regulation (Anand and Kanneganti 2013). Moreover, the NLRP12 inflammasome has recently played a protective role in colitis and colon cancer induced with dextran sulfate sodium (DSS) and azoxymethane/dextran sulfate sodium (AOM/DSS) respectively (Zaki et al. 2011). However, the specific and selective activators for these inflammasomes remain unclear.

13.2.1.4 NLRC4

The NLRC4 inflammasome is also known as ICE protease activating factor, when N-terminal domain is a CARD. NLRC4 responds to a more defined range of stimuli than the NLRP3 inflammasome. *Legionella pneumophila*, *Pseudomonas aeruginosa*, *Salmonella typhimurium*, and *Shigella flexneri* are responsible for NLRC4 inflammasome activation (Mariathasan et al. 2004; Amer et al. 2006; Miao et al. 2006, 2008; Lamkanfi et al. 2007; Sutterwala et al. 2007; Suzuki et al. 2007). NLRC4 senses bacterial flagellin and Gram-negative bacterial type III secretory system (T3SS) that are leaked into the host cell (Miao et al. 2010). These bacterial components are directly bound by NLR family apoptosis-inhibiting proteins (NAIPs), forming complexes in the cytosol. In mice, there several subtypes of NAIP proteins, whereas in humans only one NAIP protein has been characterized (Fig. 13.2), and it was discovered to bind T3SS needle protein. Murine NAIP1 binds T3SS needle protein. However, NAIP2 tends to bind T3SS rod protein, whereas NAIP5 and NAIP6 bind bacterial flagellin (Kofoed and Vance 2011). NAIPs then interact with NLRC4, triggering the complex conformation, activating caspase-1 and leading to the release of pro-inflammatory cytokines and finally, to pyroptosis.

13.2.1.5 NLRC3

The NLRs are cytoplasmic immune sensors that are involved in intestinal homeostasis (Zaki et al. 2010; Allen et al. 2012). NLRC3 (also known as NOD3) is still insufficiently characterized. Some authors classify NLRC3 as NLRs non-inflammasome-forming (Sharma and Jha 2016). However, they contribute considerably to inflammatory regulation by regulating inflammation pathways (Ting et al. 2010). Non-inflammasome-forming NLRs modulate NF-κB and other significant inflammation regulatory pathways, which are crucial in chronic inflammation and inflammation-induced tumorigenesis (Allen 2014). In fact, some studies have reported that NLRC3 expression is remarkably reduced in tumors from patients with colorectal cancer in comparison with healthy tissues (Liu et al. 2015; Karki et al. 2016). In this study, they investigate the role of NLRC3 in colorectal cancer using a mouse model of AOM/DSS colitis-induced and colorectal tumorigenesis. The conclusion is that mice lacking in Nlrc3 are significantly more susceptible to colitis and colorectal tumorigenesis. NLRC3 is presumably a protector against colorectal tumors through the inhibition of the mTOR pathway.

13.2.1.6 NLRCX

Another non-inflammasome-forming NLR is NLRCX. Unlike the other NLR proteins mentioned above, it constitutes the first noncytoplasmic NLR protein, and is localized in the mitochondria (Moore et al. 2008; Xiao and Ting 2012). The Nlrx1 expression is highest in mitochondria-rich tissues such as muscle and heart. The main functions of Nlrx1 include negative regulation of anti-viral inflammatory response via the MAVS-RIG1 signaling pathway or TLR-induced NF-κB signaling by targeting the TRAF6 and IKK signaling pathway. These data indicate that NLRX1 attenuates tumorigenesis through the negative regulation of AKT and NF-κB signaling, although it is a potential target for managing immune response in inflammation-associated diseases and cancer pathology (Allen et al. 2011). These results show specific knockdown of Nlrx1, resulting in increased gene expression of the cytokines TNF-α and IL-6 in response to LPS treatment (Xia et al. 2011). NLRX1 also plays an important role in regulating the balance between intrinsic and extrinsic apoptosis in cancer cells. NLRX1 positively regulates apoptosis in response to intrinsic apoptosis signals, and this may be why the Nlrx1 expression is down-regulated in cancer cells. Nlrx1 −/− mice develop fewer tumors than wild-type mice in the AOM-induced colorectal cancer murine model (Soares et al. 2014).

13.2.2 AIM2-Like Receptors and RIG-1-Like Receptors

The non-NLR AIM2 has an HIN-200 domain consisting of proteins that contain a PYRIN domain and the conserved DNA-binding domain hematopoietic IFN-inducible nuclear protein with 200-amino acids (HIN-200) domain (Schattgen and Fitzgerald 2011) that can directly bind its cytosolic dsDNA (Fig. 13.3). Besides, it is also able to form a caspase-1-containing inflammasome. Therefore, these proteins can theoretically bind nucleic acids and recruit ASC to trigger the conformation of an inflammasome. Indeed, AIM2 can form an inflammasome whose assembly is stimulated by recognition of cytosolic DNA of bacterial or viral origin (Fernandes-Alnemri et al. 2010; Jones et al. 2010; Rathinam et al. 2010), or self-DNA from apoptotic cells (Choubey 2012; Zhang et al. 2013). Recent studies about crystal structures of AIM2 complexes with DNA have provided an insight into the mechanism of AIM2 inflammasome activation (Jin et al. 2012). Binding of DNA to the HIN-200 domain of AIM2 results in a conformational change and AIM2 oligomerization around the DNA molecule, which then allows the recruitment of ASC and caspase-1 and inflammasome assembly (Jin et al. 2013).

Additionally, although mice have a wide range of ALRs that includes 13 members, humans have three more: IFI16, IFIX, and MNDA, but most of these ALRs remain insufficiently characterized. However, some murine ALRs were found to trigger IL-1β production, suggesting that they might form inflammasomes (Brunette et al. 2012). Activation of IFI16 in CD4 T cells during HIV infection was found to trigger pyroptosis of T cells (Monroe et al. 2014). Finally, the RIG-I-like receptor family member, which is best known as an inducer of type I IFN production in response to recognition of viral RNA, was also shown to form an inflammasome (Poeck et al. 2010). However, it remains unclear what determines when RIG-I forms an inflammasome versus when it merely triggers type I IFN production.

13.3 Inflammasomes in Age-Related Diseases

The role of inflammasomes becomes even more significant in the elderly, as they are more susceptible to infections owing to the drastic decrease in the immune system. However, a highly important event takes place during aging, starting with DAMPs accumulation (Goldberg and Dixit 2015) and the subsequent activation of the NLRP3 inflammasome due to the stimuli induced by endogenous by-products. Then, endogenous by-products are recognized by PRRs in macrophages (Schroder and Tschopp 2010; Medzhitov 2008) to trigger the singular chronic, low-grade inflammation that occurs during aging (Spadaro et al. 2016). Numerous studies have reported (Goldberg and Dixit 2015; Ferrucci et al. 2005) that systemic low-grade inflammation contributes to the onset of chronic diseases and degenerative changes as we age. Chronic inflammation also plays an essential role in the initiation and progression of metabolic disorders, such as T2D, obesity, gouty arthritis, and atherosclerosis.

Heart disease, including atherosclerosis, is the leading cause of death in the elderly (Leon and Gustafsson 2016; North and Sinclair 2012). Cholesterol crystals (Grebe and Latz 2013) and white blood cells accumulate on the arterial wall, limiting the flow of oxygen-rich blood to the organs, which can lead to life-threatening complications such as heart attack and stroke. It has long been suggested, based on evidence from mouse models (Duewell et al. 2010; Elhage et al. 2003; Mallat et al. 2001), that IL-18, a product of inflammasome activation, may play a crucial role in the initiation and progression of atherosclerosis. Furthermore, human atherosclerotic plaques have elevated concentrations of IL-18 and IL-18 receptors compared with disease-free arterial tissues. Apolipoprotein E (ApoE) is necessary for a proper cholesterol metabolism. In ApoE-deficient mice, which spontaneously develop atherosclerotic lesions, elevated IL-18 levels have been shown to cause vascular inflammation and enhance the instability of atherosclerotic plaques, whereas IL-18-deficiency resulted in reduced atherosclerotic lesion size (Tan et al. 2010; De Nooijer et al. 2004). Elevation of low-density lipoprotein and free fatty acids (FFAs) in human blood due to imbalanced lipid metabolism can induce pro-IL-1β production through TLRs, providing the first signal for inflammasome activation (Masters et al. 2011).

Another major global age-related health threat is T2D, resulting in insulin resistance and a chronic inflammatory disease characterized by elevated circulating levels of TNF, interleukins, and cytokine-like proteins known as adipokines released from adipose tissue (Donath and Shoelson 2011). IL-1β, in particular, has been strongly linked to the pathogenesis of T2D by promoting insulin resistance and causing β-cell functional impairment and apoptosis. In cell culture, IL-1β depresses insulin sensitivity by inducing JNK-dependent serine phosphorylation of insulin receptor substrate-1, resulting in the disruption of insulin-induced PI3K-Akt signaling in insulin-targeted cells. At the same time, IL-1β induces the expression of TNF-α (Wen et al. 2011), which could independently impair insulin signaling (Hotamisligil et al. 1993). Together with elevated FFAs in circulation because of imbalanced lipid metabolism, IL-1β induces metabolic stressors, such as endoplasmic reticulum stress and oxidative stress, both of which are involved in the induction of inflammation and β-cell loss, thereby leading to the pathogenesis of T2D (Robbins et al. 2014; Legrand-Poels et al. 2014). Furthermore, clinical trials in humans reported that either IL-1 receptor antagonist (IL-1RA) or anti-IL-1β-neutralizing antibody improved control of glucose levels and β-cell function. Larger-scale clinical trials have been undertaken to definitively determine the potential of this treatment strategy (Larsen et al. 2007; Böni-Schnetzler and Donath 2013). Furthermore, neuromodulatory lipids known as endocannabinoids are lipids that have recently been found to induce NLRP3 inflammasome-dependent IL-1β production by pancreatic-infiltrating macrophages through the peripheral cannabinoid receptor type 1 (CB1R), resulting in pancreatic β-cell death in a paracrine manner (Jourdan et al. 2013). Moreover, blockade of CB1R by an inhibitor delayed the progress of T2D in the Zucker diabetic fatty rat, which carries a spontaneous mutation of the leptin receptor gene and develops hyperglycemia progressively with aging accompanied by reduced β-cell apoptosis and hyperglycemia. This finding implicates CB1R as being a potential therapeutic target in T2D (Böni-Schnetzler and Donath 2013).

Apart from metabolic alterations, aging also constitutes a primary risk factor for many neurodegenerative diseases. Recent studies have suggested that misfolded protein aggregates lead to activation of the NLRP3 inflammasome in two neurodegenerative diseases: AD and amyotrophic lateral sclerosis (Masters and O'Neill 2011; Walsh et al. 2014). AD is a chronic neurodegenerative disease that mainly affects cognitive functioning and is the most common cause of dementia. Amyloid-β peptide is regularly formed in brain tissue by cleavage of the amyloid precursor protein, but it can form prion-like misfolded oligomers in the case of AD (Heneka et al. 2015). Amyloid-β was the first molecule associated with neurodegenerative disease models that was found to activate the murine NLRP3 inflammasome, resulting in IL-1β production (Halle et al. 2008). Fibrillar amyloid-β induces NLRP3-inflammasome-dependent caspase-1 activation through a mechanism dependent on endosomal rupture and cathepsin B release in LPS-primed murine macrophages (Halle et al. 2008). Interestingly, administration of cathepsin B inhibitors significantly improved memory deficit and reduced amyloid plaque load in the brain in the AD mouse model, suggesting a potential therapeutic approach for Alzheimer's treatment in which the inflammasome is targeted (Hook et al. 2008).

Parkinson's disease results in the death of dopamine-generating neurons in the substantia nigra and the presence of aggregated inclusions mainly composed of α-synuclein (αSyn) in neurons (Shulman et al. 2011). αSyn can form fibrils with a cross β-sheet structure, morphologically similar to the amyloid fibrils from AD (Chiti and Dobson 2006). Recent research has shown that in a Parkinson's disease mouse model in which Parkinson's disease is induced by the loss of nigral dopaminergic neurons caused by treatment with neurotoxin 1-methyl-4-phenyl-1,2,3,6-tetrahydropyridine, mice lacking Nlrp3 are resistant to developing PD. This provides in vivo evidence for a visible link between the NLRP3 inflammasome and Parkinson's disease (Yan et al. 2015).

13.4 Conclusions and Future Prospects

It is sufficiently clear that one of the major beneficial roles of the inflammasome is to sense microbial infection and mediate a rapid plan to host defense through the immediate secretion of cytokines. These responses are highly effective against infectious agents. However, inflammasomes can also function as sensors of nonmicrobial signals (e.g., sterile mediators of membrane damage or cellular stress). Most of these nonmicrobial triggers of the inflammasome have been studied, mainly because of their pathological roles in disease, whereas there are few examples of beneficial effects of inflammasome activation by nonmicrobial triggers. Inflammasome activation cannot all be considered harmful, and the therapeutic inhibition of this pathway has to be balanced against its beneficial contribution. As the mechanistic insight of the inflammasomes increases, opportunities for creating new therapies for patients with inflammatory diseases are expected to be enhanced proportionately. It is however possible that the beneficial effects of inflammasome activation by nonmicrobial triggers have been ignored. It is also notable that almost all known non-infectious triggers of the inflammasome mediate activation through NLRP3, which seems to be uniquely able to respond to a wide range of stimuli. So far it remains unclear whether other NLRs can also sense nonmicrobial signals of physiological stress, although there remain many NLRs whose respective roles in host defense are beginning to be understood and described, such as NLRP10 (Eisenbarth et al. 2012), NLRC5 (Cui et al. 2010; Meissner et al. 2010), and NLRC3 (Schneider et al. 2012; Zhang et al. 2014). With these NLRs, one of the major obstacles to overcome seems to be identifying the activating signals. For example, inflammasome activation by pore-forming toxins triggers caspase-1-dependent activation of membrane repair (Gurcel et al. 2006), and NLRC4 activation induced by cytosolic delivery of flagellin has been shown to trigger rapid production of inflammatory lipid mediators in a caspase-1-dependent manner (von Moltke et al. 2012). Furthermore, it is clear that inflammasomes exist in multiple cell types, including both hematopoietic and nonhematopoietic cells, and play a determinant role in the onset and development of age-related and chronic inflammatory diseases.

Bibliography

Allen IC (2014) Non-inflammasome forming NLRs in inflammation and tumorigenesis. Front Immunol 5:169

Allen IC, Moore CB, Schneider M, Lei Y, Davis BK, Scull MA et al (2011) NLRX1 protein attenuates inflammatory responses to infection by interfering with the RIG-I-MAVS and TRAF6-NF-κB signaling pathways. Immunity 34:854–865

Allen IC, Wilson JE, Schneider M, Lich JD, Roberts RA, Arthur JC et al (2012) NLRP12 suppresses colon inflammation and tumorigenesis through the negative regulation of non canonical NF-kB signaling. Immunity 36:742–754

Amer A, Franchi L, Kanneganti TD, Body- Malapel M, Ozoren N, Brady G et al (2006) Regulation of *Legionella* phagosome maturation and infection through flagellin and host Ipaf. J Biol Chem 281:35217–35223

Anand PK, Kanneganti TD (2013) NLRP6 in infection and inflammation. Microbes Infect 15:661–668

Barker BR, Taxman DJ, Ting JP (2011) Cross-regulation between the IL-1B/IL-18 processing inflammasome and other inflammatory cytokines. Curr Open Immunol 23:591–597

Bauernfeind FG et al (2009) Cutting edge: NF-kappaB activating pattern recognition and cytokine receptors license NLRP3 inflammasome activation by regulating NLRP3 expression. J Immunol 183:787–791

Böni-Schnetzler M, Donath MY (2013) How biologics targeting the IL-1 system are being considered for the treatment of type 2 diabetes. Br J Clin Pharmacol 76:263–268

Boyden ED, Dietrich WF (2006) Nalp1b controls mouse macrophage susceptibility to anthrax lethal toxin. Nat Genet 38:240–244

Brunette RL, Young JM, Whitley DG, Brodsky IE, Malik HS, Stetson DB (2012) Extensive evolutionary and functional diversity among mammalian AIM2-like receptors. J Exp Med 209:1969–1983

Chandrasekaran A, Idelchik MD, Melendez JA (2017) Redox control of senescence and age-related disease. Redox Biol 11:91–102

Chiti F, Dobson CM (2006) Protein misfolding, functional amyloid, and human disease. Annu Rev Biochem 75:333–366

Choubey D (2012) DNA-responsive inflammasomes and their regulators in autoimmunity. Clin Immunol 142:223–231

Cui J, Zhu L, Xia X, Wang HY, Legras X, Hong J et al (2010) NLRC5 negatively regulates the NF-kB and type I interferon signaling pathways. Cell 141:483–496

De Nooijer R, von der Thüsen JH, Verkleij CJ, Kuiper J, Jukema JW, van der Wall EE et al (2004) Overexpression of IL-18 decreases intimal collagen content and promote a vulnerable plaque phenotype in apolipoprotein-E-deficient mice. Aterioscler Thromb Vasc Biol 24:2313–2319

De Zoete MR, Palm NW, Zhu S, Flavell RA (2014) Inflammasomes. Cold Spring Harb Perspect Biol 6:a016287

Donath MY, Shoelson SE (2011) Type 2 diabetes as an inflammatory disease. Nat Rev Immunol 11:98–107

Dostert C, Petrilli V, Van Bruggen R, Steele C, Mossman BT, Tschopp J (2008) Innate immune activation through Nalp3 inflammasome sensing asbestos and silica. Science 320:674–677

Duewell P, Kono H, Rayner KJ, Sirois CM, Vladimer G, Bauernfeind FG et al (2010) NLRP3 inflammasomes are required for atherogenesis and activated by colesterol crystals. Nature 464:1357–1361

Eisenbarth SC, Williams A, Colegio OR, Meng H, Strowig T, Rongvaux A et al (2012) NLRP10 is a NOD-like receptor essential to initiate adaptive immunity by dendritic cells. Nature 484:510–513

Elhage R, Jawien J, Rudling M, Ljunggren HG, Takeda K, Akira S et al (2003) Reduced aterosclerosis in interleukin-18 deficient apolipoprotein E-knockout mice. Cardiovasc Res 59:234–240

Faustin B, Lartigue L, Bruey JM, Luciano F, Sergienko E, Bailly-Maitre B et al (2007) Reconstituted NALP1 inflammasome reveals two-step mechanism of caspase-1 activation. Mol Cell 25:713–724

Fernandes-Alnemri T, Yu JW, Juliana C, Solorzano L, Kang S, Wu J et al (2010) The AIM2 inflammasome is critical for innate immunity to *Francisella tularensis*. Nat Immunol 11:385–393

Ferrucci L, Corsi A, Lauretani F, Bandinelli S, Bartali B, Taub DD et al (2005) The origins of age-related pro inflammatory state. Blood 105:2294–2299

Franceschi C, Campisi J (2014) Chronic inflammation (inflammaging) and its potential contribution to age-related diseases. J Gerontol A Biol Sci Med Sci 69:S4–S9

Franceschi C, Bonafè M, Valensin S, Olivieri F, De Luca M, Ottaviani E, De Benedictis G (2000) Inflamm-aging. An evolutionary perspective on immunosenescence. Ann N Y Acad Sci 908:244–254

Goldberg EL, Dixit VD (2015) Drivers of age-related inflammation and strategies for health span extension. Immunol Rev 265:63–74

Grebe A, Latz E (2013) Cholesterol crystals and inflammation. Curr Rheumatol Rep 15:313

Grenier JM, Wang L, Manji GA, Huang WJ, Al-Garawi A, Kelly R et al (2002) Functional screening of five PYPAF family members identifies PYPAF5 as a novel regulator of NF-kB and caspase-1. FEBS Lett 530:73–78

Guarda G et al (2011) Differential expression of NLRP3 among hematopoietic cells. J Immunol 186:2529–2534

Guo H, Callaway JB, Ting JP (2015) Inflammasomes: mechanisms of action, role in disease, and therapeutics. Nat Med 21:677–687

Gurcel L, Abrami L, Girardin S, Tschopp J, van der Goot FG (2006) Caspase-1 activation of lipid metabolic pathways in response to bacterial pore-forming toxins promotes cell survival. Cell 136:1135–1145

Halle A, Hornung V, Petzold GC, Stewart CR, Monks BG, Reinheckel T et al (2008) The NALP3 inflammasome is involved in the innate immune response to amyloid-beta. Nat Immunol 9:857–865

Heneka MT, Golenbock DT, Latz E (2015) Innate immunity in Alzheimer's disease. Nat Immunol 16:229–236

Hook VY, Kindy M, Hook G (2008) Inhibitors of cathepsin B improve memory and reduce beta-amyloid in transgenic Alzheimer disease mice expressing the wild type, but not the Swedish mutant, beta-secretase site of the amyloid precursor protein. J Biol Chem 283:7745–7753

Hotamisligil GS, Shargill NS, Spiegelman BM (1993) Adipose expression of tumor necrosis factor-alpha: direct role in obesity-linked insulin resistance. Science 259:87–91

Hu B, Elinav E, Huber S, Strowig T, Had L, Hafemann A et al (2013) Microbiota-induced activation of epithelial IL-6 signaling links inflammasome-driven inflammation with transmissible cancer. Proc Natl Acad Sci U S A 110:9862–9867

Jin K (2010) Modern biological theories of aging. Aging Dis 1:72–74

Jin T, Perry A, Jiang J, Smith P, Curry JA, Unterholzner L et al (2012) Structures of the HIN domain: DNA complexes reveal ligand binding and activation mechanisms of the AIM2 inflammasome and IFI16 receptor. Immunity 36:561–571

Jin T, Perry A, Smith P, Jiang J, Xiao TS (2013) Structure of the absent in melanoma 2 (AIM2) pyrin domain provides insights into the mechanisms of AIM2 auto inhibition and inflammasome assembly. J Biol Chem 288:13225–13235

Jones JW, Kayagaki N, Broz P, Henry T, Newton K, O'Rourke K et al (2010) Absent in melanoma 2 is required for innate immune recognition of *Francisella tularensis*. Proc Natl Acad Sci U S A 107:9771–9776

Jourdan T, Godlewski G, Cinar R, Bertola A, Szanda G, Liu J et al (2013) Activation of the NLRP3 inflammasome in infiltrating macrophages by endocannabinoids mediates beta cell loss in type 2 diabetes. Nat Med 19:1132–1140

Karki R, Man SM, Malireddi RK, Kesavardhana S, Zhu Q, Burton AR et al (2016) NLRC3 is an inhibitory sensor of PI3K-mTOR pathways in cancer. Nature. https://doi.org/10.1038/nature20597

Kayagaki N, Warming S, Lamkanfi M, Vande Walle L, Louie S, Dong J et al (2011) Non-canonical inflammasome activation targets caspase-11. Nature 479:117–121

Kirkwood TB (2005) Understanding the odd science of aging. Cell 120:437–447

Kirkwood TB, Boys RJ, Gillespie CS, Proctor CJ, Shanley DP, Wilkinson DJ (2003) Towards an e-biology of aging: integrating theory and data. Nat Rev Mol Cell Biol 4:243–249

Knrone CL, Trzcinski K, Zborowski T, Sanders EA, Bogaert D (2013) Impaired innate mucosal immunity in aged mice permits prolonged Streptococcus pneumonia colonization. Infect Immun 81:4615–4625

Kofoed EM, Vance RE (2011) Innate immune recognition of bacterial ligands by NAIPs determines inflammasome specificity. Nature 477:592–595

Lamkanfi M, Amer A, Kanneganti TD, Munoz-Planillo R, Chen G, Vandenabeele P et al (2007) The Nod-like receptor family member Naip5/Bircle restricts *Legionella pneumophila* growth independently of caspase-1 activation. J Immunol 178:8022–8027

Larsen CM, Faulenbach M, Vaag A, Vølund A, Ehses JA, Seifert B et al (2007) Interleukin-1 receptor antagonist in type 2 diabetes mellitus. N Engl J Med 35:1517–1526

Legrand-Poels S, Esser N, L'homme L, Scheen A, Paquot N, Piette J (2014) Free fatty acids as modulators of the NLRP3 inflammasome in obesity/type 2 diabetes. Biochem Pharmacol 92:131–141

Leon LJ, Gustafsson AB (2016) Staying young at heart: autophagy and adaptation to cardiac aging. J Mol Cell Cardiol 95:78–85

Levinsohn JL, Newman ZL, Hellmich KA, Fattah R, Getz MA, Liu S et al (2012) Anthrax lethal factor cleavage of Nlrp1 is required for activation of the inflammasome. PLoS Pathog 8: e1002638

Lipsky MS, King M (2015) Biological theories of aging. Dis Mon 61:460–466

Liu R, Truax AD, Chen L, Hu P, Li Z, Chen J et al (2015) Expression profile of innate immune receptors, NLRs and AIM2, in human colorectal cancer: correlation with cancer stages and inflammasome components. Oncotarget 6:33456–33469

Ljubuncic P, Reznick AZ (2009) The evolutionary theories of aging revisited. A mini-review. Gerontology 55:205–216

Mallat Z, Corbaz A, Scoazec A, Graber P, Alouani S, Esposito B et al (2001) Interleukin-18/ interleukin-18 binding protein signaling modulates atherosclerotic lesion development and stability. Circ Res 89:E41–E45

Mariathasan S, Newton K, Monack DM, Vucic D, French DM, Lee WP et al (2004) Differential activation of the inflammasome by caspase-1 adaptors ASC and Ipaf. Nature 430:213–218

Mariathasan S, Weiss DS, Newton K, McBride J, O'Rourke K, Roose-Girma M et al (2006) Cryopyrin activates the inflammasome in response to toxins and ATP. Nature 440:228–232

Martinon F, Burns K, Tschopp J (2002) The inflammasome: a molecular platform triggering activation of inflammatory cascades and processing of pro-IL-B. Mol Cell 10:417–426

Martinon F, Petrilli V, Mayor A, Tardivel A, Tschopp J (2006) Gout-associated uric acid crystals activate the NALP3 inflammasome. Nature 440:237–241

Masters SL, O'Neill LA (2011) Disease-associated amyloid and misfolded protein aggregates activate the inflammasome. Trends Mol Med 17:276–282

Masters SL, Latz E, O'Neill LA (2011) The inflammasome in atherosclerosis and type 2 diabetes. Sci Transl Med 3:81ps17

Medzhitov R (2008) Origin and physiological roles of inflammation. Nature 454:428–435

Meissner TB, Li A, Biswas A, Lee KH, Liu YJ, Bayir E et al (2010) NLR family member NLRC5 is a transcriptional regulator of MHC class I genes. Proc Natl Acad Sci U S A 107:13794–13799

Miao EA, Alpuche-Aranda CM, Dors M, Clark AE, Bader MW, Miller SI et al (2006) Cytoplasmic flagellin activates caspase-1 and secretion of interleukin 1b via Ipaf. Nat Immunol 7:569–575

Miao EA, Ernst RK, Dors M, Mao DP, Aderem A (2008) *Pseudomonas aeruginosa* activates caspase 1 through Ipaf. Proc Nat Acad Sci U S A 105:2562–2567

Miao EA, Mao DP, Yudkovsky N, Bonneau R, Lorang CG, Warren SE et al (2010) Innate immune detection of type III secretion apparatus through the NLRC4 inflammasome. Proc Nat Acad Sci U S A 107:3076–3080

Monroe KM, Yang Z, Johnson JR, Geng X, Doitsh G, Krogan NJ et al (2014) IFI16 DNA sensor is required for death of lymphoid CD4 T cells abortively infected with HIV. Science 343:428–432

Moore CB, Bergstralh DT, Duncan JA, Lei Y, Morrison TE, Zimmermann AG et al (2008) NLRX1 is a regulator of mitochondrial antiviral immunity. Nature 451:573–577

North BJ, Sinclair DA (2012) The intersection between aging and cardiovascular disease. Circ Res 110:1097–1108

Poeck H, Bscheider M, Gross O, Finger K, Roth S, Rebsamen M et al (2010) Recognition of RNA virus by RIG-I results in activation of CARD9 and inflammasome signaling for interleukin 1β production. Nat Immunol 11:63 pgs

Rathinam VA, Jiang Z, Waggoner SN, Sharma S, Cole LE, Waggoner L et al (2010) The AIM2 inflammasome is essential for host defense against cytosolic bacteria and DNA viruses. Nat Immunol 11:395–402

Robbins GR, Wen H, Ting GP (2014) Inflammasomes and metabolic disorders: old genes in modern diseases. Mol Cell 54:297–308

Rock KL, Latz E, Ontiveros F, Kino H (2010) The sterile inflammatory response. Annu Rev Immunol 28:321–342

Schattgen SA, Fitzgerald KA (2011) The PYHIN protein family as mediators of host defenses. Immunol Rev 243:109–118

Schneider M, Zimmermann AG, Roberts RA, Zhang L, Swanson KV, Wen H et al (2012) The innate immune sensor NLRC3 attenuates Toll-like receptor signaling via modification of the signaling adaptor TRAF6 and transcription factor NF-kB. Nat Immunol 13:823–831

Schroder K, Tschopp J (2010) The inflammasomes. Cell 140:821–832

Sharma N, Jha S (2016) NLR-regulated pathways in cancer: opportunities and obstacles for therapeutic interventions. Cell Mol Life Sci 73:1741–1764

Shulman JM, De Jager PL, Feany MB (2011) Parkinson's disease: genetics and pathogenesis. Annu Rev Pathol 6:193–122

Soares F, Tattoli I, Rahman MA, Robertson SJ, Belcheva A, Liu D et al (2014) The mitochondrial protein NLRX1 controls the balance between extrinsic and intrinsic apoptosis. J Biol Chem 289:19317–19330

Spadaro O, Goldberg EL, Camell CD, Youm YH, Kopchick JJ, Nguyen KY et al (2016) Growth hormone receptor deficiency protects against age-related NLRP3 inflammasome activation and immune senescence. Cell Rep 14:1571–1580

Strowig T, Henao-Mejia J, Elinav E, Flavell R (2012) Inflammasomes in health and disease. Nature 481:278–286

Sutterwala FS, Mijares LA, Li L, Ogura Y, Kazmierczak BI, Flavell RA (2007) Immune recognition of *Pseudomonas aeruginosa* mediated by the IPAF/NLRC4 inflammasome. J Exp Med 204:3235–3245

Suzuki T, Franchi L, Toma C, Ashida H, Ogawa M, Yoshikawa Y et al (2007) Differential regulation of caspase-1 activation, pyroptosis, and autophagy via Ipaf and ASC in *Shigella*-infected macrophages. PLoS Pathog 3:e111

Tan HW, Liu X, Bi XP, Xing SS, Li L, Gong HP, al e (2010) IL-18 overexpression promotes vascular inflammation and remodeling in a rat model of metabolic syndrome. Atherosclerosis 208:350–357

Ting JP, Lovering RC, Alnemri ES, Bertin J, Boss JM, Davis BK et al (2008) The NLR gene family: a standard nomenclature. Immunity 28:285–287

Ting JP, Duncan JA, Lei Y (2010) How the noninflammasome NLRs function in the innate immune system. Science 327:286–290

Von Moltke J, Trinidad NJ, Moayeri M, Kintzer AF, Wang SB, van Rooijen N et al (2012) Rapid induction of inflammatory lipid mediators by the inflammasome in vivo. Nature 490:107–111

Walsh JG, Muruve DA, Power C (2014) Inflammasomes in the CNS. Nat Rev Neurosci 15:84–97

Wen H, Gris D, Lei Y, Jha S, Zhang L, Huang MT et al (2011) Fatty acid-induced NLRP3-ASC inflammasome activation interferes with insulin signaling. Nat Immunol 12:408–415

Wlodarska M, Thaiss CA, Nowarski R, Henao-Mejia J, Zhang JP, Brown EM et al (2014) NLRP6 inflammasome orchestrates the colonic host-microbial interface by regulating goblet cell mucus secretion. Cell 156:1045–1059

Xia X, Cui J, Wang HY, Zhu L, Matsueda S, Wang Q et al (2011) NLRX1 negatively regulates TLR-induced NF-κB signaling by targeting TRAF6 and IKK. Immunity 34:843–853

Xiao TS, Ting JPY (2012) NLRX1 has a tail to tell. Immunity 36:311–312

Yan Y, Jiang W, Liu L, Wang X, Ding C, Tian Z et al (2015) Dopamine controls systemic inflammation through inhibition of NLRP3 inflammasome. Cell 160:62–73

Youm YH, Grant RW, McCabe LR, Albarado DC, Nguyen KY, Ravussin A et al (2013) Canonical Nlrp3 inflammasome links systemic low-grade inflammation to functional decline in aging. Cell Metab 18:519–532

Zaki MH, Boyd KL, Vogel P, Kastan MB, Lamkanfi M, Kanneganti TD (2010) The NLRP3 inflammasome protects against loss of epithelial integrity and mortality during experimental colitis. Immunity 32:379–391

Zaki MH, Vogel P, Malireddi RK, Body-Malapel M, Anand PK, Bertin J et al (2011) The NOD-like receptor NLRP12 attenuated colon inflammation and tumorigenesis. Cancer Cell 20:649–660

Zhang W, Cai Y, Xu W, Gao X, Xiong S (2013) AIM2 facilitates the apoptotic DNA-induced systemic lupus erythematosus via arbitrating macrophage functional maturation. J Clin Immunol 33:925–937

Zhang L, Mo J, Swanson KV, Wen H, Petrucelli A, Gregory SM, et al (2014) NLRC3, a member of the NLR family of proteins, is a negative regulator of innate immune signaling induced by the DNA sensor STING. Immunity 40:329–341

Chapter 14
Genetics of Inflammasomes

Wanessa Cardoso da Silva, Edione C. Reis, Telma M. Oshiro, and Alessandra Pontillo

Contents

Abstract Mutations in inflammasome genes are responsible for rare monogenic and polygenic autoinflammatory diseases. On the other side, genetic polymorphisms in the same molecules contribute to the development of common multifactorial diseases (i.e., autoimmune diseases, cardiovascular pathologies, cancer). In this chapter we depicted the current knowledge about inflammasome genetics.

Keywords Polymorphisms · Inflammasome · IL-1β · IL-18 · Autoinflammatory syndromes · Multifactorial diseases

---

W. C. da Silva (✉) · T. M. Oshiro
Laboratório de Investigação em Dermatologia e Imunodeficiências, Hospital das Clinicas HCFMUSP, Faculdade de Medicina, Universidade de São Paulo, São Paulo, SP, Brasil

Avenida Dr. Enéas de Carvalho Aguiar, 470 – Instituto de Medicina Tropical (IMT) Prédio 2 – 3° andar, São Paulo, SP, Brasil

E. C. Reis · A. Pontillo
Laboratório de Imunogenética, Departamento de Imunologia, Instituto de Ciências Biomédicas, Universidade de São Paulo, São Paulo, SP, Brasil

Avenida Prof. Lineu Prestes, 1730 – 05508-000 Cidade Universitária, São Paulo, SP, Brasil

M. D. Cordero, E. Alcocer-Gómez (eds.), *Inflammasomes: Clinical and Therapeutic Implications*, Experientia Supplementum 108,
https://doi.org/10.1007/978-3-319-89390-7_14

14.1 Genetics of Inflammasome

Inflammasome is a cytoplasmic multiprotein complex responsible for inflammatory caspase (viz., caspase-1, caspase-4, and caspase-5) activation. It is assembled after the activation of cytosolic innate immune receptors by pathogen- or danger-associated molecular patterns (PAMPs and DAMPs, respectively). Several receptors belonging to nucleotide-binding domain and leucine-rich repeat containing receptors (NLRs) or pyrin and HIN domain-containing (PYHIN) receptor families (i.e., NLRP1, NLRP3, NLRC4, AIM2, IFI16) are able to assemble, directly or indirectly through the recruitment of the adaptor protein ASC, the inflammasome and to activate the caspases, consequent releasing several cellular substrates including the pro-inflammatory cytokines, interleukin (IL)-1β and IL-18, and the gasdermin D, responsible for the inflammatory cell death named pyroptosis (Man et al. 2017).

Great attention has been reserved for inflammasome genetics since the discovery that dominant inherited mutations in the *NLRP3* gene, coding for the cryopyrin/NLRP3 protein, cause a constitutively high production of IL1b. In fact, mutations in *NLRP3* gene are related to autoinflammatory syndromes associated with cryopyrin/NLRP3 (cryopyrin-associated periodic syndrome, CAPS), namely, familial cold-induced autoinflammatory syndrome 1 (FCAS1, OMIM:120100), Muckle-Wells syndrome (MWS, OMIM:191900), and chronic infantile neurologic cutaneous articular (CINCA) syndrome or neonatal-onset multisystem inflammatory disease (NOMID, OMIM:607115) (Hoffman et al. 2001; Dodé et al. 2002; Feldmann et al. 2002; Aksentijevich et al. 2002). Here below we depicted the current knowledge about main genes involved in inflammasome activation.

14.2 *NLRP1*

The gene codifying for the NLR PYD-containing protein 1 (*NLRP1*) is localized on chromosome 17 (17p13.2) and consists of 18 exons. Multiple alternatively spliced transcript variants encoding distinct isoforms have been found for this gene, but the biological validity of some variants has not been determined. The longest isoform (5623 bp, NM_033004.3) codes for a 1473 amino acid protein (NP_127497.1).

NLRP1 presents the typical NLRP subfamily domain architecture consisting of an N-terminal homotypic interaction domain, the pyrin domain (PYD) (residues: 9–88), a central nucleotide-binding domain (NBD), NACHT domain (residues: 328–497) with ATPase-associated activity (AAA) domain (residues: 327–416), and a series of leucine-rich repeats (LRRs) (amino acid region 810–988). In addition, NLRP1 contains two more domains at the C-terminus, namely, function to find domain (FIIND) (residues: 1100–1354) and another homotypic interaction domain, the caspase activation and recruitment domain (CARD, residues: 1380–1460) (Chavarria-Smith et al. 2016) (Fig. 14.1).

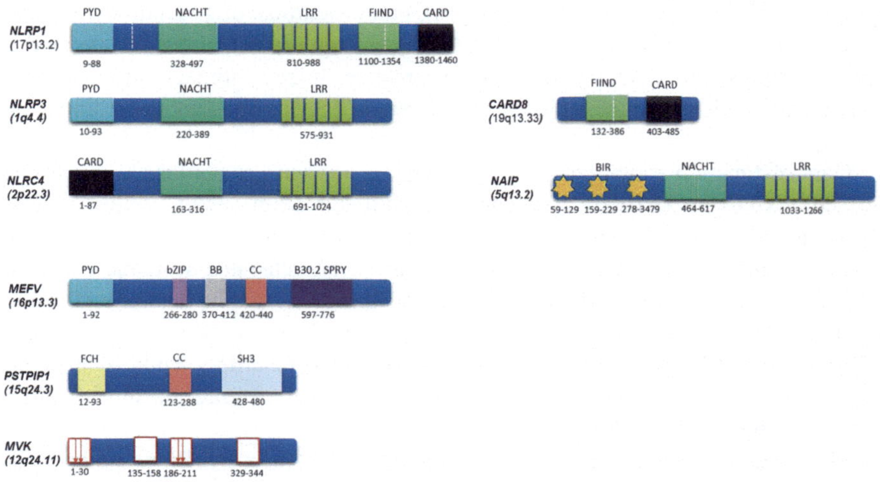

Fig. 14.1 Graphical representation of domain structure for NLRP1, NLRP3, NLRC4, NAIP, pyrin/MEFV, CARD8, PSTPIP1, and MVK

PYD is involved in PYD/PYD interaction with the PYD-CARD adaptor protein ASC. The NACHT domain is indispensable for ATP-dependent homo- and hetero-oligomerization of NLRP1. In murine orthologues, NLRP1 is activated via direct proteolysis in a specific N-terminal linker region between the PYD and NBD domain, suggesting that also in humans the N-terminal domain is auto-inhibitory for NLRP1 activity (Chavarria-Smith et al. 2016).

According to its function in Toll-like receptors (TLRs), LRRs are presumably involved in agonist sensing; however until now no agonist has been demonstrated that specifically activates NLRP1 in humans (Chavarria-Smith et al. 2016). Bacterial muramyl dipeptide (MDP) has been shown to activate NLRP1 inflammasome; however the mechanism remains unclear, with MDP being a known agonist of another NLR protein, NOD2 (Reubold et al. 2014). The possible interaction between NLRP1 and NOD2 has been postulated (Hsu et al. 2008). Recently D'Osualdo et al. described the activation of NLRP1 inflammasome by the endoplasmic reticulum (ER) stress response (or unfolded protein response), including this receptor among the sensors for ER perturbation (D'Osualdo et al. 2015).

NLRP1 has also the possibility to directly interact with inflammatory caspases through its CARD domain (Chavarria-Smith et al. 2016; Martinon et al. 2002), leading to an active debate about the role of ASC in NLRP1 inflammasome.

The FIIND domain seems to be important for NLRP1 activity, as it contains a specific auto-proteolytic cleavage site (Ser1213) and a critical residue for autolytic processing (His1186). The C-terminal cleavage of NLRP1 is necessary and sufficient to NLRP1 inflammasome activity (Finger et al. 2012).

Kummer et al. (2007) demonstrated the expression of NLRP1 protein in the lymphoid organs (T cells), skin (Langerhans cells), stomach, gut (intraepithelial T lymphocytes), lung (alveolar macrophages), brain (neurons, oligodendrocytes), and testes (spermatogonia). Moreover, RNA sequencing of human tissues showed a specific expression of *NLRP1* in the lung, prostate, spleen, stomach, thymus, pituitary gland, trachea, and uterus [as shown in public database Ensembl (Zerbino et al. 2018)].

Gain-of-function mutations in the *NLRP1* gene cause a nonfebrile syndrome termed NAIAD (NLRP1-associated autoinflammation with arthritis and dyskeratosis) (OMIM: #617388) characterized by arthritis, skin hyperplasia, and/or dyskeratosis, high transitional B-cell level, and elevated systemic levels of caspase-1 and IL-18 [(Grandemange et al. 2016), Infevers public database (Touitou et al. 2004)]. Taking into account that several single nucleotide polymorphisms (SNPs) in *NLRP1* were associated with autoimmune disorders, authors are convinced that an autoimmune dysregulation is the key feature of this disorder (Grandemange et al. 2016).

Moreover, the non-synonymous mutation of Met77Thr in *NLRP1* was identified in a dominant disorder with corneal intraepithelial dyskeratosis (Soler et al. 2013).

Moreover, since 2007, several single nucleotide polymorphisms (SNPs) in *NLRP1* have been associated with complex human diseases (Table 14.1).

The non-synonymous polymorphism rs12150220 (Leu155His) and the promoter variant rs2670660 were initially associated with autoimmune thyroid disorders in a genome-wide association study (GWAS) (Jin et al. 2007). The rs12150220 has also been associated with Addison's disease (Magitta et al. 2009), while rs2670660 with generalized vitiligo (Alkhateeb and Qarqaz 2010), systemic lupus erythematosus (Pontillo et al. 2012a), and inflammatory bowel disease (De Iudicibus et al. 2011). A haplotype containing both variants has been associated with celiac disease (Pontillo et al. 2011).

The polymorphism rs11651270 (Met1184Val) has been recently associated with HPV infection and HPV-related cervical cancer (Pontillo et al. 2016), nodular melanoma (da Silva et al. 2016), and diabetic kidney disease (Soares et al. 2018).

All these variants possibly lead to augmented gene transcription or protein function, being localized in promoter region (rs2670660) or in regions that affect the auto-inhibition of NLRP1 (rs12150220) (Chavarria-Smith et al. 2016) or NLRP1 auto-processing (rs11651270) (Finger et al. 2012), respectively; however their functional consequence is still unclear.

Levandowski et al. (2013) demonstrated that individuals containing Leu155His-Val1059Met-Met1184Val are more prone to autoimmune diseases due to an increased processing of pro-IL1β in peripheral blood mononuclear cells (PBMC).

Other variants were uniquely associated with other diseases as reported in Table 14.1.

Table 14.1 Associated studies involving genetic polymorphisms in inflammasome sensors *NLRP1*, *NLRP3*, *NLRC4*, and *MEFV* and in the NLRP3 inhibitor *CARD8*

Gene	SNP ID	Disease	Reference
NLRP1	rs1008588, rs2670660	Generalized vitiligo	Jin et al. (2007)
	rs12150220, rs2670660	Autoimmune thyroid disorders	Jin et al. (2007)
		Celiac disease	Pontillo et al. (2011)
		Generalized vitiligo	Jin et al. (2007), Alkhateeb and Qarqaz (2010), Dwivedi et al. (2013)
		Systemic lupus erythematosus	Pontillo et al. (2012a)
	rs12150220	Addison's disease	Magitta et al. (2009)
		Asbestos-associated mesothelioma	Girardelli et al. (2012)
		P. vivax malaria (severity)	Santos et al. (2016)
		Preeclampsia	Pontillo et al. (2015)
		Type 1 diabetes	Soares et al. (2018)
		HPV susceptibility	Pontillo et al. (2016)
		Sporadic malignant melanoma	da Silva et al. (2016)
		Nodular melanoma	da Silva et al. (2016)
	rs2137722, rs12150220, rs2670660	Leprosy susceptibility	Pontillo et al. (2013)
	rs2137722, rs34733791, rs11657747, rs11651595	Alzheimer's disease	Pontillo et al. (2012c)
	rs2670660	Glucocorticoid response in IBD	De Iudicibus et al. (2011)
	rs6502867, rs4790797	Vitiligo-associated autoimmune disease	Jin et al. (2007)
	rs8079034, rs878329	Psoriasis	Ekman et al. (2014)
	rs81822352	Systemic sclerosis	Dieudé et al. (2011)
	rs878329	Rheumatoid arthritis	Sui et al. (2012)
	rs878329	Partial seizures	Wang et al. (2017)
NLRP3	rs10159239, rs4925648, rs4925659	Rheumatoid arthritis	Mathews et al. (2014)
	rs10754558, rs4612666	Food-induced anaphylaxis and aspirin-induced asthma	Hitomi et al. (2009)
		Gastric cancer	Castano-Rodriguez et al. (2015)
	rs10754558, rs10925019	Ulcerative colitis	Zhang et al. (2014)
	rs10754558, rs358294199	Celiac disease	Pontillo et al. (2010a, 2011)
	rs107331113	Psoriasis	Carlström et al. (2012)

(continued)

Table 14.1 (continued)

Gene	SNP ID	Disease	Reference
	rs10754558	Acute coronary syndrome	Gonzalez-Pacheco et al. (2017)
		Anti-TNF response in patients with rheumatoid arthritis	Sode et al. (2015)
		Coronary artery disease	Zhou et al. (2016)
		HIV-1 susceptibility	Pontillo et al. (2010b, 2012b)
		HTLV-1 susceptibility	Kamada et al. (2014)
		Insulin resistance in T2D	Wang et al. (2015)
		Ischemic stroke	Zhu et al. (2016)
		Metabolic syndrome	Zhang et al. (2016)
		Type 1 diabetes	Pontillo et al. (2011)
	rs2027432, rs10754558, rs35829419	Late-onset Alzheimer's disease	Tan et al. (2013)
	rs358294199	Colorectal cancer	Ungerbäck et al. (2012)
		Crohn's disease	Schoultz et al. (2009)
		Cytokine profile in the blood	Sahdo et al. (2013)
		Delayed apoptosis of human neutrophils	Blomgran et al. (2012)
		Melanoma	Verma et al. (2012a)
		Increased IL-1β and severe inflammation	Kastbom et al. (2008)
		Mycobacterium tuberculosis growth in macrophages	Eklund et al. (2014)
		Myocardial infarction	Verma et al. (2012a)
		Pancreatic cancer, periodontitis	Miskiewicz et al. (2015)
		Rheumatoid arthritis	Ben Hamad et al. (2012), Jenko et al. (2016), Kastbom et al. (2015)
		Stroke in rheumatoid arthritis	Kastbom et al. (2015)
	rs3806265	Psoriatic juvenile idiopathic arthritis	Day et al. (2008)
	rs3806268	Primary gout	Deng et al. (2015)
	rs4353135	Oligoarticular/ polyarticular JIA	Yang et al. (2014)
	rs4612666	Pediatric severe renal parenchymal infections	Cheng et al. (2015)
CARD8	rs2043211	Abdominal aortic aneurysms	Roberts et al. (2011)
		Ankylosing spondylitis	Kastbom et al. (2013)

(continued)

Table 14.1 (continued)

Gene	SNP ID	Disease	Reference
		Cardiovascular events in rheumatoid arthritis	Garcia-Bermudez et al. (2013)
		Primary gout	McKinney et al. (2015), Chen et al. (2015)
		Inflammatory bowel disease	Yang et al. (2011), Roberts et al. (2010)
	rs7248320	Hepatocellular carcinoma and cervical cancer	Yin et al. (2015)
NLRC4	rs479333, rs212713	Increase serum levels of IL-18	Matteini et al. (2014)
MEFV	rs224204	Psoriatic juvenile idiopathic arthritis	Day et al. (2008)
	rs3743930, rs28940580	Henoch-Schönlein purpura	Xiong et al. (2017)

14.3 *NLRP3*

The gene codifying for the NLR PYD-containing protein 3 (*NLRP3*) is localized on chromosome 1 (1q44) and consists of nine exons spanning through 4470 bp. According to Ensembl public database (Zerbino et al. 2018), it has eight transcripts (splice variants) corresponding to six encoded proteins. The longest isoform (NM_004895.4) codes for a 1036 amino acid protein (NP_001073289.1). Like NLRP1, also NLRP3 presents the typical NLRP subfamily domain architecture consisting of an N-terminal pyrin domain (residues: 10–93), a central nucleotide-binding domain, a NACHT domain with a AAA motif (residues: 220–389), and a C-terminal series of LRRs (residues: 575–931) (Fig. 14.1).

The pyrin domain is involved in PYD/PYD interaction with the adaptor protein ASC. The NACHT domain is indispensable for ATP-dependent homo- and hetero-oligomerization of NLRP3 (Bae and Park 2011; Aksentijevich et al. 2007). Moreover, it has been recently demonstrated that this domain is also involved in NLRP3-CARD8 interaction leading to inhibition of NLRP3 (Ito et al. 2014). In resting conditions, LRR domain interacts with NBD, and it is presumably responsible for NLRP3 auto-inhibition (Aksentijevich et al. 2007).

NLRP3 is activated by a large number of distinct pathogen- or damage-associated molecular patterns (PAMPs or DAMPs, respectively), and it was postulated that perturbations in cellular homeostasis could be sensed by NLRP3, through three main mechanisms: K+ efflux (principally mediated by the ATP-dependent purinergic receptor P2X7), mitochondrial damage and consequent liberation of reactive oxygen species (ROS) and other mitochondrial products (i.e., mtDNA, cardiolipin), and phagolysosomal damage and rupture (cathepsins' release) (Man et al. 2017).

In "The Human Protein Atlas" public database (Uhlen et al. 2015), NLRP3 protein is reported to be expressed in the blood, bone marrow, and lymph nodes, nonkeratinizing epithelia and mucosa, the brain (cerebral cortex, hippocampus), the thyroid and adrenal glands, and the breast and female and male reproductive apparatus. It is interesting to notice that in some cells *NLRP3* is constitutively expressed, while in others it needs to be induced by TLR agonists or cytokines, through the Myd88/NF-kB signaling (Yin et al. 2009).

Since 2011, gain-of-function mutations in *NLRP3* gene have been described in patients affected by rare autosomal dominant autoinflammatory diseases, called cryopyrin/NLRP3-associated periodic syndrome (CAPS, OMIM: 606416) (Hoffman et al. 2001; Dodé et al. 2002; Feldmann et al. 2002; Aksentijevich et al. 2002). Initially described as three distinct diseases, namely, familial cold-induced autoinflammatory syndrome 1 (FCAS1, OMIM: 120100), Muckle-Wells syndrome (MWS, OMIM: 191900), and chronic infantile neurologic cutaneous articular (CINCA) syndrome or neonatal-onset multisystem inflammatory disease (NOMID) (OMIM: 607115), nowadays these overlapping inflammatory phenotypes are considered a spectrum (mild to severe) of the same syndrome (CAPS) due to the presence of NLRP3 mutations [as largely reviewed in Manthiram et al. (2017)].

CAPS are characterized by IL-1β-mediated systemic inflammation and symptoms involving the skin, joints, central nervous system (CNS), and eyes (Hoffman et al. 2001; Dodé et al. 2002; Feldmann et al. 2002; Aksentijevich et al. 2002). Patients present a neutrophil-enriched skin infiltrate (Aksentijevich et al. 2007; Manthiram et al. 2017) and showed significantly increased IL-17 serum levels as well as a higher frequency of Th-17 compared to control subjects (Lasiglie et al. 2011).

Until now, about 180 different gain-of-function mutations in *NLRP3* have been identified in CAPS patients, the majority within the central NBD domain (exon 3). In addition, some mutations within the LRR region (i.e., exon and exon 6) have also been described [from Infevers public database (Touitou et al. 2004)]. There is no apparent correlation between mutation position and disease severity, even if more rare mutations are commonly found in patients with more severe clinical presentation (Aksentijevich et al. 2007; Manthiram et al. 2017).

Aksentijevich et al. (2007) proposed a pathogenic mechanism for CAPS. In resting conditions NLRP3 is present in the cytosol in a “close” inactive conformation, ensured by the LRRs-NBD interaction. Mutations in NBD or LRRs, in turn, destabilize this inactive state, leading to a constitutively activated NLRP3 protein (“open” conformation), and to the consequent constitutive inflammasome activation and increased IL-1β production. Therefore, genetic variants in NLRP3 influence the threshold of NLRP3 inflammasome activation and disturb a balanced innate immune response.

Intriguingly, in about 40% of CAPS patients, no mutations have been found in codifying region of *NLRP3* (Aksentijevich et al. 2007; Manthiram et al. 2017). The presence of genetic variants in the promoter region of *NLRP3* has been taken into account (Anderson et al. 2008), and the effect of additional genetic factors that initiate or modulate CAPS has been postulated (Caroli et al. 2007). This fact can be attributed, at least in part, to somatic mosaicism, as observed in 70% of CINCA/NOMID cases (Tanaka et al. 2011).

Several gain-of-function SNPs in *NLRP3* gene have been associated with the pathophysiology of multifactorial diseases, as reported in Table 14.1; however the functional effect on inflammasome activity was demonstrated in few variants.

The missense variant Gln705Lys or Gln703Lys (rs35829419, Q705K or Q703K) was found to protect against the development of celiac disease (Pontillo et al. 2010a) and female myocardial infarction (Varghese et al. 2013). On the other hand, this SNP has been associated with susceptibility to melanoma (Verma et al. 2012a) and Crohn's disease (Schoultz et al. 2009). The rs35829419 is a variant with an uncertain significance localized in the NACHT domain of the *NLRP3* gene, first described in CAPS patient and later found also in healthy individuals and consequently considered a polymorphism with a low frequency in the general population (5–11%) (Touitou et al. 2004). Verma et al. (2012b) and Villani et al. (2009) showed that rs35829419, especially when in combination with the nonsense SNP Cys10* (rs2043211; C10X) in CARD8, affects NLRP3 activation leading to constitutive increased production of IL-1β (and IL-18) by human peripheral blood mononuclear cells (PBMC) or isolated monocytes.

The rs10754558 polymorphism, located in the *NLRP3* 3′-untranslated region (3′UTR), was associated with protection against several infections [i.e., HIV-1 (Pontillo et al. 2010b, 2012b), HTLV-1 (Kamada et al. 2014), tuberculosis (Souza de Lima et al. 2016)] and autoimmune development (i.e., type 1 diabetes) (Pontillo et al. 2010a) and with increased susceptibility to develop food allergy (Hitomi et al. 2009) (Table 14.1).

It has been proposed that this variant affects *NLRP3* mRNA stability (Hitomi et al. 2009) interfering with microRNA (miR)-223 binding (Bauernfeind et al. 2012).

14.4 *NLRC4*

The gene codifying for the NLR CARD-containing protein 4 (*NLRC4*) is localized on chromosome 2 (2p22.3) and consists of 12 exons spanning 41 kb. According to "Ensembl" public database (Zerbino et al. 2018), it has four transcripts (splice variants) corresponding to four encoded proteins. The longest isoform (NM_021209.4) encodes for a 1024 amino acid protein (NP_001186067).

NLRC4 presents the typical NLR structure, with the central NBD domain (residues: 163–316) and the C-terminal LRR domain-containing 13 leucine-rich repeats (residues: 691–1024), but instead of a pyrin domain at the N-terminal (such as for NLRP1 and NLRP3), it presents a caspase activation and recruitment domain (CARD) (residues: 1–87) (Vance 2015) (Fig. 14.1).

According to "The Human Protein Atlas" public database (Vance 2015), NLRC4 protein is largely expressed in the appendix, bone marrow, lymph nodes, spleen, lung, adipose tissue and placenta, and gut epithelium.

As shown for NLRP3, NBD and LRR domains interact for an auto-inhibition mechanism of NLRC4, stabilizing the NLRC4 inactive ("close") conformation (Vance 2015; Canna et al. 2014). The mechanism for NLRC4 activation is still under debate; however it was demonstrated that it results from the interaction of the receptor with another NLR protein, NAIP (neuronal apoptosis inhibitory protein), which can directly bind to bacterial flagellin (Vance 2015; Kofoed and Vance 2011). Once activated, NLRC4 recruits inflammatory caspases, directly through CARD/CARD interaction with adaptor protein ASC, leading to inflammasome activation and consequent IL-1β and/or IL-18 release, and the eventual induction of pyroptosis (Vance 2015). The presence of ASC works to increase the efficiency of caspase-1 recruitment and activation and hence enhances IL-1β production (Vance 2015).

Gain-of-function mutations in *NLRC4* have been found in patients with the familial cold autoinflammatory syndrome 4 (FCAS4, OMIM: 616115) (Romberg et al. 2014) or autoinflammation with infantile enterocolitis (AIFEC/SCAN4, OMIM: 616050) (Canna et al. 2014; Kitamura et al. 2014), two distinct described diseases, nowadays considered a unique dominant autoinflammatory syndrome associated with NLRC4 mutations. Up to now, the majority of reported mutations [i.e., Val341Ala (Romberg et al. 2014), Thr337Ser (Canna et al. 2014), and His443Pro (Kitamura et al. 2014)] are localized within the NBD domain (exon 4) of *NLRC4* gene ["Infevers" public database (Touitou et al. 2004)] and result in the constitutive activation of NLRC4 inflammasome leading to increased IL-1β, but especially IL-18 release, and abnormal pyroptosis in human monocyte-derived macrophages (Canna et al. 2014; Khameneh and Mortellaro 2014).

Similarly, to what was proposed for the mutated NLRP3, Romberg et al. (2014) and Canna et al. (2014) suggested that the described mutations (Val341Ala and Thr337Ser, respectively) could destabilize the closed/inactive conformation of NLRC4, either disrupting the LRR/NBD interactions. On the other hand, Kitamura et al. (Kitamura et al. 2014) proposed that the His443Pro substitution could increase the binding affinity of the mutated NLRC4 for other NLRC4 monomers leading to an increase NLRC4 oligomerization rate. Anyway, these defects result in constant oligomerization and constitutive activation of the inflammasome complex (Canna et al. 2014; Romberg et al. 2014; Kitamura et al. 2014; Khameneh and Mortellaro 2014).

Emphasizing the important role of NLRC4 not only in innate immune cells but also in intestinal epithelial cells and specifically with regard to IL-18 production, in AIFEC/SCAN4 patients, the intestinal involvement is prominent (Canna et al. 2014) compared to other autoinflammatory syndromes, such as NAIAD or CAPS (Manthiram et al. 2017). Moreover, two polymorphisms localized in the intronic region of *NLRC4* (rs479333 and rs212713) have been associated with serum levels of IL-18 (Matteini et al. 2014) (Table 14.1). Other study provides evidence for the functional causality of the *NLRC4* SNP rs385076 and IL-18 activation (Zeller et al. 2015) (Table 14.1).

14.5 *MEFV* (Pyrin)

The cytosolic innate immune receptor pyrin/marenostrin is codified by *MEFV* gene, which is localized on chromosome 16 (16p13.3) and consists of 10 exons (NM_000243.2). This protein does not belong to NLR or PYHIN inflammasome receptor families; however due to the presence of a PYD domain, it is able to activate the inflammasome (Heilig and Broz 2018). Pyrin is a 781 amino acid protein (NP_000234.1) constituted by an N-terminal PYD domain, a B-box zinc finger domain, a coiled-coil domain, and a C-terminal B30.2 domain (also known as the rfp/PRY/SPRY domain) (Fig. 14.1).

RNA sequencing of human tissues showed a specific expression of *MEFV* in the blood, spleen, lung, testis, and female reproductive apparatus (Zerbino et al. 2018; Uhlen et al. 2015).

Mutations in *MEFV* gene are responsible for familial Mediterranean fever (FMF, OMIM: 249100), a hereditary recurrent fever and the most prevalent monogenic autoinflammatory syndrome (about 1–5/10,000 prevalence in general population, 1/200–1/1000 in non-Ashkenazi Jews, Turks, Armenians, and Arabs) (reviewed in Özen et al. 2017), generally in a recessive pattern of transmission, even if patients with single mutation in *MEFV* gene were also recently described, or either with combined genotypes between mutations in *MEFV* and in other AID genes such as *TNFRSF1A* (as discussed in Touitou 2018). FMF is characterized by recurrent 1-day to 3-day febrile attacks accompanied by serositis, synovitis, and/or cutaneous inflammation (Özen et al. 2017). Renal amyloidosis is a common and fatal prognosis in FMF (Özen et al. 2017; Ben-Chetrit 2003).

Mutations in *MEFV* have missense or nonsense variations as well as deletions or splicing defects and are spread out in the entire gene, even if the most frequently effected is exon 10 (encoding for B30.2 domain of pyrin) (Manthiram et al. 2017; Özen et al. 2017; Touitou et al. 2004).

The exact role of pyrin has been long debated, and several models for FMF pathogenesis have been proposed throughout the years. The first model suggested that pyrin inhibits pro-inflammatory signaling by the sequestration of ASC or inhibition of the enzymatic activity of caspase-1 via its SPRY domain (Chae et al. 2006), and its loss-of-function could lead to increased inflammasome activation.

An alternative model suggested that pyrin forms a ternary complex with ASC and pro-caspase, and the formation of this complex leads to caspase-1 autoactivation. As the majority of FMF mutations are located in the B30.2 domain of pyrin, it has been suggested that this domain regulates the activity of pyrin (Yu et al. 2007).

More recently it was demonstrated that bacterial modification of host small GTPAse RhoA activates the assembling of pyrin inflammasome (Heilig and Broz 2018). In a homeostatic state, RhoA induces the phosphorylation of pyrin on its residues Ser208 and Ser242 through two protein kinases, PKN1 and 2. Once phosphorylated, pyrin interacts with the regulatory protein 14-3-3, which inhibits the recruitment of ASC and the formation of inflammasome. When RhoA is modified by certain bacterial toxins, pyrin phosphorylation does not occur, 14-3-3 proteins do not bind to pyrin, and consequently pyrin inflammasome is activated (Park et al. 2016). Taking into account these new findings, a novel pathogenic hypothesis has been made considering that mutated pyrin is less able to bind 14-3-3 proteins leading to a constitutively activated inflammasome (Park et al. 2016).

14.6 Newly Recognized Inflammasome Regulators: *PSTPIP1*, *MVK*, and *LPIN2*

The gene codifying for the proline-serine-threonine phosphatase-interacting protein 1 *(PSTPIP1)* is localized on chromosome 15 (15q24.3) and has 15 exons differently spliced into four transcripts (Zerbino et al. 2018). The full-length transcript (NM_001321136.1) codes for a 416 amino acid protein (NP_001308066.1) constituted by an N-terminal F-BAR (FCH+CC) domain, which plays a role in dimerization and membrane phospholipid binding, and a C-terminal SH3 domain, which mediates the assembly of large multiprotein complexes (Starnes et al. 2014). PSTPIP1 interacts with actin playing an important role in cytoskeleton organization (Badour et al. 2003) and with pyrin (Shoham et al. 2003)—through its F-BAR and SH3 domains—being an upstream regulator of inflammasome.

PSTPIP1 is expressed predominantly in hematopoietic tissues and the lung (Uhlen et al. 2015).

Missense mutations in F-BAR domain of *PSTPIP1* were described in patients with PAPA syndrome (pyogenic arthritis with pyoderma gangrenosum and acne), a rare genetic disorder characterized by its effects on the skin and joint (Touitou et al. 2004; Manthiram et al. 2017; Wise et al. 2002). Mutant protein is hyperphosphorylated and binds avidly to pyrin promoting inflammation (Wise et al. 2002). According to Yu et al. (2007), mutations either in *MEFV* or in *PSTPIP1* promote the assembly of pyrin inflammasome.

The hyper-IgD syndrome (HIDS, OMIM: 260920) or mevalonate kinase deficiency (MKD, OMIM: 251170) is a monogenic hereditary periodic fever characterized by early-onset repeated episodes of fever, rash, gastrointestinal symptoms, and oral ulcers (Manthiram et al. 2017). In this disease, biallelic loss-of-function mutations affect the second enzyme of the isoprenoid pathway, the so-called mevalonate kinase (*MVK*, 12q24.11) (Touitou et al. 2004), causing a reduced formation of intermediate and final metabolites (isoprenoids and cholesterol, respectively) and resulting in the unexpected activation of inflammasome and caspase-1. The first etiologic evidence was that the lower level of isoprenoids leads to a reduced amount of protein prenylation especially of the small GTP proteins Rac1 and RhoA geranylgeranylation (Kuijk et al. 2008). Two independent studies have linked Rac1 with the activation of caspase-1. GTP-bound active form of Rac1 activates caspase-1 through the p21-activated kinase (PAK)-1 (Normand et al. 2009). Decrease of Rac1 geranylgeranylation leads to caspase-1 activation through the Rac1/PI3K/PKB pathway (Kuijk et al. 2008). Finally, according to Park et al. (2016), defect in RhoA geranyl-geranylation affects pyrin phosphorylation and inhibited state, leading to a constitutively activated pyrin inflammasome.

Concluding, Majeed syndrome (OMIM: 609628) is an autoinflammatory disease characterized by multifocal sterile osteomyelitis, dyserythropoietic anemia, and neutrophilic skin lesions and was associated with recessive mutations in *LPIN2* gene (18p11.31), which encodes lipin 2, an enzyme and transcription factor involved in lipid metabolism. Recently it has been demonstrated that lipin 2 regulates also NLRP3 inflammasome by affecting potassium efflux (Lordén et al. 2017), suggesting that genetic defect in *LPIN2* could promote constitutive NLRP3 inflammasome activation. SNPs in LPIN2 were previously associated with obesity and metabolic diseases (Aulchenko et al. 2007; Meidtner et al. 2014).

References

Aksentijevich I, Nowak M, Mallah M, Chae JJ, Watford WT, Hofmann SR, Stein L, Russo R, Goldsmith D, Dent P, Rosenberg HF, Austin F, Remmers EF, Balow JE, Rosenzweig S, Komarow H, Shoham NG, Wood G, Jones J, Mangra N, Carrero H, Adams BS, Moore TL, Schikler K, Hoffman H, Lovell DJ, Lipnick R, Barron K, O'Shea JJ, Kastner DL, Goldbach-Mansky R (2002) De novo CIAS1 mutations, cytokine activation, and evidence for genetic heterogeneity in patients with neonatal-onset multisystem inflammatory disease (NOMID) – A new member of the expanding family of pyrin-associated autoinflammatory diseases. Arthritis Rheum 46(12):3340–3348. https://doi.org/10.1002/art.10688

Aksentijevich I, Putnam CD, Remmers EF, Mueller JL, Le J, Kolodner RD, Moak Z, Chuang M, Austin F, Goldbach-Mansky R, Hoffman HM, Kastner DL (2007) The clinical continuum of cryopyrinopathies: novel CIAS1 mutations in North American patients and a new cryopyrin model. Arthritis Rheum 56(4):1273–1285. https://doi.org/10.1002/art.22491

Alkhateeb A, Qarqaz F (2010) Genetic association of NALP1 with generalized vitiligo in Jordanian Arabs. Arch Dermatol Res 302(8):631–634. https://doi.org/10.1007/s00403-010-1064-1

Anderson JP, Mueller JL, Misaghi A, Anderson S, Sivagnanam M, Kolodner RD, Hoffman HM (2008) Initial description of the human NLRP3 promoter. Genes Immun 9(8):721–726. https://doi.org/10.1038/gene.2008.66

Aulchenko YS, Pullen J, Kloosterman WP, Yazdanpanah M, Hofman A, Vaessen N, Snijders PJ, Zubakov D, Mackay I, Olavesen M, Sidhu B, Smith VE, Carey A, Berezikov E, Uitterlinden AG, Plasterk RH, Oostra BA, van Duijn CM (2007) LPIN2 is associated with type 2 diabetes, glucose metabolism, and body composition. Diabetes 56(12):3020–3026. https://doi.org/10.2337/db07-0338

Badour K, Zhang JY, Shi F, McGavin MKH, Rampersad V, Hardy LA, Field D, Siminovitch KA (2003) The Wiskott-Aldrich syndrome protein acts downstream of CD2 and the CD2AP and PSTPIP1 adaptors to promote formation of the immunological synapse. Immunity 18 (1):141–154. https://doi.org/10.1016/s1074-7613(02)00516-2

Bae JY, Park HH (2011) Crystal structure of NALP3 protein pyrin domain (PYD) and its implications in inflammasome assembly. J Biol Chem 286(45):39528–39536. https://doi.org/10.1074/jbc.M111.278812

Bauernfeind F, Rieger A, Schildberg FA, Knolle PA, Schmid-Burgk JL, Hornung V (2012) NLRP3 inflammasome activity is negatively controlled by miR-223. J Immunol 189(8):4175–4181. https://doi.org/10.4049/jimmunol.1201516

Ben Hamad M, Cornelis F, Marzouk S, Chabchoub G, Bahloul Z, Rebai A, Fakhfakh F, Ayadi H, Petit-Teixeira E, Maalej A (2012) Association study of CARD8 (p.C10X) and NLRP3 (p.

Q705K) variants with rheumatoid arthritis in French and Tunisian populations. Int J Immunogenet 39(2):131–136. https://doi.org/10.1111/j.1744-313X.2011.01070.x

Ben-Chetrit E (2003) Familial Mediterranean fever (FMF) and renal AA amyloidosis–phenotype-genotype correlation, treatment and prognosis. J Nephrol 16(3):431–434

Blomgran R, Brodin VP, Verma D, Bergstrom I, Soderkvist P, Sjowall C, Eriksson P, Lerm M, Stendahl O, Sarndahl E (2012) Common genetic variations in the NALP3 inflammasome are associated with delayed apoptosis of human neutrophils. PLoS One 7(3). https://doi.org/10.1371/journal.pone.0031326

Canna SW, de Jesus AA, Gouni S, Brooks SR, Marrero B, Liu Y, DiMattia MA, Zaal KJM, Sanchez GAM, Kim H, Chapelle D, Plass N, Huang Y, Villarino AV, Biancotto A, Fleisher TA, Duncan JA, O'Shea JJ, Benseler S, Grom A, Deng Z, Laxer RM, Goldbach-Mansky R (2014) An activating NLRC4 inflammasome mutation causes autoinflammation with recurrent macrophage activation syndrome. Nat Genet 46(10):1140–1146. https://doi.org/10.1038/ng.3089

Carlström M, Ekman AK, Petersson S, Söderkvist P, Enerbäck C (2012) Genetic support for the role of the NLRP3 inflammasome in psoriasis susceptibility. Exp Dermatol 21(12):932–937. https://doi.org/10.1111/exd.12049

Caroli F, Pontillo A, D'Osualdo A, Travan L, Ceccherini I, Crovella S, Alessio M, Stabile A, Gattorno M, Tommasini A, Martini A, Lepore L (2007) Clinical and genetic characterization of Italian patients affected by CINCA syndrome. Rheumatology 46(3):473–478. https://doi.org/10.1093/rheumatology/kel269

Castano-Rodriguez N, Kaakoush NO, Goh KL, Fock KM, Mitchell HM (2015) The NOD-like receptor signalling pathway in Helicobacter pylori infection and related gastric cancer: a case-control study and gene expression analyses (vol 9, e98899, 2014). PLoS One 10(1). https://doi.org/10.1371/journal.pone.0117870

Chae JJ, Wood G, Masters SL, Richard K, Park G, Smith BJ, Kastner DL (2006) The B30.2 domain of pyrin, the familial Mediterranean fever protein, interacts directly with caspase-1 to modulate IL-1beta production. Proc Natl Acad Sci U S A 103(26):9982–9987

Chavarria-Smith J, Mitchell PS, Ho AM (2016) Functional and evolutionary analyses identify proteolysis as a general mechanism for NLRP1 inflammasome activation. PLoS Pathog 12(12): e1006052. https://doi.org/10.1371/journal.ppat.1006052

Chen Y, Ren XF, Li CG, Xing SC, Fu ZJ, Yuan Y, Wang R, Wang YA, Lv WS (2015) CARD8 rs2043211 polymorphism is associated with gout in a Chinese male population. Cell Physiol Biochem 35(4):1394–1400. https://doi.org/10.1159/000373960

Cheng CH, Lee YS, Chang CJ, Lin JC, Lin TY (2015) Genetic polymorphisms in inflammasome-dependent innate immunity among pediatric patients with severe renal parenchymal infections. PLoS One 10(10). https://doi.org/10.1371/journal.pone.0140128

D'Osualdo A, Anania VG, Yu K, Lill JR, Kaufman RJ, Matsuzawa S, Reed JC (2015) Transcription factor ATF4 induces NLRP1 inflammasome expression during endoplasmic reticulum stress. PLoS One 10(6):e0130635. https://doi.org/10.1371/journal.pone.0130635

da Silva WC, Oshiro TM, de Sa DC, Franco D, Neto CF, Pontillo A (2016) Genotyping and differential expression analysis of inflammasome genes in sporadic malignant melanoma reveal novel contribution of CARD8, IL1B and IL18 in melanoma susceptibility and progression. Cancer Genet 209(10):474–480. https://doi.org/10.1016/j.cancergen.2016.09.004

Day TG, Ramanan AV, Hinks A, Lamb R, Packham J, Wise C, Punaro M, Donn RP (2008) Autoinflammatory genes and susceptibility to psoriatic juvenile idiopathic arthritis. Arthritis Rheum 58(7):2142–2146. https://doi.org/10.1002/art.23604

De Iudicibus S, Stocco G, Martelossi S, Londero M, Ebner E, Pontillo A, Lionetti P, Barabino A, Bartoli F, Ventura A, Decorti G (2011) Genetic predictors of glucocorticoid response in pediatric patients with inflammatory bowel diseases. J Clin Gastroenterol 45(1):e1–e7. https://doi.org/10.1097/MCG.0b013e3181e8ae93

Deng JP, Lin W, Chen YP, Wang X, Yin Z, Yao CH, Liu TB, Lv YH (2015) rs3806268 of NLRP3 gene polymorphism is associated with the development of primary gout. Int J Clin Exp Pathol 8 (10):13747–13752

Dieudé P, Guedj M, Wipff J, Ruiz B, Riemekasten G, Airo P, Melchers I, Hachulla E, Cerinic MM, Diot E, Hunzelmann N, Caramaschi P, Sibilia J, Tiev K, Mouthon L, Riccieri V, Cracowski JL, Carpentier PH, Distler J, Amoura Z, Tarner I, Avouac J, Meyer O, Kahan A, Boileau C, Allanore Y (2011) NLRP1 influences the systemic sclerosis phenotype: a new clue for the contribution of innate immunity in systemic sclerosis-related fibrosing alveolitis pathogenesis. Ann Rheum Dis 70(4):668–674. https://doi.org/10.1136/ard.2010.131243

Dodé C, Le Dû N, Cuisset L, Letourneur F, Berthelot JM, Vaudour G, Meyrier A, Watts RA, Scott DG, Nicholls A, Granel B, Frances C, Garcier F, Edery P, Boulinguez S, Domergues JP, Delpech M, Grateau G (2002) New mutations of CIAS1 that are responsible for Muckle-Wells syndrome and familial cold urticaria: a novel mutation underlies both syndromes. Am J Hum Genet 70(6):1498–1506

Dwivedi M, Laddha NC, Mansuri MS, Marfatia YS, Begum R (2013) Association of NLRP1 genetic variants and mRNA overexpression with generalized vitiligo and disease activity in a Gujarat population. Br J Dermatol 169(5):1114–1125. https://doi.org/10.1111/bjd.12467

Eklund D, Welin A, Andersson H, Verma D, Soderkvist P, Stendahl O, Sarndahl E, Lerm M (2014) Human gene variants linked to enhanced NLRP3 activity limit intramacrophage growth of Mycobacterium tuberculosis. J Infect Dis 209(5):749–753. https://doi.org/10.1093/infdis/jit572

Ekman AK, Verma D, Fredrikson M, Bivik C, Enerback C (2014) Genetic variations of NLRP1: susceptibility in psoriasis. Br J Dermatol 171(6):1517–1520. https://doi.org/10.1111/bjd.13178

Feldmann J, Prieur AM, Quartier P, Berquin P, Certain S, Cortis E, Teillac-Hamel D, Fischer A, de Saint Basile G (2002) Chronic infantile neurological cutaneous and articular syndrome is caused by mutations in CIAS1, a gene highly expressed in polymorphonuclear cells and chondrocytes. Am J Hum Genet 71(1):198–203

Finger JN, Lich JD, Dare LC, Cook MN, Brown KK, Duraiswami C, Bertin J, Gough PJ (2012) Autolytic proteolysis within the function to find domain (FIIND) is required for NLRP1 inflammasome activity (vol 287, pg 25030, 2012). J Biol Chem 287(37):31456–31456. https://doi.org/10.1074/jbc.A112.378323

Garcia-Bermudez M, Lopez-Mejias R, Gonzalez-Juanatey C, Corrales A, Castaneda S, Ortiz AM, Miranda-Filloy JA, Gomez-Vaquero C, Fernandez-Gutierrez B, Balsa A, Pascual-Salcedo D, Blanco R, Llorca J, Martin J, Gonzalez-Gay MA (2013) CARD8 rs2043211 (p.C10X) polymorphism is not associated with disease susceptibility or cardiovascular events in Spanish rheumatoid arthritis patients. DNA Cell Biol 32(1):28–33. https://doi.org/10.1089/dna.2012.1836

Girardelli M, Maestri I, Rinaldi RR, Tognon M, Boldorini R, Bovenzi M, Crovella S, Comar M (2012) NLRP1 polymorphisms in patients with asbestos-associated mesothelioma. Infect Agents Cancer 7. https://doi.org/10.1186/1750-9378-7-25

Gonzalez-Pacheco H, Vargas-Alarcon G, Angeles-Martinez J, Martinez-Sanchez C, Perez-Mendez-O, Herrera-Maya G, Martinez-Rios MA, Peña-Duque MA, Posadas-Romero C, Fragoso JM (2017) The NLRP3 and CASP1 gene polymorphisms are associated with developing of acute coronary syndrome: a case-control study. Immunol Res. https://doi.org/10.1007/s12026-017-8924-0

Grandemange S, Sanchez E, Louis-Plence P, Tran Mau-Them F, Bessis D, Coubes C, Frouin E, Seyger M, Girard M, Puechberty J, Costes V, Rodière M, Carbasse A, Jeziorski E, Portales P, Sarrabay G, Mondain M, Jorgensen C, Apparailly F, Hoppenreijs E, Touitou I, Geneviève D (2016) A new autoinflammatory and autoimmune syndrome associated with NLRP1 mutations: NAIAD (NLRP1-associated autoinflammation with arthritis and dyskeratosis). Ann Rheum Dis 0:1–8. https://doi.org/10.1136/annrheumdis-2016-210021

Heilig R, Broz P (2018) Function and mechanism of the pyrin inflammasome. Eur J Immunol 48 (2):230–238. https://doi.org/10.1002/eji.201746947

Hitomi Y, Ebisawa M, Tomikawa M, Imai T, Komata T, Hirota T, Harada M, Sakashita M, Suzuki Y, Shimojo N, Kohno Y, Fujita K, Miyatake A, Doi S, Enomoto T, Taniguchi M, Higashi N, Nakamura Y, Tamari M (2009) Associations of functional NLRP3 polymorphisms with susceptibility to food-induced anaphylaxis and aspirin-induced asthma. J Allergy Clin Immunol 124(4):779–785. https://doi.org/10.1016/j.jaci.2009.07.044

Hoffman HM, Mueller JL, Broide DH, Wanderer AA, Kolodner RD (2001) Mutation of a new gene encoding a putative pyrin-like protein causes familial cold autoinflammatory syndrome and Muckle-Wells syndrome. Nat Genet 29(3):301–305. https://doi.org/10.1038/ng756

Hsu LC, Ali SR, McGillivray S, Tseng PH, Mariathasan S, Humke EW, Eckmann L, Powell JJ, Nizet V, Dixit VM, Karin M (2008) A NOD2-NALP1 complex mediates caspase-1-dependent IL-1beta secretion in response to Bacillus anthracis infection and muramyl dipeptide. Proc Natl Acad Sci U S A 105(22):7803–7808. https://doi.org/10.1073/pnas.0802726105

Ito S, Hara Y, Kubota T (2014) CARD8 is a negative regulator for NLRP3 inflammasome, but mutant NLRP3 in cryopyrin-associated periodic syndromes escapes the restriction. Arthritis Res Ther 16(1):R52. https://doi.org/10.1186/ar4483

Jenko B, Praprotnik S, Tomsic M, Dolzan V (2016) NLRP3 and CARD8 polymorphisms influence higher disease activity in rheumatoid arthritis. J Med Biochem 35(3):319–323. https://doi.org/10.1515/jomb-2016-0008

Jin Y, Mailloux CM, Gowan K, Riccardi SL, LaBerge G, Bennett DC, Fain PR, Spritz RA (2007) NALP1 in vitiligo-associated multiple autoimmune disease. N Engl J Med 356(12):1216–1225. https://doi.org/10.1056/NEJMoa061592

Kamada AJ, Pontillo A, Guimaraes RL, Loureiro P, Crovella S, Cavalcanti Brandao LA (2014) NLRP3 polymorphism is associated with protection against human T-lymphotropic virus 1 infection. Memorias Do Instituto Oswaldo Cruz 109(7):960. https://doi.org/10.1590/0074-0276140154

Kastbom A, Verma D, Eriksson P, Skogh T, Wingren G, Söderkvist P (2008) Genetic variation in proteins of the cryopyrin inflammasome influences susceptibility and severity of rheumatoid arthritis (the Swedish TIRA project). Rheumatology (Oxford) 47(4):415–417. https://doi.org/10.1093/rheumatology/kem372

Kastbom A, Klingberg E, Verma D, Carlsten H, Forsblad-d'Elia H, Wesamaa J, Cedergren J, Eriksson P, Söderkvist P (2013) Genetic variants in CARD8 but not in NLRP3 are associated with ankylosing spondylitis. Scand J Reumatol 42(6). https://doi.org/10.3109/03009742.2013.779020

Kastbom A, Arlestig L, Rantapaa-Dahlqvist S (2015) Genetic variants of the NLRP3 inflammasome are associated with stroke in patients with rheumatoid arthritis. J Rheumatol 42(10):1740–1745. https://doi.org/10.3899/jrheum.141529

Khameneh HJ, Mortellaro A (2014) NLRC4 gets out of control. Nat Genet 46(10):1048–1049. https://doi.org/10.1038/ng.3100

Kitamura A, Sasaki Y, Abe T, Kano H, Yasutomo K (2014) An inherited mutation in NLRC4 causes autoinflammation in human and mice. J Exp Med 211(12):2385–2396. https://doi.org/10.1084/jem.20141091

Kofoed EM, Vance RE (2011) Innate immune recognition of bacterial ligands by NAIPs determines inflammasome specificity. Nature 477(7366):592–595. https://doi.org/10.1038/nature10394

Kuijk LM, Beekman JM, Koster J, Waterham HR, Frenkel J, Coffer PJ (2008) HMG-CoA reductase inhibition induces IL-1beta release through Rac1/PI3K/PKB-dependent caspase-1 activation. Blood 112(9):3563–3573. https://doi.org/10.1182/blood-2008-03-144667

Kummer JA, Broekhuizen R, Everett H, Agostini L, Kuijk L, Martinon F, van Bruggen R, Tschopp J (2007) Inflammasome components NALP 1 and 3 show distinct but separate expression profiles in human tissues suggesting a site-specific role in the inflammatory response. J Histochem Cytochem 55(5):443–452. https://doi.org/10.1369/jhc.6A7101.2006

Lasiglie D, Traggiai E, Federici S, Alessio M, Buoncompagni A, Accogli A, Chiesa S, Penco F, Martini A, Gattorno M (2011) Role of IL-1 beta in the development of human T(H)17 cells: lesson from NLPR3 mutated patients. PLoS One 6(5):e20014. https://doi.org/10.1371/journal.pone.0020014

Levandowski CB, Mailloux CM, Ferrara TM, Gowan K, Ben S, Jin Y, McFann KK, Holland PJ, Fain PR, Dinarello CA, Spritz RA (2013) NLRP1 haplotypes associated with vitiligo and autoimmunity increase interleukin-1β processing via the NLRP1 inflammasome. Proc Natl Acad Sci U S A 110(8):2952–2956. https://doi.org/10.1073/pnas.1222808110

Lordén G, Sanjuán-García I, de Pablo N, Meana C, Alvarez-Miguel I, Pérez-García MT, Pelegrín P, Balsinde J, Balboa MA (2017) Lipin-2 regulates NLRP3 inflammasome by affecting P2X7 receptor activation. J Exp Med 214(2):511–528. https://doi.org/10.1084/jem.20161452

Magitta NF, Wolff ASB, Johansson S, Skinningsrud B, Lie BA, Myhr KM, Undlien DE, Joner G, Njolstad PR, Kvien TK, Forre O, Knappskog PM, Husebye ES (2009) A coding polymorphism in NALP1 confers risk for autoimmune Addison's disease and type 1 diabetes. Genes Immun 10 (2):120–124. https://doi.org/10.1038/gene.2008.85

Man SM, Karki R, Kanneganti TD (2017) Molecular mechanisms and functions of pyroptosis, inflammatory caspases and inflammasomes in infectious diseases. Immunol Rev 277(1):61–75. https://doi.org/10.1111/imr.12534

Manthiram K, Zhou Q, Aksentijevich I, Kastner DL (2017) The monogenic autoinflammatory diseases define new pathways in human innate immunity and inflammation. Nat Immunol 18 (8):832–842. https://doi.org/10.1038/ni.3777

Martinon F, Burns K, Tschopp J (2002) The inflammasome: a molecular platform triggering activation of inflammatory caspases and processing of pro IL-beta. Mol Cell 10(2):417–426. https://doi.org/10.1016/s1097-2765(02)00599-3

Mathews RJ, Robinson JI, Battellino M, Wong C, Taylor JC, Eyre S, Churchman SM, Wilson AG, Isaacs JD, Hyrich K, Barton A, Plant D, Savic S, Cook GP, Sarzi-Puttini P, Emery P, Barrett JH, Morgan AW, McDermott MF, Braggss (2014) Evidence of NLRP3-inflammasome activation in rheumatoid arthritis (RA); genetic variants within the NLRP3-inflammasome complex in relation to susceptibility to RA and response to anti-TNF treatment. Ann Rheum Dis 73 (6):1202–1210. https://doi.org/10.1136/annrheumdis-2013-203276

Matteini AM, Li J, Lange EM, Tanaka T, Lange LA, Tracy RP, Wang Y, Biggs ML, Arking DE, Fallin MD, Chakravarti A, Psaty BM, Bandinelli S, Ferrucci L, Reiner AP, Walston JD (2014) Novel gene variants predict serum levels of the cytokines IL-18 and IL-1ra in older adults. Cytokine 65(1):10–16. https://doi.org/10.1016/j.cyto.2013.10.002

McKinney C, Stamp LK, Dalbeth N, Topless RK, Day RO, Kannangara DRW, Williams KM, Janssen M, Jansen TL, Joosten LA, Radstake TR, Riches PL, Tausche AK, Liote F, So A, Merriman TR (2015) Multiplicative interaction of functional inflammasome genetic variants in determining the risk of gout. Arthritis Res Ther 17. https://doi.org/10.1186/s13075-015-0802-3

Meidtner K, Fisher E, Ängquist L, Holst C, Vimaleswaran KS, Boer JMA, Halkjær J, Masala G, Østergaard JN, Mortensen LM, van der A DL, Tjønneland A, Palli D, Overvad K, Wareham NJ, Loos RJF, Sørensen TIA, Boeing H (2014) Variation in genes related to hepatic lipid metabolism and changes in waist circumference and body weight. Genes Nutr 9(2):385. https://doi.org/10.1007/s12263-014-0385-7

Miskiewicz A, Szparecki G, Durlik M, Rydzewska G, Ziobrowski I, Gorska R (2015) The Q705K and F359L single-nucleotide polymorphisms of NOD-like receptor signaling pathway: association with chronic pancreatitis, pancreatic cancer, and periodontitis. Arch Immunol Ther Exp 63 (6):485–494. https://doi.org/10.1007/s00005-015-0355-9

Normand S, Massonnet B, Delwail A, Favot L, Cuisset L, Grateau G, Morel F, Silvain C, Lecron JC (2009) Specific increase in caspase-1 activity and secretion of IL-1 family cytokines: a putative link between mevalonate kinase deficiency and inflammation. Eur Cytokine Netw 20 (3):101–107. https://doi.org/10.1684/ecn.2009.0163

Özen S, Batu ED, Demir S (2017) Familial Mediterranean fever: recent developments in pathogenesis and new recommendations for management. Front Immunol 8:253. https://doi.org/10.3389/fimmu.2017.00253

Park YH, Wood G, Kastner DL, Chae JJ (2016) Pyrin inflammasome activation and RhoA signaling in the autoinflammatory diseases FMF and HIDS. Nat Immunol 17(8):914. https://doi.org/10.1038/ni.3457

Pontillo A, Brandao L, Guimaraes R, Segat L, Araujo J, Crovella S (2010a) Two SNPs in NLRP3 gene are involved in the predisposition to type-1 diabetes and celiac disease in a pediatric

population from northeast Brazil. Autoimmunity 43(8):583–589. https://doi.org/10.3109/08916930903540432
Pontillo A, Brandao LA, Guimaraes RL, Segat L, Athanasakis E, Crovella S (2010b) A 3'UTR SNP in NLRP3 gene is associated with susceptibility to HIV-1 infection. J Acquir Immune Defic Syndr 54(3):236–240. https://doi.org/10.1097/QAI.0b013e3181dd17d4
Pontillo A, Vendramin A, Catamo E, Fabris A, Crovella S (2011) The missense variation Q705K in CIAS1/NALP3/NLRP3 gene and an NLRP1 haplotype are associated with celiac disease. Am J Gastroenterol 106(3):539–544. https://doi.org/10.1038/ajg.2010.474
Pontillo A, Girardelli M, Kamada AJ, Pancotto JAT, Donadi EA, Crovella S, Sandrin-Garcia P (2012a) Polymorphisms in inflammasome genes are involved in the predisposition to systemic lupus erythematosus. Autoimmunity 45(4):271–278. https://doi.org/10.3109/08916934.2011.637532
Pontillo A, Oshiro TM, Girardelli M, Kamada AJ, Crovella S, Duarte AJS (2012b) Polymorphisms in inflammasome genes and susceptibility to HIV-1 infection. J Acquir Immune Defic Syndr 59 (2):121–125. https://doi.org/10.1097/QAI.0b013e3182392ebe
Pontillo A, Catamo E, Arosio B, Mari D, Crovella S (2012c) NALP1/NLRP1 genetic variants are associated with Alzheimer disease. Alzheimer Dis Assoc Disord 26(3):277–281. https://doi.org/10.1097/WAD.0b013e318231a8ac
Pontillo A, Laurentino W, Crovella S, Pereira AC (2013) NLRP1 haplotypes associated with leprosy in Brazilian patients. Infect Genet Evol 19:274–279. https://doi.org/10.1016/j.meegid.2013.06.006
Pontillo A, Reis EC, Bricher PN, Vianna P, Diniz S, Fernandes KS, Chies JA, Sandrim V (2015) NLRP1 L155H polymorphism is a risk factor for preeclampsia development. Am J Reprod Immunol 73(6):577–581. https://doi.org/10.1111/aji.12353
Pontillo A, Bricher P, Leal VNC, Lima S, Souza PRE, Crovella S (2016) Role of inflammasome genetics in susceptibility to HPV infection and cervical cancer development. J Med Virol 88 (9):1646–1651. https://doi.org/10.1002/jmv.24514
Reubold TF, Hahne G, Wohlgemuth S, Eschenburg S (2014) Crystal structure of the leucine-rich repeat domain of the NOD-like receptor NLRP1: implications for binding of muramyl dipeptide. FEBS Lett 588(18):3327–3332. https://doi.org/10.1016/j.febslet.2014.07.01
Roberts RL, Topless RKG, Phipps-Green AJ, Gearry RB, Barclay ML, Merriman TR (2010) Evidence of interaction of CARD8 rs2043211 with NALP3 rs35829419 in Crohn's disease. Genes Immun 11(4):351–356. https://doi.org/10.1038/gene.2010.11
Roberts RL, Van Rij AM, Phillips LV, Young S, McCormick SPA, Merriman TR, Jones GT (2011) Interaction of the inflammasome genes CARD8 and NLRP3 in abdominal aortic aneurysms. Atherosclerosis 218(1):123–126. https://doi.org/10.1016/j.atherosclerosis.2011.04.043
Romberg N, Al Moussawi K, Nelson-Williams C, Stiegler AL, Loring E, Choi M, Overton J, Meffre E, Khokha MK, Huttner AJ, West B, Podoltsev NA, Boggon TJ, Kazmierczak BI, Lifton RP (2014) Mutation of NLRC4 causes a syndrome of enterocolitis and autoinflammation. Nat Genet 46(10):1135–1139. https://doi.org/10.1038/ng.3066
Sahdo B, Fransen K, Idosa BA, Eriksson P, Soderquist B, Kelly A, Sarndahl E (2013) Cytokine profile in a cohort of healthy blood donors carrying polymorphisms in genes encoding the NLRP3 inflammasome. PLoS One 8(10). https://doi.org/10.1371/journal.pone.0075457
Santos MLS, Reis EC, Bricher PN, Sousa TN, Brito CFA, Lacerda MVG, Fontes CJF, Carvalho LH, Pontillo A (2016) Contribution of inflammasome genetics in Plasmodium vivax malaria. Infect Genet Evol 40:162–166. https://doi.org/10.1016/j.meegid.2016.02.038
Schoultz I, Verma D, Halfvarsson J, Torkvist L, Fredrikson M, Sjoqvist U, Lordal M, Tysk C, Lerm M, Soderkvist P, Soderholm JD (2009) Combined polymorphisms in genes encoding the inflammasome components NALP3 and CARD8 confer susceptibility to Crohn's disease in Swedish men. Am J Gastroenterol 104(5):1180–1188. https://doi.org/10.1038/ajg.2009.29
Shoham NG, Centola M, Mansfield E, Hull KM, Wood G, Wise CA, Kastner DL (2003) Pyrin binds the PSTPIP1/CD2BP1 protein, defining familial Mediterranean fever and PAPA

syndrome as disorders in the same pathway. Proc Natl Acad Sci U S A 100(23):13501–13506. https://doi.org/10.1073/pnas.2135380100

Soares JLS, Fernandes FP, Patente TA, Monteiro MB, Parisi MC, Giannella-Neto D, Correa-Giannella ML, Pontillo A (2018) Gain-of-function variants in NLRP1 protect against the development of diabetic kidney disease: NLRP1 inflammasome role in metabolic stress sensing? Clin Immunol (Orlando, Fla) 187:46–49. https://doi.org/10.1016/j.clim.2017.10.003

Sode J, Vogel U, Bank S, Andersen PS, Hetland ML, Locht H, Heegaard NHH, Andersen V (2015) Genetic variations in pattern recognition receptor loci are associated with anti-TNF response in patients with rheumatoid arthritis. PLoS One 10(10). https://doi.org/10.1371/journal.pone.0139781

Soler VJ, Tran-Viet KN, Galiacy SD, Limviphuvadh V, Klemm TP, St Germain E, Fournie PR, Guillaud C, Maurer-Stroh S, Hawthorne F, Suarez C, Kantelip B, Afshari NA, Creveaux I, Luo XY, Meng WH, Calvas P, Cassagne M, Arne JL, Rozen SG, Malecaze F, Young TL (2013) Whole exome sequencing identifies a mutation for a novel form of corneal intraepithelial dyskeratosis. J Med Genet 50(4):246–254. https://doi.org/10.1136/jmedgenet-2012-101325

Souza de Lima D, Ogusku MM, Sadahiro A, Pontillo A (2016) Inflammasome genetics contributes to the development and control of active pulmonary tuberculosis. Infect Genet Evol 41:240–244. https://doi.org/10.1016/j.meegid.2016.04.015

Starnes TW, Bennin DA, Bing X, Eickhoff JC, Grahf DC, Bellak JM, Seroogy CM, Ferguson PJ, Huttenlocher A (2014) The F-BAR protein PSTPIP1 controls extracellular matrix degradation and filopodia formation in macrophages. Blood 123(17):2703–2714. https://doi.org/10.1182/blood-2013-07-516948

Sui J, Li H, Fang Y, Liu Y, Li M, Zhong B, Yang F, Zou Q, Wu Y (2012) NLRP1 gene polymorphism influences gene transcription and is a risk factor for rheumatoid arthritis in han chinese. Arthritis Rheum 64(3):647–654. https://doi.org/10.1002/art.33370

Tan MS, Yu JT, Jiang T, Zhu XC, Wang HF, Zhang W, Wang YL, Jiang W, Tan L (2013) NLRP3 polymorphisms are associated with late-onset Alzheimer's disease in Han Chinese. J Neuroimmunol 265(1–2):91–95. https://doi.org/10.1016/j.jneuroim.2013.10.002

Tanaka N, Izawa K, Saito MK, Sakuma M, Oshima K, Ohara O, Nishikomori R, Morimoto T, Kambe N, Goldbach-Mansky R, Aksentijevich I, de Saint Basile G, Neven B, van Gijn M, Frenkel J, Arostegui JI, Yague J, Merino R, Ibanez M, Pontillo A, Takada H, Imagawa T, Kawai T, Yasumi T, Nakahata T, Heike T (2011) High incidence of NLRP3 somatic mosaicism in patients with chronic infantile neurologic, cutaneous, articular syndrome results of an international multicenter collaborative study. Arthritis Rheum 63(11):3625–3632. https://doi.org/10.1002/art.30512

Touitou I (2018) Inheritance of autoinflammatory diseases: shifting paradigms and nomenclature. J Med Genet 50(6):349–359. https://doi.org/10.1136/jmedgenet-2013-101577

Touitou I, Lesage S, McDermott M, Cuisset L, Hoffman H, Dode C, Shoham N, Aganna E, Hugot JP, Wise C, Waterham H, Pugnere D, Demaille J, Sarrauste de Menthiere C (2004) Infevers: an evolving mutation database for auto-inflammatory syndromes. Hum Mutat 24(3):194–198

Uhlen M, Fagerberg L, Hallstrom BM, Lindskog C, Oksvold P, Mardinoglu A, Sivertsson A, Kampf C, Sjostedt E, Asplund A, Olsson I, Edlund K, Lundberg E, Navani S, Szigyarto CA, Odeberg J, Djureinovic D, Takanen JO, Hober S, Alm T, Edqvist PH, Berling H, Tegel H, Mulder J, Rockberg J, Nilsson P, Schwenk JM, Hamsten M, von Feilitzen K, Forsberg M, Persson L, Johansson F, Zwahlen M, von Heijne G, Nielsen J, Ponten F (2015) Proteomics. Tissue-based map of the human proteome. Science (New York, NY) 347 (6220):1260419. doi: https://doi.org/10.1126/science.1260419

Ungerbäck J, Belenki D, Jawad ul-Hassan A, Fredrikson M, Fransén K, Elander N, Verma D, Söderkvist P (2012) Genetic variation and alterations of genes involved in NFκB/TNFAIP3- and NLRP3-inflammasome signaling affect susceptibility and outcome of colorectal cancer. Carcinogenesis 33(11):2126–2134. https://doi.org/10.1093/carcin/bgs256

Vance RE (2015) The NAIP/NLRC4 inflammasomes. Curr Opin Immunol 32:84–89. https://doi.org/10.1016/jcoi201501.010

Varghese GP, Fransén K, Hurtig-Wennlöf A, Bengtsson T, Jansson JH, Sirsjö A (2013) Q705K variant in NLRP3 gene confers protection against myocardial infarction in female individuals. Biomed Rep 1(6):879–882. https://doi.org/10.3892/br.2013.155

Verma D, Bivik C, Farahani E, Synnerstad I, Fredrikson M, Enerback C, Rosdahl I, Soderkvist P (2012a) Inflammasome polymorphisms confer susceptibility to sporadic malignant melanoma. Pigment Cell Melanoma Res 25(4):506–513. https://doi.org/10.1111/j.1755-148X.2012.01008.x

Verma D, Sarndahl E, Andersson H, Eriksson P, Fredrikson M, Jonsson JI, Lerm M, Soderkvist P (2012b) The Q705K polymorphism in NLRP3 is a gain-of-function alteration leading to excessive interleukin-1 beta and IL-18 production. PLoS One 7(4). https://doi.org/10.1371/journal.pone.0034977

Villani A-C, Lemire M, Fortin G, Louis E, Silverberg MS, Collette C, Baba N, Libioulle C, Belaiche J, Bitton A, Gaudet D, Cohen A, Langelier D, Fortin PR, Wither JE, Sarfati M, Rutgeerts P, Rioux JD, Vermeire S, Hudson TJ, Franchimont D (2009) Common variants in the NLRP3 region contribute to Crohn's disease susceptibility. Nat Genet 41(1):71–76. https://doi.org/10.1038/ng.285

Wang S, Fang F, Jin WB, Wang X, Zheng XS (2015) Investigation into the association between NLRP3 gene polymorphisms and susceptibility to type 2 diabetes mellitus. Genet Mol Res 14 (4):17447–17452. https://doi.org/10.4238/2015.December.21.15

Wang HD, Xu PF, Liao DH, Dang RL, He X, Guo YJ, Jiang P (2017) Association between NLPR1, NLPR3, and P2X7R gene polymorphisms with partial seizures. Biomed Res Int. https://doi.org/10.1155/2017/9547902

Wise CA, Gillum JD, Seidman CE, Lindor NM, Veile R, Bashiardes S, Lovett M (2002) Mutations in CD2BP1 disrupt binding to PTP PEST and are responsible for PAPA syndrome, an autoinflammatory disorder. Hum Mol Genet 11(8):961–969. https://doi.org/10.1093/hmg/11.8.961

Xiong SJ, Xiong Y, Huang Q, Wang JR, Zhang XF (2017) The association between MEFV gene polymorphisms and Henoch-Schonlein purpura, and additional SNP-SNP interactions in Chinese Han children. Rheumatol Int 37(3):455–460. https://doi.org/10.1007/s00296-016-3596-y

Yang SK, Kim H, Hong M, Lim J, Choi E, Ye BD, Park SK, Song K (2011) Association of CARD8 with inflammatory bowel disease in Koreans. J Hum Genet 56(3):217–223. https://doi.org/10.1038/jhg.2010.170

Yang CA, Huang ST, Chiang BL (2014) Association of NLRP3 and CARD8 genetic polymorphisms with juvenile idiopathic arthritis in a Taiwanese population. Scand J Rheumatol 43 (2):146–152. https://doi.org/10.3109/03009742.2013.834962

Yin Y, Yan Y, Jiang X, Mai J, Chen NC, Wang H, Yang XF (2009) Inflammasomes are differentially expressed in cardiovascular and other tissues. Int J Immunopathol Pharmacol 22 (2):311–322

Yin J, Wen J, Hang D, Han J, Jiang J, Song C, Liu Y, Liu JB, Liu L, Zhu LG, Chen JG, Zhai XJ, Xie SH, Hu ZB, Shen HB, Dai M, Li N (2015) Expression quantitative trait loci for CARD8 contributes to risk of two infection-related cancers-hepatocellular carcinoma and cervical cancer. PLoS One 10(7). https://doi.org/10.1371/journal.pone.0132352

Yu JW, Fernandes-Alnemri T, Datta P, Wu J, Juliana C, Solorzano L, McCormick M, Zhang Z, Alnemri ES (2007) Pyrin activates the ASC pyroptosome in response to engagement by autoinflammatory PSTPIP1 mutants. Mol Cell 28(2):214–227

Zeller T, Haase T, Muller C, Riess H, Lau D, Zeller S, Krause J, Baumert J, Pless O, Dupuis J, Wild PS, Eleftheriadis M, Waldenberger M, Zeilinger S, Ziegler A, Peters A, Tiret L, Proust C, Marzi C, Munzel T, Strauch K, Prokisch H, Lackner KJ, Herder C, Thorand B, Benjamin EJ, Blankenberg S, Koenig W, Schnabel RB (2015) Molecular characterization of the NLRC4 expression in relation to interleukin-18 levels. Circ Cardiovasc Genet 8(5):717–726. https://doi.org/10.1161/circgenetics.115.001079

Zerbino DR, Achuthan P, Akanni W, Amode MR, Barrell D, Bhai J, Billis K, Cummins C, Gall A, Girón CG, Gil L, Gordon L, Haggerty L, Haskell E, Hourlier T, Izuogu OG, Janacek SH,

Juettemann T, To JK, Laird MR, Lavidas I, Liu Z, Loveland JE, Maurel T, McLaren W, Moore B, Mudge J, Murphy DN, Newman V, Nuhn M, Ogeh D, Ong CK, Parker A, Patricio M, Riat HS, Schuilenburg H, Sheppard D, Sparrow H, Taylor K, Thormann A, Vullo A, Walts B, Zadissa A, Frankish A, Hunt SE, Kostadima M, Langridge N, Martin FJ, Muffato M, Perry E, Ruffier M, Staines DM, Trevanion SJ, Aken BL, Cunningham F, Yates A, Flicek P (2018) Ensembl 2018. Nucleic Acids Res 46(D1):D754–D761. https://doi.org/10.1093/nar/gkx1098

Zhang HX, Wang ZT, XX WYG, Zhong J, Liu J (2014) NLRP3 gene is associated with ulcerative colitis (UC), but not Crohn's disease (CD), in Chinese Han population. Inflamm Res 63(12):979–985. https://doi.org/10.1007/s00011-014-0774-9

Zhang L, Zhang D, Zheng Y, Hu Z, Zeng Y, Yue F (2016) Association of NLRP3 gene single nucleotide polymorphisms with metabolic syndrome. Zhonghua Yi Xue Yi Chuan Xue Za Zhi 33(4):530–534. https://doi.org/10.3760/cma.j.issn.1003-9406.2016.04.024

Zhou D, Wang XH, Chen T, Wen W, Liu Y, Wu Y, Yuan ZY (2016) The NLRP3 rs10754558 polymorphism is associated with the occurrence and prognosis of coronary artery disease in the Chinese Han population. Biomed Res Int. https://doi.org/10.1155/2016/3185397

Zhu ZS, Yan J, Geng CS, Wang DG, Li CY, Feng S, Wang HP (2016) A polymorphism within the 3'UTR of NLRP3 is associated with susceptibility for ischemic stroke in Chinese population. Cell Mol Neurobiol 36(6):981–988. https://doi.org/10.1007/s10571-015-0288-1

Chapter 15
Inhibiting Inflammasomes with Small Molecules

Avril A. B. Robertson

Contents

A. A. B. Robertson (✉)
School of Chemistry and Molecular Biosciences, The University of Queensland, Brisbane, QLD, Australia
e-mail: a.robertson3@uq.edu.au

M. D. Cordero, E. Alcocer-Gómez (eds.), *Inflammasomes: Clinical and Therapeutic Implications*, Experientia Supplementum 108,
https://doi.org/10.1007/978-3-319-89390-7_15

Abstract Modulation of inflammasomes has tremendous therapeutic potential and is hotly pursued by industry and academia alike. Indeed a growing number of patents are emerging to protect the intellectual property in valuable compound classes. This chapter focusses specifically on the suite of small-molecule NLRP3 inflammasome inhibitors published, as specific modulation of other inflammasomes is not yet well established. Synthetic molecules, known drugs and natural product NLRP3 modulators will be detailed. Some of the molecular classes discussed have been extensively characterised through cell-based screening, pharmacokinetic profiling and therapeutic proof of concept animal models. However, many inhibitors lack rigorous studies and/or have multiple activities of which NLRP3 modulation is only one. While this is not intended as an exhaustive list, it should give an impression of the range of structures and strategies that are being used, alongside challenges encountered, in an effort to exploit the significant therapeutic benefits of targeting inflammasomes.

Keywords Inflammasome · NLRP3 · Drug discovery · Anti-inflammatory · Immunomodulation

15.1 NLRP3 Inflammasome Inhibition

Inflammasome-mediated secretion of interleukin (IL)-1β and IL-18 is emerging as a key disease driver in a surprisingly wide variety of acquired conditions. These range from diseases of metabolic dysregulation (e.g. type 2 diabetes, gout, NASH) to T cell-mediated or organ-specific autoimmune diseases (e.g. multiple sclerosis, psoriasis, type 1 diabetes, rheumatoid arthritis), to systemic autoimmune disease (e.g. Muckle-Wells, NOMID, Behcet's disease, Sjogren's syndrome, Schnitzler syndrome), to inflammatory reactions in the skin (e.g. contact hypersensitivity, sunburn), joints (e.g. rheumatoid arthritis, osteoarthritis, systemic juvenile idiopathic arthritis, adult-onset Still's disease, relapsing polychondritis), muscle (e.g. polymyositis), heart (e.g. post-infarction cardiac remodelling) and brain (e.g. Alzheimer's and Parkinson's disease, multiple sclerosis, depression, ischaemic stroke) (Latz et al. 2013; Schroder and Tschopp 2010; Dinarello 2010). The therapeutic potential of inflammasome inhibition, particularly NLRP3 inflammasome, is both exciting and commercially valuable (Sheridan 2017). There are an ever-increasing number of molecules reported as inflammasome inhibitors including natural products, synthetic molecules and known drugs which will be described in this chapter.

Inflammasome inhibitors have commonly been discovered on the basis of phenotypic screening, not structure-based drug design, typically using murine bone marrow-derived macrophages (BMDM) or THP-1 cell lines with IL-1β as a readout. These assays involve two steps in activating the inflammasome; ***Signal 1*** is a priming signal resulting in expression of inflammasome protein components and the inactive pro-inflammatory cytokines pro-IL-1β and pro-IL-18. This priming phase occurs via the activation of the transcription factor nuclear factor-κB (NF-κB) pathway. Pro-IL-18 is constitutively expressed and also increased after activation (He et al. 2016a). The most commonly used priming signal in cell-based screening assays is lipopolysaccharide (LPS) where in vivo this would be provided by similar microbial molecules or endogenous cytokines. ***Signal 2*** is required to trigger assembly of the inflammasome components to form the protein-processing platform, resulting in cleavage of pro-caspase-1 to active caspase-1. The active caspase-1 can then catalyse proteolytic activation of pro-IL-1 and pro-IL-18 to their active secreted forms. Caspase-1 also triggers pyroptotic cell death via proteolytic cleavage of gasdermin D which migrates to the cell membrane forming pores (He et al. 2016a). NLRP3 is the most extensively characterised inflammasome and indeed the most complex due to the plethora of stimuli which initiate activation: adenosine triphosphate (ATP), nigericin, elevated glucose, saturated free fatty acids, ceramide, amyloid deposits formed by islet amyloid polypeptide, α-synuclein and particulates such as crystals of cholesterol, monosodium urate or silica amongst many others. Other inflammasomes are much more specific with a narrower spectrum of activators. In vitro assays commonly use NLRC4 activator *Salmonella typhimurium*, AIM2-trigger double-stranded DNA analog Poly(dA:dT) (Coll et al. 2015), and NLRP1 inflammasome activator *Bacillus anthracis* lethal toxin or muramyl dipeptide (Chavarria-Smith and Vance 2015).

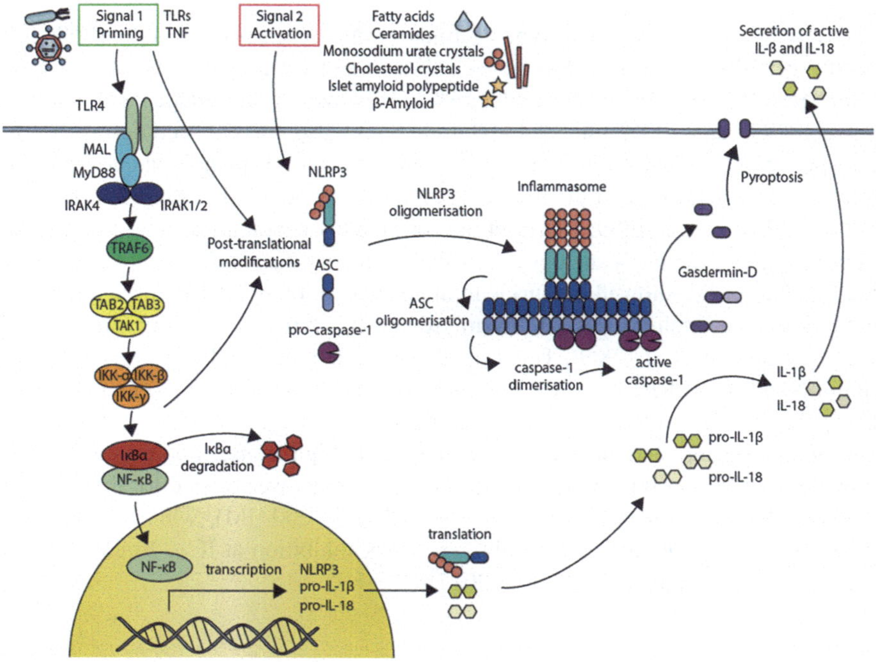

NLRP3 inflammasome pathway: image by Dr Rebecca C. Coll

After hit identification through phenotypic testing, it is necessary to interrogate these compounds carefully to identify their mode of action and specificity. Some of the inflammasome modulatory compounds described herein are very potent and well characterised with respect to their effect on inflammasome pathways, while others are tentative and need much more work. In some cases, particularly in the natural products field and for reactive small molecules, identified compounds are non-specific with multiple modes of action, and inflammasome inhibition has been added to the ever-expanding list. The potency of inflammasome inhibition also needs to be considered in context. For inhibitors active in the micromolar range, it may not be realistic to reach these concentrations in vivo thereby hindering translation into clinical use. If high compound concentrations are required, this could also be accompanied by significant off target effects and toxicity in vivo.

In this chapter, new synthetic molecules and known drugs, identified as inflammasome inhibitors, are discussed followed by natural product-based studies. While this is not intended as an exhaustive list, it should give an impression of the range of structures and strategies that are being used, alongside challenges encountered, in an effort to exploit the therapeutic benefits of targeting inflammasomes.

15.2 New Synthetic Molecules and Known Drugs as Inflammasome Inhibitors

15.2.1 Sulfonylureas

Glyburide (also known as glybenclamide) was developed in the 1960s by Boehringer Mannheim and Hoechst as an insulin secretagogue treatment for type 2 diabetes (Marble 1971; Ashcroft 2005). However, four independent studies, between 1997 and 2005, indicated glyburide could also prevent secretion of IL-1β, in response to LPS and ATP, from murine and human macrophages and also murine Schwann cells with IC_{50} in the low μM range (Perregaux et al. 2001; Hamon et al. 1997; Laliberte et al. 1999; Marty et al. 2005). This gave an exciting potential to repurpose glyburide or its analogs as anti-inflammatory agents. It was not until 2009 that Lamkanfi et al. showed glyburide inhibited activation of NLRP3 inflammasome and that this accounted for its ability to prevent IL-1β release from murine macrophages (Lamkanfi et al. 2009). The drug also had a degree of selectivity as it did not inhibit NLRC4, NLRP1 or AIM2 inflammasomes (Lamkanfi et al. 2009; Coll et al. 2011). While glyburide clearly had inhibitory activity against NLRP3, it was not particularly potent, and the ability to gain such high concentrations in vivo is unlikely. Ten further type 2 diabetes sulfonylurea drugs were tested for their ability to inhibit NLRP3: seven showed no effect (IC_{50} > 200 μM), while glimepiride, glisoxepide and gliquidone gave only very weak inhibition at IC_{50} 75 μM, 73 μM and 116 μM, respectively (Hill et al. 2017).

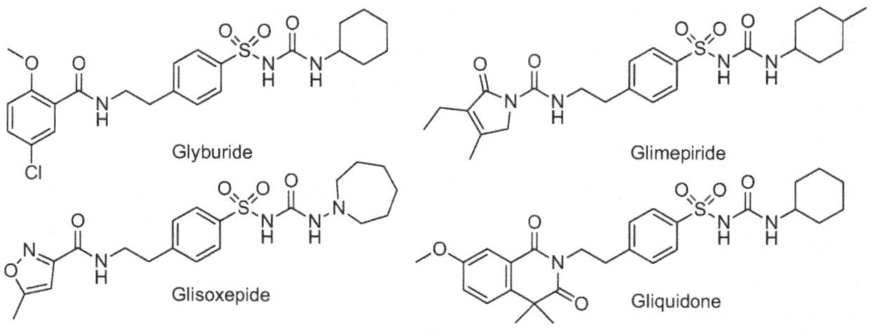

Type 2 diabetes patients have an increased susceptibility to sepsis triggered by bacterial infection. The benefit of inflammatory response in fighting this infection is somewhat controversial: some believe the response is essential, while others suggest modulation is required (Wiersinga and van der Poll 2007). In a clinical study of sepsis, caused by *Burkholderia pseudomallei*-induced melioidosis, involving 1160 patients (410 diabetic), those patients taking glyburide showed improved survival, while metformin- or insulin-treated patients did not (Koh et al. 2011). Glyburide-treated patients had differential expression of 63 immune-related genes in their peripheral whole blood leukocytes linking to an anti-inflammatory effect (Koh et al. 2011). The authors speculated this may be due to NLRP3 inflammasome inhibition by glyburide, but there are clearly multiple effects and further investigation that would be required.

Pfizer identified an interesting set of potent IL-1β inhibitory sulfonylureas, termed cytokine release inhibitory drugs or CRIDs, through a phenotypic screen of ion channel inhibitors (Perregaux et al. 2001). These diarylsulfonylureas were patented in 1998 (Dombroski and Eggler 1998). Although structure activity relationships were not specified in the patent, more detailed studies were published on three of these molecules: CP-412-245, CP-424,174 and CP-456,773 (the latter molecule is also known as CRID3 or MCC950). This work demonstrated the nanomolar inhibitory activity of these compounds against IL-1β secretion from cells and also in vivo for CP-424,174 which was dosed orally in mice (ED_{50} around 15 mg/kg) (Perregaux et al. 2001). The compounds were relatively selective as secretion of IL-6 and TNFα remained unaffected (Perregaux et al. 2001). Two related molecules, epoxides CRID1 and 2, were synthesised for use as affinity labels in mode of action studies identifying glutathione *S*-transferase (GST) Omega 1-1 (Laliberte et al. 2003) as a possible target of the CRID molecules; however, Pfizer has since published studies which dismiss this, leaving the precise mode of action of these molecules unknown (Primiano et al. 2016).

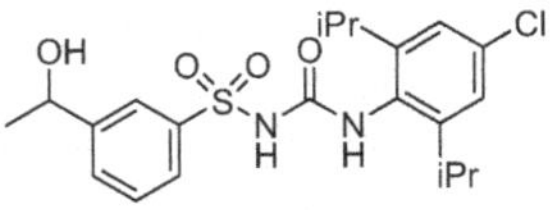

CP-412,245

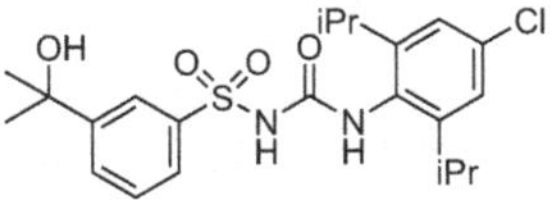

CP-424,174

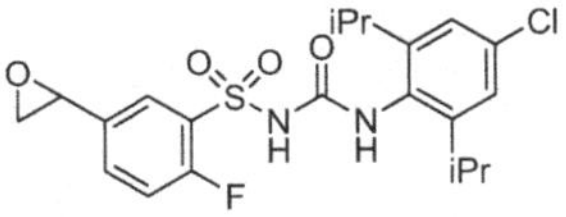

CRID1

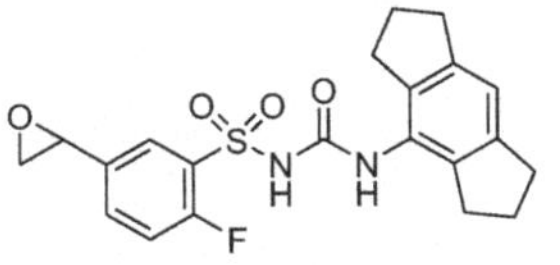

CRID2

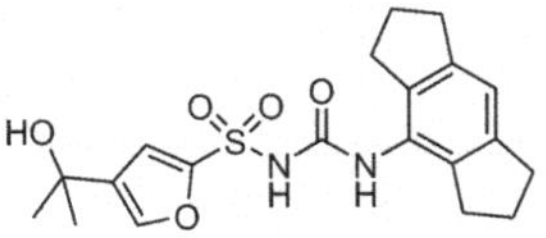

CRID3
CP-456,773
MCC950

At the time of the diarylsulfonylurea IL-1β inhibitor discovery, the field of inflammasome biology was only just beginning in the laboratory of Jurg Tschopp. It was therefore many years before the link between these molecules and NLRP3 inflammasome was finally established. Coll et al. (2015) profiled MCC950 to find it had an IC_{50} of 7.5 nM against NLRP3 inflammasome, exquisite selectivity over AIM2, NLRC4 and NLRP1 and no effect on TLR signalling. Efforts were also made to fully elucidate the mechanism through which NLRP3 inhibition occurred. No effect was identified on common NLRP3 activating mechanisms such as potassium ion efflux or calcium ion signalling. MCC950 did not inhibit interactions between NLRP3 proteins *nor* with ASC (Coll et al. 2015). Pfizer later added further insight from results of commercially available off target screens covering 196 different targets to find little or no inhibition with MCC950 at concentration of 10 μM simultaneously ruling out putative targets such as SUR1, SUR2a, SUR2b, caspase-1, SYK, JNK, GPR40 and GPR120 (Primiano et al. 2016). Furthermore ToxCast/Tox21 data indicate MCC950 had activity in only 18 out of 410 screening assays, and even where activity was observed the response was weak ($AC_{50} >$ 20 μM; Emax $< 50\%$) (Primiano et al. 2016). Most recently Jiang et al. (2017) suggested MCC950 might target chloride efflux to suppress inflammasome activation based on their observation that this compound could dose dependently inhibit chloride efflux in $NLRP3^{-/-}$ BMDMs. Some have speculated MCC950 may act on NEK7 (Van Hauwermeiren and Lamkanfi 2016), a mitotic kinase recently identified (Schmid-Burgk et al. 2016) to directly interact with the leucine-rich repeat domain of NLRP3, in a kinase-independent manner. NEK7 is required for inflammasome assembly (Shi et al. 2016; He et al. 2016b) and it is specific as it is not involved in NLRC4 or AIM2 activation *nor* TLR responses. Moreover the roles of NEK7 as mitotic kinase and NLRP3 regulator are mutually exclusive, i.e. mitosis and NLRP3 inflammasome activation cannot occur simultaneously. While this is a plausible theory, there is no evidence, to date, that this is the target of MCC950. Despite intense interest, the precise mode of action and binding site(s) of MCC950 remains elusive.

The in vitro ADME studies of MCC950 (Coll et al. 2015) indicated it had excellent stability to mouse or human liver microsomes, no significant inhibition of five major cytochrome P450 isoforms (1A2, 2C9, 2C19, 2D6, 3A4) at 10 μM and did not affect the hERG ion channel (IC_{50} >30 μM) when tested using an automated patch-clamp method. The fraction unbound for MCC950 in mouse plasma was extremely small (Primiano et al. 2016), in keeping with the sulfonylurea drug class which typically shows high plasma protein binding. Metabolism of MCC950 when exposed to human liver microsomes identified a single metabolite with much weaker NLRP3 inhibitory activity than the parent compound (Salla et al. 2016).

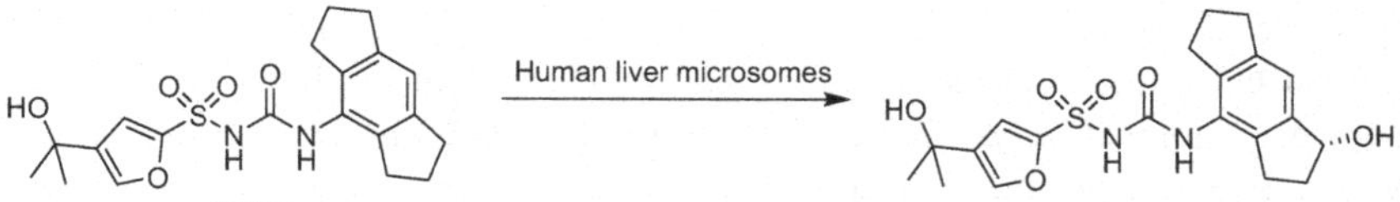

CRID3
CP-456,773
MCC950

NLRP3 IC_{50} 1.2 μM

The in vivo pharmacokinetic profile of MCC950, based on a single dose of 3 mg/kg i.v. and 20 mg/kg p.o. in C57BL/6 mice, showed impressive oral bioavailability of 68%, C_{max} 25,333 ng/ml, AUC 163,410 ng h/ml and half-life ~3 h (Coll et al. 2015). In repeat dosing of MCC950 over 5 days at 200 mg/kg p.o., no accumulation was observed in mouse serum; however, detailed distribution studies have not been published (Primiano et al. 2016). Levels of IL-1β in vivo after MCC950 dosing in mice were reduced by 50% at 0.4 mg/kg, 90% at 1.2 mg/kg and >90% above 4 mg/kg illustrating the in vivo potency of the molecule (Primiano et al. 2016). Many in vivo disease models have been conducted with oral doses as high as 200 mg/kg without any noted toxicity concerns. Nevertheless, the compounds inclusion in the ToxCast set should be noted, and details of Pfizer's clinical trials on this molecule have not been made publically available. These data could prove key to future development of the series to benefit human health in a multitude of disease areas considering the very substantial, and ever-expanding, body of data on this tool compound.

MCC950 had efficacy and selectivity in a murine model of Muckle-Wells syndrome (Coll et al. 2015). This model provides perfect proof of concept for any NLRP3 inflammasome inhibitor as the basis of the disease lies with an activating mutation in NLRP3. This mutation leads to elevated concentrations of the pro-inflammatory cytokines IL-1β and IL-18, and, in the murine model, mice die in the neonatal period (usually < 14 days) due to excessive inflammation. Treatment with MCC950 extended survival to an impressive 38–45 days (until dosing was stopped), and correspondingly low levels of IL-1β and IL-18 were measured during treatment. When treatment was halted, these cytokines increased, and the mice died within the usual 14-day period. A similar murine model involving an NLRP1-activating mutation was also investigated; however, MCC950 was ineffective in this case, as would be expected based on the established selectivity profile of the compound (Coll et al. 2015).

Inflammasome inhibition will undoubtedly be beneficial in many diseases, but the question of whether response to infection would be diminished is an important one. Murine models where IL-1 signalling is defective lead to elevated risk of developing an infection. In the case of approved IL-1β-targeted biologics, IL-1β signalling is entirely blocked, and predisposition to serious infection has been noted (Corporation 2011). Moreover, biologics typically have a long half-life in vivo with canakinumab only requiring administration twice per month, so full response to infection will take time to return after dosing is halted (Corporation 2011). Inflammasome inhibitors have potential to be far more selective, for example, MCC950 blocks only NLRP3-mediated IL-1β (and IL-18) secretion leaving other inflammasome pathways fully responsive. Small-molecule inhibitors of these pathways are unlikely to have the extended half-life of the biologics and should clear from the body quickly when treatment is stopped. A detailed analysis of the US Food and Drug Administration Adverse Event Reporting of drugs after market approval indicated use of IL-1β inhibitors conferred only moderate risk of infection (LaRock et al. 2016). However, such patients had exceptionally high chance of severe invasive group A *Streptococcal* infection (GAS), even when compared to other immunomodulatory approaches. The reasons for this apparent disparity with other infections were investigated revealing that GAS cleaves pro-IL-1β, independently of NLRP3, through its own protease SpeB; the subsequent release of IL-1β restricts its invasive capacity (LaRock et al. 2016). However, in severe GAS infections, SpeB is mutated and cannot function allowing the infection to progress unabated (LaRock et al. 2016).

Earlier studies on the role of NLRP3 in uropathogenic *Escherichia coli* infection showed reference strains CFT073- and UTI89-triggered NLRP3-dependent pyroptotic cell death in human macrophages (Schaale et al. 2016). In contrast, the multidrug-resistant sequence type 131 strains EC958 and MS3179 and asymptomatic strains 83972 and VR50 did not cause pyroptotic cell death. As with the GAS studies, it appears some strains of UPEC may have adapted to avoid inflammasome activation and therefore detection by the host. When the CFT073-induced macrophage death was further investigated, significant differences were observed between the murine and human cells (Schaale et al. 2016). In murine macrophages the IL-1β release and cell death is completely dependent on NLRP3 and the pore forming toxin α-haemolysin. In human macrophages, cell death was triggered by an, as yet unidentified, NLRP3-independent mechanism, and α-haemolysin was not required. There is clearly a complex interplay between pathogens and the innate immune system which remains to be untangled.

Fungal pathogens such as *Candida albicans* does not evade NLRP3 inflammasome but uses it to trigger macrophage pyroptosis and therefore escape from the macrophage into the extracellular milieu where it replicates (Tucey et al. 2016). Moreover there is an ensuing phase of non-pyroptotic fungal induced macrophage killing designed to fully eradicate the macrophage population. The endoplasmic reticulum-mitochondria encounter structure (ERMES) tethering complex is common to fungi and required to retain the mitochondrial morphology (Tucey et al. 2016). An *mmm1* ERMES mutant showed a 10 h delay in triggering NLRP3-mediated macrophage pyroptosis compared to the wild-type strain. The mutation in ERMES led to altered mitochondrial morphology, reduced fungal virulence, altered cell wall structure, stunted hyphal growth and gene expression. After the 10 h period, a pronounced NLRP3 activation was triggered, by an unknown mechanism, but the liberated fungal pathogen lacked capacity to initiate the non-pyroptotic phase of macrophage death. In keeping with these discoveries, MCC950 effectively prevented *C. albicans* induced macrophage pyroptosis for at least 9 h (Tucey et al. 2016). There appears to be a critical balance between the innate immune response to infection and the ability of the infectious agent to exploit its defence mechanisms to persist or even thrive in the host.

The inflammatory response mounted to clear pathogens from the host can lead to extensive tissue damage and poor prognosis. In a murine study of NLRP3's role during influenza A infection, any deficiency in NLRP3 inflammasome early in the infection whether through genetic mutation or pharmacological inhibition using MCC950 (5 mg/kg i.n.) led to expedited mortality (Pinar et al. 2017). However if MCC950 was used at the peak of infection, increased survival rate was observed with much less severe disease outcomes (Pinar et al. 2017). There is clearly much to learn about the interplay between infectious agents and the immune response such that potential to manipulate the system therapeutically can be understood. This is likely to be particularly difficult where multiple infectious agents are present as they may exploit host immunity in different ways. Furthermore differences even exist for various strains of the same pathogen, and these can behave differently depending on the cell type or host species studied.

Researchers at Trinity College, Dublin, first showed efficacy of MCC950 in models of inflammatory brain disease (Coll et al. 2015). Multiple sclerosis is a disease where myelin sheath of nerve cells in the brain and spinal cord is damaged. In the murine model of multiple sclerosis termed experimental autoimmune encephalomyelitis (EAE), NLRP3-mediated production of IL-1β and IL-18 has a critical role in disease induction (Gris et al. 2010). These cytokines, together with IL-23, drive production of the pro-inflammatory cytokine IL-17 by $CD4^+$ T helper 17 cells and $\gamma\delta^+$ T cells (Lalor et al. 2011). Mice with NLRP3 deficiency do not develop EAE (Gris et al. 2010). Pharmacological inhibition of NLRP3 was tested by administration of MCC950 (10 mg/kg i.p., on days 0, 1 and 2 then every second day thereafter) which both delayed EAE and reduced severity of the disease (Coll et al. 2015). In this model MCC950 was present from before disease induction (prophylactic), rather than dosing once disease had been established; however, it did show promise for treatment of inflammatory brain disease with MCC950.

In Alzheimer's disease, an association with NLRP3 inflammasome activation in response to amyloid-β was identified over a decade ago, but specific NLRP3 inhibitors had not been identified (Halle et al. 2008). In the APP/PS1 transgenic mouse model of Alzheimer's disease, accelerated deposition of amyloid-β is observed alongside associated neuroinflammation and impaired cognitive capacity. MCC950 was administered in this model at 10 mg/kg i.p. every second day for a period of 3 months with behavioural tests (T-maze and novel object recognition) at 8th and 9th weeks of treatment (Dempsey et al. 2017). Cognitive function improved significantly accompanied by reduced activation of microglia, while phagocytosis of amyloid-β was promoted ameliorating amyloid-β pathology. These results are impressive, particularly since 98% of all known therapeutics cannot penetrate the blood-brain barrier.

In the EAE and Alzheimer's studies, the levels of MCC950 in the brain were not reported, and peripheral effects may also contribute to the efficacy observed. Early theories of Heneka et al. (2013) raised the interesting possibility that CNS diseases (Alzheimer's disease, Parkinson's disease, Amyotrophic lateral sclerosis, Huntington's disease, Creutzfeldt-Jakob disease) linked with misfolded protein aggregates (amyloid-β, synuclein, huntingtin, TDP-43, prion protein, etc.) may be innate immunity disorders, and their protein triggers would be considered components of this system. Whether or not this is the case, early studies with MCC950 in inflammatory disorders of the CNS certainly give potential to investigate application of this compound and close analogs in other such diseases.

Imiquimod triggers NEK7-dependent NLRP3 inflammasome activation through inhibition of quinone oxidoreductases NQO2 and mitochondrial complex I leading to generation of reactive oxygen species (Gross et al. 2016). In a model of imiquimod-induced mouse ear dermal inflammation, MCC950 (200 mg/kg, p.o., bid for 5 days) (Primiano et al. 2016) treatment reduced ear swelling by >40%, and levels of IL-22, IL-17A and IL-17F were also reduced by approximately 50%. No topical treatment of MCC950 has yet been published, and it would be interesting to see if topical administration would be efficacious in this model.

Myelodysplastic syndromes (MDSs) are characterised by defective differentiation of haematopoietic stem cells to form new blood cells in the body. This is a complex disease but cytokine profile, abundance of the alarmin S100A9 (a known TLR4 and CD33 agonist) and overexpression of TLRs led to investigations into the role of the innate immune system in these diseases (Basiorka et al. 2016). As part of this detailed study, alarmins and reactive oxygen species were found to elicit pyroptotic cell death, and activate β-catenin, by triggering NLRP3 inflammasome formation. Inhibition of NLRP3 inflammasome using MCC950, in MDS patient bone marrow mononuclear cells, prevented pyroptotic death of haematopoietic stem and progenitor cells and allowed functional haematopoiesis to proceed. This exciting development provides a new avenue to investigate in treatment of this disease class.

Inflammasomes play a major role in asthma and other inflammatory lung diseases where noxious stimuli such as allergens, pathogens, smoke, particulates, *etc.* can activate the pathway (Pinkerton et al. 2017). Efficacy of MCC950 (200 mg/kg, bid) was demonstrated in a murine model of allergic asthma triggered by house dust mite challenge (Primiano et al. 2016). Mice were treated before challenge and after, then sacrificed at 24 h. Analysis of bronchiolar lavage fluid showed significant reduction of neutrophil infiltration and normalisation of the small lymphocyte increase observed. These results were recapitulated in a dexamethasone corticosteroid treatment group. Nine out of thirty-six different inflammatory mediators measured were increased in response to the house dust mite challenge. Four inflammatory mediators (CXCL1, CXCL5, CXCL10, CCL7) were reduced by MCC950 treatment while eight, including those suppressed by MCC950, were suppressed by dexamethasone (Primiano et al. 2016). While allergic asthma is the most common type of asthma, between 5 and 20% of asthmatics are refractory to corticosteroid treatment. These resistant patents typically have more severe disease, account for >50% of health-care costs for asthma, and remain without effective treatment (Ito et al. 2006). Severe steroid-resistant asthma is commonly characterised by increased Th1 and/or Th17 responses and a non-eosinophilic endotype (Green et al. 2002). Substantial increases in IL-1β expression occurs in this non-eosinophilic asthma compared to mild to moderate asthma(Wanderer 2009), and this positively correlates with increased neutrophilic inflammation in these patients (Hastie et al. 2010). MCC950 treatment was investigated in two murine models of infection-induced (Chlamydia or Haemophilus) severe steroid-resistant allergic airway disease (Kim et al. 2017). Disease was first established, then on days 32–34 mice were treated with MCC950 (10 mg/kg i.n., bd), and effects were assessed on day 35. Both mouse models showed impressive reduction in airway hyper-responsiveness, inflammation and IL-1β levels. These responses were equivalent to those observed by dexamethasone treatment in steroid-sensitive allergic airway disease. Together these studies open a new field of investigation for treatment of different asthma subtypes. Future studies will likely show if other inflammatory lung diseases such as COPD, silicosis and asbestosis may also be alleviated using inflammasome inhibitors.

NLRP3 inflammasome activation has been observed in obese individuals and is strongly linked to pathogenesis of metabolic diseases such as type 2 diabetes and fatty liver disease. MCC950 has been investigated in two murine models of non-alcoholic steatohepatitis (NASH), the foz/foz "fat aussie" model and the methionine- and choline-deficient diet (MCD) model (Mridha et al. 2017). In the foz/foz model, mice are fed an atherogenic diet, over 16 weeks, and develop elevated alanine transaminase and severe steatohepatitis including hepatocyte ballooning, inflammation and fibrosis. Treatment with MCC950 (20 mg/kg, p.o., qd) from 16 to 24 weeks provided no benefit to metabolic features of disease (weight gain, plasma insulin, fasting blood glucose and other markers) which was perhaps surprising based on prior hypothesis (Henao-Mejia et al. 2012; Wen et al. 2012). However, liver inflammation was almost completely abrogated showing remarkable reduction in levels of pro-IL-1β, IL-1β, IL-6 and MCP1 alongside suppressed macrophage and neutrophil infiltration. The typical markers of liver damage alanine transaminase and aspartate transaminase were also significantly reduced. One of the most important findings was the ability of MCC950 to reverse established liver fibrosis. This is highly significant as liver fibrosis is predictive of disease-specific mortality (Ekstedt et al. 2015). Results of liver pathology showed a reduction from 64% of mice categorised with definite NASH to only 18% (Mridha et al. 2017). In the MCD model, liver fibrosis is normally severe, but treatment with MCC950 (10 mg/kg p.o., qd for 5 days, followed by 20 mg/kg p.o. every second day up to 6 weeks) almost completely protected against development of fibrosis as compared with vehicle-treated controls. These data indicate MCC950 was effective both therapeutically and prophylactically in treatment of liver fibrosis.

Fibrosis is problematic in a wide array of crystal-related kidney diseases ultimately leading to renal failure. These crystalline deposits can form naturally, as a result of metabolic disease, or may be drug induced, and they trigger NLRP3 (Mulay et al. 2014). Inhibition of NLRP3 by MCC950 (~200 mg/kg, p.o., administered in chow) in a murine model of crystal nephropathy (Ludwig-Portugall et al. 2016) reduced inflammasome activation and associated formation of IL-1β and IL-18. Critically, the development of fibrosis was prevented; however, in contrast to the aforementioned NASH study (Mridha et al. 2017), established fibrosis could not be reversed; therefore early treatment would be recommended. An elegant means to visualise effects on NLRP3 in vivo was used employing transfer of bone marrow cells harbouring an inflammasome-specific luciferase reporter (interleukin-1-β-Gaussia luciferase) into the mice. When inflammasomes were activated, the special reporter system produced light which was detected in vivo by bioluminescent imaging (Ludwig-Portugall et al. 2016). This study also investigated the effects of MCC950 inhibition on the adaptive immune response through ovalbumin challenge in the crystal nephropathy murine model. No significant difference was observed between treated and untreated mice as measured by count of ovalbumin-specific cytotoxic T cells or ovalbumin-specific IgG antibody serum titres. These studies indicate a promising future for treatment of crystal-induced kidney disease with first in class NLRP3 inflammasome inhibitors.

Individuals who suffer hypertension typically have high levels of IL-1β and IL-18 where the IL-1β levels show positive correlation with blood pressure (Barbaro et al. 2015; Dalekos et al. 1997; Rabkin 2009). In addition there is significant accompanying inflammatory response and fibrosis in the kidney. Recent murine studies showed this inflammatory response is NLRP3 mediated (Krishnan et al. 2016). MCC950 was tested in hypertensive C57BL/6 mice where the hypertension was induced by the removal of one kidney, implant of a deoxycorticosterone acetate pellet and 0.9% saline to drink (Krishnan et al. 2016). These hypertensive mice showed systolic blood pressure increase of −30 mmHg and increased gene expression in the kidney of inflammasome components: NLRP3, ASC, pro-caspase-1, pro-IL-1β but not pro-IL-18. Inhibiting inflammasome activity in this salt-based model by use of mice deficient in inflammasome components (caspase-1$^{-/-}$ or ASC$^{-/-}$) alleviated inflammation, and significantly reduced fibrosis and hypertension (Krishnan et al. 2016). More specific pharmacological NLRP3 inhibition using MCC950 (10 mg/kg/day via osmotic pump) also showed these beneficial responses, and blood pressure was almost normalised; however, effects on fibrosis were unfortunately not assessed. Notably, MCC950 was effective even when administered after the hypertension had been established (Krishnan et al. 2016). These results raised the exciting possibility that hypertension may be treated using inflammasome inhibitors. They also suggest that inflammasome formation in the kidney may be linked with hypertension in other diseases such as the crystal-related hyperuricemia and gout where hypertension is commonplace. Indeed efficacy of MCC950 has already been shown in crystallopathy mouse models (Ludwig-Portugall et al. 2016), but hypertension was not the focus of those studies, and the wider context should be explored further. Another model of hypertension can be initiated using the vasoconstrictor angiotensin II. In agreement with the salt-based model already described, ASC$^{-/-}$ mice had 40% less hypertension, as assessed by blood pressure measurement, in comparison to wild-type controls. This model used young adult male mice 10–12 weeks old, and unfortunately MCC950 was not examined (Krishnan et al. 2016). Later studies in this model focussed on aged male mice 23–31 months old where inflammation is already established prior to induction of hypertension (Dinh et al. 2017). In these aged mice a much stronger response to angiotensin II was measured in comparison to young mice attributed to increased expression of the angiotensin 1 receptor (AT1R) while AT2R was suppressed. In the aged mice, the known NLRP3 activator, reactive oxygen species was significantly increased alongside renal expression of NLRP3 inflammasome components (NLRP3, caspase-1 and IL-1β). It was surprising, given the previous data, that MCC950 (10 mg/kg/day via osmotic pump) was not effective in lowering hypertension in this angiotensin II model indicating, in this case, the hypertension was not mediated by NLRP3. The effect of MCC950 on the inflammatory response in this model was not pursued and would be informative. Availability of MCC950 as a tool compound allows elucidation of NLRP3 involvement in disease beyond that previously gained from genetically modified animal models. The relevance of this cellular signalling platform can now be understood at various stages of disease and during the ageing process.

NLRP3 inflammasome is active in cardiac fibroblasts, circulating inflammatory cells and cardiomyocytes (Grundmann et al. 2011; Takahashi 2014). In myocardial infarction, an inflammatory response is mounted in order to allow healing, but elevated IL-1β and IL-18 are linked with increased infarct size, reduced cardiac contractility and are predictive of heart failure (van Hout et al. 2016). In a clinically relevant pig model of myocardial infarction, MCC950 was used to study the therapeutic benefit of NLRP3 inflammasome inhibition (van Hout et al. 2016). Just prior to the end of a 75 min transluminal occlusion, MCC950 was dosed at either 3 mg/kg or 6 mg/kg intravenously and repeated daily for 1 week in a randomised blinded study. MCC950 treatment led to reduced infarct size, improved left ventricular function, elevated reserve capacity in the myocardium, suppressed neutrophil infiltration and decreased angiogenesis. The effects observed were dose dependent with the maximum response given in the 6 mg/kg treatment group. NLRP3 inhibition in this cardiac model by the sulfonylurea MCC950 was clearly beneficial. The authors suggest NLRP3 inhibition provides a selective means to modulate the inflammatory response rather than blocking it completely, allowing effective tissue repair mechanisms to remain.

The sulfonylurea class of NLRP3 inhibitors are the most extensively studied to date with impressive results in multiple disease models. This is also generating significant commercial interest with multiple companies such as Inflazome, IFM, Nodthera and Jecure all publishing patents in this space.

15.2.2 Dual Action Sulfonylureas

An interesting set of nine hybrid molecules have been synthesised based on known type 2 diabetes drugs and exploiting the nanomolar NLRP3 inhibitory activity of MCC950 (CRID3) (Hill et al. 2017).

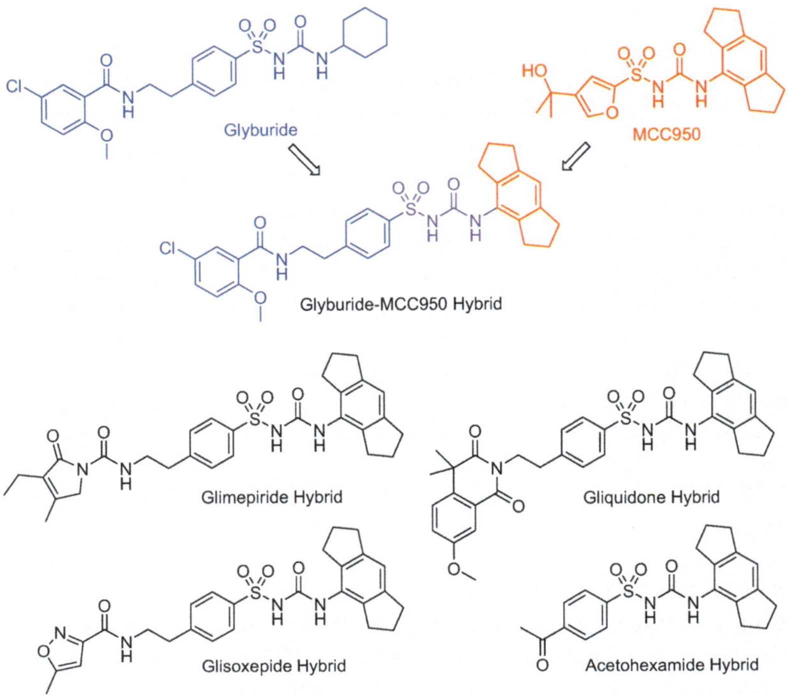

These hybrids all show nanomolar potency as NLRP3 inhibitors, and a subset (hybrids of glyburide, glimepiride, gliquidone, glisoxepide and acetohexamide) also retained insulin secretory activity of the parent sulfonylurea (Hill et al. 2017). Results of in vivo studies have not yet been communicated; however, multiple uses can be envisioned. Pancreatic β-cells are continually depleted as type 2 diabetes progresses, and therefore sulfonylurea drugs become less efficacious. This pancreatic β-cell death has been attributed to NLRP3-mediated IL-1β release, indicating dual action molecules which can prevent β-cell death while also stimulating insulin release may prove valuable. There are also a multitude of additional disease complications, such as nephropathy, coronary atherosclerosis, neuroinflammation and wound healing, and many of these may benefit from the NLRP3 inhibitory effect of these hybrid molecules.

Glyburide is in clinical trials as an intravenous formulation for treatment of ischaemic stroke due to its inhibition of SUR1-TRPM4 ion channels which alleviates oedema and haemorrhagic transformation (Sheth et al. 2016). NLRP3 inflammasome activation is also evident after haemorrhagic stroke leading to tissue damage (Yang et al. 2017). The dual action glyburide-MCC950 hybrid may therefore prove even more efficacious than glyburide in treatment of stroke.

15.2.3 Sulfonamides

The glyburide synthetic precursor sulfonamide 16673-34-0, more recently called JC-21, was reported as an NLRP3 inhibitor (Marchetti et al. 2014) effective in murine models of myocardial infarction and ischaemia reperfusion (Marchetti et al. 2014, 2015; Toldo et al. 2016). However, this was refuted by Hill et al. (2017) who tested this compound along with sulfonylurea precursors of other type 2 diabetes drugs for NLRP3 inhibitory activity using LPS-primed murine BMDM stimulated with ATP.

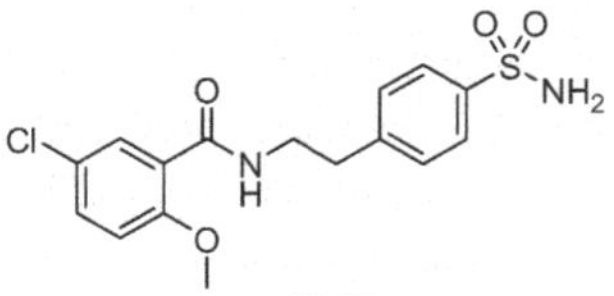

JC-21
16673-34-0

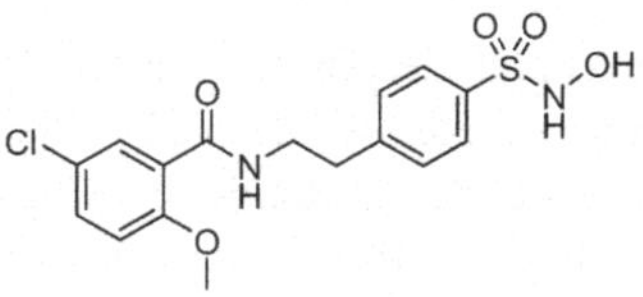

JC-171

In order to improve solubility of JC-21, a hydroxysulfonamide analog was synthesised, JC-171 (Guo et al. 2017). JC-171 showed NLRP3 inhibitory IC_{50} of 8.45 μM in LPS-primed J774A.1 macrophages stimulated with ATP, while in the same assay JC-21 had IC_{50} of 3.25 μM. JC-171 appeared more potent when tested in murine BMDM and did not inhibit secretion of TNF-α or IL-6 indicating a degree of selectivity. Furthermore, expression of inflammasome protein components was not affected by JC-171 leading to the assumption that inflammasome activation was inhibited rather than the priming step. JC-171 (100 mg/kg) was effective in suppressing IL-1β in vivo in mice challenged with LPS. In the murine EAE disease model of multiple sclerosis, JC-171 (100 mg/kg i.p. every second day) gave an impressive delay of disease onset and reduced symptoms. When dosed therapeutically, after onset of disease in the EAE model, and compared directly to MCC950, both compounds (10 mg/kg i.p. every second day) proved equally efficacious in preventing disease progression.

15.2.4 Vinyl Sulfones

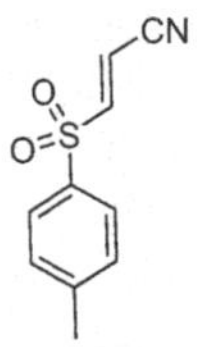

BAY11-7082

Investigation of different NF-κB inhibitors and their effect on NLRP3 inflammasome inhibition identified the vinyl sulfone small molecule BAY11-7082 as an NLRP3 inhibitor with IC_{50} of 12 μM (Juliana et al. 2010). Although this compound targeted the NF-κB pathway via IKKβ kinase (and thereby modulated inflammasome priming), it also had an independent activity through inhibition of inflammasome assembly. This was demonstrated by treatment of LPS-stimulated bone marrow macrophages with BAY11-7082. Further evidence was gained using stable NLRP3$^{-/-}$ bone marrow macrophages where NLRP3 was constitutively expressed without the need for LPS priming. In these cells inflammasome activation using ATP, nigericin or MSU was blocked by BAY11-7082. In this manner BAY11-7082 behaved very similarly to parthenolide which was identified in the same study (and is described in Sect. 15.3.3). However, BAY11-7082 proved to be much more selective over NLRP1, NLRC4 (Juliana et al. 2010) and AIM2 (Jiang et al. 2017) inflammasomes and, despite its non-specific cysteine modifying capacity, could not directly inhibit caspase-1. BAY11-7082 is a Michael acceptor and reacts with biological nucleophiles glutathione and L-cysteine (Strickson et al. 2013). BAY11-7082 also reacts with other biological targets such as tyrosine phosphatases and the ubiquitin system (Strickson et al. 2013).

An interesting study was conducted to characterise the structural elements of BAY11-7082 needed for this irreversible binding (Juliana et al. 2010). Both priming in response to LPS and activation in response to ATP were investigated with Bay11-7082 and its seven structural analogs measuring pro-IL-1β expression and pro-caspase-1 processing, respectively (Juliana et al. 2010).

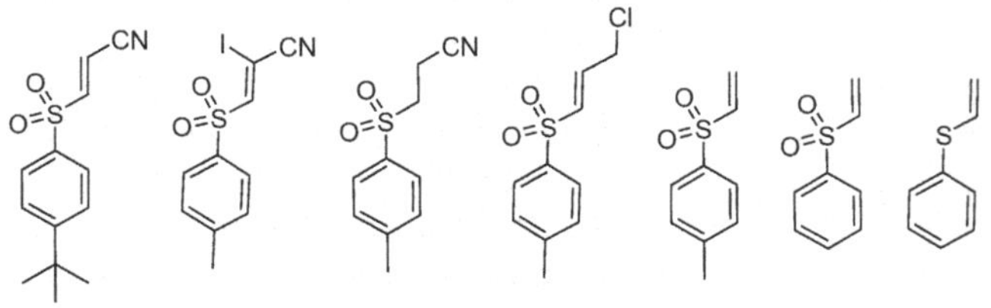

(1) (2) (3) (4) (5) (6) (7)

Seven Structural Analogues of BAY11-7082

The bulky *t*-butyl and iodo of analogs **1** and **2** did not affect inhibition indicating these were not the sites of nucleophilic attack. As anticipated, the alkene reduction in analog **3** abrogated activity. Introduction of a chloromethylene in place of nitrile, analog **4**, or simple replacement with hydrogen (analogs **5** and **6**) gave weak retention of NLRP3 inhibitory activity through both priming and activation. These modifications provided evidence of nucleophilic attack at the C3 position ultimately resulting in scission of the carbon-sulfur bond. This was facilitated by the electron withdrawing sulfone group which ultimately acts as a leaving group in *p*-toluenesulfonic acid. Reduction of the sulfone, analog **7**, therefore caused complete loss of inhibitory activity. BAY11-7082 inhibited pyroptosome formation and pyroptotic cell death (Juliana et al. 2010).

BAY11-7082 was tested in a murine model of lupus nephritis, where both NF-κB and NLRP3 inflammasome activation are dysregulated (Zhao et al. 2013).

Proteinurea, renal malfunction, cytokine release and neutrophil infiltration were all suppressed by BAY11-7082 alongside a reduced mortality. In complex diseases such as systemic lupus erythematosus, multiaction drugs may have an important role. This feature of BAY11-7082 was also exploited in a rat model of neuropathic pain using lumbar disc herniation (Zhang et al. 2017). NF-κB was identified as a pain mediator while also acting to prime NLRP3 inflammasome. Subsequent activation of the inflammasome perpetuates the inflammatory phenotype. BAY11-7082 was dosed at 5 mg/kg via i.p. injection three times per week over 4 weeks in the neuropathic pain model (Zhang et al. 2017). This resulted in statistically significant reduction of NF-κB, IL-1β and IL-18, and pain was attenuated. Given the reactive nature of BAY11-7082, it is possible that other effects also contributed to the positive outcome; nevertheless, this small molecule has an impressive activity in vivo, and it would be interesting to see if it can cross the blood-brain barrier to have direct effect in CNS disease models.

NLRP3 activation is triggered in response to burn-induced acute lung injury (Han et al. 2015). At a cellular level, alveolar macrophages upregulate NLRP3 expression and activation in response to burn serum, and this can be inhibited by BAY11-7082. This effect was also observed in vivo where IL-1β and IL-18 levels peak at 24–48 h post injury (Han et al. 2015). BAY11-7082 3 mg/kg was dosed via i.p. injection immediately after initiating burn-induced acute lung injury. Approximately two- to threefold reduction of NLRP3-related inflammatory cytokines and proteins was achieved along with a reduction in myeloperoxidase. More importantly the histopathologic features of the injury, neutrophil infiltration, oedema, alveolar wall thickening and haemorrhage, were all reduced.

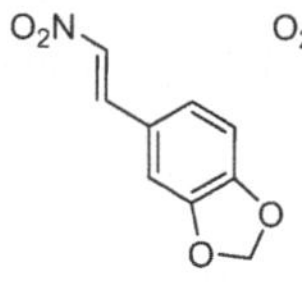

MNS

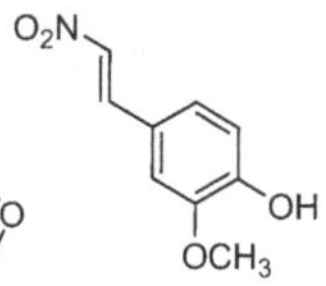

HMNS

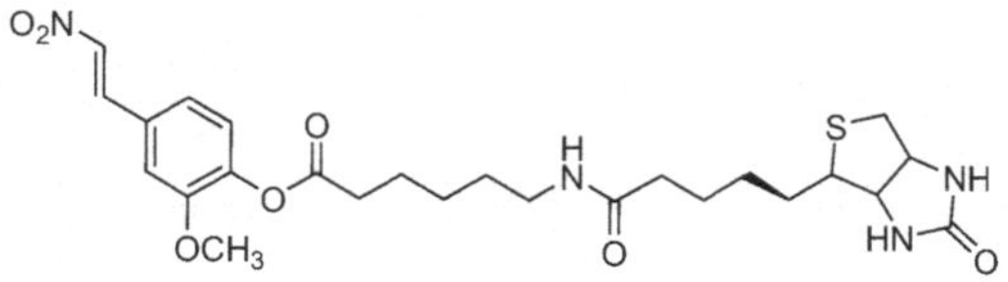

Biotin HMNS

15.2.5 *β-Nitrostyrenes*

3,4-Methylenedioxy-β-nitrostyrene (MNS) was identified from screening 160 kinase targeted compounds against LPS-primed murine BMDM (He et al. 2014). Such libraries typically have molecules which target kinase ATP-binding sites, and it was therefore possible that hits would be identified which could inhibit the ATP-binding site of NLRP3. MNS, a known Syk kinase inhibitor, was identified with an IC_{50} of 2 μM (He et al. 2014). Secretion of IL-1β, IL-18 and active caspase-1 formation was inhibited by MNS but not mRNA levels of inflammasome components. In a similar manner to BAY11-7082, MNS acts as a Michael acceptor, and the nitrovinyl side chain is therefore essential for its biological activity as an inhibitor of the NLRP3 pathway. Modification to the benzodioxole ring was tolerated but decreased compound potency. MNS directly targets NLRP3 inflammasome reacting with the leucine-rich repeat and nucleotide-binding oligomerisation domains while also targeting the ATP-binding site. Given this reactive nature, it is surprising that MNS did not inhibit other inflammasomes NLRC4 or AIM2.

A biotinylated probe, based on the active MNS analog, HMNS, successfully pulled down NLRP3 protein from cell lysate and could also pull down recombinantly expressed NLRP3 (He et al. 2014). This interaction was abrogated in the presence of excess HMNS. In contrast, NLRC4 could not be isolated using biotin-HMNS. Full-length NLRP3 and mutants of pyrin or LRR domain could all be isolated using biotin-HMNS. Likewise, the NOD and LRR domain could be isolated whereas the pyrin domain could not. There have not been a large number of in vivo studies with MNS after its identification as an NLRP3 modulator. One study was conducted in a rat model of wound healing after burn injury where MNS promoted healing (Xiao et al. 2016). Nitro-containing compounds and styrenes are not generally considered drug-like and are well-known toxicophores.

15.2.6 *Acrylate Derivatives*

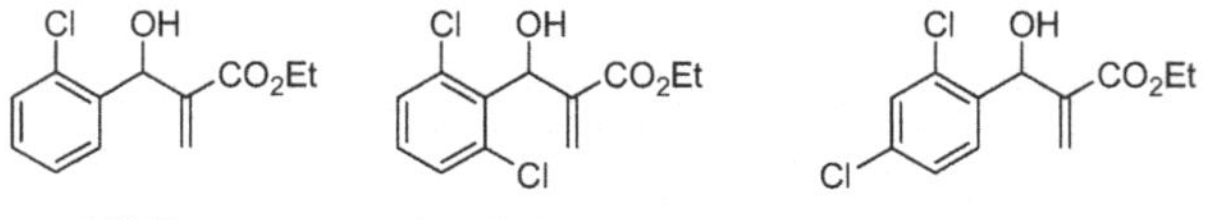

INF4E INF4E close structural analogs

Cocco et al. (2014) observed that inflammasome inhibitors commonly contained Michael acceptor functionality. They hypothesised this allowed inhibition of the pathway through reaction with cysteine residues in caspase-1, NLRP3 or other relevant proteins in the cascade. A library was therefore designed around this electrophilic pharmacophore with the intent of discovering new irreversible NLRP3 inhibitors which could be optimised for specificity and minimal toxicity. In common with other known irreversible NLRP3 inhibitors, many of the new compounds inhibited caspase-1 and NLRP3 ATPase activity in line with expectation. Out of 36 compounds, INF4E and its 2 close structural analogs were most promising for further work (Cocco et al. 2014).

INF4E and its analogs were not ideal; a degree of cellular cytotoxicity was noted. Importantly INF4E also irreversibly bound to human serum albumin forming three covalent adducts which may trigger idiosyncratic adverse reactions in vivo (Cocco et al. 2016). This lead compound was systematically modified (Cocco et al. 2017) through removal or substitution of the alcohol moiety, hydrolysis of the ester to the carboxylic acid, and a small number of compounds also reduced the alkene, thereby removing the likelihood of covalent mode of action. All compounds were triaged through assays for Michael acceptor reactivity, cytotoxicity, NLRP3 activity and pyroptosis to ultimately identify the improved molecule INF39. INF39 has a potency in the micromolar range; and while a full IC_{50} was not recorded, a 10 μM concentration reduced IL-1β secretion by around 50% from LPS-primed mouse BMDM. Unlike INF4E, no activity was observed on caspase-1 showing that INF39 was more selective. INF39 inhibited NLRP3 ATPase activity by 52% at 100 μM and also gave partial inhibition of NLRP3 inflammasome priming. Rapid metabolism, via INF39 ester hydrolysis to the carboxylic acid, was found during permeability through rat intestine (ex vivo) and also during in vitro microsomal stability. Both the acid and ester are active as NLRP3 inhibitors, and neither were cytotoxic against THP-1 cell line (MTT assay) up to 100 μM.

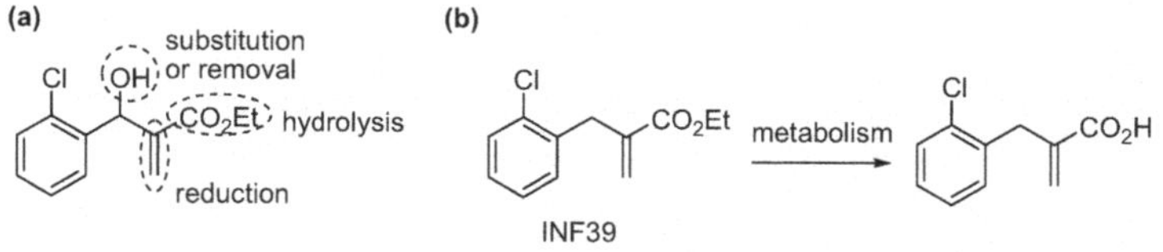

INF39 was highly insoluble and lipophilic, leading to formulation as a suspension in olive oil for oral dosing in vivo (Cocco et al. 2017). A rat model of 2,4-dinitrobenzenesulfonic acid-induced colitis was investigated where INF39 was dosed at 12.5, 25 and 50 mg/kg/day for 6 days commencing at induction of colitis (Cocco et al. 2017). Positive outcomes were observed on body and spleen weight, colonic length and macroscopic damage, while reduction in levels of IL-1β, TNF-α and myeloperoxidase was also observed.

Despite early results, which were promising, it is difficult to ascertain the wider application of INF39 without further study. INF39 pharmacokinetic profile and distribution, either as the ester prodrug or the acid form, need to be evaluated alongside more extensive investigation of toxicity profile. Alternative prodrugs or salt forms may help modulate the solubility issues and facilitate use of this compound in other disease models where direct administration to the site of action is more challenging.

As an extension of the acrylate containing covalent drug investigation, hybrid analogs were synthesised and tested for their NLRP3 inhibitory activity (Cocco et al. 2016). The type 2 diabetes sulfonylurea drug glyburide, a weakly NLRP3 active insulin secretagogue, and its precursor sulfonamide, 16673-34-0, were conjugated to the INF39 acrylate warhead. Both hybrids retained Michael acceptor reactivity, were not cytotoxic at 100 μM to THP-1 cells and did not react with serum albumin. Inhibition of THP-1 pyroptotic cell death at 10 μM concentration of test compound was more effective for the 16673-34-0 hybrid at 46% versus 17% for the glyburide hybrid. While IL-1β release from wild-type mouse BMDM was approximately 50% at 20 μM concentration of the 16673-34-0 hybrid, the effect was much less in macrophages bearing NLRP3-activating mutations, typical of NLRP3-specific genetic diseases such as Muckle-Wells. Inhibition of NLRP3 ATPase activity by the 16673-34-0 hybrid was not very potent with IC_{50} of 74 μM.

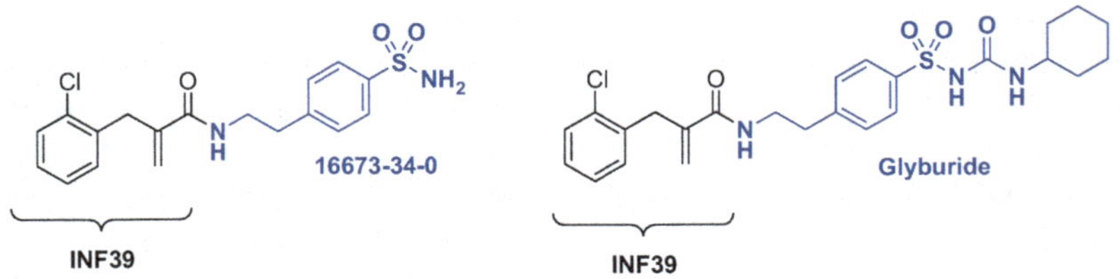

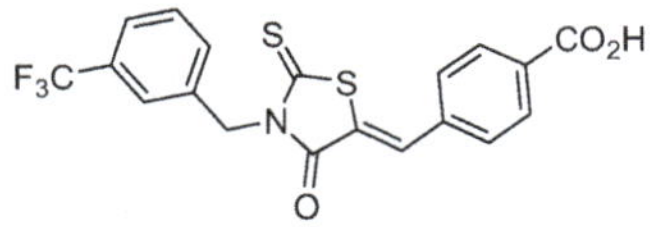
CY-09

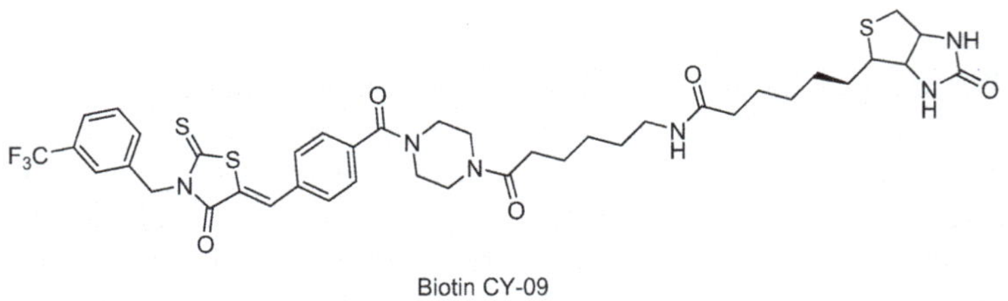
Biotin CY-09

15.2.7 *Glitazones*

In a screening campaign using a proprietary library of bioactive compounds, CY-09 was identified and characterised as an inhibitor of NLRP3 inflammasome (Jiang et al. 2017). CY-09 is from the glitazone molecular class and contains a Michael acceptor moiety which might be prone to conjugation with nucleophilic species such as cysteine side chains. CY-09 blocked IL-1β secretion in response to nigericin, ATP and MSU crystals in LPS-primed BMDM with no effect observed in response to NLRC4 or AIM2 triggers. CY-09 did not disturb the priming phase with levels of pro-IL-1β and NLRP3 expression unchanged. An interesting set of comparative data indicated CY-09 was similar in potency to BAY11-072 and the natural products parthenolide, sulforaphane and isoliquiritigenin, with IC_{50} in the 5 μM range. CY-09 was however more selective than parthenolide (described in Sect. 15.3.3) or sulforaphane (Sect. 15.3.2) over AIM2 and NLRC4. No effect was observed on chloride channels, but NLRP3 oligomerisation was inhibited.

Through synthesis of a biotinylated version of CY-09, pull-down experiments could be conducted using the affinity of streptavidin beads for the biotin tag (Jiang et al. 2017). The probe was incubated with cell lysates of LPS-primed BMDM then extracted using streptavidin beads. Using this technique NLRP3 was identified as attached to the probe and could be removed from the matrix using competition with non-labelled CY-09, illustrating reversible binding kinetics. NEK7 was not pulled down in these experiments, and other experiments designed towards alternative inflammasomes did not pull down those proteins. Direct affinity was also observed between the biotinylated CY-09 and purified NLRP3, and this was further characterised by microscale thermophoresis to establish a K_D of 500 nM. To more accurately identify the binding site, the LRR, NACHT and pyrin domains were studied to find only the NACHT domain bound to CY-09. Mutational studies showed CY-09 bound to the ATP site of the Walker A motif and prevented subsequent ATPase activity, oligomerisation and activation of NLRP3.

CY-09 had good pharmacokinetic properties (Jiang et al. 2017) with stability in plasma and no inhibition of the hERG ion channel up to 10 μM. No potent inhibition of the five major cytochromes P450 enzymes was found; however, it should be considered the potency of the compound in vitro is around IC_{50} 5 μM; therefore, inhibition of CYP1A2 at 18.9, CYP2C9 at 8.18 μM and CYP3A4 at 26 μM may prove problematic in vivo. Moreover, hERG channel effects should be ascertained at a higher top concentration to give a more convincing safety margin. CY-09 had an in vivo half-life of 2.4 h in mice with bioavailability of 72% and AUC of 8232 (h ng)/mL, suitable to determine efficacy in murine models. CY-09 did indeed prove highly efficacious in models of MSU-induced peritonitis and Muckle-Wells where both CY-09 and MCC950 were comparable at 40 mg/kg i.p. dose. CY-09 also showed reversal of metabolic disorder in diabetic mice, but MCC950 was not included as a comparator in that experiment. CY-09 therefore looks promising as a lead for further investigation of NLRP3 inhibitors which bind the ATP site.

15.2.8 Edaravone

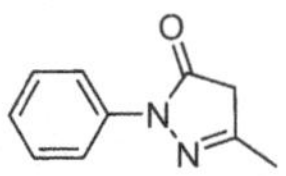

Edaravone

Edaravone (Radicava®, Radicut® developed in the 1980s) is now a generic treatment for ALS (approved in 2015 in Japan and 2017 in the USA) and stroke (approved in 2011) (Administration 2017b; Miyaji et al. 2015). Although the mode of action of edaravone is largely unknown, it has free radical scavenging and anti-oxidant properties and is neuroprotective. Edaravone's reactive oxygen species scavenging ability blocks NLRP3-mediated IL-1β secretion from amyloid-β-treated microglia (Wang et al. 2017a). These studies are only preliminary, and much more work would be needed to fully confirm effects on NLRP3. Edaravone does have side effects including skin inflammation, hypersensitivity and gait disturbance (Administration 2017a).

15.2.9 *Antidepressants*

Major depressive disorder pathogenesis has recently been linked with NLRP3 inflammasome-mediated IL-1β and neuroinflammation and has triggered significant interest in therapeutic modulation of these pathways (Kaufmann et al. 2017). A selection of antidepressant drugs in clinical use to treat major depressive disorder were investigated for their potential effect on NLRP3 inflammasome (Alcocer-Gomez et al. 2017). This was an interesting study employing a structurally diverse set of compounds (fluoxetine, paroxetine, mianserin, mirtazapine, venlafaxine, desvenlafaxine, amitriptyline, imipramine and agomelatine). When tested at 1 μM concentration on LPS-stimulated THP-1 cell line with ATP as NLRP3 inflammasome trigger, all compounds caused 50–60% reduction in secretion of IL-1β and IL-18. NLRP3 expression was also reduced. Mirtazapine, fluoxetine and agomelatine were marginally more potent than the other analogs. In a murine model of depression, where forced swimming test was used, all mice showed elevation of IL-1β which was alleviated by the antidepressants (Alcocer-Gomez et al. 2017).

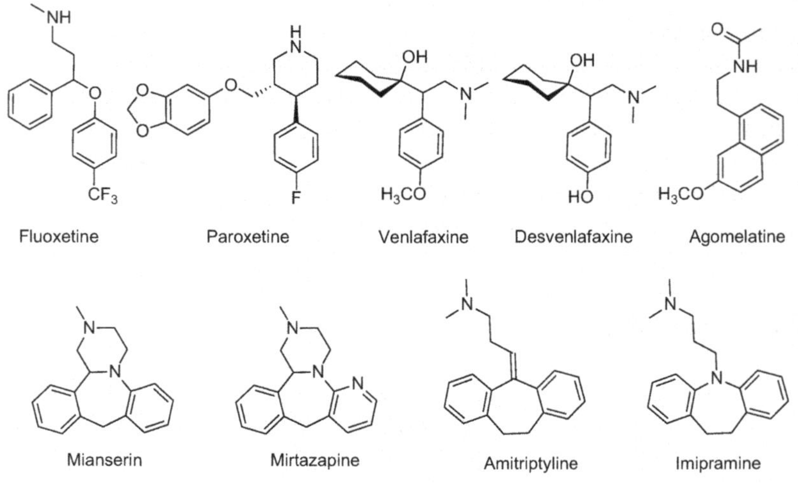

A cohort of 214 patients (aged 18–60) diagnosed with major depressive disorder were recruited, and the effect of 9 different antidepressant medications (fluoxetine, paroxetine, mianserin, mirtazapine, venlafaxine, desvenlafaxine, amitriptyline, imipramine and agomelatine) on NLRP3 inflammasome was examined (Alcocer-Gomez et al. 2017). In comparison to untreated controls, all treated patients showed reduction of IL-1β and IL-18 in serum with reduced expression of NLRP3 mRNA in blood mononuclear cells. Genes typical of autophagy, BECLIN and MAP-LC3, were significantly upregulated indicating enhanced autophagy pathways may be protective. Further examination in vitro employed autophagy-deficient mouse embryonic fibroblast cells, none of the antidepressants inhibited NLRP3 inflammasome pathways in these cells. In agreement with this, no increase of LC3B-11 (autophagy readout) and no reduction of cleaved caspase-1 was observed. The wild-type mouse embryonic fibroblast cells did show strong inhibition of ATP-triggered NLRP3 pathways in the presence of antidepressants. This was accompanied by the increase of LC3B-11 and reduction of cleaved caspase-1. Overall, these results have led to the intriguing possibility that NLRP3 inflammasome could act as a biomarker for treatment response and therefore aid in drug selection.

15.2.10 Acylhydrazone EMD638683

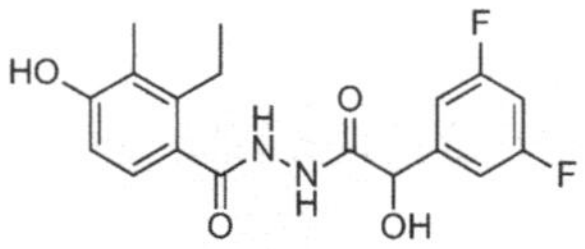

EMD638683

EMD638683 is a known inhibitor of serum- and glucocorticoid-inducible kinase 1 and reported to inhibit NLRP3 inflammasome (Gan et al. 2017). A high dose of EMD638683 (600 mg/kg/day in chow) administered to mice, alleviated cardiac inflammation and fibrosis induced by angiotensin II. This was also compared to MCC950 treatment (10 mg/kg i.p.) which had similar efficacy. Further investigation of EMD638683 treatment indicated an inhibition of NLRP3 and IL-1β expression as measured by mRNA alongside a reduction in cleaved caspase-1 and IL-1β in cardiac tissues of the treated mice. Murine BMDM were used to compare effect of EMD638683 to NLRP3 siRNA and also MCC950 (alone and also in combination) on reduction of IL-1β levels, and in each case results were comparable. It should, however, be noted that EMD638683 has, thus far, only been investigated as a tool compound in animal studies. EMD638683 is of the acylhydrazone class of compounds which are commonly regarded in medicinal chemistry as toxicophores. Indeed in an earlier study EMD638683 triggered increased fluid intake and urination alongside a marked reduction in body weight (Ackermann et al. 2011). Extensive toxicity profiling would be required were this compound to be pursued further.

15.2.11 Benzimidazoles

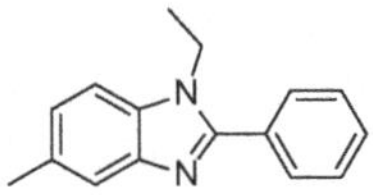

FC11a-2

Benzimidazole Fc11a-2 inhibited secretion of IL-1β and IL-18 from LPS-primed THP-1 cells stimulated with ATP showing an IC_{50} of around 10 μM (Liu et al. 2013). However total inhibition of secretion was not observed at the highest tested concentration (30 μM). Further studies indicated Fc11a-2 did not inhibit the priming phase but interfered with the cleavage of pro-caspase-1 and hence proteolytic processing of pro- IL-1β and pro-IL-18. In a murine model of dextran sodium sulfate-induced colitis, Fc11a-2 (10–30 mg/kg, intragastrically) dose-dependent improvement was evident in body weight, colon length, histopathologic scoring and myeloperoxidase activity (Liu et al. 2013). Macrophage infiltration and active caspase-1 were similarly reduced alongside a notable decrease in mRNA for IL-1β, IL-18, TNF-α, IL-17A and IFN-γ. These results prompted further investigation into analogs to understand the SAR. However, none of the analogs were significantly more potent than FC-11a-2 which at 10–30 μM showed inhibition of only 40% in the aforementioned cell-based assay (Pan et al. 2017).

15.2.12 Organoboron NLRP3 Inhibitors

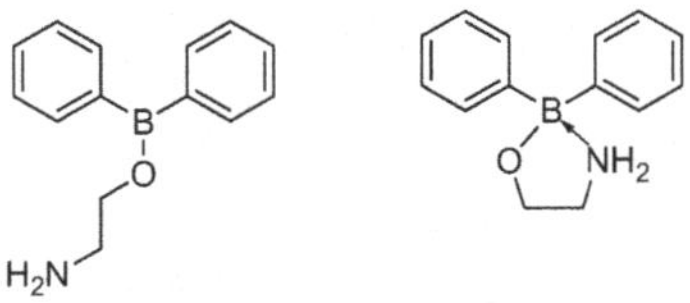

2-aminoethoxy diphenylborinate (2APB)
acyclic and cyclic forms

Boron is a largely overlooked element in medicinal chemistry and not common in pharmaceutical drugs with only Velcade and Tavaborole on the market. An interesting series of NLRP3 inhibitors have been both published (Baldwin et al. 2017) and patented (Brough et al. 2017) based on the boron semimetal scaffold 2-aminoethoxy diphenylborinate (2APB). This parent scaffold is known to disrupt cellular Ca^{2+} homeostasis via a variety of mechanisms, and, until recently, this explained its NLRP3 inhibitory effect. However Kastnelson et al. recently revealed 2APB inflammasome inhibition was independent of its effects on Ca^{2+} channels (Katsnelson et al. 2015).

This work prompted a search for similar boron-based NLRP3 inhibitory compounds which were devoid of the Ca^{2+} modulatory activity. During these studies the cyclic nature of 2APB was used to search the zinc database for compounds with similar pharmacophore and shape; additional boron-containing compounds were also identified through searches of SciFinder Scholar (Baldwin et al. 2017). Two of the most potent early hits were BC7 (NLRP3 IC_{50} 1.16 μM) and BC23 (NLRP3 IC_{50} 2.29 μM). During SAR studies the ring oxygen, boron, NH and substituent CCl_3 were all identified as essential to compound activity. The bisphenyl was not altered, while the remaining two ring substituents could be used to modulate activity. Compounds NBC6, 18 and 24 had improved potency alongside more attractive physicochemical properties. The lead compound NBC6 had NLRP3 IC_{50} 0.57 μM, which was substantially more potent than the parent 2APB (NLRP3 IC_{50} 67 μM). Gratifyingly no effect was observed on calcium homeostasis in cell-based assays. NBC6 was well characterised as a specific inhibitor of NLRP3 inflammasome up to top concentration of 30 μM in murine BMDM or neutrophils. Both canonical and non-canonical inflammasome pathways could be blocked to prevent IL-1β secretion, but IL-1α was not affected. NLRC4 or AIM2 inflammasomes were not inhibited by NBC6 indicating the compound had a degree of specificity across inflammasomes. Importantly, NBC6 was tested in washout cell-based assays and was identified as an irreversible inhibitor of NLRP3.

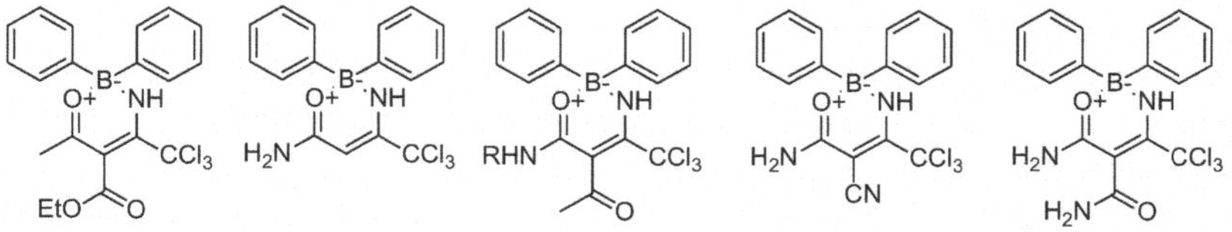

BC7 BC23 NBC6 R=H NBC13 R=Ph NBC18 NBC24

One murine model of peritonitis was investigated with a close analog of NBC6 called NBC13, reported as easier to formulate but equipotent (Baldwin et al. 2017). Both wild-type and NLRP3 knockout mice were used, and MCC950 was tested simultaneously, as a comparator compound. NBC13 did indeed inhibit LPS-induced IL-1β in vivo but not to the same degree as MCC950. This may be due to issues of pharmacokinetics, but associated data was not presented. Nevertheless, this series proves to be an interesting novel compound set which may find future application in inflammatory disease.

15.2.13 Anthranilic Acid NSAIDs

The widely used fenamate non-steroidal anti-inflammatory drugs (NSAIDs) typically target COX enzymes, but recent research found these also target NLRP3 (Daniels et al. 2016). Four clinically used fenamates (diclofenac, flufenamic acid, meclofenamic acid, mefenamic acid) were tested on LPS-primed immortalised murine BMDM cells. These drugs inhibited IL-1β secretion when the cells were stimulated with ATP. The most potent inhibitor in this assay was meclofenamic acid with an IC_{50} around 25 μM. The structurally unrelated NSAID ibuprofen had no effect. Similar result was observed using MSU as NLRP3 activator. Testing of the fenamates using NLRC4 stimuli *S. typhimurium* or AIM2 stimuli dsDNA showed the compounds did not inhibit these inflammasomes and therefore had a degree of selectivity. The fenamates also inhibited ASC speck formation.

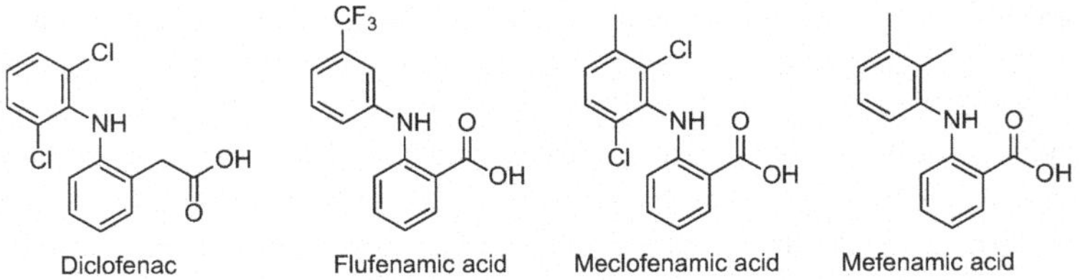

Investigation into the mode of action showed no evidence of cysteine modification. The compounds were also readily washed out of cells reversing the observed inhibition. Whole-cell patch-clamp testing was used to examine ion currents showing no effect on ATP-induced cation currents. However, inhibition of VRAC was identified preventing chloride ion transport in a similar manner to known chloride channel and VRAC inhibitors.

In vivo models of Alzheimer's disease were conducted using mefenamic acid (Daniels et al. 2016). The first showed a protective effect with prophylactic treatment using mefenamic acid (5 mg/kg/day, i.p.) in rats which had been injected intracerebroventricularly with soluble oligomeric amyloid β_{1-42}. The second therapeutic model was investigated using 13–14-month-old 3 × TgAD transgenic mice where mefenamic acid was delivered continuously (25 mg/kg/day) via osmotic mini pump (Daniels et al. 2016). The treatment completely reversed neuroinflammation with concomitant reduction in activated microglia and IL-1β. This finding is exciting as these drugs are already in clinical use and could be readily repurposed.

15.3 Natural Products as Inflammasome Modulators

15.3.1 β-Hydroxybutyrate

β-Hydroxybutyrate (BHB) is a ketone metabolite increased in response to various forms of caloric restriction or energy deficiency to provide an alternative fuel source, it also functions in cell signalling pathways (Newman and Verdin 2014). Under these conditions there is an anti-inflammatory effect recently linked to the innate immune response (Youm et al. 2015). Youm et al. showed BHB effectively blocked K^+ efflux, a known NLRP3 inflammasome trigger, while also preventing ASC polymerisation and speck formation. BHB was effective at blocking activation of NLRP3 inflammasome with an IC_{50} ~1 mM, a physiologically relevant concentration (Youm et al. 2015). A characteristic dose-dependent reduction of IL-1β and

IL-18 was observed from LPS-primed human monocytes treated with BHB, while TNF-α levels were unchanged. This selectivity was further illustrated as NLRC4 and AIM2, and the non-canonical inflammasome pathway was not affected by BHB. Further investigation of the mode of action is required; current investigations were not conclusive albeit ruling out AMP-activated protein kinase, reactive oxygen species, autophagy and glycolytic inhibition. It is interesting to note other ketogenic species such as acetoacetate and related short chain fatty acids butyrate and acetate did not recapitulate the effects observed with BHB, and also the chirality of BHB did not affect its ability to inhibit NLRP3-mediated pathways.

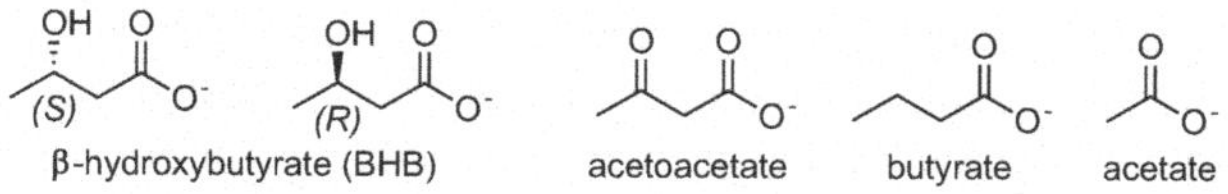

A murine model of the NLRP3-specific human disease familial cold autoinflammatory syndrome (NLRP3 (L351P) Cre^{+}) compared ketogenic (to increase BHB) versus chow diet (Youm et al. 2015). In this model NLRP3 is activated causing excessive neutrophilia (Brydges et al. 2009). The ketogenic diet prevented neutrophilia and hyperglycaemia. This prophylactic approach suggested elevated BHB levels might prove beneficial and widely applicable in the context of inflammatory disease. To dose BHB in vivo, formulation to extend its half-life was required and therefore a nanolipogel complex was used. Two published models indicate promise for BHB in gout, a disease characterised by presence of monosodium urate crystals which cause macrophage activation and, inflammasome dependent, neutrophil infiltration resulting in pain and swelling. The first model administered an intraperitoneal injection of monosodium urate crystals to mice; upon dosing with the aforementioned nanolipogel formulation of BHB (125 mg/kg i.p.), levels of circulating IL-1β and influx of neutrophils into the peritoneum were normalised (Youm et al. 2015). This held significant promise, and a subsequent rat model was investigated. Ketogenic or chow diet was fed to outbred Sprague-Dawley rats, injected with MSU crystals in the knee. In agreement with earlier models, remarkable reduction of inflammatory phenotype and tissue damage was observed in the ketogenic diet group, and circulating IL-1β levels were not elevated. BHB was equally effective in both adult and elderly mice; moreover *Staphylococcus aureus* infection was not exacerbated by BHB giving early indication that immune response to infection has not been unduly compromised although this may be pathogen dependent.

Efficacy of BHB has also been investigated in a rat model of Parkinson's disease (Fu et al. 2015) induced by treatment of the right substantia nigra pars compacta with LPS. BHB was administered continuously via subcutaneous osmotic pump at three doses (0.4, 0.8, 1.6 mmol/kg/day = 41.6, 83, 166 mg/kg/day) from 3 days prior to LPS treatment and for 3 weeks posttreatment. Analysis of motor function through amphetamine-induced rotational behaviour did show a dose-dependent improvement but, however, did not completely alleviate symptoms. Levels of both dopamine and its metabolite DOPAC showed a dose-dependent increase, while results from

Western blot and immunohistochemistry analysis also supported a protective effect of BHB on dopaminergic neurons. Cell-based work, in support of this study, suggested the effects of BHB were mediated by GPR109A in microglia; however, this contrasts with the work of Youm et al. (2015) who studied BMDM from GPR109A competent and deficient mice to conclude BHB mode of action was not via this receptor.

The most recent murine study of BHB and NLRP3 inflammasome was focussed on major depressive disorder after elevated IL-1β, IL-6 and TNF-α were identified as part of the pathophysiology (Yamanashi et al. 2017). IL-1β was linked to anti-neurogenic and anhedonic behaviour, and symptoms were ameliorated in response to IL-1β receptor antagonist, IL-1Ra, or knockout of the IL-1β receptor IL-1RI (Koo and Duman 2008). In contrast administration of IL-1β arrested the cell cycle via IL-1RI and activation of the NF-κB pathway, preventing proliferation of hippocampal cells (Koo and Duman 2008). BHB was administered subcutaneously (250 mg/kg) in mice showing peak brain concentration of around 8 μg/g of hippocampus (Yamanashi et al. 2017), which is well below the IC_{50} of 1 mM, and although brains were rinsed prior to processing, it was not clear whether these were fully perfused, so blood may remain in the blood vessels affecting the analytical results. Administration of BHB in a murine immobilisation stress model gave a small but statistically significant reduction in levels of IL-1β. However, in a model of chronic unpredictable stress, no change in IL-1β was found with BHB treatment (Yamanashi et al. 2017). Examination of the BHB administration method or perhaps use of ketogenic diet to increase endogenous BHB in the murine model may prove more conclusive. Indeed, observation of BHB antidepressant effect may not be mediated via NLRP3, and further study would be required to clarify this point.

15.3.2 Glucosinolate-Derived Isothiocyanates

Glucosinolates are a class of more than 130 bioactive compounds naturally occurring in all plants of the mustard family. Many plants of this family are referred to as "superfoods" due to the beneficial biological effects conferred by the glucosinolate system (Greaney et al. 2016). Upon plant damage, glucosinolates are hydrolysed by the atypical β-glucosidase "myrosinase", and usually a subsequent Lössen-type rearrangement occurs to give the corresponding isothiocyanates (Sturm and Wagner 2017). Isothiocyanates are reactive and in vivo can modify thiol moieties, such as cysteine side chains, amongst other nucleophilic functional groups. Perhaps unsurprisingly, isothiocyanates have numerous biological effects including inhibition of phase I metabolic enzymes (specifically cytochromes P450), induction of phase II enzymes and antimicrobial and anticancer properties. Isothiocyanates also have anti-inflammatory properties typically mediated through the nuclear factor erythroid-derived 2-like factor 2 (Nrf2) and the anti-oxidant response-element (ARE) pathways (Sturm and Wagner 2017).

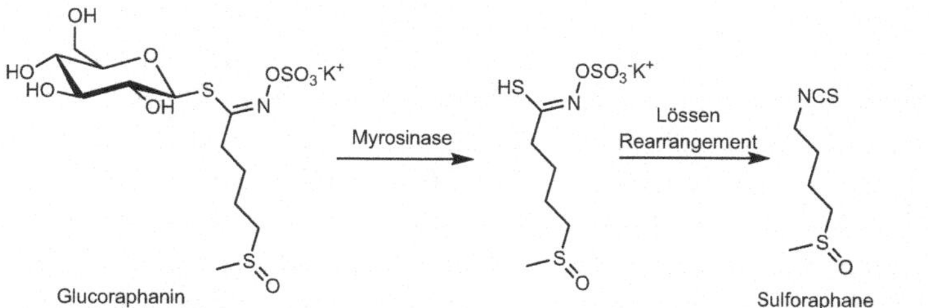

Sulforaphane, generated from the glucosinolate glucoraphanin, was studied at the National Institute for Health in Bethesda for inflammasome inhibitory effects (Greaney et al. 2016). Multiple inflammasomes were inhibited, NLRP1b, NLRP3, NAIP5/NLRC4 and AIM2 post-priming, and, despite the presence of a cysteine residue in the active site, sulforaphane was not a direct inhibitor of caspase-1. Several pathways were investigated to identify the mode of inhibition including investigation of the heat-shock response, effects on Nrf2, and tubulin modification, but these did not reveal a unifying hypothesis, and further work is required. Nevertheless, sulforaphane (25 mg/kg, i.p. at 0 and 4 h) was tested in an acute murine (C57BL/6) model of monosodium urate crystal-induced gout (Greaney et al. 2016). In this model NLRP3 inflammasome is typically activated by the crystalline deposits resulting in release of IL-1β and influx of inflammatory cells. Peritoneal lavage fluid of sulforaphane MSU-treated mice indicated a significant reduction of both IL-1β and recruited cell count compared to vehicle MSU-treated controls. These data are intriguing given the popular consumption of the *Cruciferae*, including the *Brassica* vegetables, in which these parent glucosinolates exist. Evgen Ltd. are developing Sulforadex®, a synthetic formulation of sulforaphane which successfully completed phase I clinical trial (Co 19th Nov 2016). One dose of Sulforadex® is equivalent to consuming 2.5 kg of broccoli. There are reports of testing Sulforadex® in diseases where NLRP3 inflammasome is implicated such as osteoarthritis, COPD, AMD, diabetes and cardiovascular disease (Co 19th Nov 2016).

15.3.3 Sesquiterpene Lactones

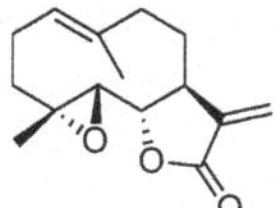

Parthenolide

Parthenolide is a naturally occurring germacranolide sesquiterpene lactone first isolated, as a principle active component, from Feverfew and generally found throughout the *Tanacetum* genus (*Studies in Natural Products Chemistry*: Volume 52 Bioactive natural products 2017; Dey et al. 2016). This natural product has been studied for efficacy against a plethora of different diseases such as hypoadiponectinemia, infection (bacterial or parasitic), cancers and inflammatory conditions (Studies in natural products chemistry: Volume 52 Bioactive natural products 2017; Dey et al. 2016). Clinical trials as a migraine therapeutic have been conducted with the natural source, Feverfew (Murphy et al. 1988).

Parthenolide has multiple modes of action and can ablate NLRP3 inflammasome-mediated cytokine release with an IC_{50} of 2.6 μM in THP-1 cell line (Li et al. 2015). Parthenolide inhibits the NF-κB pathway which regulates transcription of NLRP3 inflammasome components in the cell-priming phase (Juliana et al. 2010). The NF-κB pathway is triggered in response to toll-like receptor activation, commonly with LPS. NLRP3 Inflammasome assembly can then be activated by a second signal, such as ATP. In 2010, parthenolide was found to also inhibit the NLRP3 inflammasome assembly step (Juliana et al. 2010). Bone marrow macrophages were primed with LPS then treated with parthenolide, but no inflammasome assembly could be observed. This experiment showed inhibitory activity of parthenolide was not restricted to effects on the NF-κB pathway. Further evidence was gained using stable NLRP3$^{-/-}$ bone marrow macrophages where NLRP3 was constitutively expressed without the need for LPS priming. In these cells inflammasome activation using ATP, nigericin or MSU was blocked by parthenolide. These inhibitory effects were recapitulated in NLRP1 and NLRC4 inflammasome-specific assays. The protein common to these inflammasomes is caspase-1; parthenolide reacts with the active-site cysteine residue of caspase-1, Cys^{285}, preventing its proteolytic activation of IL-1β. Parthenolide also modified Cys residues at the ATP site of NLRP3 preventing ASC pyroptosome assembly and NLRP3 inflammasome formation. The reactivity of parthenolide seems to be responsible for its inhibitory activity on multiple proteins. Therefore it is interesting to note that 10 μM parthenolide did not inhibit AIM2-dependent caspase-1 activation or related pyroptotic cell death, so there remains an unexplained degree of selectivity (Coll et al. 2011).

Unfortunately parthenolide has low aqueous solubility, poor bioavailability and is sensitive to pH, which is a significant challenge for its use in vivo. These challenges have led to investigation of analogs with increased bioavailability and aqueous solubility (Guzman et al. 2007; Shanmugam et al. 2011; D'Anneo et al. 2013; Yang et al. 2015). Also, the α,β-unsaturated carbonyl "Michael acceptor" functionality of sesquiterpene lactones commonly correlates with cytotoxicity particularly if other alkylating centres are present (Siedle et al. 2004). The parthenolide mode of action in inflammation depends on these reactive centres, and therefore modification to alleviate cytotoxicity, while retaining anti-inflammatory effects of NLRP3 inhibition may prove challenging.

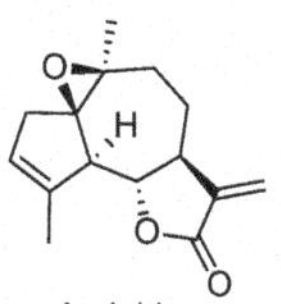

Arglabin

Arglabin is of the guaianolide class and occurs in *Artemisia glabella*, a type of wormwood, found only in Kazakhstan (Abderrazak et al. 2016). Recent investigations have attempted to delineate how arglabin may interact with innate immune pathways. Although the carbon skeleton is slightly different from parthenolide, a similar system of reactive functional groups are present: α,β-unsaturated carbonyl, lactone, epoxide and alkene. The strained lactone ring is particularly susceptible to nucleophilic attack in vitro and in vivo (Abderrazak et al. 2016). Comparatively little investigation of mode of action/binding to NLRP3 has been conducted with arglabin, and this remains to be confirmed. Arglabin inhibited NLRP3 inflammasome in LPS-primed murine BMDM treated with cholesterol crystals, using IL-1β and IL-18 secretion as readout (Abderrazak et al. 2016). The effect was exceptionally potent at IC_{50} ~10 nM. Cholesterol crystals are phagocytosed and activate NLRP3 after lysosomal rupture of the phagosome. In contrast using ATP to promote NLRP3 assembly, the effect was much less dramatic; 50 nM arglabin gave only 20% decrease in IL-1β secretion. Without testing significantly higher concentrations of arglabin, in this assay system, an IC_{50} cannot be determined. There was no inhibitory effect (up to 50 nM) on NLRP1, AIM2 or NLRC4 inflammasome in response to their respective activating stimuli (Abderrazak et al. 2016); again testing higher concentrations of arglabin would be necessary to show true selectivity. Arglabin induced microtubule-associated protein 1 light chain 3 II (LC3-II) protein accumulation at autophagosomal membranes, in the presence and absence of NLRP3 activators, inducing autophagy which the authors indicate increases degradation of NLRP3 and pro-IL-1β. In this regard multiple NLRP3 inflammasome activators such as silica and MSU should be inhibited by arglabin; however, this was not tested.

Cholesterol crystals are a common inflammasome trigger in atherosclerosis. The observed cytokine inhibitory effect of arglabin in response to cholesterol crystals therefore prompted investigation of in vivo efficacy in $ApoE_2$.Ki mice expressing human $ApoE_2$ (2/2) (Abderrazak et al. 2015). These mice are predisposed to develop atherosclerotic plaques particularly when fed an atherogenic diet. A twice daily low dose of arglabin (2.5 ng/g i.p.) over 13 weeks effectively reduced the plasma levels of IL-1β by approximately 50%. A 59% reduction of total cholesterol, 42% reduction in triglycerides and 44% reduction in autoantibodies against oxLDL were measured relative to vehicle-treated mice. Moreover, a switch in macrophage phenotype from pro-inflammatory M1 to the anti-inflammatory M2 was evident in spleen and arterial lesions. Both aortic sinus and whole aorta en face in arglabin-treated mice were reduced to the same level as was observed in $ApoE_2$. Ki/NLRP3 $^{-/-}$ mice fed atherogenic diet. No differences were measured in LDL receptor expression, hepatic steatosis or cholesterol biosynthesis. In an additional study, in the same murine model, arglabin attenuated plasma levels of glucose (20% reduction) and insulin (50% reduction) (Abderrazak et al. 2016). Overall, these models had a positive outcome, but much more work remains to fully delineate the mechanisms behind the observations.

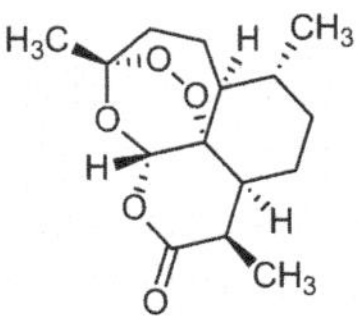

Artemisinin is a sesquiterpene lactone antimalarial drug which occurs naturally in *Artemisia annua* (sweet wormwood) and has formed the basis for many similar analogs (Dai et al. 2017). The drug is generally well tolerated and active in multiple models of disease (Dai et al. 2017) including inflammatory conditions: post-infarct myocardial remodelling, EAE murine model of multiple sclerosis and lupus nephritis. One identified mode of action is via inhibition of the NF-κB pathway to modulate the immune response, as observed in human astrocytoma T67 cells (Aldieri et al. 2003) and also in microglia (Zhu et al. 2012). This mode of action commonly results in NLRP3 inhibitory activity as recognised in a transgenic mouse (APPswe/PS1dE9) model of Alzheimer's disease (Shi et al. 2013). Once daily, administration of artemisinin (40 mg/kg i.p.) diminished amyloid plaques in the cortex by 48% and in the hippocampus by 61%. Examining the underlying mechanisms established no impairment of amyloid-β transport across the blood-brain barrier, but amyloid precursor peptide cleavage was impaired via inhibition of the required enzyme β-secretase (BACE1) (Shi et al. 2013). NF-κB expression and translocation to the nucleus is integral to the expression of BACE1, and this was significantly impaired by artemisinin. The NLRP3 priming phase also involves NF-κB nuclear translocation, and accordingly NLRP3 pathways were suppressed alongside a subsequent reduction in cleaved caspase-1 and IL-1β production. The NLRP3 modulatory activity of artemisinin was also identified in a murine model of burn sepsis (Long et al. 2016). The artemisinin class of molecules may find an expanded range of therapeutic application in the future.

15.3.4 *Flavonoids*

Flavones are widely distributed throughout the human diet and may contribute to the health benefits of fruit and vegetable consumption. This molecular class has therefore attracted much interest. After consumption, typical peak plasma flavone concentrations vary, and isoflavones tend to be more bioavailable, for example, soy isoflavones (and citrus flavaonones) peaked at 10 μM; however, this is exceptional, and <1 μM is more common (Higdon et al. 2005). Flavones are poorly absorbed, extensively metabolised and readily excreted leading to low bioavailability (Higdon et al. 2005). Moreover, there are numerous effects on cell signalling. In recent years there have been an increasing number of studies on inflammasome modulation by flavones in vitro and in vivo. Previous studies of the anti-inflammatory properties of flavonoids indicated the C2–C3 alkene, and the position of the hydroxyl groups is important to the anti-inflammatory potency of these compounds.

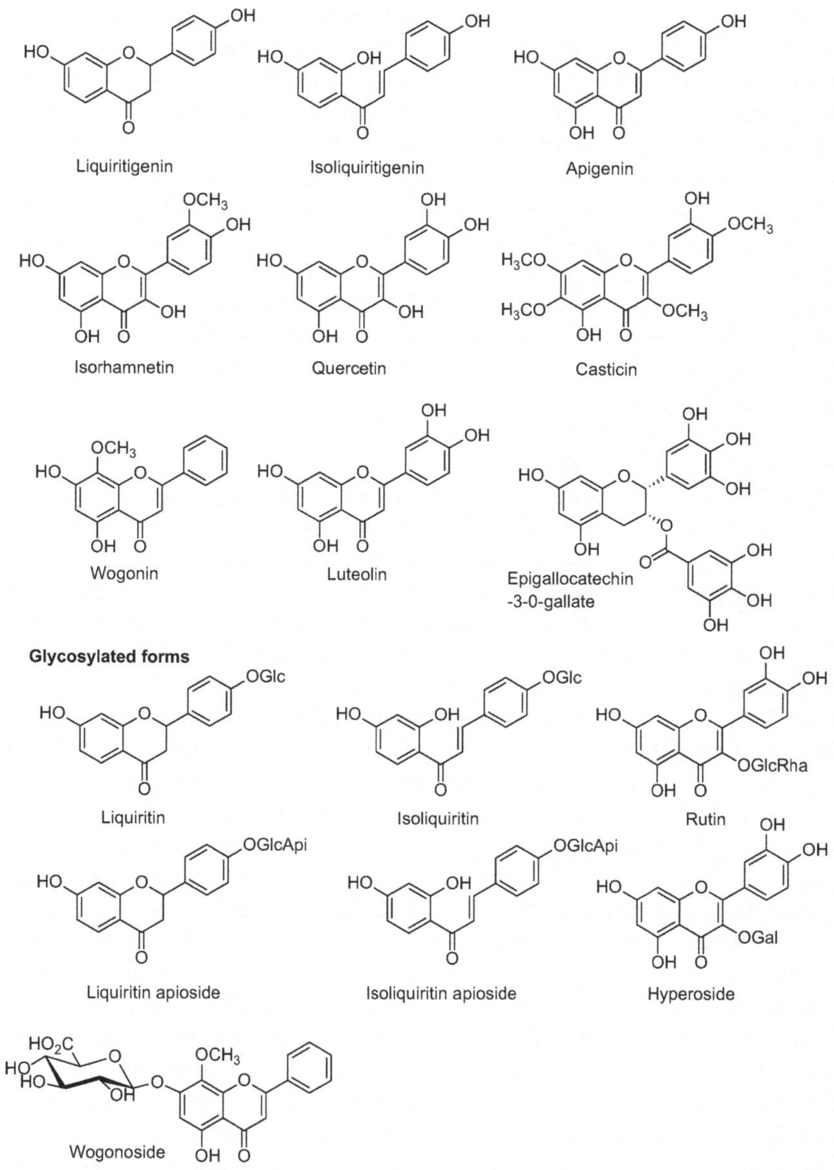

The flavonoid liquiritigenin, the closely related chalcone isoliquiritigenin, and a saponin, glycyrrhizin, were isolated from *Glycyrrhiza Uralensis* of the licorice plant family. Both isoliquiritigenin and glycyrrhizin blocked TLR4/MD-2 complex and also IKK, preventing NF-κB activation and therefore inflammasome priming. Isoliquiritigenin was more potent than glycyrrhizin and showed selectivity over AIM2 inflammasome which glycyrrhizin did not. Both compounds inhibit priming and activation steps required for NLRP3 inflammasome formation. Using murine

BMDM with inflammasome stimulation (LPS priming followed by ATP, MSU or nigericin trigger), NLRP3-derived IL-1β secretion was reduced by around 50%, as compared to vehicle control, by 1–10 μM isoliquiritigenin, while 1 mM glycyrrhizin was required to elicit the same response. Parthenolide was included as a control in these assays, showing approximately similar response to isoliquiritigenin and the related flavonoid liquiritigenin. In contrast, glycosylated forms of either isoliquiritigenin or liquiritigenin (liquirtin, isoliquirtin, liquirtin etoposide, isoliquirtin etoposide), also found in *G. uralensis*, were ineffective. This agrees with reports that deglycosylation of flavones led to an improvement of anti-inflammatory activity (Hostetler et al. 2012). Other studies on the anti-inflammatory effect of isoliquiritigenin found an induction of regulatory T cells in vitro and in vivo (Guo et al. 2015). In adipose tissues inflammation was suppressed by both inflammasome-dependent and inflammasome-independent mechanisms (Watanabe et al. 2016). Therefore, effects of flavones in modulating inflammation are complex and not purely inflammasome driven.

A related flavonoid, apigenin, commonly found in fruit, vegetables and herbs such as camomile, has structural similarity to the aforementioned liquiritigenin. It is perhaps unsurprising that the anti-inflammatory effects on the NLRP3 pathway were similar, although it is difficult to compare relative potency as these were not examined in the same cell lines. Apigenin inhibited IL-1β secretion from THP-1 cells with a potency of around 25 μM (Zhang et al. 2014). Both NLRP3 activation and priming were inhibited by apigenin in THP-1 and J774A.1 cells. Additional studies, based on results of quantitative real-time PrimePCR array, showed a multifactorial effect with more than 24 genes upregulated, while TLR4, CCL5, ICAM1 and VACM1 amongst others were downregulated. Notably apigenin prevented an LPS-induced reduction of anti-inflammatory cytokine IL-10. Apigenin inhibited activation of NF-κB and prevented ASC speck formation and ultimately caspase-1 maturation. Cytokine mRNA stability was also impaired through modulation of post-translational processing. Recent research has indicated a likely link between NLRP3 activation, stress and major depressive disorder (Kaufmann et al. 2017). Apigenin was tested in vivo to determine efficacy in a rat model of chronic unpredictable mild stress (Li et al. 2016b). A dose of 20 mg/kg intragastrically normalised sucrose consumption, indicative of antidepressant effect, improved behavioural symptoms and reduced microglial activation. Furthermore, oxidative stress, IL-1β, IL-18 and expression of NLRP3, caspase-1 and ASC proteins were all significantly reduced to levels close to the vehicle-only group.

Isoliquiritigenin was successfully tested in a rat model of intracerebral haemorrhage, and attempts were made to elucidate the mechanism of action (Zeng et al. 2017). There are currently no effective therapeutics for this debilitating disorder, but anti-inflammatory approaches hold significant promise. NLRP3 knockout or inhibition prevents brain damage, while the Nrf2 anti-oxidant pathway can prevent release of reactive oxygen species, a known NLRP3 inflammasome trigger. The rat model of collagenase type IV-induced intracerebral haemorrhage was conducted using three experiments (Zeng et al. 2017). The first experiment used

5 groups of 36 rats each: sham, vehicle only and isoliquiritigenin 10, 20, 40 mg/kg. The 10 mg/kg group generally showed no significant effect, whereas 20 and 40 mg/kg were efficacious to a similar degree. Behavioural deficits (motor, sensory, balance and reflex) and histological effects were improved, while haematoma volume, brain oedema and BBB permeability reduced. The second experiment was designed to examine the mechanisms of therapeutic effect using 4 groups of 30 rats: sham, intracerebral haemorrhage only, vehicle treated and isoliquiritigenin 20 mg/kg treated. This experiment confirmed isoliquiritigenin activated Nrf2 expression and nuclear translocation. The authors speculate that isoliquiritigenin may alkylate reactive cysteine residues responsible for stress sensing in the Nrf2 partner protein kelchlike ECH-associated protein 1 (Keap1) (Zeng et al. 2017). Given the reactive nature of these flavonoid molecules, it is possible they also alkylate proteins of the inflammasome pathway, but this has not been investigated thus far. Isoliquiritigenin suppressed both the NF-κB pathway and NLRP3 inflammasome proteins (NLRP3, ASC, pro-caspase-1, pro-IL-1β and pro-IL-18) (Zeng et al. 2017). In agreement with downregulation of the inflammasome pathway, the secretion of active IL-1β and IL-18 was significantly ablated. The third experiment used siRNA for Nrf2 and also co-administration with isoliquiritigenin. The Nrf2 siRNA group were notably impaired with exacerbation of all aforementioned measures of brain injury and inflammatory markers. This was alleviated in the isoliquiritigenin treatment arm (Zeng et al. 2017). This experiment provided a detailed examination of isoliquiritigenin's therapeutic potential in this CNS disorder giving strong evidence on which to base additional studies. The authors acknowledge several limitations which will need to be addressed and that they mainly considered inflammasome pathways in these experiments. Also, collagenase may itself trigger inflammatory responses.

Isoliquiritigenin (0.5% wt/wt in chow) showed significant promise in models of adipose tissue inflammation (Honda et al. 2014; Watanabe et al. 2016), typically observed in type 2 diabetes, where C57BL/6 mice were fed a high-fat diet over a 20-week period (Honda et al. 2014). Serum cholesterol, triglycerides, leptin and insulin levels were reduced, and an improvement in insulin sensitivity was evident. Although isoliquiritigenin-treated mice consumed slightly more food, they gained less weight and hepatic steatosis was alleviated. Considering adipose tissue inflammation, there were significantly less crown-like structures formed around adipocytes, indicative of a reduction in inflammatory cells. Accompanying this, there was also a reduction in the expression of inflammatory genes TNF-α, IL-6 and MCP-1 and an increase in adiponectin. Looking at modulation of inflammasomes, IL-1β and caspase-1 production were markedly reduced at the 20-week time point but also at the much earlier 4-week stage where significant elevation of these components was observed in the vehicle-only control groups. Isoliquiritigenin had clearly beneficial effects in this murine obesity model; however, not all of these are due to inhibition of NLRP3 inflammasome. In similar overnutrition model of NASH, NLRP3 inhibition by MCC950, did not alleviate hepatic steatosis and had no effect on the metabolic aspects of disease (Mridha et al. 2017).

Wogonoside, isolated from the flowering plant *Scutellaria baicalensis Georgi* (Chinese skullcap), inhibits IL-1β secretion from LPS-treated THP-1 cells triggered with ATP with an IC_{50} around 50 μM (Sun et al. 2015). NF-κB nuclear translocation and DNA binding was inhibited along with expression of pro-IL-1β and NLRP3. In addition, active caspase-1 was suppressed which coincided with reduction in secreted IL-1β. Wogonoside (12.5, 25, 50 mg/kg, intragastrically) suppressed dextran sodium sulfate-induced colitis in a dose-dependent manner (Sun et al. 2015). A similar dosing level was also investigated in a rat model of spinal cord injury and reported to alleviate associated inflammation via suppression of NF-κB and inhibition of NLRP3 inflammasome (Yonglin et al. 2017). Wogonoside (10, 20, 40 mg/kg) dose dependently increased survival of BALB/c mice in a model of lipopolysaccharide (LPS), and D-galactosamine induced liver injury by activating Nrf2 and inhibiting NLRP3 in a similar manner to other flavonoids (Gao et al. 2016).

Casticin (vitexicarpin) alleviated LPS-induced acute lung injury; similar to other flavonoids, the mechanism was attributed to suppression of NF-κB and inhibition of NLRP3 inflammasome (Wang et al. 2016). Rutin, a disaccharide form of quercetin, proved to modulate NLRP3 inflammasome in three models of inflammatory disease: pancreatitis, spinal cord injury and endothelial dysfunction (Wu et al. 2016; Wang et al. 2017c; Aruna et al. 2014). Quercetin, luteolin and epigallocatechin gallate also modulated NLRP3 inflammasome in vivo (Wu et al. 2014; Jiang et al. 2016b; Wang et al. 2013; Fu et al. 2017). Epigallocatechin gallate was additionally reported to inhibit NLRP1 inflammasome in a study considering melanoma growth (Ellis et al. 2011).

Isorhamnetin and hyperoside (the 3-*O*-galactoside of quercetin) isolated from water dropwort (*Oenanthe javanica*) were tested using LPS-primed murine BMDM against inflammasome triggers for NLRP3, NLRC4 and AIM2 (Ahn and Lee 2017). Isorhamnetin attenuated IL-1β, IL-18 and cleaved caspase-1 with NLRP3 (ATP, nigericin, alum) and AIM2 (dsDNA) triggers, but these were enhanced in the case of NLRC4 trigger (flagellin). Pro-IL-1β, TNF-α, IL -6 and NLRP3 expression, in response to LPS, was prevented by isorhamnetin. Hyperoside however, did not affect the expression of inflammasome proteins; this molecule inhibited the activation of NLRC4 inflammasome and AIM2 but not NLRP3 inflammasome (Ahn and Lee 2017). This work raises the possibility that other glycosylated flavones may be active against AIM2 and NLRC4.

Efficacy of isoliquiritigenin, and its close structural analogs, in various disease models was undoubtedly due to intriguing multimodal action. In some cases the molecules were studied mainly for effect on NLRP3 inflammasome and much remains to be elucidated about the effects on other inflammasome pathways. It is difficult to compare activity of these molecules across the varied studies. Extensive metabolism of the flavones (Higdon et al. 2005) has not been taken into account and it may be that some of the activity observed is due to metabolites. This would be particularly relevant in the liver and intestine where extensive metabolism of flavonoids occurs. It would be interesting to investigate inflammasome inhibitory activity of the flavone class more fully to understand the preliminary SAR arising from this series. Even at a cellular level, to test these compounds and their metabolites side by side would give an indication of their relative potency on inflammasome pathways. However, cellular and in vivo conclusions need to be carefully viewed in the context of bioavailability and likely in vivo concentrations (Cassidy and Minihane 2017). Given the worldwide human consumption of flavonoid natural sources, this class of molecules is certainly of interest.

15.3.5 Quinones

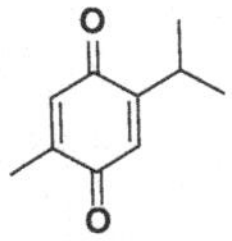

Thymoquinone

Thymoquinone is a component of black cumin seed (*Nigella sativa*) and commonly studied as an anticancer phytochemical. As is common with many natural products, thymoquinone has multiple modes of action and associated biological effects including anti-inflammatory and immunomodulatory properties (Khan et al. 2017). In a study of mouse (B16F10) and human (A375) metastatic melanoma cell migration, thymoquinone dose dependently retarded migratory ability in both cell lines (Ahmad et al. 2013). At 40 μM thymoquinone, cell proliferation was ablated; therefore a top concentration of 20 μM was used. Cell migration of B16F10 cells was increased in the presence NLRP3 inflammasome stimuli LPS and ATP. Thymoquinone suppressed NLRP3 inflammasome protein expression and NF-κB activity, while proteolytic cleavage of caspase-1 and subsequent secretion of IL-1β and IL-18 were dose dependently reduced. Thymoquinone can function as both an oxidant and anti-oxidant with some studies suggesting the anti-oxidant activity is prevalent at low concentrations and oxidant activity is promoted at high concentrations. The anti-oxidant effect of thymoquinone on NLRP3 inflammasome activity, and cell migration was explored by testing the cell migration assay in the presence of a reactive oxygen species inhibitor *N*-acetyl-L-cysteine. This did decrease cell migration but to a lesser extent than thymoquinone the authors therefore concluded that thymoquinone (anti-inflammatory) activity was not purely due to anti-oxidant effect. While these experiments show thymoquinone did perturb NLRP3 pathways and retard cell migration, the link between the two is not unequivocal. Thymoquinone was also examined in vivo for NLRP3 related effects in a murine model of cancer cell migration. B16F10 cells were injected into the tail, and tumour nodule formation in the lungs was used as a readout. Treatment with thymoquinone dose dependently reduced nodules from 15 to 1 (Periyanayagam et al. 2015).

The effect of thymoquinone on pancreatitis has also been investigated with respect to NLRP3 mediated effects (Periyanayagam et al. 2015). Thymoquinone administration (100 mg/kg p.o. for 60 days), in a murine model of pancreatic inflammation, resulted in a decrease in oxidative stress markers and IL-1β, IL-18 and TNF-α mRNA levels. Disappointingly, NLRP3 was not measured. Evidence including interstitial oedema inflammation and inflammatory cell infiltration, parenchymal cell necrosis and haemorrhage also indicated improvement of pancreatitis.

15.3.6 Stilbenoids and Close Analogs

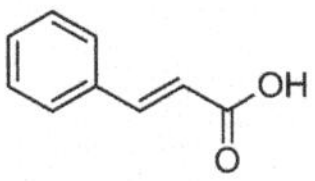

Cinnamic acid

Cinnamic acid is naturally found in cinnamon and has a Michael acceptor functionality commonly found in many synthetic NLRP3 inhibitory molecules (BAY11-7082, MNS, acrylate derivatives). It would be reasonable to assume cinnamic acid would have potential to covalently modify proteins on the NLRP3 inflammasome pathways, but this remains to be proven. Cinnamic acid was tested for efficacy in LPS-induced endotoxin-poisoned mice (Xu et al. 2017). Neutrophil infiltration was reduced significantly, alongside reduction of IL-1β, IL-18 and TNF-α in serum. NLRP3, caspase-1, IL-1β mRNA, protein of NLRP3 and cleaved caspase-1 were similarly reduced. These data gave early indication that cinnamic acid may have NLRP3 modulatory effects.

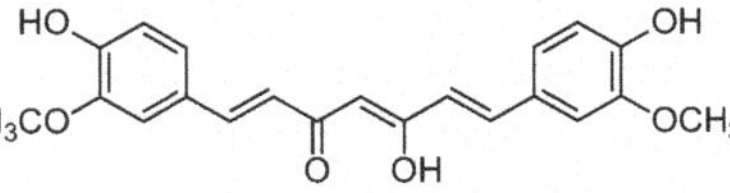

Curcumin

Curcumin is a major bioactive constituent of turmeric which has seen an upsurge in popularity in the superfoods arena. Like so many natural products, multiple modes of action have been identified and many disease models studied. One recent study of the biological mode of action used a murine model of osteoarthritis, where destabilisation of the medial meniscus (DMM) surgery was used and addressed to some degree in vivo modulation of inflammasome (Sun et al. 2017). Curcumin (50 μM, i.p.) retarded disease progression and reduced expression of IL-1β, IFN-γ, IL-17A, IL-18, TNF-α and VCAM1 mRNA. In LPS-primed THP-1 cells, activated with ATP, curcumin (10 μM) suppressed IL-1β and TNF-α at both mRNA and protein level, while cleaved caspase-1 levels were reduced to baseline. An earlier study found curcumin suppressed the TLR4, MD88, NFκB and P2X7R pathways in murine macrophages (Kong et al. 2016). Other murine models have also been reported with investigation of NLRP3 involvement such as chronic kidney disease (Bugyei-Twum et al. 2016), LPS-induced septic shock (Gong et al. 2015) and diabetic nephropathy (Lu et al. 2017) amongst others. Whether the reactive Michael acceptor functionality is integral to the NLRP3 activity, as it is in many other NLRP3 inhibitory compounds, has not yet been defined.

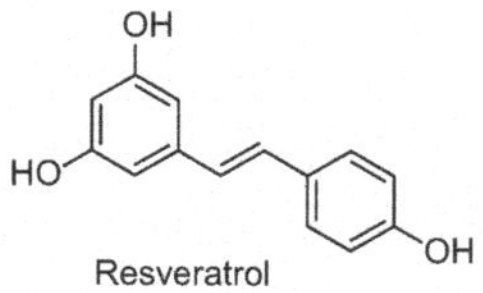

Resveratrol

Resveratrol is a stilbenoid which commonly occurs in the skin of grapes and other berries. This polyphenolic compound has been subjected to a plethora of studies illustrating its bioactivity in a wide range of disease models. However, while resveratrol is well absorbed (~70%), it has poor oral bioavailability, at just 0.5%, due to extensive metabolism (Walle et al. 2004). In 2013, a role for resveratrol in modulation of NLRP3 inflammasome was identified while studying pathways involved in radiation injury (Fu et al. 2013). Resveratrol activated Sirt1 deacetylase preventing its activity in transactivation of NF-κB and subsequent effect on NLRP3 transcription (Fu et al. 2013). These effects were also observed in murine microglia (Sui et al. 2016). Additional studies indicated resveratrol inhibited NLRP3 inflammasome assembly as well as priming. NLRP3 cell-based inhibition for resveratrol had an IC_{50} ~15 μM using J774A.1 macrophages (Chang et al. 2015). However resveratrol is non-selective as it also inhibits NLRP1 and NLRC4 inflammasomes (Chang et al. 2015). Resveratrol prevented mitochondrial damage and therefore also averted release of mitochondrial reactive oxygen species and translocation of mitochondrial DNA into the cytosol (Chang et al. 2015). This protective mechanism prevented activation of NLRP3. In addition, the autophagy marker LC3B-11 was increased by resveratrol; this pathway increases degradation of NLRP3 and pro-IL-1β, hence suppressing NLRP3 inflammasome activity (Chang et al. 2015). NLRP3 activity has been assessed in vivo using murine models of renal inflammation (Chang et al. 2015), LPS-induced acute lung injury (Jiang et al. 2016a), adipose tissue inflammation (Li et al. 2016a) and sepsis-associated encephalopathy (Chang et al. 2015; Sui et al. 2016) amongst many others. Efficacy in human disease has also been tested; indeed 130 clinical trials are listed in clinicaltrials.gov with 17 of these at phase III.

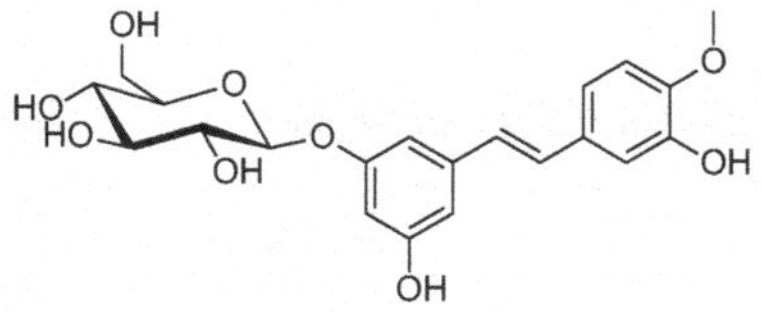

Rhaponticin

Rhaponticin is a glycosylated stilbenoid naturally occurring in rhubarb rhizomes. The aglycone is closely related to resveratrol and has very similar reported activity as an activator of Sirt1 (Wei et al. 2017). No investigation has yet been made with this compound into the mitochondrial protective effects or autophagy. Rhaponticin (20–100 mg/kg, p.o.) was protective in a dextran sodium sulfate-induced murine model of colitis (Wei et al. 2017). These results should be confirmed through cell-based studies to support the in vivo findings, and establish potency and selectivity.

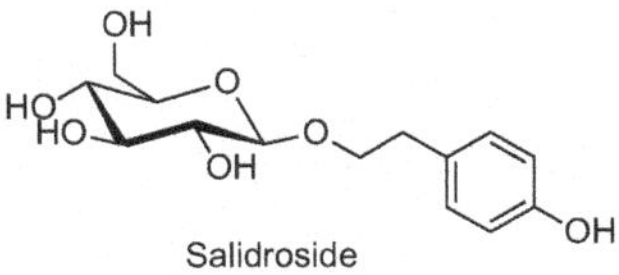

Salidroside

Salidroside is a glycosylated form of tyrosol which occurs in Rhodiola rosea (golden root) and is typically regarded as antidepressant and anxiolytic. Recent studies have shown beneficial effects in diabetic nephropathy (Wang et al. 2017b) and ventilation induced lung injury (Wang et al. 2017d). Similar to resveratrol and rhapontin, salidroside is an activator of Sirt1 thereby preventing its activity in transactivation of NF-κB and subsequent effect on NLRP3 transcription (Wang et al. 2017d). These results should be confirmed through cell-based studies to support the in vivo findings and establish potency and selectivity.

15.3.7 Steroids

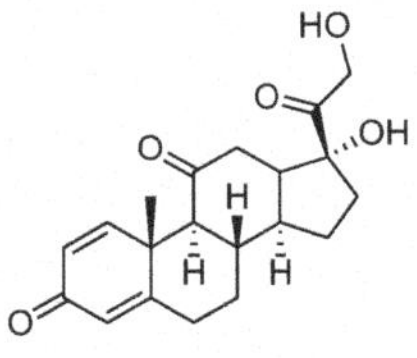

Prednisone

Prednisone (10 mg/kg, p.o.), a potent anti-inflammatory drug, in a mouse model of cuprizone-induced demyelination gave early indications of NLRP3 modulatory effects (Yu et al. 2017a). Prednisone reduced microglial and astrocyte activation and protein levels of NLRP3, IL-1β and active caspase-1. Further work is required to support the in vivo findings, establish potency and selectivity.

15.3.8 Pentacyclic Natural Products

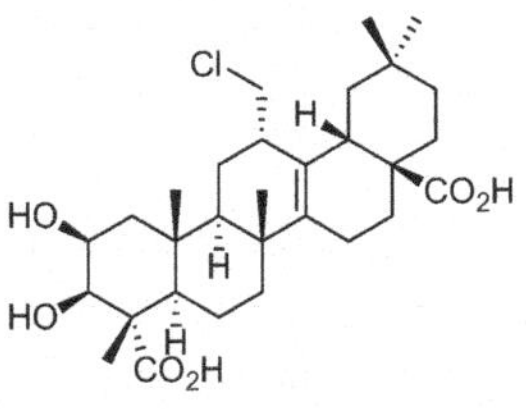

Senegenin

Senegenin (tenuigenin) is isolated from the traditional Chinese medicinal herb *Polygala tenuifolia* known to act as an antidepressant; effects were confirmed in a murine model of chronic unpredictable mild stress, and for the first time, these effects were linked to NLRP3 inhibition (Li et al. 2017). NLRP3-mediated IL-1β secretion was prevented through inhibition of the NF-κB pathway as determined through hippocampal phosphorylation levels of NF-κB by Western blot and immunohistochemistry. These early results require further investigation.

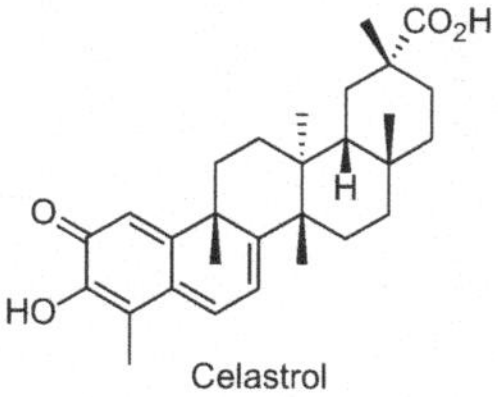

Celastrol

Celastrol is isolated from the roots of *Tripterygium wilfordii* (thunder god vine) and in 2009 was successfully tested, as an extract, in clinical trials for rheumatoid arthritis (Goldbach-Mansky et al. 2009). Reports of celastrol's IL-1β inhibitory properties prompted investigation into potential NLRP3 inhibitory action in an, LPS-stimulated, J774A.1 macrophage cell line with ATP as NLRP3 inflammasome trigger (Xin et al. 2017). IL-1β, IL-18 and TNF-α were all inhibited by celastrol in a dose-dependent manner, but the effect was not particularly potent with IC_{50} >50 μM. Levels of NLRP3, caspase-1 p10 and pro-IL-1β were decreased by celastrol, but again the effect was not potent. Pyroptotic cell death is typical of NLRP3 activation and was reduced by celastrol as measured by LDH release assay. Celastrol inhibited reactive oxygen species, through its well-known anti-oxidant activity, and also prevented NF-κB activation (Xin et al. 2017). A subsequent study also identified celastrol as an NLRP3 inhibitor and provided considerably more detail (Yu et al. 2017b). It is interesting to note this study used LPS-primed peritoneal macrophages stimulated with ATP where the inhibitory effect was much more convincing with IC_{50} <125 nM. No effect was observed on TNF-α if celastrol was added after LPS stimulation. This study also found celastrol could enhance autophagy (which is known to inhibit NLRP3 inflammasome activation) and prevent ASC oligomerisation and formation of the NLRP3 complex. Unfortunately, there was little assessment of compound selectivity in vitro. Promising activity was observed in vivo in murine models of LPS-induced septic shock where celastrol was administered at 1 mg/kg i.p. and also in dextran sodium sulfate-induced colitis (compound dose was not recorded). These results are promising and warrant further investigation.

15.3.9 Alkaloids

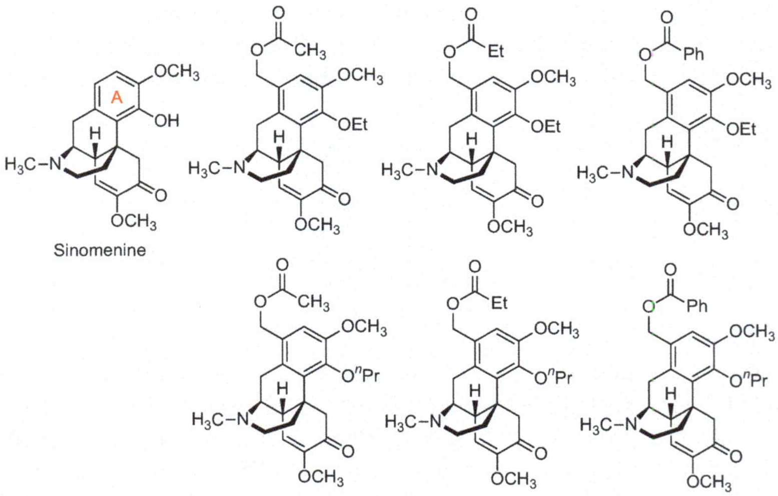

Sinomenine (cocculine) is an alkaloid, of similar structure to morphine, isolated from *Sinomenium acutum* and traditionally used to treat rheumatism, arthritis and neuralgia. Sinomenine (10–20 mg/kg i.p.) proved protective in a mouse model of ischaemic stroke. In this model, a decreased expression of NLRP3 and ASC was found, while cleaved caspase-1 and IL-1β levels were reduced, along with IL-6, IL-18 and TNF-α (Qiu et al. 2016). Sinomenine (30 mg/kg) was also efficacious in a mouse model of traumatic brain injury where efficacy was attributed to activation of the Nrf2 anti-oxidant response-element pathways known to negatively regulate NF-κB and NLRP3 inflammasome (Yang et al. 2016). A number of analogs have been synthesised which were modified in the A ring of sinomenine leaving the remainder of the molecule unchanged; these proved to be more potent inhibitors of cellular IL-1β release when tested at 10 μg/mL (Zhao et al. 2015).

15.4 Other Possibilities for Indirect NLRP3 Inhibitors

15.4.1 Kinase Inhibitors as Indirect NLRP3 Inhibitors

Multiple kinases have been identified which regulate NLRP3 inflammasome components and hence NLRP3 complex formation (Neumann and Ruland 2013). Untangling and understanding this inflammasome kinome network is only just beginning. Given the strong drug discovery efforts in cancer therapy with kinase inhibitors, there is a promising foundation on which to find advanced leads for indirectly targeting inflammasomes.

In 2013, Hara et al. (2013) identified Syk and Jnk kinases were important to NLRP3 inflammasome activation. Syk kinase is thought to phosphorylate ASC, an event which is crucial for oligomerisation (Lin et al. 2015). Hara et al. used a chemical biology approach, where a series of kinase inhibitors were tested to identify which prevented release of IL-1β. R406, the active form of prodrug fostamatinib, and BAY61-3606 were Syk kinase inhibitors identified in this study (Hara et al. 2013). Prior to this finding, fostamatinib had already been in clinical development as a rheumatoid arthritis therapeutic, but trials were terminated after phase IIb where primary and secondary endpoints were not achieved (Ron Leuty 2013). It is worth noting the covalent modifier of NLRP3 inflammasome components 3,4-methylenedioxy-β-nitrostyrene is also a known inhibitor of Syk kinase and inhibits NLRP3 however it is not particularly drug-like (as previously discussed) (He et al. 2014). Hara also discovered the small molecule AP600125 and peptidic TAT-JI_TIP Jnk inhibitors were effective inflammasome inhibitors (Hara et al. 2013). Both the Syk and Jnk inhibitor effects were supported through extensive studies including mutation of active-site residues in each kinase. There are many other published and commercial Syk and Jnk kinase inhibitors [nilvadipine (Paris et al. 2014), TAK659 (Liu and Mamorska-Dyga 2017), bentamapimod (Gehringer et al. 2015), tanzisertib (Gehringer et al. 2015)] which may now be re-examined for possible effects on inflammasome activity.

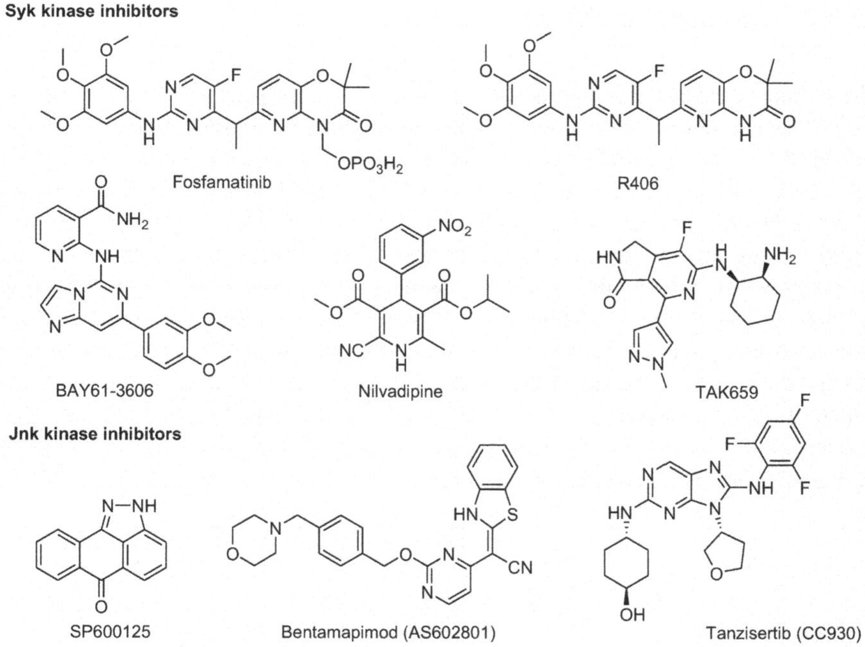

Further kinases have been found that regulate NLRP3 inflammasome assembly: PKR (Boriushkin et al. 2016; Yim and Williams 2014), Lyn (Shio et al. 2009), BTK (Ito et al. 2015), TAK1-Jnk (Okada et al. 2014), PI(3)K (Ives et al. 2015), DAPK (Lai and Chen 2014), IRAK1/4 (Lin et al. 2014; Fernandes-Alnemri et al. 2013) and, in 2016, Pyk2 (Chung et al. 2016) and NEK7 (Schmid-Burgk et al. 2016). NEK7 is particularly intriguing as its catalytic phosphorylation activity is dispensable for its action on NLRP3. Shi et al. (2016) reported on the NLRP3-NEK7 axis in more detail showing NEK7 is a selective upstream regulator of NLRP3 inflammasome activation. NEK7 interacts directly with the leucine-rich repeat domain of NLRP3, in a kinase-independent manner, and this association is required for inflammasome assembly. NEK7 is very specific as it is not involved in NLRC4 (another LRR domain containing inflammasome) or AIM2 (lacks an LRR domain) activation nor toll-like receptor (TLR) responses. Moreover the roles of NEK7 as mitotic kinase and NLRP3 regulator are mutually exclusive, i.e. mitosis and NLRP3 inflammasome activation cannot occur simultaneously. Understanding and disrupting the NEK7-NLRP3 interaction could give a novel means by which to target NLRP3 activity. Little is known about how most kinases are interacting with NLRP3, and other inflammasomes have been even less studied. Through understanding the kinase networks surrounding these systems, their full potential can be realised and appropriate kinase inhibitors may be developed or repurposed as anti-inflammatory drugs.

15.4.2 Caspase Inhibitors

The ultimate goal of the inflammasome is activation of caspase-1 such that cytokine processing can occur. It is a viable therapeutic strategy to inhibit caspase-1 directly; indeed there are a significant number of caspase-1-targeted molecules in the published and patent literature (MacKenzie et al. 2010; Lee et al. 2017). Examples of these include pralnacasan (VX740), emricasan and VX765. Human caspases 4 and 5 (murine ortholog is caspase-11) are also of significant interest as drug targets; these directly interact with LPS via their CARD domain to trigger activation of their protease function (Shi et al. 2014); this is termed the non-canonical inflammasome pathway. However, despite many years of research, caspase inhibitors have not yet succeeded in the clinic. There are significant challenges beyond selectivity issues; caspase inhibitors have typically suffered from toxicity and poor pharmacokinetics, but a keen interest remains, and it is likely a drug will emerge from this effort, perhaps with the aid of novel methods of delivery (Lee et al. 2017).

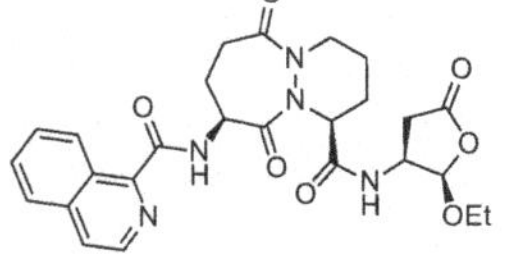
Pralnacasan

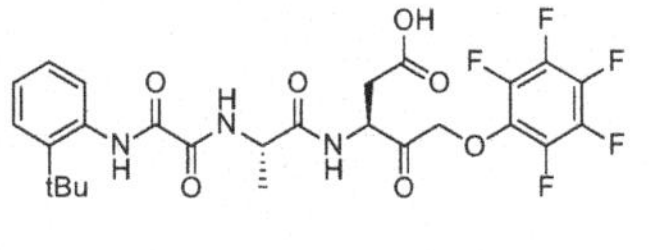
Emricasan

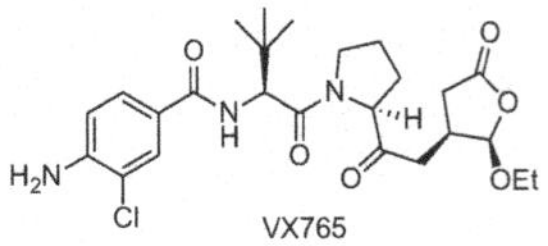
VX765

15.4.3 Ion Channels, Reactive Oxygen Species and Lysosomal Destabilisation

There are a number of studies linking Ca^{2+} to inflammasome triggering. Interestingly, increase in cellular Ca^{2+} can trigger NLRP3 activation via an unknown mechanism; in contrast, blocking calcium channels prevents NLRP3 activation (Yaron et al. 2015; Rada et al. 2014). This is a possible avenue for therapeutic intervention where calcium channels could be targeted using known inhibitors such as nilvadipine (also a known Syk inhibitor).

Lysosomal rupture leads to generation of reactive oxygen species (suggested to activate NLRP3), activating calcium-dependent ion channels. Calcium-dependent protein kinases (CaMkII) are also triggered upstream of the TAK1-JNK pathway, ultimately resulting in NLRP3 activation (Okada et al. 2014). Anti-oxidants and radical scavengers may prevent the impact of reactive oxygen species in triggering the inflammasome. There are many dietary examples of anti-oxidants in natural products as already covered. Lysosomal damage also involves release of protease cathepsin B, and this is believed to trigger NLRP3 inflammasome. There are inhibitors of cathepsin B such as E-64 (Terada et al. 2010) or Ca-074 (Terada et al. 2010) which may provide promising leads as NLRP3 inflammasome inhibitors. Although it is worth noting, these epoxide-containing compounds are likely to react, non-specifically, with biological nucleophiles.

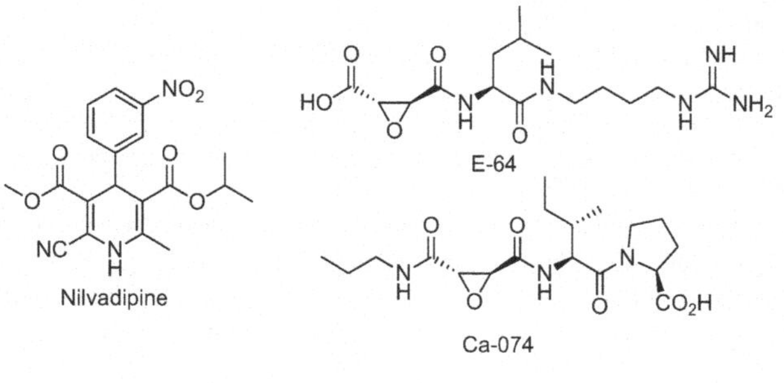

15.5 Conclusions

As the knowledge of inflammasome biology forges ahead, chemists are beginning to use this to build a substantial pipeline of inhibitory molecules as tools and also as therapeutics. There are clearly many avenues towards inhibition of inflammasome activity, and the molecules thus far published tend to exploit more than one. This is especially true of the natural products and reactive compounds which covalently modify biological nucleophiles. The most well-characterised NLRP3 inhibitors are sulfonylureas and the number of publications and patents in this class continually increase. In contrast, many of the inhibitors described herein require significantly more work to characterise their mode of action, increase their potency and optimise their pharmacokinetic properties before hope of successful translation. That said, the illustrated preliminary efficacy in models of inflammatory disease gives promise for the future. The recent success for Novartis with Ilaris (canakinumab), a biologic which target IL-1β, in meeting the endpoints of the 6-year 10,061 patient CANTOS trial has further excited the inflammasome field. Indeed a number of established large pharmaceutical companies and new companies are in active pursuit of inflammasome inhibitors including Inflazome, IFM Therapeutics, Olatec, Nodthera and Selvita, and the race to bring molecules to the clinic is well underway.

References

Abderrazak A, Couchie D, Mahmood DF, Elhage R, Vindis C, Laffargue M, Mateo V, Buchele B, Ayala MR, El Gaafary M, Syrovets T, Slimane MN, Friguet B, Fulop T, Simmet T, El Hadri K, Rouis M (2015) Anti-inflammatory and antiatherogenic effects of the NLRP3 inflammasome inhibitor arglabin in ApoE2.Ki mice fed a high-fat diet. Circulation 131(12):1061–1070. https://doi.org/10.1161/CIRCULATIONAHA.114.013730

Abderrazak A, El Hadri K, Bosc E, Blondeau B, Slimane MN, Buchele B, Simmet T, Couchie D, Rouis M (2016) Inhibition of the inflammasome NLRP3 by arglabin attenuates inflammation, protects pancreatic beta-Cells from apoptosis, and prevents type 2 diabetes mellitus development in ApoE2Ki mice on a chronic high-fat diet. J Pharmacol Exp Ther 357(3):487–494. https://doi.org/10.1124/jpet.116.232934

Ackermann TF, Boini KM, Beier N, Scholz W, Fuchss T, Lang F (2011) EMD638683, a novel SGK inhibitor with antihypertensive potency. Cell Physiol Biochem 28(1):137–146. https://doi.org/10.1159/000331722

Administration USFaD (2017a) Drugs@FDA: FDA approved drug products. New drug application 209176. https://www.accessdata.fda.gov/scripts/cder/daf/index.cfm?event=overview.process&ApplNo=209176. Accessed 15 Nov 2017

Administration USFaD (2017b) FDA approves drug to treat ALS. https://www.fda.gov/NewsEvents/Newsroom/PressAnnouncements/ucm557102.htm. Accessed 15 Nov 2017

Ahmad I, Muneer K, Tamimi I, Chang M, Ata M, Yusuf N (2013) Thymoquinone suppresses metastasis of melanoma cells by inhibition of NLRP3 inflammasome. Toxicol Appl Pharmacol 270:70–76. https://doi.org/10.1016/j.taap.2013.03.027

Ahn H, Lee G-S (2017) Isorhamnetin and hyperoside derived from water dropwort inhibits inflammasome activation. Phytomedicine 24:77–86. https://doi.org/10.1016/j.phymed.2016.11.019

Alcocer-Gomez E, Casas-Barquero N, Williams MR, Romero-Guillena SL, Canadas-Lozano D, Bullon P, Sanchez-Alcazar JA, Navarro-Pando JM, Cordero MD (2017) Antidepressants induce autophagy dependent-NLRP3-inflammasome inhibition in major depressive disorder. Pharmacol Res 121:114–121. https://doi.org/10.1016/j.phrs.2017.04.028

Aldieri E, Atragene D, Bergandi L, Riganti C, Costamagna C, Bosia A, Ghigo D (2003) Artemisinin inhibits inducible nitric oxide synthase and nuclear factor NF-kB activation. FEBS Lett 552(2–3):141–144. https://doi.org/10.1016/S0014-5793(03)00905-0

Aruna R, Geetha A, Suguna P (2014) Rutin modulates ASC expression in NLRP3 inflammasome: a study in alcohol and cerulein-induced rat model of pancreatitis. Mol Cell Biochem 396:269–280. https://doi.org/10.1007/s11010-014-2162-8

Ashcroft FM (2005) ATP-sensitive potassium channelopathies: focus on insulin secretion. J Clin Invest 115(8):2047–2058. https://doi.org/10.1172/JCI25495

Baldwin AG, Rivers-Auty J, Daniels MJD, White CS, Schwalbe CH, Schilling T, Hammadi H, Jaiyong P, Spencer NG, England H, Luheshi NM, Kadirvel M, Lawrence CB, Rothwell NJ, Harte MK, Bryce RA, Allan SM, Eder C, Freeman S, Brough D (2017) Boron-based inhibitors of the NLRP3 inflammasome. Cell Chem Biol 24(11):1321–1335. https://doi.org/10.1016/j.chembiol.2017.08.011

Barbaro NR, Fontana V, Modolo R, De Faria AP, Sabbatini AR, Fonseca FH, Anhe GF, Moreno H (2015) Increased arterial stiffness in resistant hypertension is associated with inflammatory biomarkers. Blood Press 24(1):7–13. https://doi.org/10.3109/08037051.2014.940710

Basiorka AA, McGraw KL, Eksioglu EA, Chen X, Johnson J, Zhang L, Zhang Q, Irvine BA, Cluzeau T, Sallman DA, Padron E, Komrokji R, Sokol L, Coll RC, Robertson AA, Cooper MA, Cleveland JL, O'Neill LA, Wei S, List AF (2016) The NLRP3 inflammasome functions as a driver of the myelodysplastic syndrome phenotype. Blood 128(25):2960–2975. https://doi.org/10.1182/blood-2016-07-730556

Boriushkin E, Wang JJ, Li J, Bhatta M, Zhang SX (2016) p58(IPK) suppresses NLRP3 inflammasome activation and IL-1beta production via inhibition of PKR in macrophages. Sci Rep 6:25013. https://doi.org/10.1038/srep25013

Brough D, Allan SM, Freeman S, Baldwin AG (2017) Cyclic diarylboron derivatives as NLRP3 inflammasome inhibitors. University of Manchester. WO/2017/017469

Brydges SD, Mueller JL, McGeough MD, Pena CA, Misaghi A, Gandhi C, Putnam CD, Boyle DL, Firestein GS, Horner AA, Soroosh P, Watford WT, O'Shea JJ, Kastner DL, Hoffman HM (2009) Inflammasome-mediated disease animal models reveal roles for innate but not adaptive immunity. Immunity 30(6):875–887. https://doi.org/10.1016/j.immuni.2009.05.005

Bugyei-Twum A, Abadeh A, Thai K, Zhang Y, Mitchell M, Kabir G, Connelly KA (2016) Suppression of NLRP3 inflammasome activation ameliorates chronic kidney disease-induced cardiac fibrosis and diastolic dysfunction. Sci Rep 6:39551. https://doi.org/10.1038/srep39551

Cassidy A, Minihane AM (2017) The role of metabolism (and the microbiome) in defining the clinical efficacy of dietary flavonoids. Am J Clin Nutr 105(1):10–22. https://doi.org/10.3945/ajcn.116.136051

Chang YP, Ka SM, Hsu WH, Chen A, Chao LK, Lin CC, Hsieh CC, Chen MC, Chiu HW, Ho CL, Chiu YC, Liu ML, Hua KF (2015) Resveratrol inhibits NLRP3 inflammasome activation by preserving mitochondrial integrity and augmenting autophagy. J Cell Physiol 230(7):1567–1579. https://doi.org/10.1002/jcp.24903

Chavarria-Smith J, Vance RE (2015) The NLRP1 inflammasomes. Immunol Rev 265(1):22–34. https://doi.org/10.1111/imr.12283

Chung IC, OuYang CN, Yuan SN, Li HP, Chen JT, Shieh HR, Chen YJ, Ojcius DM, Chu CL, Yu JS, Chang YS, Chen LC (2016) Pyk2 activates the NLRP3 inflammasome by directly phosphorylating ASC and contributes to inflammasome-dependent peritonitis. Sci Rep 6:36214. https://doi.org/10.1038/srep36214

Co Ha (19 Nov 2016) Harnessing the clinical potential of sulforaphane. http://www.directorstalkinterviews.com/evgen-pharma-plc-harnessing-clinical-potential-sulforaphane/412717321. Accessed 09 Nov 2017

Cocco M, Garella D, Di Stilo A, Borretto E, Stevanato L, Giorgis M, Marini E, Fantozzi R, Miglio G, Bertinaria M (2014) Electrophilic warhead-based design of compounds preventing NLRP3 inflammasome-dependent pyroptosis. J Med Chem 57(24):10366–10382. https://doi.org/10.1021/jm501072b

Cocco M, Miglio G, Giorgis M, Garella D, Marini E, Costale A, Regazzoni L, Vistoli G, Orioli M, Massulaha-Ahmed R, Detraz-Durieux I, Groslambert M, Py BF, Bertinaria M (2016) Design, synthesis, and evaluation of acrylamide derivatives as direct NLRP3 inflammasome inhibitors. ChemMedChem 11(16):1790–1803. https://doi.org/10.1002/cmdc.201600055

Cocco M, Pellegrini C, Martinez-Banadocha H, Giorgis M, Marini E, Costale A, Miglio G, Fornai M, Antonioli L, Lopez-Castejon G, Tapia-Abellan A, Angosto D, Hafner-Bratkovic I, Regazzoni L, Blandizzi C, Pelegrin P, Bertinaria M (2017) Development of an acrylate derivative targeting the NLRP3 inflammasome for the treatment of inflammatory bowel disease. J Med Chem 60(9):3656–3671. https://doi.org/10.1021/acs.jmedchem.6b01624

Coll RC, Robertson A, Butler M, Cooper M, O'Neill LA (2011) The cytokine release inhibitory drug CRID3 targets ASC oligomerisation in the NLRP3 and AIM2 inflammasomes. PLoS One 6(12):e29539. https://doi.org/10.1371/journal.pone.0029539

Coll RC, Robertson AA, Chae JJ, Higgins SC, Munoz-Planillo R, Inserra MC, Vetter I, Dungan LS, Monks BG, Stutz A, Croker DE, Butler MS, Haneklaus M, Sutton CE, Nunez G, Latz E, Kastner DL, Mills KH, Masters SL, Schroder K, Cooper MA, O'Neill LA (2015) A small-molecule inhibitor of the NLRP3 inflammasome for the treatment of inflammatory diseases. Nat Med 21(3):248–255. https://doi.org/10.1038/nm.3806

Corporation NP (2011) Highlights of prescribing information: Ilaris (canakinumab). https://www.accessdata.fda.gov/drugsatfda_docs/label/2012/125319s047lbl.pdf. Accessed 15 Nov 2017

D'Anneo A, Carlisi D, Lauricella M, Puleio R, Martinez R, Di Bella S, Di Marco P, Emanuele S, Di Fiore R, Guercio A, Vento R, Tesoriere G (2013) Parthenolide generates reactive oxygen species and autophagy in MDA-MB231 cells. A soluble parthenolide analogue inhibits tumour growth and metastasis in a xenograft model of breast cancer. Cell Death Dis 4:e891. https://doi.org/10.1038/cddis.2013.415

Dai YF, Zhou WW, Meng J, Du XL, Sui YP, Dai L, Wang PQ, Huo HR, Sui F (2017) The pharmacological activities and mechanisms of artemisinin and its derivatives: a systematic review. Med Chem Res 26(5):867–880. https://doi.org/10.1007/s00044-016-1778-5

Dalekos GN, Elisaf M, Bairaktari E, Tsolas O, Siamopoulos KC (1997) Increased serum levels of interleukin-1beta in the systemic circulation of patients with essential hypertension: additional risk factor for atherogenesis in hypertensive patients? J Lab Clin Med 129(3):300–308. https://doi.org/10.1016/S0022-2143(97)90178-5

Daniels MJ, Rivers-Auty J, Schilling T, Spencer NG, Watremez W, Fasolino V, Booth SJ, White CS, Baldwin AG, Freeman S, Wong R, Latta C, Yu S, Jackson J, Fischer N, Koziel V, Pillot T, Bagnall J, Allan SM, Paszek P, Galea J, Harte MK, Eder C, Lawrence CB, Brough D (2016) Fenamate NSAIDs inhibit the NLRP3 inflammasome and protect against Alzheimer's disease in rodent models. Nat Commun 7:12504. https://doi.org/10.1038/ncomms12504

Dempsey C, Rubio Araiz A, Bryson KJ, Finucane O, Larkin C, Mills EL, Robertson AAB, Cooper MA, O'Neill LAJ, Lynch MA (2017) Inhibiting the NLRP3 inflammasome with MCC950 promotes non-phlogistic clearance of amyloid-beta and cognitive function in APP/PS1 mice. Brain Behav Immun 61:306–316. https://doi.org/10.1016/j.bbi.2016.12.014

Dey S, Sarkar M, Giri B (2016) Anti-inflammatory and anti-tumor activities of parthenolide: an update. J Chem Biol Therp 1(2):107. https://doi.org/10.4172/2572-0406.1000107

Dinarello CA (2010) Anti-inflammatory agents: present and future. Cell 140(6):935–950. https://doi.org/10.1016/j.cell.2010.02.043

Dinh QN, Drummond GR, Kemp-Harper BK, Diep H, De Silva TM, Kim HA, Vinh A, Robertson AAB, Cooper MA, Mansell A, Chrissobolis S, Sobey CG (2017) Pressor response to angiotensin II is enhanced in aged mice and associated with inflammation, vasoconstriction and oxidative stress. Aging (Albany NY) 9(6):1595–1606. https://doi.org/10.18632/aging.101255

Dombroski MA, Eggler JF (1998) Sulfonyl urea derivatives and their use in the control of interleukin-1 activity. Pfizer Inc. WO1998032733

Ekstedt M, Hagstrom H, Nasr P, Fredrikson M, Stal P, Kechagias S, Hultcrantz R (2015) Fibrosis stage is the strongest predictor for disease-specific mortality in NAFLD after up to 33 years of follow-up. Hepatology 61(5):1547–1554. https://doi.org/10.1002/hep.27368

Ellis LZ, Liu W, Luo Y, Okamoto M, Qu D, Dunn JH, Fujita M (2011) Green tea polyphenol epigallocatechin-3-gallate suppresses melanoma growth by inhibiting inflammasome and IL-1beta secretion. Biochem Biophys Res Commun 414(3):551–556. https://doi.org/10.1016/j.bbrc.2011.09.115

Fernandes-Alnemri T, Kang S, Anderson C, Sagara J, Fitzgerald KA, Alnemri ES (2013) Cutting edge: TLR signaling licenses IRAK1 for rapid activation of the NLRP3 inflammasome. J Immunol 191(8):3995–3999. https://doi.org/10.4049/jimmunol.1301681

Fu Y, Wang Y, Du L, Xu C, Cao J, Fan T, Liu J, Su X, Fan S, Liu Q, Fan F (2013) Resveratrol inhibits ionising irradiation-induced inflammation in MSCs by activating SIRT1 and limiting NLRP-3 inflammasome activation. Int J Mol Sci 14(7):14105–14118. https://doi.org/10.3390/ijms140714105

Fu SP, Wang JF, Xue WJ, Liu HM, Liu BR, Zeng YL, Li SN, Huang BX, Lv QK, Wang W, Liu JX (2015) Anti-inflammatory effects of BHBA in both in vivo and in vitro Parkinson's disease models are mediated by GPR109A-dependent mechanisms. J Neuroinflammation 12:9. https://doi.org/10.1186/s12974-014-0230-3

Fu J, Sun H, Zhang Y, Xu W, Wang C, Fang Y, Zhao J (2017) Neuroprotective effects of luteolin against spinal cord ischemia-reperfusion injury by attenuation of oxidative stress, inflammation, and apoptosis. J Med Food 21(1):13–20. https://doi.org/10.1089/jmf.2017.4021

Gan W, Ren J, Li T, Lv S, Li C, Liu Z, Yang M (2017) The SGK1 inhibitor EMD638683, prevents Angiotensin II-induced cardiac inflammation and fibrosis by blocking NLRP3 inflammasome activation. Biochim Biophys Acta 1864(1):1–10. https://doi.org/10.1016/j.bbadis.2017.10.001

Gao Y-Z, Zhao L-F, Ma J, Xue W-H, Zhao H (2016) Protective mechanisms of wogonoside against Lipopolysaccharide/D-galactosamine-induced acute liver injury in mice. Eur J Pharmacol 780:8–15. https://doi.org/10.1016/j.ejphar.2016.02.040

Gehringer M, Muth F, Koch P, Laufer SA (2015) c-Jun N-terminal kinase inhibitors: a patent review (2010-2014). Expert Opin Ther Pat 25(8):849–872. https://doi.org/10.1517/13543776.2015.1039984

Goldbach-Mansky R, Wilson M, Fleischmann R, Olsen N, Silverfield J, Kempf P, Kivitz A, Sherrer Y, Pucino F, Csako G, Costello R, Pham TH, Snyder C, van der Heijde D, Tao X, Wesley R, Lipsky PE (2009) Comparison of Tripterygium wilfordii Hook F versus sulfasalazine in the treatment of rheumatoid arthritis: a randomized trial. Ann Intern Med 151(4):229–240 W249-251

Gong Z, Zhou J, Li H, Gao Y, Xu C, Zhao S, Chen Y, Cai W, Wu J (2015) Curcumin suppresses NLRP3 inflammasome activation and protects against LPS-induced septic shock. Mol Nutr Food Res 59(11):2132–2142. https://doi.org/10.1002/mnfr.201500316

Greaney AJ, Maier NK, Leppla SH, Moayeri M (2016) Sulforaphane inhibits multiple inflammasomes through an Nrf2-independent mechanism. J Leukoc Biol 99(1):189–199. https://doi.org/10.1189/jlb.3A0415-155RR

Green RH, Brightling CE, Woltmann G, Parker D, Wardlaw AJ, Pavord ID (2002) Analysis of induced sputum in adults with asthma: identification of subgroup with isolated sputum neutrophilia and poor response to inhaled corticosteroids. Thorax 57(10):875–879

Gris D, Ye Z, Iocca HA, Wen H, Craven RR, Gris P, Huang M, Schneider M, Miller SD, Ting JP (2010) NLRP3 plays a critical role in the development of experimental autoimmune encephalomyelitis by mediating Th1 and Th17 responses. J Immunol 185(2):974–981. https://doi.org/10.4049/jimmunol.0904145

Gross CJ, Mishra R, Schneider KS, Medard G, Wettmarshausen J, Dittlein DC, Shi H, Gorka O, Koenig PA, Fromm S, Magnani G, Cikovic T, Hartjes L, Smollich J, Robertson AA, Cooper MA, Schmidt-Supprian M, Schuster M, Schroder K, Broz P, Traidl-Hoffmann C, Beutler B, Kuster B, Ruland J, Schneider S, Perocchi F, Gross O (2016) K+ Efflux-Independent NLRP3 Inflammasome Activation by Small Molecules Targeting Mitochondria. Immunity 45 (4):761–773. https://doi.org/10.1016/j.immuni.2016.08.010

Grundmann S, Bode C, Moser M (2011) Inflammasome activation in reperfusion injury: friendly fire on myocardial infarction? Circulation 123(6):574–576. https://doi.org/10.1161/CIRCULATIONAHA.111.018176

Guo A, He D, Xu HB, Geng CA, Zhao J (2015) Promotion of regulatory T cell induction by immunomodulatory herbal medicine licorice and its two constituents. Sci Rep 5:14046. https://doi.org/10.1038/srep14046

Guo C, Fulp JW, Jiang Y, Li X, Chojnacki JE, Wu J, Wang XY, Zhang S (2017) Development and characterization of a hydroxyl-sulfonamide analogue, 5-chloro-N-[2-(4-hydroxysulfamoyl-phenyl)-ethyl]-2-methoxy-benzamide, as a novel NLRP3 inflammasome inhibitor for potential treatment of Multiple Sclerosis. ACS Chem Neurosci 8(10):2194–2201. https://doi.org/10.1021/acschemneuro.7b00124

Guzman ML, Rossi RM, Neelakantan S, Li X, Corbett CA, Hassane DC, Becker MW, Bennett JM, Sullivan E, Lachowicz JL, Vaughan A, Sweeney CJ, Matthews W, Carroll M, Liesveld JL, Crooks PA, Jordan CT (2007) An orally bioavailable parthenolide analog selectively eradicates acute myelogenous leukemia stem and progenitor cells. Blood 110(13):4427–4435. https://doi.org/10.1182/blood-2007-05-090621

Halle A, Hornung V, Petzold GC, Stewart CR, Monks BG, Reinheckel T, Fitzgerald KA, Latz E, Moore KJ, Golenbock DT (2008) The NALP3 inflammasome is involved in the innate immune response to amyloid-beta. Nat Immunol 9(8):857–865. https://doi.org/10.1038/ni.1636

Hamon Y, Luciani MF, Becq F, Verrier B, Rubartelli A, Chimini G (1997) Interleukin-1beta secretion is impaired by inhibitors of the Atp binding cassette transporter, ABC1. Blood 90 (8):2911–2915

Han SC, Cai WX, Yang XK, Jia YH, Zheng Z, Wang HT, Li J, Li Y, Gao JX, Fan L, Hu DH (2015) ROS-,ediated NLRP3 inflammasome activity is essential for burn-induced acute lung injury. Mediat Inflamm. Artn 720457. https://doi.org/10.1155/2015/720457

Hara H, Tsuchiya K, Kawamura I, Fang R, Hernandez-Cuellar E, Shen Y, Mizuguchi J, Schweighoffer E, Tybulewicz V, Mitsuyama M (2013) Phosphorylation of the adaptor ASC acts as a molecular switch that controls the formation of speck-like aggregates and inflammasome activity. Nat Immunol 14(12):1247–1257. https://doi.org/10.1038/ni.2749

Hastie AT, Moore WC, Meyers DA, Vestal PL, Li H, Peters SP, Bleecker ER (2010) Analyses of asthma severity phenotypes and inflammatory proteins in subjects stratified by sputum granulocytes. J Allergy Clin Immunol 125(5):1028–1036 e1013. https://doi.org/10.1016/j.jaci.2010.02.008

He Y, Varadarajan S, Munoz-Planillo R, Burberry A, Nakamura Y, Nunez G (2014) 3,4-methylenedioxy-beta-nitrostyrene inhibits NLRP3 inflammasome activation by blocking assembly of the inflammasome. J Biol Chem 289(2):1142–1150. https://doi.org/10.1074/jbc.M113.515080

He Y, Hara H, Nunez G (2016a) Mechanism and regulation of NLRP3 inflammasome activation. Trends Biochem Sci 41(12):1012–1021. https://doi.org/10.1016/j.tibs.2016.09.002

He Y, Zeng MY, Yang D, Motro B, Nunez G (2016b) NEK7 is an essential mediator of NLRP3 activation downstream of potassium efflux. Nature 530(7590):354–357. https://doi.org/10.1038/nature16959

Henao-Mejia J, Elinav E, Jin C, Hao L, Mehal WZ, Strowig T, Thaiss CA, Kau AL, Eisenbarth SC, Jurczak MJ, Camporez JP, Shulman GI, Gordon JI, Hoffman HM, Flavell RA (2012) Inflammasome-mediated dysbiosis regulates progression of NAFLD and obesity. Nature 482 (7384):179–185. https://doi.org/10.1038/nature10809

Heneka MT, Kummer MP, Stutz A, Delekate A, Schwartz S, Vieira-Saecker A, Griep A, Axt D, Remus A, Tzeng TC, Gelpi E, Halle A, Korte M, Latz E, Golenbock DT (2013) NLRP3 is activated in Alzheimer's disease and contributes to pathology in APP/PS1 mice. Nature 493 (7434):674. https://doi.org/10.1038/nature11729

Higdon J, Drake V, Delage B, Crozier A (2005) Flavonoids. Linus Pauling Institute, Oregon State University. http://lpi.oregonstate.edu/mic/dietary-factors/phytochemicals/flavonoids. Accessed 12 Nov 2017

Hill JR, Coll RC, Sue N, Reid JC, Dou J, Holley CL, Pelingon R, Dickinson JB, Biden TJ, Schroder K, Cooper MA, Robertson AAB (2017) Sulfonylureas as concomitant insulin secretagogues and NLRP3 inflammasome inhibitors. ChemMedChem 12(17):1449–1457. https://doi.org/10.1002/cmdc.201700270

Honda H, Nagai Y, Matsunaga T, Okamoto N, Watanabe Y, Tsuneyama K, Hayashi H, Fujii I, Ikutani M, Hirai Y, Muraguchi A, Takatsu K (2014) Isoliquiritigenin is a potent inhibitor of NLRP3 inflammasome activation and diet-induced adipose tissue inflammation. J Leukoc Biol 96(6):1087–1100. https://doi.org/10.1189/jlb.3A0114-005RR

Hostetler G, Riedl K, Cardenas H, Diosa-Toro M, Arango D, Schwartz S, Doseff AI (2012) Flavone deglycosylation increases their anti-inflammatory activity and absorption. Mol Nutr Food Res 56(4):558–569. https://doi.org/10.1002/mnfr.201100596

Ito K, Chung KF, Adcock IM (2006) Update on glucocorticoid action and resistance. J Allergy Clin Immunol 117(3):522–543. https://doi.org/10.1016/j.jaci.2006.01.032

Ito M, Shichita T, Okada M, Komine R, Noguchi Y, Yoshimura A, Morita R (2015) Bruton's tyrosine kinase is essential for NLRP3 inflammasome activation and contributes to ischaemic brain injury. Nat Commun 6:7360. https://doi.org/10.1038/ncomms8360

Ives A, Nomura J, Martinon F, Roger T, LeRoy D, Miner JN, Simon G, Busso N, So A (2015) Xanthine oxidoreductase regulates macrophage IL1beta secretion upon NLRP3 inflammasome activation. Nat Commun 6:6555. https://doi.org/10.1038/ncomms7555

Jiang L, Zhang L, Kang K, Fei DS, Gong R, Cao YH, Pan SH, Zhao MR, Zhao MY (2016a) Resveratrol ameliorates LPS-induced acute lung injury via NLRP3 inflammasome modulation. Biomed Pharmacother 84:130–138. https://doi.org/10.1016/j.biopha.2016.09.020

Jiang W, He F, Li M, Bian Z, Zhu L, Huang Y, Han N, Liu J, Sun T (2016b) Quercetin suppresses NLRP3 inflammasome activation and attenuates histopathology in a rat model of spinal cord injury. Spinal Cord 54(6):592–596. https://doi.org/10.1038/sc.2015.227

Jiang H, He H, Chen Y, Huang W, Cheng J, Ye J, Wang A, Tao J, Wang C, Liu Q, Jin T, Jiang W, Deng X, Zhou R (2017) Identification of a selective and direct NLRP3 inhibitor to treat inflammatory disorders. J Exp Med 214(11):3219–3238. https://doi.org/10.1084/jem.20171419

Juliana C, Fernandes-Alnemri T, Wu J, Datta P, Solorzano L, Yu JW, Meng R, Quong AA, Latz E, Scott CP, Alnemri ES (2010) Anti-inflammatory compounds parthenolide and Bay 11-7082 are direct inhibitors of the inflammasome. J Biol Chem 285(13):9792–9802. https://doi.org/10.1074/jbc.M109.082305

Katsnelson MA, Rucker LG, Russo HM, Dubyak GR (2015) K+ Efflux agonists induce NLRP3 inflammasome activation independently of Ca2+ signaling. J Immunol 194(8):3937. https://doi.org/10.4049/jimmunol.1402658

Kaufmann FN, Costa AP, Ghisleni G, Diaz AP, Rodrigues ALS, Peluffo H, Kaster MP (2017) NLRP3 inflammasome-driven pathways in depression: clinical and preclinical findings. Brain Behav Immun 64:367–383. https://doi.org/10.1016/j.bbi.2017.03.002

Khan MA, Tania M, Fu SY, Fu JJ (2017) Thymoquinone, as an anticancer molecule: from basic research to clinical investigation. Oncotarget 8(31):51907–51919. https://doi.org/10.18632/oncotarget.17206

Kim RY, Pinkerton JW, Essilfie AT, Robertson AA, Baines KJ, Brown AC, Mayall JR, Ali MK, Starkey MR, Hansbro NG, Hirota JA, Wood LG, Simpson JL, Knight DA, Wark PA, Gibson PG, O'Neill LA, Cooper MA, Horvat JC, Hansbro PM (2017) Role for NLRP3 inflammasome-mediated, IL-1beta-dependent responses in severe, steroid-resistant asthma. Am J Respir Crit Care Med. https://doi.org/10.1164/rccm.201609-1830OC

Koh GC, Maude RR, Schreiber MF, Limmathurotsakul D, Wiersinga WJ, Wuthiekanun V, Lee SJ, Mahavanakul W, Chaowagul W, Chierakul W, White NJ, van der Poll T, Day NP, Dougan G, Peacock SJ (2011) Glyburide is anti-inflammatory and associated with reduced mortality in melioidosis. Clin Infect Dis 52(6):717–725. https://doi.org/10.1093/cid/ciq192

Kong F, Ye B, Cao J, Cai X, Lin L, Huang S, Huang W, Huang Z (2016) Curcumin represses NLRP3 inflammasome activation via TLR4/MyD88/NF-kappaB and P2X7R signaling in PMA-induced macrophages. Front Pharmacol 7:369. https://doi.org/10.3389/fphar.2016.00369

Koo JW, Duman RS (2008) IL-1beta is an essential mediator of the antineurogenic and anhedonic effects of stress. Proc Natl Acad Sci U S A 105(2):751–756. https://doi.org/10.1073/pnas.0708092105

Krishnan SM, Dowling JK, Ling YH, Diep H, Chan CT, Ferens D, Kett MM, Pinar A, Samuel CS, Vinh A, Arumugam TV, Hewitson TD, Kemp-Harper BK, Robertson AA, Cooper MA, Latz E, Mansell A, Sobey CG, Drummond GR (2016) Inflammasome activity is essential for one kidney/deoxycorticosterone acetate/salt-induced hypertension in mice. Br J Pharmacol 173(4):752–765. https://doi.org/10.1111/bph.13230

Lai MZ, Chen RH (2014) Regulation of inflammation by DAPK. Apoptosis 19(2):357–363. https://doi.org/10.1007/s10495-013-0933-4
Laliberte RE, Eggler J, Gabel CA (1999) ATP treatment of human monocytes promotes caspase-1 maturation and externalization. J Biol Chem 274(52):36944–36951
Laliberte RE, Perregaux DG, Hoth LR, Rosner PJ, Jordan CK, Peese KM, Eggler JF, Dombroski MA, Geoghegan KF, Gabel CA (2003) Glutathione s-transferase omega 1-1 is a target of cytokine release inhibitory drugs and may be responsible for their effect on interleukin-1beta posttranslational processing. J Biol Chem 278(19):16567–16578. https://doi.org/10.1074/jbc.M211596200
Lalor SJ, Dungan LS, Sutton CE, Basdeo SA, Fletcher JM, Mills KH (2011) Caspase-1-processed cytokines IL-1beta and IL-18 promote IL-17 production by gammadelta and CD4 T cells that mediate autoimmunity. J Immunol 186(10):5738–5748. https://doi.org/10.4049/jimmunol.1003597
Lamkanfi M, Mueller JL, Vitari AC, Misaghi S, Fedorova A, Deshayes K, Lee WP, Hoffman HM, Dixit VM (2009) Glyburide inhibits the Cryopyrin/Nalp3 inflammasome. J Cell Biol 187(1):61–70. https://doi.org/10.1083/jcb.200903124
LaRock CN, Todd J, LaRock DL, Olson J, O'Donoghue AJ, Robertson AAB, Cooper MA, Hoffman HM, Nizet V (2016) IL-1beta is an innate immune sensor of microbial proteolysis. Sci Immunol 1(2):eaah3539. https://doi.org/10.1126/sciimmunol.aah3539
Latz E, Xiao TS, Stutz A (2013) Activation and regulation of the inflammasomes. Nat Rev Immunol 13(6):397–411. https://doi.org/10.1038/nri3452
Lee H, Shin EA, Lee JH, Ahn D, Kim CG, Kim JH, Kim SH (2017) Caspase inhibitors: a review of recently patented compounds (2013-2015). Expert Opin Ther Pat:1–13. https://doi.org/10.1080/13543776.2017.1378426
Li S, Gao X, Wu X, Wu Z, Cheng L, Zhu L, Shen D, Tong X (2015) Parthenolide inhibits LPS-induced inflammatory cytokines through the toll-like receptor 4 signal pathway in THP-1 cells. Acta Biochim Biophys Sin Shanghai 47(5):368–375. https://doi.org/10.1093/abbs/gmv019
Li A, Zhang S, Li J, Liu K, Huang F, Liu B (2016a) Metformin and resveratrol inhibit Drp1-mediated mitochondrial fission and prevent ER stress-associated NLRP3 inflammasome activation in the adipose tissue of diabetic mice. Mol Cell Endocrinol 434:36–47. https://doi.org/10.1016/j.mce.2016.06.008
Li R, Wang X, Qin T, Qu R, Ma S (2016b) Apigenin ameliorates chronic mild stress-induced depressive behavior by inhibiting interleukin-1beta production and NLRP3 inflammasome activation in the rat brain. Behav Brain Res 296:318–325. https://doi.org/10.1016/j.bbr.2015.09.031
Li H, Lin S, Qin T, Li H, Ma Z, Ma S (2017) Senegenin exerts anti-depression effect in mice induced by chronic un-predictable mild stress via inhibition of NF-kappaB regulating NLRP3 signal pathway. Int Immunopharmacol 53:24–32. https://doi.org/10.1016/j.intimp.2017.10.001
Lin KM, Hu W, Troutman TD, Jennings M, Brewer T, Li XX, Nanda S, Cohen P, Thomas JA, Pasare C (2014) IRAK-1 bypasses priming and directly links TLRs to rapid NLRP3 inflammasome activation. Proc Natl Acad Sci U S A 111(2):775–780. https://doi.org/10.1073/pnas.1320294111
Lin YC, Huang DY, Wang JS, Lin YL, Hsieh SL, Huang KC, Lin WW (2015) Syk is involved in NLRP3 inflammasome-mediated caspase-1 activation through adaptor ASC phosphorylation and enhanced oligomerization. J Leukoc Biol 97(5):825–835. https://doi.org/10.1189/jlb.3HI0814-371RR
Liu D, Mamorska-Dyga A (2017) Syk inhibitors in clinical development for hematological malignancies. J Hematol Oncol 10(1):145. https://doi.org/10.1186/s13045-017-0512-1
Liu W, Guo W, Wu J, Luo Q, Tao F, Gu Y, Shen Y, Li J, Tan R, Xu Q, Sun Y (2013) A novel benzo[d]imidazole derivate prevents the development of dextran sulfate sodium-induced murine experimental colitis via inhibition of NLRP3 inflammasome. Biochem Pharmacol 85(10):1504–1512. https://doi.org/10.1016/j.bcp.2013.03.008

Long HB, Xu BC, Luo YL, Luo KQ (2016) Artemisinin protects mice against burn sepsis through inhibiting NLRP3 inflammasome activation. Am J Emerg Med 34(5):772–777. https://doi.org/10.1016/j.ajem.2015.12.075

Lu M, Yin N, Liu W, Cui X, Chen S, Wang E (2017) Curcumin ameliorates diabetic nephropathy by suppressing NLRP3 inflammasome signaling. Biomed Res Int 2017:1516985. https://doi.org/10.1155/2017/1516985

Ludwig-Portugall I, Bartok E, Dhana E, Evers BD, Primiano MJ, Hall JP, Franklin BS, Knolle PA, Hornung V, Hartmann G, Boor P, Latz E, Kurts C (2016) An NLRP3-specific inflammasome inhibitor attenuates crystal-induced kidney fibrosis in mice. Kidney Int 90(3):525–539. https://doi.org/10.1016/j.kint.2016.03.035

MacKenzie SH, Schipper JL, Clark AC (2010) The potential for caspases in drug discovery. Curr Opin Drug Discov Devel 13(5):568–576

Marble A (1971) Glibenclamide, a new sulphonylurea: whither oral hypoglycaemic agents? Drugs 1(2):109–115

Marchetti C, Chojnacki J, Toldo S, Mezzaroma E, Tranchida N, Rose SW, Federici M, Van Tassell BW, Zhang S, Abbate A (2014) A novel pharmacologic inhibitor of the NLRP3 inflammasome limits myocardial injury after ischemia-reperfusion in the mouse. J Cardiovasc Pharmacol 63 (4):316–322. https://doi.org/10.1097/FJC.0000000000000053

Marchetti C, Toldo S, Chojnacki J, Mezzaroma E, Liu K, Salloum FN, Nordio A, Carbone S, Mauro AG, Das A, Zalavadia AA, Halquist MS, Federici M, Van Tassell BW, Zhang S, Abbate A (2015) Pharmacologic inhibition of the NLRP3 inflammasome preserves cardiac function after ischemic and nonischemic injury in the mouse. J Cardiovasc Pharmacol 66(1):1–8. https://doi.org/10.1097/FJC.0000000000000247

Marty V, Medina C, Combe C, Parnet P, Amedee T (2005) ATP binding cassette transporter ABC1 is required for the release of interleukin-1beta by P2X7-stimulated and lipopolysaccharide-primed mouse Schwann cells. Glia 49(4):511–519. https://doi.org/10.1002/glia.20138

Miyaji Y, Yoshimura S, Sakai N, Yamagami H, Egashira Y, Shirakawa M, Uchida K, Kageyama H, Tomogane Y (2015) Effect of edaravone on favorable outcome in patients with acute cerebral large vessel occlusion: subanalysis of RESCUE-Japan Registry. Neurol Med Chir (Tokyo) 55 (3):241–247. https://doi.org/10.2176/nmc.ra.2014-0219

Mridha AR, Wree A, Robertson AA, Yeh MM, Johnson CD, Van Rooyen DM, Haczeyni F, Teoh NC, Savard C, Ioannou GN, Masters SL, Schroder K, Cooper MA, Feldstein AE, Farrell GC (2017) NLRP3 inflammasome blockade reduces liver inflammation and fibrosis in experimental NASH in mice. J Hepatol 66(5):1037–1046. https://doi.org/10.1016/j.jhep.2017.01.022

Mulay SR, Evan A, Anders HJ (2014) Molecular mechanisms of crystal-related kidney inflammation and injury. Implications for cholesterol embolism, crystalline nephropathies and kidney stone disease. Nephrol Dial Transplant 29(3):507–514. https://doi.org/10.1093/ndt/gft248

Murphy JJ, Heptinstall S, Mitchell JR (1988) Randomised double-blind placebo-controlled trial of feverfew in migraine prevention. Lancet 2(8604):189–192. https://doi.org/10.1016/S0140-6736(88)92289-1

Neumann K, Ruland J (2013) Kinases conquer the inflammasomes. Nat Immunol 14 (12):1207–1208. https://doi.org/10.1038/ni.2763

Newman JC, Verdin E (2014) beta-hydroxybutyrate: much more than a metabolite. Diabetes Res Clin Pr 106(2):173–181. https://doi.org/10.1016/j.diabres.2014.08.009

Okada M, Matsuzawa A, Yoshimura A, Ichijo H (2014) The lysosome rupture-activated TAK1-JNK pathway regulates NLRP3 inflammasome activation. J Biol Chem 289(47). https://doi.org/10.1074/jbc.M114.579961

Pan L, Hang N, Zhang C, Chen Y, Li S, Sun Y, Li Z, Meng X (2017) Synthesis and biological evaluation of novel benzimidazole derivatives and analogs targeting the NLRP3 inflammasome. Molecules 22(2):213–228. https://doi.org/10.3390/molecules22020213

Paris D, Ait-Ghezala G, Bachmeier C, Laco G, Beaulieu-Abdelahad D, Lin Y, Jin C, Crawford F, Mullan M (2014) The spleen tyrosine kinase (Syk) regulates Alzheimer amyloid-beta production and Tau hyperphosphorylation. J Biol Chem 289(49):33927–33944. https://doi.org/10.1074/jbc.M114.608091

Periyanayagam S, Arumugam G, Ravikumar A, Ganesan VS (2015) Thymoquinone ameliorates NLRP3-mediated inflammation in the pancreas of albino Wistar rats fed ethanol and high-fat diet. J Basic Clin Physiol Pharmacol 26(6):623–632. https://doi.org/10.1515/jbcpp-2014-0109

Perregaux DG, McNiff P, Laliberte R, Hawryluk N, Peurano H, Stam E, Eggler J, Griffiths R, Dombroski MA, Gabel CA (2001) Identification and characterization of a novel class of interleukin-1 post-translational processing inhibitors. J Pharmacol Exp Ther 299(1):187–197

Pinar A, Dowling JK, Bitto NJ, Robertson AA, Latz E, Stewart CR, Drummond GR, Cooper MA, McAuley JL, Tate MD, Mansell A (2017) PB1-F2 peptide derived from avian influenza a virus H7N9 induces inflammation via activation of the NLRP3 inflammasome. J Biol Chem 292 (3):826–836. https://doi.org/10.1074/jbc.M116.756379

Pinkerton JW, Kim RY, Robertson AAB, Hirota JA, Wood LG, Knight DA, Cooper MA, O'Neill LAJ, Horvat JC, Hansbro PM (2017) Inflammasomes in the lung. Mol Immunol 86:44–55. https://doi.org/10.1016/j.molimm.2017.01.014

Primiano MJ, Lefker BA, Bowman MR, Bree AG, Hubeau C, Bonin PD, Mangan M, Dower K, Monks BG, Cushing L, Wang S, Guzova J, Jiao A, Lin LL, Latz E, Hepworth D, Hall JP (2016) Efficacy and pharmacology of the NLRP3 inflammasome inhibitor CP-456,773 (CRID3) in murine models of dermal and pulmonary inflammation. J Immunol 197(6):2421–2433. https://doi.org/10.4049/jimmunol.1600035

Qiu J, Wang M, Zhang J, Cai Q, Lu D, Li Y, Dong Y, Zhao T, Chen H (2016) The neuroprotection of Sinomenine against ischemic stroke in mice by suppressing NLRP3 inflammasome via AMPK signaling. Int Immunopharmacol 40:492–500. https://doi.org/10.1016/j.intimp.2016.09.024

Rabkin SW (2009) The role of interleukin 18 in the pathogenesis of hypertension-induced vascular disease. Nat Clin Pract Cardiovasc Med 6(3):192–199. https://doi.org/10.1038/ncpcardio1453

Rada B, Park JJ, Sil P, Geiszt M, Leto TL (2014) NLRP3 inflammasome activation and interleukin-1 beta release in macrophages require calcium but are independent of calcium-activated NADPH oxidases. Inflamm Res 63(10):821–830. https://doi.org/10.1007/s00011-014-0756-y

Ron Leuty SFBT (2013) Rigel to cut 30 jobs, focus on three drug programs. https://www.bizjournals.com/sanfrancisco/blog/biotech/2013/09/rigel-fostamatinib-itp-rigl-layoffs.html. Accessed 14 Nov 2017

Salla M, Butler MS, Pelingon R, Kaeslin G, Croker DE, Reid JC, Baek JM, Bernhardt PV, Gillam EM, Cooper MA, Robertson AA (2016) Identification, synthesis, and biological evaluation of the major human metabolite of NLRP3 inflammasome inhibitor MCC950. ACS Med Chem Lett 7(12):1034–1038. https://doi.org/10.1021/acsmedchemlett.6b00198

Schaale K, Peters KM, Murthy AM, Fritzsche AK, Phan MD, Totsika M, Robertson AA, Nichols KB, Cooper MA, Stacey KJ, Ulett GC, Schroder K, Schembri MA, Sweet MJ (2016) Strain- and host species-specific inflammasome activation, IL-1beta release, and cell death in macrophages infected with uropathogenic Escherichia coli. Mucosal Immunol 9(1):124–136. https://doi.org/10.1038/mi.2015.44

Schmid-Burgk JL, Chauhan D, Schmidt T, Ebert TS, Reinhardt J, Endl E, Hornung V (2016) A genome-wide CRISPR (clustered regularly interspaced short palindromic repeats) screen identifies NEK7 as an essential component of NLRP3 inflammasome activation. J Biol Chem 291 (1):103–109. https://doi.org/10.1074/jbc.C115.700492

Schroder K, Tschopp J (2010) The inflammasomes. Cell 140(6):821–832. https://doi.org/10.1016/j.cell.2010.01.040

Shanmugam R, Kusumanchi P, Appaiah H, Cheng L, Crooks P, Neelakantan S, Peat T, Klaunig J, Matthews W, Nakshatri H, Sweeney CJ (2011) A water soluble parthenolide analog suppresses in vivo tumor growth of two tobacco-associated cancers, lung and bladder cancer, by targeting NF-kappaB and generating reactive oxygen species. Int J Cancer 128(10):2481–2494. https://doi.org/10.1002/ijc.25587

Sheridan C (2017) Novartis trial validates inflammasome as chronic disease driver. Nat Biotechnol 35(10)

Sheth KN, Simard JM, Elm J, Kronenberg G, Kunte H, Kimberly WT (2016) Human data supporting glyburide in ischemic stroke. Acta Neurochir Suppl 121:13–18. https://doi.org/10.1007/978-3-319-18497-5_3

Shi JQ, Zhang CC, Sun XL, Cheng XX, Wang JB, Zhang YD, Xu J, Zou HQ (2013) Antimalarial drug artemisinin extenuates amyloidogenesis and neuroinflammation in APPswe/PS1dE9 transgenic mice via inhibition of nuclear factor-kappaB and NLRP3 inflammasome activation. CNS Neurosci Ther 19(4):262–268. https://doi.org/10.1111/cns.12066

Shi JJ, Zhao Y, Wang YP, Gao WQ, Ding JJ, Li P, Hu LY, Shao F (2014) Inflammatory caspases are innate immune receptors for intracellular LPS. Nature 514(7521):187–192. https://doi.org/10.1038/nature13683

Shi H, Wang Y, Li X, Zhan X, Tang M, Fina M, Su L, Pratt D, Bu CH, Hildebrand S, Lyon S, Scott L, Quan J, Sun Q, Russell J, Arnett S, Jurek P, Chen D, Kravchenko VV, Mathison JC, Moresco EM, Monson NL, Ulevitch RJ, Beutler B (2016) NLRP3 activation and mitosis are mutually exclusive events coordinated by NEK7, a new inflammasome component. Nat Immunol 17(3):250–258. https://doi.org/10.1038/ni.3333

Shio MT, Eisenbarth SC, Savaria M, Vinet AF, Bellemare MJ, Harder KW, Sutterwala FS, Bohle DS, Descoteaux A, Flavell RA, Olivier M (2009) Malarial hemozoin activates the NLRP3 inflammasome through Lyn and Syk kinases. PLoS Pathog 5(8):e1000559. https://doi.org/10.1371/journal.ppat.1000559

Siedle B, Garcia-Pineres AJ, Murillo R, Schulte-Monting J, Castro V, Rungeler P, Klaas CA, Da Costa FB, Kisiel W, Merfort I (2004) Quantitative structure-activity relationship of sesquiterpene lactones as inhibitors of the transcription factor NF-kappaB. J Med Chem 47(24):6042–6054. https://doi.org/10.1021/jm049937r

Strickson S, Campbell DG, Emmerich CH, Knebel A, Plater L, Ritorto MS, Shpiro N, Cohen P (2013) The anti-inflammatory drug BAY 11-7082 suppresses the MyD88-dependent signalling network by targeting the ubiquitin system. Biochem J 451(3):427–437. https://doi.org/10.1042/BJ20121651

Studies in natural products chemistry: Volume 52 Bioactive natural products (2017). Elsevier.

Sturm C, Wagner AE (2017) Brassica-derived plant bioactives as modulators of chemopreventive and inflammatory signaling pathways. Int J Mol Sci 18(9):1890–1991. https://doi.org/10.3390/ijms18091890

Sui DM, Xie Q, Yi WJ, Gupta S, Yu XY, Li JB, Wang J, Wang JF, Deng XM (2016) Resveratrol protects against sepsis-associated encephalopathy and inhibits the NLRP3/IL-1beta axis in microglia. Mediat Inflamm 2016:1045657. https://doi.org/10.1155/2016/1045657

Sun Y, Zhao Y, Yao J, Zhao L, Wu Z, Wang Y, Pan D, Miao H, Guo Q, Lu N (2015) Wogonoside protects against dextran sulfate sodium-induced experimental colitis in mice by inhibiting NF-kappaB and NLRP3 inflammasome activation. Biochem Pharmacol 94(2):142–154. https://doi.org/10.1016/j.bcp.2015.02.002

Sun Y, Liu W, Zhang H, Li H, Liu J, Zhang F, Jiang T, Jiang S (2017) Curcumin prevents osteoarthritis by inhibiting the activation of inflammasome NLRP3. J Interf Cytokine Res 37(10):449–455. https://doi.org/10.1089/jir.2017.0069

Takahashi M (2014) NLRP3 inflammasome as a novel player in myocardial infarction. Int Heart J 55(2):101–105. https://doi.org/10.1536/ihj.13-388

Terada K, Yamada J, Hayashi Y, Wu Z, Uchiyama Y, Peters C, Nakanishi H (2010) Involvement of cathepsin B in the processing and secretion of interleukin-1beta in chromogranin A-stimulated microglia. Glia 58(1):114–124. https://doi.org/10.1002/glia.20906

Toldo S, Marchetti C, Mauro AG, Chojnacki J, Mezzaroma E, Carbone S, Zhang S, Van Tassell B, Salloum FN, Abbate A (2016) Inhibition of the NLRP3 inflammasome limits the inflammatory injury following myocardial ischemia-reperfusion in the mouse. Int J Cardiol 209:215–220. https://doi.org/10.1016/j.ijcard.2016.02.043

Tucey TM, Verma-Gaur J, Nguyen J, Hewitt VL, Lo TL, Shingu-Vazquez M, Robertson AA, Hill JR, Pettolino FA, Beddoe T, Cooper MA, Naderer T, Traven A (2016) The endoplasmic reticulum-mitochondrion tether ERMES orchestrates fungal immune evasion, illuminating inflammasome responses to hyphal signals. mSphere 1(3). https://doi.org/10.1128/mSphere.00074-16

Van Hauwermeiren F, Lamkanfi M (2016) The NEK-sus of the NLRP3 inflammasome. Nat Immunol 17(3):223–224. https://doi.org/10.1038/ni.3391

van Hout GP, Bosch L, Ellenbroek GH, de Haan JJ, van Solinge WW, Cooper MA, Arslan F, de Jager SC, Robertson AA, Pasterkamp G, Hoefer IE (2016) The selective NLRP3-inflammasome inhibitor MCC950 reduces infarct size and preserves cardiac function in a pig model of myocardial infarction. Eur Heart J 18(9):828–836. https://doi.org/10.1093/eurheartj/ehw247

Walle T, Hsieh F, DeLegge MH, Oatis JE Jr, Walle UK (2004) High absorption but very low bioavailability of oral resveratrol in humans. Drug Metab Dispos 32(12):1377–1382. https://doi.org/10.1124/dmd.104.000885

Wanderer AA (2009) Corticosteroid resistance in pulmonary neutrophilic inflammatory disorders and rationale for adjunct IL-1beta targeted therapy. Am J Respir Cell Mol Biol 41(2):246–247

Wang WW, Wang C, Ding X-Q, Pan Y, Gu T-T, Wang M-X, Liu Y-L, Wang F-M, Wang S-J, Kong L-D (2013) Quercetin and allopurinol reduce liver thioredoxin-interacting protein to alleviate inflammation and lipid accumulation in diabetic rats. Br J Pharmacol 169 (6):1352–1371. https://doi.org/10.1111/bph.12226

Wang C, Zeng L, Zhang T, Liu J, Wang W (2016) Casticin inhibitrs lipopolysaccharide-induced acute lung injury in mice. Eur J Pharmacol 789:172–178. https://doi.org/10.1016/j.ejphar.2016.07.035

Wang HM, Zhang T, Huang JK, Xiang JY, Chen JJ, Fu JL, Zhao YW (2017a) Edaravone attenuates the proinflammatory response in amyloid-beta-treated microglia by inhibiting NLRP3 inflammasome-mediated IL-1beta secretion. Cell Physiol Biochem 43(3):1113–1125. https://doi.org/10.1159/000481753

Wang S, Zhao X, Yang S, Chen B, Shi J (2017b) Salidroside alleviates high glucose-induced oxidative stress and extracellular matrix accumulation in rat glomerular mesangial cells by the TXNIP-NLRP3 inflammasome pathway. Chem Biol Interact 278:48–53. https://doi.org/10.1016/j.cbi.2017.10.012

Wang W, Wu QH, Sui Y, Wang Y, Qiu X (2017c) Rutin protects endothelial dysfunction by disturbing Nox4 and ROS-sensitive NLRP3 inflammasome. Biomed Pharmacother 86:32–40. https://doi.org/10.1016/j.biopha.2016.11.134

Wang Y, Xu CF, Liu YJ, Mao YF, Lv Z, Li SY, Zhu XY, Jiang L (2017d) Salidroside attenuates ventilation induced lung injury via SIRT1-dependent inhibition of NLRP3 inflammasome. Cell Physiol Biochem 42(1):34–43. https://doi.org/10.1159/000477112

Watanabe Y, Nagai Y, Honda H, Okamoto N, Yamamoto S, Hamashima T, Ishii Y, Tanaka M, Suganami T, Sasahara M, Miyake K, Takatsu K (2016) Isoliquiritigenin attenuates adipose tissue inflammation in vitro and adipose tissue fibrosis through inhibition of innate immune responses in mice. Sci Rep 6:23097. https://doi.org/10.1038/srep23097

Wei W, Wang L, Zhou K, Xie H, Zhang M, Zhang C (2017) Rhapontin ameliorates colonic epithelial dysfunction in experimental colitis through SIRT1 signaling. Int Immunopharmacol 42:185–194. https://doi.org/10.1016/j.intimp.2016.11.024

Wen H, Ting JP, O'Neill LA (2012) A role for the NLRP3 inflammasome in metabolic diseases – did Warburg miss inflammation? Nat Immunol 13(4):352–357. https://doi.org/10.1038/ni.2228

Wiersinga WJ, van der Poll T (2007) Is the septic response good or bad? Curr Infect Dis Rep 9 (5):366–373. https://doi.org/10.1007/s11908-007-0057-5

Wu J, Xu X, Li Y, Kou J, Huang F, Liu B, Liu K (2014) Quercetin, luteolin and epigallocatechin gallate alleviate TXNIP and NLRP3-mediated inflammation and apoptosis with regulation of AMPK in endothelial cells. Eur J Pharmacol 745:59–68. https://doi.org/10.1016/j.ejphar.2014.09.046

Wu J, Maoqiang L, Fan H, Zhenyu B, Qifang H, Xuepeng W, Liulong Z (2016) Rutin attenuates neuroinflammation in spinal cord injury rats. J Surg Res 203(2):331–337. https://doi.org/10.1016/j.jss.2016.02.041

Xiao M, Li L, Li C, Liu L, Yu Y, Ma L (2016) 3,4-Methylenedioxy-beta-nitrostyrene ameliorates experimental burn wound progression by inhibiting the NLRP3 inflammasome activation. Plast Reconstr Surg 137(3):566e–575e. https://doi.org/10.1097/01.prs.0000479972.06934.83

Xin W, Wang Q, Zhang D, Wang C (2017) A new mechanism of inhibition of IL-1beta secretion by celastrol through the NLRP3 inflammasome pathway. Eur J Pharmacol 814:240–247. https://doi.org/10.1016/j.ejphar.2017.08.036

Xu F, Wang F, Wen T, Sang W, He X, Li L, Zeng N (2017) Protective effect of cinnamic acid in endotoxin-poisoned mice. Phytother Res 31(12):1946–1953. https://doi.org/10.1002/ptr.5944

Yamanashi T, Iwata M, Kamiya N, Tsunetomi K, Kajitani N, Wada N, Iitsuka T, Yamauchi T, Miura A, Pu S, Shirayama Y, Watanabe K, Duman RS, Kaneko K (2017) Beta-hydroxybutyrate, an endogenic NLRP3 inflammasome inhibitor, attenuates stress-induced behavioral and inflammatory responses. Sci Rep 7(1):7677. https://doi.org/10.1038/s41598-017-08055-1

Yang ZJ, Ge WZ, Li QY, Lu Y, Gong JM, Kuang BJ, Xi X, Wu H, Zhang Q, Chen Y (2015) Syntheses and biological evaluation of costunolide, parthenolide, and their fluorinated analogues. J Med Chem 58(17):7007–7020. https://doi.org/10.1021/acs.jmedchem.5b00915

Yang Y, Wang H, Li L, Li X, Wang Q, Ding H, Wang X, Ye Z, Wu L, Zhang X, Zhou M, Pan H (2016) Sinomenine provides neuroprotection in model of traumatic brain injury via the Nrf2-ARE pathway. Front Neurosci 10:580. https://doi.org/10.3389/fnins.2016.00580

Yang SJ, Shao GF, Chen JL, Gong J (2017) The NLRP3 inflammasome: an important driver of neuroinflammation in hemorrhagic stroke. Cell Mol Neurobiol 38(3):595–603. https://doi.org/10.1007/s10571-017-0526-9

Yaron JR, Gangaraju S, Rao MY, Kong X, Zhang L, Su F, Tian Y, Glenn HL, Meldrum DR (2015) K^+ regulates Ca^{2+} to drive inflammasome signaling: dynamic visualization of ion flux in live cells. Cell Death Dis 6:e1954. https://doi.org/10.1038/cddis.2015.277

Yim HC, Williams BR (2014) Protein kinase R and the inflammasome. J Interf Cytokine Res 34(6):447–454. https://doi.org/10.1089/jir.2014.0008

Yonglin Z, Zhaojie W, Fengguang G, Jingsheng W, Hanzhong Z, Wenqiang Z (2017) Wogonoside alleviates inflammation induced by traumatic spinal cord injury by suppressing NF-κB and NLRP3 inflammasome activation. Exp Ther Med 14(4):3304–3308. https://doi.org/10.3892/etm.2017.4904

Youm YH, Nguyen KY, Grant RW, Goldberg EL, Bodogai M, Kim D, D'Agostino D, Planavsky N, Lupfer C, Kanneganti TD, Kang S, Horvath TL, Fahmy TM, Crawford PA, Biragyn A, Alnemri E, Dixit VD (2015) The ketone metabolite beta-hydroxybutyrate blocks NLRP3 inflammasome-mediated inflammatory disease. Nat Med 21(3):263–269. https://doi.org/10.1038/nm.3804

Yu H, Wu M, Lu G, Cao T, Chen N, Zhang Y, Jiang H, Fan H, Yao R (2017a) Prednisone alleviates demyelination through regulation of the NLRP3 inflammasome in a C57BL/6 mouse model of cuprizone-induced demyelination. Brain Res 1678:75–84. https://doi.org/10.1016/j.brainres.2017.09.034

Yu X, Zhao Q, Zhang X, Zhang H, Liu Y, Wu X, Li M, Li X, Zhang J, Ruan X, Zhang H (2017b) Celastrol ameliorates inflammation through inhibition of NLRP3 inflammasome activation. Oncotarget 8(40):67300–67314. https://doi.org/10.18632/oncotarget.18619

Zeng J, Chen Y, Ding R, Feng L, Fu Z, Yang S, Deng X, Xie Z, Zheng S (2017) Isoliquiritigenin alleviates early brain injury after experimental intracerebral hemorrhage via suppressing ROS- and/or NF-kappaB-mediated NLRP3 inflammasome activation by promoting Nrf2 antioxidant pathway. J Neuroinflammation 14(1):119. https://doi.org/10.1186/s12974-017-0895-5

Zhang X, Wang G, Gurley EC, Zhou H (2014) Flavonoid apigenin inhibits lipopolysaccharide-induced inflammatory response through multiple mechanisms in macrophages. PLoS One 9(9): e107072. https://doi.org/10.1371/journal.pone.0107072

Zhang A, Wang K, Ding L, Bao X, Wang X, Qiu X, Liu J (2017) Bay11-7082 attenuates neuropathic pain via inhibition of nuclear factor-kappa B and nucleotide-binding domain-like receptor protein 3 inflammasome activation in dorsal root ganglions in a rat model of lumbar disc herniation. J Pain Res 10:375–382. https://doi.org/10.2147/JPR.S119820

Zhao J, Zhang H, Huang Y, Wang H, Wang S, Zhao C, Liang Y, Yang N (2013) Bay11-7082 attenuates murine lupus nephritis via inhibiting NLRP3 inflammasome and NF-kappaB activation. Int Immunopharmacol 17(1):116–122. https://doi.org/10.1016/j.intimp.2013.05.027

Zhao Z, Xiao J, Wang J, Dong W, Peng Z, An D (2015) Anti-inflammatory effects of novel sinomenine derivatives. Int Immunopharmacol 29(2):354–360. https://doi.org/10.1016/j.intimp.2015.10.030

Zhu CS, Xiong ZJ, Chen XH, Peng FH, Hu XQ, Chen YM, Wang Q (2012) Artemisinin attenuates lipopolysaccharide-stimulated proinflammatory responses by inhibiting NF-kappa B pathway in microglia cells. PLoS One 7(4):e35125. https://doi.org/10.1371/journal.pone.0035125

The manufacturer's authorised representative in the EU is Springer
Nature Customer Service Centre GmbH, Europaplatz 3, 69115 Heidelberg,
Germany. If you have any concerns regarding our products, please
contact ProductSafety@springernature.com

Printed and bound by CPI Group (UK) Ltd, Croydon, CR0 4YY
05/10/2026
02228138-0015